THE U.S. MODEL 1917 RIFLE

"AMERICA'S ENFIELD"

by Bruce N. Canfield

Bruce N. Canfield
The U.S. Model 1917 RIfle – "America's Enfield"

Woonsocket, R.I.: MOWBRAY PUBLISHING

300 pages

ISBN: 978-1-931464-92-5

To order additional copies of this book, call Mowbray Publishing at 800-999-4697 or go to www.gunandswordcollector.com
The design and layout of this book was conceived by Jo-Ann Langlois.

Printed in China

On the cover: *Montage of a Model 1917 rifle atop a WWI U.S. Army Reserve Captain's tunic.*
(American Rifleman. Courtesy of Jonathan Krisko)

1 2 3 4 5 6 7 8 9 10

This book is dedicated to the men and women
of our Armed Forces who put their lives on the line daily
to protect us from the forces of evil in this world.

Foreword

The U.S. Rifle Model of 1917 has been the "red-headed stepchild" of U.S. military arms — chronically ignored, misunderstood, or denigrated by users, arms collectors and historians. The Model 1917 actually has a complex and interesting history that deserves to be understood, along with its significant place in history. Collectors will be delighted that Bruce Canfield has successfully taken up that mission.

People collect things for varying reasons. My personal interest in the M1917 began during a high school job near the old Eddystone factory where the majority of M1917s were made. Already on a quest to own examples of U.S. military longarms illustrated in James E. Hicks' wonderful book *U.S. Military Firearms, 1776–1956*, the M1917 became a bit special for me. At the time, there was little written about them, so I accumulated bits of M1917 information and related collectibles as I found them, always hoping for more.

Gun collecting enthusiasm grew and flourished in the 1960s, and books appeared about many topics but seldom anything noteworthy on the M1917. In 1983, Ian Skennerton, author of books on the Lee-Enfield, wrote *The American Enfield*, providing quite a bit of information on the M1917. But, it was an incomplete story, and it is now outdated and out of print. In 2004, S.C. "Nick" Ferris wrote *The United States Rifle Model of 1917*, adding significant production and marking information from archival sources and specimen analysis, but it was narrowly focused on hardware details. Daniel Morrison's *The U.S. Enfield Bayonet*, written in 2017, shared previously unknown insights into the widespread use of the bayonets and the rifles. But many mysteries have remained unexplored, and there has not been a truly comprehensive and solidly researched book devoted to the M1917 until now.

The Model 1917 family history involves several makers in two countries spanning about 30 years, although nearly all manufacturing activity was between 1913 and 1919. The story began with British trials of a Mauser-style action and a new .276 caliber cartridge, designed by the Royal Small Arms Factory at Enfield, with a small test batch of rifles made by Vickers in 1913. That was aborted in 1914 when World War I broke out, and the trials Pattern 1913 rifle was quickly redesigned in .303 caliber as the Pattern 1914 rifle.

Overwhelmed British manufacturing assets necessitated contracting with three American firms to make these rifles in 1914. About 1.2 million were made by Winchester, Remington, and Eddystone in 1916 and 1917. Fortuitously, British orders were completed just about the time America entered WWI in April of 1917. This left a "hot production line" for military rifles available to fill the U.S. Army's sudden need for millions of rifles. While the U.S. Army wanted M1903 Springfields, their only feasible option was to reluctantly use a redesigned version of the British rifle, adapted to standard U.S. ".30-06" ammunition, and thus the U.S. Rifle Model of 1917 was born.

The problems of getting each of these variations approved and into production have never been fully told before and are instructive as to wartime challenges and solutions, often from unlikely sources. The ability of budgets and bureaucracy to unnecessarily or needlessly obfuscate or hinder production never ends.

American civilian industry amazingly turned out over two million M1917 rifles in less than 19 months, compared to only about 900,000 M1903 rifles from the two government arsenals already tooled up for that model. Thus, the M1917 became the de facto main U.S. Army rifle used in WWI.

Following WWI, the M1917 almost replaced the M1903 as the standard rifle on account of its many good features, but "not invented here" and ingrained preference for windage adjustable sights for target use resulted in the M1917 being relegated to war reserve status.

Actually, this turned out to be a good thing, as it resulted in more than a million rifles in storage at the outbreak of WWII. Many were rebuilt before or during WWII at various locations, including one which has eluded positive identification. A few went to arm U.S. troops for a second world war, but most went to allied nations around the world who were desperate for arms to fight the Axis menace. During and after WWII, they were used by Britain, Canada, China, France, Denmark, the Philippines, and even the Viet Cong, as well as other places.

After the end of the "War to End All Wars," M1917 tooling and parts on hand at Remington and Winchester were seen as potential foundations for new sporting arms, or even some military sales, but these had limited production with little success in the marketplace. Hunters did transform many surplus M1917 service rifles into sporting rifles, ranging from ugly hatchet jobs by Bubba to fine custom sporting arms by talented gunsmiths, making them a common, albeit unattractive, sporting arm. One little-known fact discovered by Canfield shows that, in 2023, more than a century after they were made, M1917 rifles that were modified to fit local needs remain in use in Greenland with dog sled patrols!

Collectors and historians have come to know and trust Bruce Canfield for his thorough research on U.S. martial arms, appearing in numerous magazine articles and more than a dozen successful books. His *U.S. Military Combat Shotguns* and *Johnson Rifles and Machine Guns*, like this book on the Model 1917 rifle, provide definitive references on overlooked topics. For more than two decades, his classics works, *U.S. Infantry Weapons of World War One* and *U.S. Infantry Weapons of WWII*, have been essential references on those specialties. Recently, he updated and massively expanded the latter into *U.S. Small Arms of WWII*, an exceptional and comprehensive guide over a much broader scope, which is the new definitive reference for that collecting field.

This new book raises the bar in terms of the breadth and depth of coverage of the M1917 rifle. It is a balanced and comprehensive account of the origins, manufacture, use, refurbishment, modifications, and worldwide distribution of the M1917. The excellent illustrations, most not previously published, add context and confirmation to the story. His information, as usual, is especially useful for collectors, and historians will appreciate the source documentation.

This is a welcome and valuable addition to my library, and I am sure that other collectors will find it informative, readable, and useful as well.

John H. Spangler
Captain, U.S. Navy (Ret.)
Salt Lake City, Utah

Receiver marking on an early-production Winchester Model 1917 rifle. Interestingly, the serial number is the same as the rifle model number! *(Photo courtesy of Joe Salter)*

Preface

There are a number of 20th-century U.S. military weapons that have reached iconic status either through their historical significance, their use in military conflicts and/or interest by martial arms collectors and gun enthusiasts. These include such weapons as the Model 1903 Springfield, M1 Garand, M14 and M16 rifles, M1 carbines, Thompson and M3 submachine guns, Browning Automatic Rifles and M1911/M1911A1 pistols. Production totals of these arms ranged from several hundred thousand to over six million.

One weapon conspicuously absent from this list was used by the majority of American troops in the First World War, and although it was only in production for less than two years, well over two million were made. While the existence of the rifle is certainly no secret, for whatever reasons, the important, if often unheralded role that it played in two world wars has never been adequately acknowledged.

Most of the weapons cited above have been the subject of numerous books and articles. However, by comparison, not a great deal has been written about the Model 1917 rifle, undoubtedly due, at least in some measure, to the rather lukewarm interest shown by collectors, shooters, and arms fanciers.

While "U.S. Rifle, Caliber .30, Model of 1917" was the official nomenclature for the weapon, there are a number of other terms used today to refer to the rifle. Some are accepted shorthand such as "Model 1917" and "M1917," or even "M17." The terms "Model 1917" and "M1917" will be used interchangeably in this book. Interestingly, the rifle has often been referred to since its adoption by the U.S. military as the "Enfield," or the "American Enfield" both informally and in numerous official government documents and correspondence. It could be said, seemingly with some justification, that this appellation is incorrect because the term "Enfield" is synonymous with the standard British service rifle for over a century, the Short Magazine Lee-Enfield (SMLE), and since the Model 1917 was not made by Enfield, it cannot be properly referred to as such.

As will be seen, however, the rifle on which the Model 1917 was originally based was the .276 caliber British Pattern 1913, which was indeed designed and developed at the Royal Small Arms Factory in Enfield and made in limited numbers in Great Britain. The Pattern 1913 morphed into the .303 caliber Pattern 1914, which was manufactured in the United States under British contract. The subsequent U.S. Model of 1917 was, in reality, only a slightly modified Pattern 1914. Even though the Model 1917 rifle was nowhere near the Enfield factory in England, it was clearly based on a design that originated there. In one sense, this is analogous to the U.S. Model of 1903 rifle, which was designed, developed, and manufactured by the Springfield Armory. However, Model 1903 rifles were also made by Rock Island Arsenal and Remington Arms Company and are routinely referred to today as "Springfields," even though they were not manufactured at the government armory in Massachusetts. In any event, given the widespread use of "Enfield" in reference to the Model 1917 rifle in a number of official U.S. government documents and publications from 1917 to 1945, one would be hard-pressed to categorically claim it is an incorrect term that should not be used. The term "American Enfield" will occasionally be used in this book but with quotation marks to indicate it is non-standard nomenclature.

There is, however, one term that is unquestionably incorrect with no justification for use other than the fact it has been bandied about for years — "Pattern 1917," "P17" or some derivation thereof. Unlike the British, the U.S. military did not use the term "Pattern" to designate small arms. It's a bit of a mystery why these erroneous terms are so ubiquitous. It certainly isn't in the interest of brevity since it takes no more effort to write "Model 1917" or "M17" than it does "Pattern 1917" or "P17." Much like the pervasive use of "clip" when "magazine" would be correct, the use of such incorrect terms as "P17" should be avoided lest they become even more entrenched in the lexicon of military weapons collecting and study.

Substantial numbers of Model 1917 rifles were released as surplus following WWII. The rifles sold for modest cost and possessed strong nickel steel actions, which made them popular candidates for sporterizing to be used as low-cost hunting and sporting rifles. The same was true of the Model 1903 Springfield but not to the same extent as the Model 1917. Unlike today, back in the late 1940s to the mid-60s, there were few avid collectors of U.S. martial arms, which undoubtedly was a major reason for the previously mentioned lack of reference material on the Model 1917 rifle. While there was occasionally a rather cursory mention of the weapon in some books, most authors are not particularly interested in expending the time and effort to write about a subject in which few people are keenly interested. Such was the case until the 1980s.

One of the first books specifically written on the Model 1917 rifle was *The U.S. Enfield* by Ian Skennerton, published in 1983, which covered the Model 1917 and its predecessors, the British Pattern 1913 and Pattern 1914 rifles, using a lot of archival documentation. A couple of WWI British War Office manuals on the Pattern 1914 rifle as well as a 1917-era U.S. military manual on the Model 1917 rifle were included and made up a substantial portion of the book. It is an interesting and informative book but with little collector-oriented information. For almost two decades, this was essentially the standard work on the subject. There was one other small book on the subject that came out in the late 1990s, but it had numerous flaws and erroneous information and was not generally well received.

Interest in U.S. military weapons of all types began to attract more and more collectors in the late 1980s and into the 1990s, including such arms as the Model 1903 and M1 Garand rifles, M1 carbines and M1911 pistols. However, the Model 1917 remained something of a "red-headed stepchild" when it came to most collectors and some shooters. This started to change a bit in 2000 when Charles R. Stratton penned an excellent book, *The Pattern 1914 and U.S. Model 1917 Rifles*. Mr. Stratton covered the various components of both rifles in minute detail, illustrated mostly by helpful line drawings.

This was followed four years later by another first-rate book, *United States Rifle Model of 1917*, by C.S. Ferris. Mr. Ferris took a more historical approach and thoroughly discussed many aspects of the Model 1917's inception, production, issuance, and other related topics in a through and scholarly manner using a great deal of primary documentation. Both are excellent works, and for the next almost twenty years, were about the extent of credible books on the Model 1917 rifle.

As the more popular U.S. military weapons cited above continued to grow in interest, helped along by new reference books and collector associations, this invariably led to less availability and higher prices, as more and more collectors were chasing fewer and fewer pecimens. In response, some collectors turned to "under-collected" arms such as the Model 1917 rifle, which could be found for much more attractive prices. There is no doubt the books written by the three authors cited above were helpful to the collectors who decided that the Model 1917 was worthy of being collected after all.

In contemplating my next project, I considered several potential topics before deciding on writing a book on the Model 1917 rifle. I believe a book that incorporates the latest available information and covers the historical background of the Model 1917 rifle, its development, and subsequent employment, along with some collector-oriented material would be helpful, especially with good-quality photographs (most in color).

With the topic decided, my next step was to determine what the proposed book should look like. My last two books, *The M1 Garand Rifle* and *U.S. Small Arms of World War II* were both voluminous tomes of almost 900 pages each but were near polar opposites in their scope. The *M1 Garand Rifle* book covered the period of time from the dawn of the 20th century to the present day but was essentially a single-topic work. The *U.S. Small Arms of WWII* book, on the other hand, concentrated on the weapons used by American armed forces from c.1941 to 1945 and covered a wide gamut of weaponry from knives to flame throwers and everything in between.

While a book on the Model 1917 rifle would obviously be a single topic, I opted not to do another extremely detailed and weighty book (literally and figuratively). This book will not cover the various parts in minute detail as did my *M1 Garand Rifle* book, since not every potential reader is interested in such esoterica. Also, I chose only briefly to profile the Model 1917's immediate predecessor, the Pattern of 1914, and its development. If one is so inclined to learn more about such matters, the books listed above are recommended. My goal with this book is to have a work of sufficient scope and detail to impart enough interesting information and data on the Model 1917 rifle that would be of interest only to a minority of potential reader, without going into extreme detail. This book contains a lot of historical information regarding the background, development and utilization of the Model 1917 rifle. This is my 14th book to have been published by Mowbray Publishing, and I am indeed fortunate to have had an excellent relationship with Stuart Mowbray and his fine staff for over 30 years.

I always strive to use unimpeachable primary documentation whenever possible. The dean of researchers, the late Frank Mallory and his Springfield Research Service, had graciously provided me with a great deal of invaluable information and data over the years for many of my previous book projects, and much of this information has been used in this book as well. In the last 20 years or so, quite a bit of additional archival information has surfaced thanks, in large measure to the efforts of several excellent researchers. Special thanks go to the Archival Research Group founded by Andrew Stolinski and Alec Tulkoff / Tulkoff Military Archives, who are talented and dedicated researchers and continue to discover very interesting primary documents, many from the National Archives. Numerous documents cited in this book were provided by these gentlemen and credited in the endnotes.

It is often necessary to use secondary information from other books and articles, and I have attempted to cross-reference such sources whenever possible. It is an unfortunate fact that incorrect or unproven statements are often repeated by subsequent writers who don't bother to verify the reliability of the information. Sometimes, speculation and conjecture are appropriate but should clearly be labeled as such. The "References" section of this book gives the primary and secondary sources used so readers can ascertain the veracity of the information presented.

Clearly, the Model 1917 rifle has not been given its rightfully deserved place in the pantheon of U.S. military weapons. In many ways, the story of the Model 1917 rifle is very different from that of the venerable Model 1903 and the M1 Garand rifles, which were developed at the Springfield Armory. The '03 and variants were in production at varying levels for over 40 years and the M1 for over 20 years, which is in stark contrast with the Model 1917's production run of less than two years. Even so, the Model 1917 rifle served our armed forces and some of our allies for over 25 years during some of the most critical times in our history. That's quite a legacy!

Bruce N. Canfield
Erie, Colorado, 2023

Acknowledgements

It goes without saying that a book of this type would not be possible without the help of a number of knowledgeable individuals who graciously provided information and photographs. Particular thanks go to two individuals for their assistance with this project: John Spangler and Hayes Otoupalik.

Hayes has been a good friend for many years and is a long-time, highly regarded dealer of U.S. weapons and related militaria. He has encyclopedic knowledge on the subject and generously shares his accumulated information and insight with interested parties. Hayes has a literal world-class collection of amazing weaponry, artifacts, uniforms, equipment, ad infinitum. A number of rare and interesting items from Hayes' collection are featured in this book, photographed by his associate Bill O'Donnell.

John Spangler, who wrote the foreword to this book, is also one of the premier dealers of firearms and associated items in the country and is highly respected in the collecting and historical militaria community. John is a dedicated researcher in his own right and has extensive knowledge on all aspects of weaponry and militaria, which he shares with many specialized collecting organizations. Both he and Hayes are members of the prestigious American Society of Arms Collectors. His website, oldguns.net, always has an extensive array of interesting items for sale and the Q&A section of the site has provided accurate answers to numerous inquiries over the years. John offered to review this manuscript prior to publication, and his comments, criticisms, and corrections were very helpful and much appreciated. I am very grateful to both of these gentlemen for their contributions to this book.

A special note of thanks goes to those who pioneered studies regarding the Model 1917 rifle: Ian Skennerton, Charles Stratton, and C.S. "Nick" Ferris. And a thank you to James E. Hicks and his illustrator Andre Jandot whose 1962 book *U.S. Military Firearms, 1776–1956* inspired a generation of collectors and introduced the Model 1917 as a legitimate part of the U.S. military arms family.

Many other individuals listed below (in no particular order) were very helpful in providing documentation, photographs and other information for this book. Several of them, regrettably, are no longer with us. My sincerest apologies if I've left anyone off the list.:

Roy Marcot	Jo-Ann Langlois
Herbert Houze	Jathan Clark
David Albert	Kris Gayman Leinicke
Gary Cunningham	Kevin Braafladt
Frank Mallory	Maureen Denfeld
Alex MacKenzie	David Labrozzi
Alec Tulkoff	Mark Keefe
Andrew Stolinksi	Brian Sheetz
Danny Michael	Joe Salter
Jonathan Krisko	Cheairs Aguilar
Tom Laemlein	Marc Wade
Stuart Mowbray	Daniel Jay Morrison

Table of Contents

Martini-Henry Mk I Rifle, .577-450 caliber. *(Rock Island Auctions)*

Lee-Metford Mk I Rifle. *(Rock Island Auctions)*

Rifle, Magazine, Lee-Enfield, Mk I. *(National Firearms Museum)*

Short Magazine Lee-Enfield No. 1 Mk III. *(National Firearms Museum)*

British .276 caliber cartridge. *(Militarycartridges.com)*

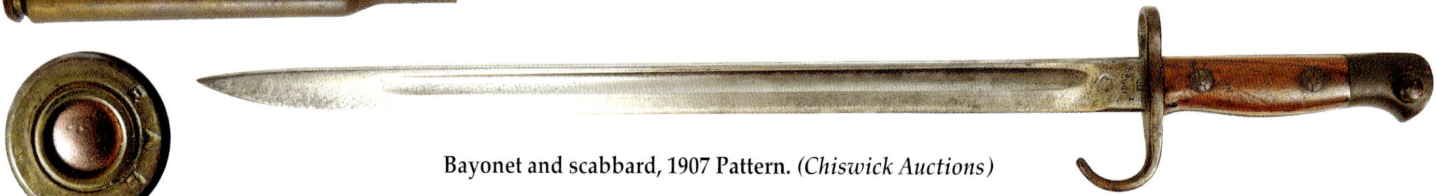

Bayonet and scabbard, 1907 Pattern. *(Chiswick Auctions)*

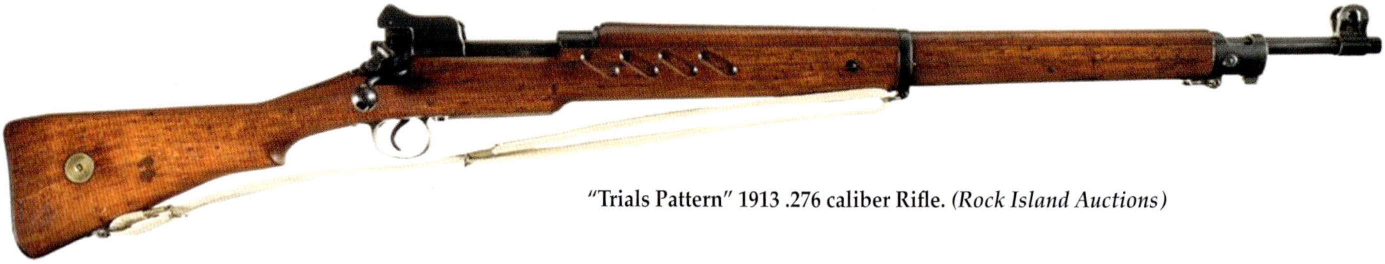

"Trials Pattern" 1913 .276 caliber Rifle. *(Rock Island Auctions)*

Pattern 1913 Rifle (left side). *(National Firearms Museum)*

Chapter 1

The Forerunners
of the U.S. Model 1917 Rifle

The standard British Army rifle from the years 1871 through 1888 was the "falling block" single-shot Martini-Henry, which was chambered for a .577-450 caliber cartridge using a paper-patched lead bullet.[1] By the mid-to-late 1880s, most of the industrial powers had adopted, or were developing, repeating bolt-action, reduced-caliber service rifles. Understandably, Great Britain did not wish to lag behind.

In 1888, the British "Rifle, Magazine, Lee-Metford Mark I" was introduced. It featured a bolt-action mechanism developed by the American inventor James Paris Lee with a rifling pattern designed by William Metford. The rifle initially used a black powder version of the .303 caliber British cartridge. Several relatively minor changes in the design resulted in the later Mark 1*, Mark II and Mark II* variants.

In 1893, the smokeless powder, cordite .303 caliber cartridge was developed to replace the black powder round. The rifles were manufactured by the Royal Small Arms Factory in Sparkbrook, Birmingham, England and the Birmingham Small Arms Co. (BSA) in Small Heath, Birmingham, England.[2]

In November of 1895, the Lee-Metford was superseded by the "Rifle, Magazine, Lee-Enfield Mark I."[3] It differed from its predecessor only in that it had deeper Enfield-type rifling rather than the shallow Metford type. The MK I*, which omitted the cleaning rod and was fitted with improved sights, was approved in 1899. It continued to be chambered for the smokeless powder .303 caliber cartridge.

Circa 1900, the British War Office established the Small Arms Committee to explore the development of improved weaponry. This led to the adoption in December of 1902, of the "Short Magazine Lee-Enfield Rifle Mark I." The rifle and its numerous later variants were referred to as the "SMLE." Early SMLE rifles used the 1907 pattern bayonet with a hooked quillon. Later versions would have a straight cross guard.

There were a number of subsequent variations of the SMLE rifle that were developed, and it remained the standard British service rifle into the late 1950s.

Pattern 1913 Rifle — While the short magazine Lee-Enfield proved to be a reliable and effective weapon, the British War Office was interested in the development of a rifle with even better performance. It was felt by many within the British Army that the Mauser rifles had shown their superiority over the Lee-action rifles during the recently concluded Boer War. It was believed that a rifle with a Mauser-style action, teamed with a cartridge having a smaller caliber, higher muzzle velocity, and a flatter trajectory than the standard .303 cartridge, would be an excellent choice for an improved service rifle.

In 1910, the Small Arms Committee established the criteria for the proposed rifle, and several prototypes were fabricated at the Royal Small Arms Factory in Enfield. By 1913, the final design was tested. The cartridge developed for the rifle was .276 caliber, similar to the Canadian Ross .280 caliber cartridge.

The new rifle, with a modified Mauser action, was originally designated as the "Rifle, Magazine, Enfield, .276 inch" and was initially referred to as a "Trials Pattern." The designation was eventually changed to the "Pattern 1913." A total of 1,251 rifles were manufactured by the Royal Small Arms Factory.[4] The fore-end of the stock had four rearward-sloping vertical grasping grooves on either side. As with several previous British service rifles, the Pattern 1913 was fitted with volley sights.

The bayonet adopted for use with the "Trials Pattern" rifle was also designated as the Pattern 1913 and was very similar to the Pattern 1907 sword bayonet used with the short magazine Lee-Enfield rifle.[5]

In comparison to the standardized British Army rifle, the Lee-Enfield Mark III testing revealed a number of perceived advantages of the new Pattern 1913, including:

- Increased muzzle velocity and flatter trajectory
- Greater strength with the bolt's front locking design
- The one-piece stock was stronger than the SMLE's two-piece stock
- Better sights
- The magazine was enclosed entirely in the stock,

which was less prone to damage and had better handling characteristics.

- The rifle was easier to make than the SMLE since it had fewer parts and required less machining.[6]

However, despite these positive attributes, there were some problems encountered with the .276 caliber cartridge, as it was found to cause excessive barrel erosion and generated heavy recoil, which could result in parts breakage. Some minor changes were made in the design of the Pattern 1913 rifle and the .276 cartridge but, by this time, it was becoming increasingly likely that Great Britain would soon become embroiled in a war that was brewing in Europe. This resulted in further development of the Pattern 1913 rifle and the .276 cartridge being suspended.[7]

Pattern 1914 Rifle — With the outbreak of war in 1914, it was apparent that Great Britain would need a large number of rifles with which to equip the rapidly growing armed forces. The Lee-Enfield remained the standard army rifle, and production was increased as quickly as possible. However, additional rifles would undoubtedly be needed. Since the Pattern 1913 performed reasonably well in testing, and since the manufacturing drawings and specifications were already available, it made sense to adopt the weapon as a supplement to the SMLE. However, some changes were obviously in order. The most important was the type of cartridge that would be used with the rifle. As stated, the .276 caliber round possessed some ballistic advantages over the .303, but there were problems with the cartridge that needed to be corrected. Even if these issues were satisfactorily addressed, the introduction of another type of rifle cartridge into the military's inventory would be undesirable from a logistics standpoint. The logical course of action was to modify the Pattern 1913 to be chambered for the standardized .303 caliber round instead of the problematic .276 cartridge.

The proposed new rifle was initially designated as the "276-.303 Magazine Rifle," and the Birmingham Small Arms Company (BSA) was approached regarding possible manufacture of the rifle, but they declined the contract, as the firm was heavily involved in manufacture of the SMLE rifle.

The firm of Vickers Limited agreed to take on the project and fabricated six pre-production samples by the end of April 1914. These pilot models made by Vickers were not totally acceptable, and additional modifications were made. Specifications for the rifle were approved by the Small Arms Committee on October 22, 1914, and the official designation was changed to "Caliber .303, Pattern 1914 Mk I, Rifle."[8]

The bayonet adopted for the Pattern 1914 rifle retained the original Pattern 1913 designation and was very similar to the "Trials Pattern" bayonets, but the hooked quillon of the earlier model was eliminated. Contracts for production of the Pattern 1913 bayonets for the Pattern 1914 rifles were eventually granted to Vickers, Enfield, and Wilkinson, with a combined total of less than 11,000 being produced.[9]

Vickers was still struggling with rectifying the problems encountered with the pre-production models, and it was not until late March of 1915 that the situation was sufficiently improved for production contracts to be finalized. On June 3, 1915, a contract for 100,000 rifles with bayonets was granted. It called for weekly delivery of 2,000 rifles slated to begin on July 31, 1915, with an increase of 3,000 per week by November 27, 1915. Unfortunately, this turned out to be totally unrealistic, as Vickers was only able to fabricate a handful of pre-production sample rifles along with about 1,500 Pattern 1913 bayonets. Needless to say, this turn of events resulted in the contract being cancelled on July 6, 1916.[10]

The American Gambit — The British sought in vain to find other domestic arms makers to manufacture the Pattern 1914 rifle, but all were either producing SMLE rifles or other war material and were unable to take on the new task. In September of 1914, the British Council of the War Department looked to a friendly nation that had the capacity and expertise to manufacture the rifles. The J.P. Morgan Company of New York was engaged to arrange production contracts, and eventually, four established American arms makers were given contracts for British SMLE Mk III and Pattern 1914 Mk I Enfield rifles. On March 17, 1915, the Hopkins & Allen Arms Company of Norwich, Connecticut, was granted a contract for 400,000 SMLE rifles, but the company was unable to start production, and the contract was cancelled in July of 1915 with no rifles being made.[11]

Winchester Repeating Arms Company of New Haven, Connecticut, and Remington Arms – Union Metallic Cartridge Company of Bridgeport, Connecticut, were solicited by representatives of the British War Office to manufacture the Pattern 1914 Mk I rifle. The British War Office finalized contractual arrangements to have the rifle manufactured in the United States by these firms. Winchester was tasked

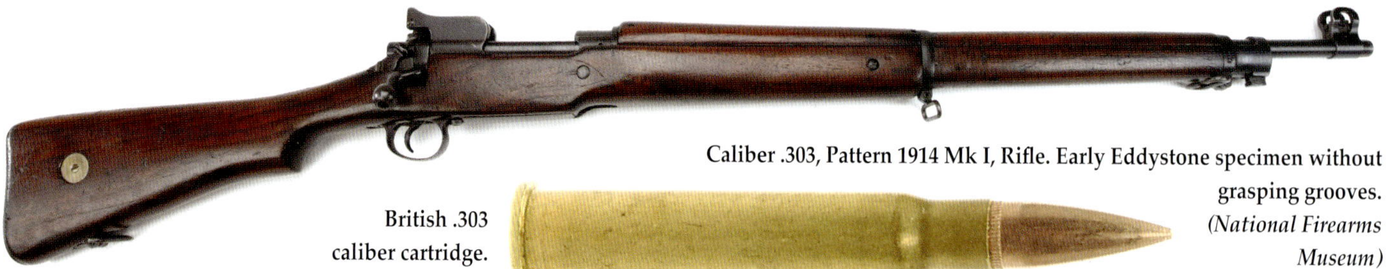

Caliber .303, Pattern 1914 Mk I, Rifle. Early Eddystone specimen without grasping grooves. *(National Firearms Museum)*

British .303 caliber cartridge.

Pattern 1913 bayonet and scabbard adopted for use with the Pattern 1914 rifle.

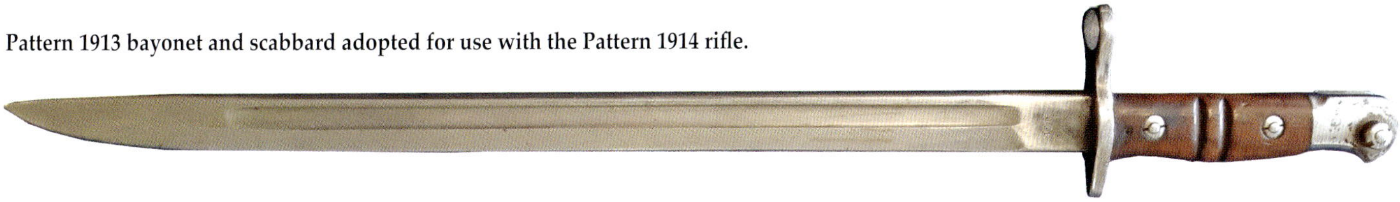

Winchester "production sample" Pattern 1914 Mk 1 rifle.
(Rock Island Auctions)

with developing the modifications required for efficient mass production. The company was initially optimistic that converting the Pattern 1913 Enfield to .303 caliber Pattern 1914 specifications wasn't going to be difficult, but this optimism soon proved to be misplaced. Many internal Winchester documents detailed myriad delays and obstacles, some of which were due to the British War Department continually changing specifications. As related by author Harold Williamson in the authoritative book *Winchester – The Gun That Won the West*:

> "Although the British also had American contracts for the Enfield rifle with Remington… Winchester was chosen to work out the final model of the rifle to be put into production… This was not accomplished without difficulty. It was soon discovered that there were discrepancies between the dimensions of the model and those on the detailed drawings received from England. Winchester gun experts were concerned over the tendency of the British rimmed .303 to jam, and recommended the use of a rimless cartridge instead, a recommendation that was not accepted."[12]

Final approval of the proposed production model developed by Winchester was given on March 22, 1915, and the contractors began to tool up for mass production. Although Winchester was selected to develop the required modifications to convert the Pattern 1913 rifle to .303 caliber, Remington Arms Company apparently wanted some of the proprietary credit for this work as well since internal Remington factory documents referred to the modified rifle as the "Remington-Enfield Bolt-Action Military Rifle, Cal. .303 inch."[13]

The Pattern 1914 manufacturing tooling and equipment that was used by Vickers was shipped to the United States to facilitate the American company's Pattern 1914 production programs. Author Luke Mercaldo in his excellent and well-researched book *Allied Rifle Contracts in America, 1914–1918* described the initial contracts:

> "The first two contracts in the United States for British rifles were negotiated by Sir Courtenay W. Bennett, His Majesty's Consul-General in New York, acting on behalf of the War Office, with Winchester and Remington-UMC (Remington-Union Cartridge Company) on November 24, 1914. Each of these contracts was for 200,000 'magazine Enfield .303 pattern 1914 rifles,' bayonets and scabbards. Remington-UMC was represented by Marcellus Hartley Dodge, Remington's chairman. The contracted price was $30.00 per set and delivery was to begin nine to twelve months from the date of acceptance at a rate of 1,000 rifles per day. Rifles ready for shipment before July 31, 1915 would earn a $1.00 premium per rifle. The contract was to close on May 31, 1916, and could be cancelled with three months' notice. The Winchester contract was similar to Remington's except the price was $32.50 per set and an advance payment of 25 percent was made."[14]

A report dated August 27, 1927, authored by J.E. Brennan, Vice President of Remington Arms Company, gave a detailed background of the beginnings of that company's Pattern 1914 rifle production program:

> "Negotiations between the British Government and the Remington Arms Company were begun in the fall of 1914, and the initial order for rifles was received in November of that year. Plans were laid to equip for a production of 2,000 rifles in a 10-hour day, which called for new buildings, new machine tools and other incidental plant facilities. Property to the east of the plant was purchased on November 13, 1914, and it is a significant fact, in connection that machine tool equipment was being installed in the new building, which had been erected on the 1st day

of March 1915 — quite an accomplishment, especially when it is considered that dwelling and other building occupying the site on November 13, 1914, had to be razed or moved to other locations, and the building construction carried on through the winter months. Other buildings were constructed with like celerity. Building 52, the long building on the north side off Main Street, was being occupied with machine equipment in less than four months from the time of breaking ground…

The sample rifle that the British submitted was, so-to-speak, only in an embryo state. But a few had been made – and 'handmade' at that – by the designer, Mr. Edward Reavill, at one of the British Arsenal, and these had been adapted to use a new cartridge which the British were developing – in .276 caliber. However, the British had no equipment to produce the .276 cartridge, in fact, no equipment existed, so they had to resort to the .303 caliber cartridge which could be obtained in the quantities needed, not only from the British ammunition factories, but from factories in the United States as well. The British had no drawings of the rifle; had made no studies of dimensions, or limits of tolerance; and in fact, that nothing but the sample .276 rifle to go by. This necessitated, on our part, a tremendous amount of work. Sample rifles to handle the .303 cartridge were made, hurriedly, by our expert model makers. These were subjected to tests by our own people, as well as by the British engineers, and as we reached results that were satisfactory, dimensions and limits were established, and the work of producing the tools, fixtures, gauges, etc., required for a production of 2,000 rifles in 10 hours, began. Simultaneously with the erection of the new buildings in Ilion, preparation for the rifle works at Eddystone were made, and as rapidly as data was established at Ilion, it was forwarded to Eddystone for their guidance. Another copy of this data was furnished to the British representatives and it is assumed, though we have no proof to substantiate this belief, was forwarded to the Winchester Repeating Arms Company at New Haven, who, likewise, had received an order for similar rifles…a total of 403,193 Remington-Enfield Rifles, caliber .303-inch, were delivered to the British Government, final deliveries taking place in midyear of 1917."[15]

According to the above memorandum, the internal Remington factory designation for the Pattern 1914 rifle was the "Model 28."[16]

As the contractors geared up for mass production, problems immediately cropped up, which Winchester and Remington had to address. The lack of machinery specifica-

Pattern 1914 rifle assembly line at the Remington, Ilion, plant, c.1916. (Remington Society, courtesy of Roy Marcot

tions, manufacturing drawings and engineering data for the sample Pattern 1914 rifles fabricated by the British hindered the efforts of the American firms to tool up for mass production. Other issues that hampered production by both firms were labor unrest and raw materials shortages.[17] The companies worked diligently to overcome these obstacles and were eventually able to start manufacturing the rifles at a satisfactory rate. Production was progressing sufficiently for the British to award additional contracts to Remington and Winchester for another 200,000 Pattern 1914 rifles from each manufacturer.[18]

Even with these additional contracts, the British War Office wanted more rifles. Since the two companies were now contractually obligated to deliver a total of 800,000 rifles to the British, both were wary about taking on any more commitments. Winchester had accepted sizable contracts for the Russian Model 1895 rifles, and Remington was heavily involved in manufacturing Model 1891 Mosin-Nagant rifles for the czarist government. Neither firm was inclined to accept any further orders from the British at this time without more easily manageable delivery schedules and sizable advance payments to obtain the required tooling and equipment. There were also problems to overcome regarding recruiting and maintaining suitable work forces and battling the continual headaches of raw material shortages. Revised contract terms and other concessions would enable production of the Pattern 1914 rifle to be substantially increased and gave Winchester and Remington-UMC some breathing room to handle their additional Pattern 1914 production commitments.

Pattern 1914 rear sight assembly (top view). Note the rifle serial number stamped on the bolt handle on a Winchester-made rifle. *(Great War forum)*

Pattern 1914 rifle (left side). ***(National Firearms Museum)***

Sectionalized Pattern 1914 rifle. ***(Cody Firearms Museum)***

In order to be able to manufacture even more Pattern 1914 rifles for the British, the chairman of Remington-UMC, Marcellus Hartley Dodge, arranged for a new corporation, "Remington Arms Company of Delaware," to be formed. Despite the name, it was a separate legal entity from Remington-UMC. The newly formed corporation leased the Baldwin Locomotive Works plant in Eddystone, Pennsylvania. The Eddystone facility was a huge complex in excellent condition and ideal for the intended purpose. With this additional manufacturing source, the Remington Arms Company of Delaware entered into a contract on April 30, 1915, with the British government for the manufacture of 1,500,000 Pattern 1914 rifles, bayonets and scabbards at a cost of $30.00 per set. This was followed on August 2, 1915, by another order for 500,000 additional rifles to be made at the Eddystone plant. The British still wanted additional rifles, and on September 27, 1915, Remington-UMC accepted an order for 600,000 more Pattern 1914 rifles. The total orders for American-made Pattern 1914 rifles was a somewhat astonishing 3,400,000![19]

Colonel John T. Thompson, who would later gain fame for a submachine gun he developed, resigned from the U.S. Army to oversee the conversion of the massive Baldwin Locomotive plant at Eddystone into a manufacturing facility for the British rifle, and it eventually became one of the largest firearms manufacturing plants in the world. Thompson remained as Works Supervisor. He later went back into service and was promoted to Brigadier General and Chief of the Small Arms Division of the Army Ordnance Department.

The Eddystone plant was referred to as the Eddystone Rifle Works but was (and is) sometimes informally called "Eddystone Arsenal." Later in WWI, it was merged into a new corporate entity, the "Midvale Steel and Ordnance Company."

As events transpired, the Eddystone plant did not manufacture any of the Pattern 1913 bayonets even though it was part of their initial production contracts. As stated by author Ian Skennerton, "...it is likely that all the Remington bayonets were produced at Bridgeport, even though on paper they were divided up between Ilion and Eddystone."[20]

The probable reason was that Remington management wanted both the Ilion and Eddystone plants to concentrate on rifle production and not be sidetracked by making bayonets as well. Since the company's Bridgeport plant had sufficient capacity, it made sense to have the bayonets manufactured there to supply both the Eddystone and Ilion plants. This will be further discussed in Chapter 10 of this book.

Eventually, the inevitable glitches and unforeseen obstacles were addressed, and all three plants concentrated on manufacture of the Pattern 1914 rifles. The first rifles were delivered by Winchester the week ending February 15, 1916, followed by Remington-UMC the week ending February 29th and Eddystone the week ending March 31, 1916.[21] However, another problem soon cropped up that threatened to seriously reduce production of the rifles. As related by author Ian Skennerton:

> "The first rifles turned out in 1916 were rejected by the British Inspection Department who had been sent to America to control quality, and although it was considered that the improvements in the standard were achieved, the Inspectorate failed a large percentage of these early batches. Continued large scale rejection by the Inspectorate resulted in the American companies' refusal to continue the contracts at the quoted prices, and their management indicated that the additional time and effort required to meet the British Inspectorate's standards would result in an increase in the unit price.
>
> Difficulties in meeting the required standards were encountered particularly by the Eddystone factory, and certain relaxations in the standards were permitted, and extended to all the factories."[22]

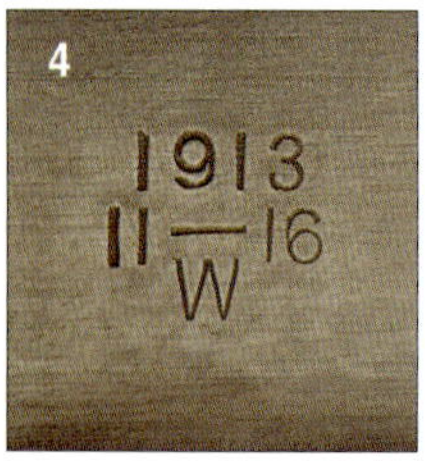

1 Pattern 1914 rear sight assembly (left side). Note the British proof/inspection markings on the side of the receiver. *(National Firearms Museum)*

2 Pattern 1914 rifle volley sight. *(National Firearms Museum)*

3 Remington Pattern 1914 receiver markings. Note the asterisk "✱" marking denoting the rifle is a modified Pattern 1914 Mk 1*. *(Great War forum)*

4 Blade markings on Pattern 1913 bayonet made by Winchester. Note the "1913" designation and "11 — 16" (Nov. 1913) production date. *(Arundel Militaria)*

5 Blade markings on Pattern 1913 bayonet made by Remington. Note the April 1917 production date. *(Tenniswood Militaria)*

6 British property and inspection markings on Pattern 1913 blade. *(Tenniswood Militaria)*

7 Winchester Pattern 1914 receiver markings. *(Great War forum)*

8 Eddystone Pattern 1914 receiver markings, low serial number "624".

9 Typical manufacturer's "roundel" on Pattern 1914 stock. This example is on a Winchester-made rifle. *(Rock Island Auctions)*

Pattern 1914 Mk I* (T) Sniper rifle with BSA telescopic sight. *(Rock Island Auctions)*

Prototype Winchester Pattern 1914 rifle with 20-round extension magazine. *(Royal Armories, Nat Edwards, Director General and Master of the Armouries)*

Pattern 1913 bayonet and British-pattern leather scabbard as issued with Pattern 1914 rifles.

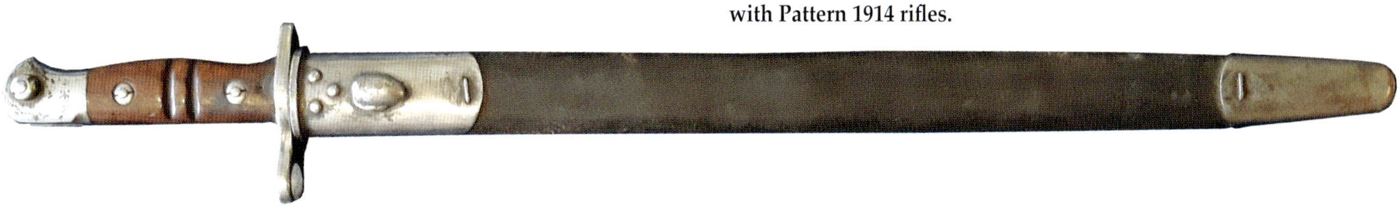

The relaxation of the stringent inspection standards resulted in more of the rifles being accepted. Clearly, the British were willing to sacrifice some degree of interchangeability in order to facilitate production of the rifles. However, the British were not completely comfortable with the lessened standards of interchangeably and designated the rifle produced by the three manufacturers as a distinct model — Pattern 1914 Mark I [W], Mark I [R]and Mark I [E], with the "*" marking on most.[23]

Due to the delays, however, it was not until May of 1916 that the American-made rifles reached England.[24] As will be discussed in a subsequent chapter, the easing of the British inspection standards later resulted in some controversy regarding the early Model 1917 rifles.

The Pattern 1914 Mk I rifle weighed 9.13 pounds. The barrel was 26 inches long with an overall length of 46.16 inches. Unlike the later U.S. Model 1917 rifles, the Pattern 1914 barrel was not marked with the initials of the manufacturer along with the date of production behind the front sight. The rear sight had a folding leaf peep aperture adjustable for elevation but not windage. The front sight was a blade with protective "ears."

Rather than the four rearward-sloping vertical grasping grooves found on the Pattern 1913 rifle, the Pattern 1914 had horizontal grooves on the fore-end of the stock with the exception of early-production Eddystone rifles, which had no grooves. The rifles were fitted with "volley sights," which had been commonly found on many British rifles for decades including the Pattern 1913. These sights were used chiefly for units firing in unison at very long range. The Pattern 1914 volley sights were graduated from 1,500 yards to 2,600 yards but were becoming obsolete by the time of the First World War.

The receiver ring was marked with the serial number and the identity of the manufacturer: "ERA" for Eddystone and "RE" for Remington. The Winchester-made rifles had a "W" prefix to the serial number. The left side of the receiver was stamped with British proof markings. The maker and year of production were stamped on the bottom of the barrel adjacent to the receiver ring. The bolt and rear sight leaf were stamped with the rifle's serial number. Winchester-made rifles typically had the serial number stamped on the outside of the bolt handle, while Remington and Eddystone stamped the serial number on the underside.[25] Stamping the rifle's serial number on various components was commonly done on many European military firearms, but the only serial number generally found on standardized United States military rifles was applied to the receiver. Most of the parts of sufficient size to do so were marked with the initial of the manufacturer.

The metal was finished in rust blue, called "browning," and the stock and handguards were made of good-quality black walnut. The right side of the stock was stamped with the manufacturer's "roundel" with the initials of the maker and, generally, either "1914" (stamped inside) or "PATT.'14" (stamped below).[26]

The initial Pattern 1914 Mk I rifle had some functioning issues, including magazine feeding and cartridge ejection. These were addressed by lengthening the bolt lock lugs and recess in the chamber. On December 21, 1916, the modified rifle was designated as the Pattern 1914 Mk I*. The modified Pattern 1914 Mk 1* rifles had the identifying asterisk "*" marking stamped on top of the receiver ring. The majority of the rifles made by the three American firms were Pattern 1914 Mk I*. The unmodified Pattern 1914 Mk I rifles are uncommon today.[27]

By mid-1917, increasing numbers of Pattern 1914 rifles began to see combat action in France where they proved to be reliable, rugged and accurate rifles. Many were also used in training. The rifle's inherent accuracy made it a candidate for use as a sniping weapon, and various types of telescopic sights were fitted to select specimens for testing and evaluation.[28] In 1918, the "Pattern 1914 Mk I"* (T) was introduced into service with the Commonwealth forces. The newly introduced sniper rifle was generally fitted with a Birmingham Small Arms (BSA) telescopic sight. Another variant was the "Pattern 1914 Mk1" (F) fitted with a "fine adjustment" rear sight with screw-type knob, similar to the later SMLE No. 4 rear sight.[29]

Variations of telescopic-sighted Pattern 1914 rifle remained in use well into the Second World War.[30] Most, if not all, of the Pattern 1914 rifles converted for sniper use in WWI were reportedly made by Winchester.

There is a Winchester Pattern 1914 rifle fitted with a 20-round magazine in the Royal Armories National Museum Collection in Leeds, England. Little is known about this interesting variant, and it was likely a prototype "one-off." Similar extension magazines were fabricated during the First World War for the American Model 1903 rifle and German Mauser, ostensibly for trench warfare, but few were actually fielded.

The Pattern 1913 bayonets made by Remington and Winchester had an overall length of 21.75 inches with a 17-inch blade and two-piece walnut grips. Winchester bayonets were stamped with a small "W" on the left side of the blade between the ricasso along with the "1913" (the pattern designation) and the month and year of manufacture. The Remington bayonets had "Remington" enclosed in a circle on the left side along with "1913" and the month and year of production. The right side was stamped with British proof/acceptance markings including the "broad arrow" British government ownership marking.[31] The Pattern 1907 scabbards made for the British Pattern 1913 bayonets were manufactured under subcontract by the Jewell Belting Company (Hartford, Conn.) and Graton & Knight Company (Worchester, Mass.).[32]

By the fall of 1916, the pressing need for service rifles lessened a bit, and after some negotiations, an agreement was reached between the British government and the three American manufacturers to reduce the total production numbers for the Pattern 1914 rifles to 1,250,000. However, even this revised total was never reached.[33] Most of the production machinery to make the new rifles was purchased by the British government and provided to the three manufacturers.[34]

While there may be some slight variations on published sources, based on reliable documentation, the final total of Pattern 1914 rifle production in the United States was:

Eddystone: 604,940 – Final delivery June 30, 1917
Winchester: 225,508 – Final delivery July 7, 1917
Remington-UMC: 403,117 – Final delivery August 11, 1917[35]

While these numbers total 1,243,565, British government records indicate that the final number of these rifles accepted in Great Britain was 1,117,850, The difference is explained by author Luke Mercaldo, "The difference between the number of rifles delivered in the U.S. and accepted in Britain was a result of losses at sea (approximately 25,715) and 100,000 Pattern 1914 rifles sent directly to India from the Eddystone Plant. This shipment to India was authorized in March, 1917."[36]

When the last rifles left the Remington factory on August 11, 1917, manufacture of the Pattern 1914 rifle came to a close. However, the story of the "U.S. Rifle, Cal. 30, Model of 1917" was just beginning.

Chapter 2

Genesis of the Model 1917 Rifle

"One of the Great Decisions of the Executive Prosecution of the War"

By 1916, it was apparent to all but the most myopic that the United States could very likely be drawn into the war that had been raging in Europe for two years. Should such an event occur, the American military would be faced with a serious shortage of modern armaments, including service rifles. The standard rifle at the time was the "U.S. Rifle, Caliber .30, Model of 1903," which had been manufactured by two government-owned and operated facilities. Springfield Armory was still manufacturing '03 rifles in 1916, but Rock Island Arsenal had ceased production of the weapon in November of 1913, although the production machinery remained in place at the arsenal.

At the time, the United States had about 600,000 Model 1903 rifles in inventory, along with some 160,000 .30-40 Krag rifles.[1] The Krags were judged to be obsolete and only suitable primarily for training and security duties. The number of Model 1903 rifles in inventory was insufficient for the Ordnance Department's projected requirements of at least one million more rifles in the event of war. On September 19, 1916, Rock Island Arsenal was directed to reestablish its Model 1903 production line, but the workforce that had previously manufactured the rifle was gone, and many of the experienced machinists had taken jobs in other industries. Also, due to the demands of the war in Europe, raw materials and machine tools were in short supply. It would be several months before Rock Island could get production underway even at a reduced level. However, even operating at full capacity, the anticipated demand would soon outstrip the combined production capabilities of Springfield and Rock Island. This was due, in some measure, to the outdated manufacturing equipment at the national armories and the seemingly inefficient production procedures.

With this in mind, in 1916, Secretary of War Newton D. Baker requested that Fred H. Colvin, editor of the widely respected publication *American Machinist* and a recognized expert on machine tooling and manufacturing, prepare a comprehensive treatise to assist civilian contractors with manufacturing the Model 1903 rifle. This culminated with Colvin and an associate, Ethan Viall, publishing a book containing the extremely detailed engineering drawings, directions and specifications needed for a manufacturer to go into production of the '03 rifle should such production become necessary.[2]

Colvin discussed his work at Springfield during this period in his classic book *60 Years with Men and Machines*:

> "In August of 1916 I was accordingly given office space in the Armory by its able commandant, Col. W.S. Peirce, who was assisted by Maj. G.H. Stewart and Capt. R.R. Nix. All of whom gave me their utmost cooperation in the work... I was able to complete the research end of the project in about four months. Beginning with the barrel, we studied the rifle piece by piece and meticulously followed the operations being performed. We reported on the kinds of material used, the tools and cutters, feeds and operating speeds, cutting lubricant, heat-treatment, and the time required for each separate operation. This detailed information then appeared in the *American Machinist*, first in successive issues, then in a special edition, and finally came out in book form."[3]

Colvin's detailed findings did not reflect well on the work being done in Springfield, as explained in his book:

> "Much of the machine equipment at the Springfield Armory, unfortunately, was very antiquated in comparison with contemporary standards. The speeds and the feeds were almost leisurely and the methods employed were of about...1890 vintage. Perhaps the greatest handicap on some of the parts, however, was some of the design and the materials indicated in their specifications. Regardless of the use of a particular part in the operation of the rifle, or the stresses to which that part was subjected, the designers of small arms insisted on having forgings throughout. Such parts as the band that held the barrel on the stock, the butt plate, or the swivel, which carried the strap for slinging the rifle on the shoulder were required to be machined from deep forgings, which meant machining them all over."

These and other obvious inefficiencies began to bother me and I lay awake nights mulling over the problems. I requested and obtained an interview with the Secretary of War Newton D. Baker, who happened to be staying at the same hotel in Springfield while he visited the Armory. I told Secretary Baker of some of the practices I had observed, which didn't seem quite right to me, and pointed out where they could be improved upon… Moreover, if these West Point Officers, who use the rifles and train other men how to use them, do not know how to make the rifles, who in the name of reason does?"[4]

Secretary Baker listened to Colvin's concerns and came to the conclusion that the situation required more investigation. Baker requested that Colvin prepare a detailed report covering all of the deficiencies he pointed out and give recommendations on how the situation could be improved. He closed with the following instructions: "Oh — one very important point — make sure you mark your report 'Confidential' and see that it is addressed to me personally."[5]

Colvin did as instructed, but unfortunately, when the report was received in Secretary Baker's office, his secretary inadvertently gave it to General Crozier instead. After reading the pointed criticisms in the report, Crozier was livid and sent a message to the *American Machinist* saying, in effect, "Colvin doesn't know what he's talking about," and ordered that he be barred from any future visits to the armory. When Secretary Baker learned of this turn of events, he apologized to Colvin and said he "…regretted sincerely the position" in which he had been placed. Apparently, Baker had a conversation with General Crozier as well. A few days later, Crozier issued an order rescinding the first one and stated that Colvin would be permitted to visit Springfield as well as any other arsenals. This news leaked out and was reported in the local press, which resulted in Colvin becoming something of a *cause celebre*. In a subsequent meeting with Secretary Baker, Colvin was still angry about his treatment by Crozier and made the scathing observation, "…there is a great need for officers around here who know more about modern production methods than they know military regulations."[6]

Newton D. Baker, Secretary of War, 1917–1921. *(Library of Congress)*

Brig. General William Crozier, Chief of Ordnance, 1901–1917. *(Library of Congress)*

Benedict Crowell, Asst. Secretary of War and Director of Munitions, 1917–1920. *(Library of Congress)*

On April 6, 1917, the United States declared war, and the projected shortage of service rifles with which to arm our burgeoning military went from a theoretical problem to a stark reality. To this end, the War Department considered possible options for addressing the serious situation.

OPTION #1 — Make arrangements for the Model 1903 rifle to be manufactured by commercial arms manufacturers and to increase the output of Springfield Armory and Rock Island Arsenal.

This was exactly the scenario envisioned by Secretary Baker when Fred Colvin was tasked with evaluating Springfield Armory's rifle production program and then prepare and publish the necessary drawings and specifications required for commercial firms to manufacture the '03.

However, there were several obstacles making this course of action unachievable. As evidenced by Colvin's report, the serious deficiencies with Springfield Armory's rifle production program (and by default Rock Island's as well), would preclude any likelihood of significantly increased rifle production by either entity that would remotely meet the demand.

It was also seen that engaging commercial manufacturers to produce the Model 1903 rifle would result in lengthy time delays that were required to find suitable manufacturers, negotiate contracts, acquire the necessary production machinery and inspection gauges and recruit and train workforces. This would result in a crippling shortage of rifles when they would be most needed.

OPTION #2 — Adopt the British Pattern 1914 rifle "as is." This option was available due to a fortuitous circumstance that was later referred to by the Assistant Secretary of War as "well-nigh providential."[7] The three plants manufacturing the Pattern 1914 rifle, Winchester, Remington, and Eddystone, were winding down production of the British rifle at precisely the time America was coming into the war. All of these plants had modern manufacturing equipment and experienced workforces still intact and could continue manufacturing the Pattern 1914 rifle for the United States government in large numbers in short order.

The War Department explored the positive and negative aspects of adopting the British Pattern 1914 with no alterations. The most obvious benefit would be the ability to manufacture large numbers of serviceable military rifles that would be available to the American military with little delay in production.

There were, however, several inherent disadvantages. Adoption of the Pattern 1914 rifle, chambered for the British .303 caliber cartridge, would result in two different types of ammunition having to be produced and issued since the Model 1903 rifle and American machine guns were chambered for the caliber .30 Springfield (.30-06) round. This would severely complicate logistics and supply problems. Also, it was widely believed that the rimmed .303 round was markedly inferior to the cannelured .30-06 in several respects.

OPTION #3 — Adopt a version of the Pattern 1914 rifle modified for the U.S. .30-06 cartridge. This would supply the United States military with a robust and serviceable rifle that could be manufactured using the existing Pattern 1914 machinery (slightly modified) and work forces. Use of the excellent .30-06 cartridge would eliminate the logistical problems inherent in having to manufacture and issue .303 caliber ammunition as well.

The major negative was that it would take some time to reengineer the Pattern 1914 rifle for the American cartridge. Also, there were a number of individuals, in and out of the military, who weren't fans of the British rifle and thought the United States should concentrate all efforts on producing more Model 1903 rifles.

The authoritative book *America's Munitions 1917–1918*, by Benedict Crowell, the Assistant Secretary of War and Director of Munitions, addressed some of the issues involved and comparisons of the Pattern 1914 to the Model 1903:

> "The rifle (Pattern 1914) had many characteristics of the 1903 Springfield, but it was not so good as the Springfield in its proportions, and its sights lacked some of the refinements to which Americans were accustomed. Yet even so it was a weapon obviously superior to either the French or Russian rifles. The ammunition it fired was out of the question for us. Not only was it inferior, but, since we expected to continue to build the Springfields at the Government arsenals, we should if we adopted the Enfield as it was, be forced to produce to two sizes of rifle ammunition, a condition leading to delay and unsatisfactory output. The rifle had been designed originally for rimless ammunition and later modified; so it could be modified readily back again to shoot our standard .30-caliber Springfield cartridges."[8]

Actually, this option was something of a *fait accompli*. The Ordnance Department had already decided on this course of action, pending approval by the War Department. This was the subject of the following memo dated April 11, 1917, just four days after the declaration of war:

> "From: The Ordnance Office
> To: The Adjutant General of the Army
> Subject: Emergency procurement of small arms other than of U.S. design.
>
> 1. In the present emergency it is necessary that this Department secure such arms as may be readily available during the period, which must elapse before the manufacturing concerns can reach a state of quantity production of small arms of the adopted design.
> 2. It is therefore recommended that authority be given for this Department to procure such small arms within the limits of authorized expenditures, with the understanding that serviceable small arms, and in so far as is possible such as are closely related to our types and use our standard ammunition, shall be procured, and in that procurement of this class of small arms shall continue only until it can be replaced without pause by production of adopted types."[9]

A subsequent memo from the Chief of Ordnance requested more specific information regarding the proposed small arms in question. A reply dated April 19, 1917, contained the following:

> "1. The small arms under consideration by this Department at the present are as follows:
> a) Enfield rifle, modified to take the service caliber .30 ammunition.
> b) Smith & Wesson revolver, caliber .45, designed to take the service flangeless cartridge.
> c) Colt revolver, caliber .45, designed to take the service flangeless cartridge.
>
> 2. It is contemplated that placing initial orders for 1,000,000 modified Enfield rifles and 100,000 each Smith & Wesson and Colt revolvers."

This memo had the following handwritten notation at the bottom: "Taken up personally with the Secretary of War and the Acting Chief of Staff."[10]

Needless to say, this recommendation (Option #3) was approved and resulted in the adoption of a Pattern 1914 rifle, modified to take the U.S. .30-06 cartridge, and designated as the "Rifle, Caliber 30, Model of 1917." In hindsight, this was unquestionably the most reasonable course of action. Secretary Crowell commented on the decision to go with the modified Enfield rifle despite the inevitable disapproval from some quarters that would follow: "The decision to modify the Enfield was one of the great decisions of the executive prosecution of the war — all honor to the men who made it."[11]

The bayonet adopted for use with the Model 1917 rifle was the same Pattern 1913 bayonet used with the Pattern 1914 rifle, which had been manufactured by Winchester and Remington. When adopted, it was given the "Model 1917" designation, and plans were made for Winchester and

Remington to manufacture bayonets for the new rifle. The Pattern 1913 bayonets not yet delivered to the British were acquired by the United States and redesignated "Model 1917." The only modification was that the British property markings on the blade were defaced and overstamped with "U.S." markings. The Model 1917 bayonet will be discussed in detail later in this book.

Almost immediately, the new rifle was frequently referred to as an "Enfield," but not everyone concurred that this was a proper designation for the Model 1917 rifle. As related in an *American Rifleman* article:

> "Lt. Col. Webley Hope, a British officer responsible for inspecting Pattern 14 rifles in America, implored Brig. Gen. Crozier to shun the word 'Enfield' for three reasons. First, it was not the official British nomenclature for the Pattern 14, and the attachment of the term 'Enfield' was an error made in the early days of the war. Second, he felt 'it would be anomalous to give this name to a rifle, which requires modification in so many ways in order to take the U.S. cartridge.' And third, was 'the importance of the new rifle not being confused with the British Lee-Enfield rifle...' "[12]

Regardless of Lt. Col. Hope's admonitions, the term "Enfield" to refer to the Model 1917 rifle continued unabated and was frequently cited in official War Department, Army, and Ordnance documents. Collectors today routinely refer to the rifle as "Model 1917 Enfields" or "American Enfields."

Almost immediately after the U.S. declared war on Germany, General Crozier made overtures to Winchester about the feasibility of the company manufacturing such rifles for the government. As will be discussed in greater detail in Chapter 4 of this book, Winchester was not immediately offered a production contract but was authorized to develop a Pattern 1914 rifle modified for the .30-06 cartridge.[13] The reason(s) why Winchester was selected to develop the modified rifle rather than Remington is not known.

Secretary of War Baker announced the decision to adopt the modified British rifle in an April 28, 1917 speech in which he stated that "...an unlimited supply of rifles..." had been found. However, possibly anticipating that the decision might be criticized by factions that still favored the '03, Baker added "...these rifles would serve until production of the American Springfield could catch up."[14] This was a bit disingenuous since Baker knew that finding additional sources of Model 1903 Springfield production was unlikely.

Despite Secretary Baker's sanguine comments about adopting the modified Pattern 1914 rifle, this decision was far from universally praised. Even before any of the newly adopted Model 1917 rifles had left the factories, an article appeared in the *New York Sun* titled "Why our Forces in France Must Use an Inferior Rifle." The article contained numerous erroneous facts and half-truths. There were other stories in the media questioning the wisdom of adopting any rifle other than the Model 1903 Springfield. In order to set the record straight, *Arms and the Man* magazine (predecessor to the NRA's *American Rifleman*) ran a feature article on July 28, 1917, refuting the *New York Sun* story and outlining the valid reasons for adopting the Model 1917 rifle. The *Arms and the Man* article pointed out that the Model 1917 rifle actually had sights that were superior to the Model 1903 for combat use.[15]

A follow-up *Arms and the Man* article made several more positive statements about the Model 1917 rifle and some other interesting observations, including a jab at the isolationist factions in the United States prior to the nation's entry into the war: "Had it not been for the opposition to preparedness, which prevailed until recently we might have been able to provide our troops with the Springfield rifle. However, the modified Enfield will shoot as accurately as the Springfield rifle and in the hands of the average marksman will be in every way as efficient."[16]

The negative publicity notwithstanding, with the decision to adopt the Model 1917 rifle, arrangements had to be made with the British for acquisition of the Pattern 1914 manufacturing equipment. Great Britain financed and held title to much of the machinery and tooling used by the three American contractors. Contractual agreements stated that the equipment and other manufacturing implements and raw materials still on hand when production of the rifle ended would revert to Great Britain. The contracts were scheduled to be completed in the spring of 1917, but with the decision to adopt the Model 1917 rifle, the deck was reshuffled and new arrangements made regarding disposition of the Pattern 1914 manufacturing equipment and related material. With this in mind, the following agreement was reached:

- The British government would sell the machinery, equipment, tools, dies, jigs, fixtures, cutters and gauges for $900,000 (these cost Great Britain $20,000,000).
- Raw materials and those in process to be taken over by the United States.
- All new small tools, which were unused to be sold at cost to the United States.
- The transfer to take place prior to the completion of the British contracts, so that the initial 1 million U.S. rifle contract could begin as soon as possible, even though it would mean simultaneous production for a short time.
- The British government would assign (an existing) contract for bayonets with Remington at $2.75 per bayonet."[17]

Since the U.S. wouldn't have any need for Pattern 1914 components that were still on hand, those were returned to England.

Plans were formulated for Winchester, Remington, and Eddystone to switch gears and begin manufacturing the Model 1917 instead of the Pattern 1914. It was soon discovered that this was much easier said than done. Interchangeability of parts for military firearms had been a touchstone of the U.S. Ordnance Department since the late 1790s when Eli Whitney began making muskets under government contract. It was made clear from the beginning that the Model 1917 rifles made by the three contractors would have to have a very high degree of standardization of the components. This was in contrast with the British policy, which required only a modicum of standardization of parts for the Pattern 1914 rifle.

One of Fred Colvin's major criticisms when he had submitted his report back in the fall of 1916 was Springfield Armory's insistence on absolutely perfect specifications with no allowable deviations. Of course, it would not be possible to have anything close to total interchangeability without adhering to exact specifications. This was the topic of discussion during a meeting between officials of Winchester, Remington, and Eddystone held in New York City on April 4, 1917:

> "The object of the meeting was to standardize as far as possible the materials from which the Enfield Rifle Pattern #14 is to be made for the United States Government. Also, to standardize as far as possible the changes necessary to adapt the Enfield Pattern #14 to take the .30 Government 1906 cartridge.
>
> ...It was the concensus [sic] of opinion of those present that inasmuch as there had never been any attempt during the life of the English contract to make the product of the three companies interchangeable that it would be impossible at this late date to attempt interchangeability between the three companies. It is further the opinion of the Conference that whatever specifications were laid down that they be made as broad as possible in order to give the individual manufacturers the greatest amount of latitude consistent with producing properly functioning arms."[18]

The manufacturers apparently understood that some acceptable level of interchangeability would be required by the Ordnance Department but endeavored to go on record that it should be limited so as not to unduly hinder the production rate. There was undoubtedly some thought that "If minimal interchangeability was good enough for the British, it should be good enough for Uncle Sam." Fred Colvin concurred, and he was once again critical of the Ordnance Department's insistence on total interchangeability. Colvin pointed out in a *New York Times* article that the British required only minimal parts interchangeability, yet well over a million rifles, that functioned perfectly well, were manufactured on essentially the same machinery that would make the Model 1917 rifle.[19]

Despite, these objections, the U.S. Ordnance Department wasn't terribly concerned with what the British did or didn't do and considered interchangeability to be vitally important for two major reasons:

1. Manufacture would be much faster and more efficient since hand-fitting of individual components would not be necessary.
2. When rifles needed repair "in the field" or when overhauled / rebuilt at ordnance facilities, non-interchangeable parts would severely hamper efficient replacement of broken or worn parts.

It was determined that the following alterations were required to adapt the .303-caliber Pattern 1914 rifle for use with the American .30-06 cartridge:

"**Receiver**

(1) Cartridge opening in bottom changed in shape and size to suit different shape of cartridge.
(2) Bullet ramp altered for same reason.
(3) Ejector slot changed to suit altered ejector
(4) Magazine engagement surfaces altered to fit new magazine.
(5) Clip guides altered to fit U.S. clip.

Bolt

(1) Front face altered to suit new diameter cartridge head.
(2) Firing pin hole enlarged to suit larger diameter pin.

Firing Pin

Diameter increased to same as Springfield rifle M1903.

Extractor

Front end changed to provide proper engagement with Model 1906 ammunition.

Main Spring

Made stronger.

Trigger Guard

(1) Magazine opening in bottom altered to suit new magazine.
(2) Front end changed to conform closely to stock contour.

Stock

(1) Long range dial sight eliminated, thereby changing form of stock at point where dial sight was attached.
(2) Magazine slot to suit new magazine.
(3) Barrel bed enlarged to suit larger diameter barrel.
(4) Marking disc cut eliminated.

Hand Guards

Barrel clearance cut enlarged in diameter to suit change in barrel.

Barrel

(1) Diameter of bore and dimensions of rifling changed to suit Model 1906 ammunition.
(2) Chamber dimensions changed for same reason.
(3) Bullet ramp changed to facilitate feeding due to new cartridge.
(4) Extractor cut changed to suit new extractor

Rear Sight

(1) Reduced in height due to flatter trajectory.
(2) Graduation changed to suit U.S. cartridge ballistics.

Aperture Sight

(1) Eliminated (this was the rear volley sight used in conjunction with the long-range dial sight eliminated from stock).
(2) Bolt stop spring abutment added to take place of that part of aperture sight that formed abutment for bolt stop spring.

Magazine Platform

Complete new design.

Ejector

Length and shape changed
Ejector marking disc eliminated."[20]

On May 10, 1917, the three companies each fabricated

Right side and top view of the Model 1917 rifle, identified as "Tool Room Sample" manufactured by Winchester. Unlike the production rifles, this sample gun had no markings. The rifle was retained by a Winchester production line superintendent and was subsequently donated to the National Firearms Museum. *(National Firearms Museum)*

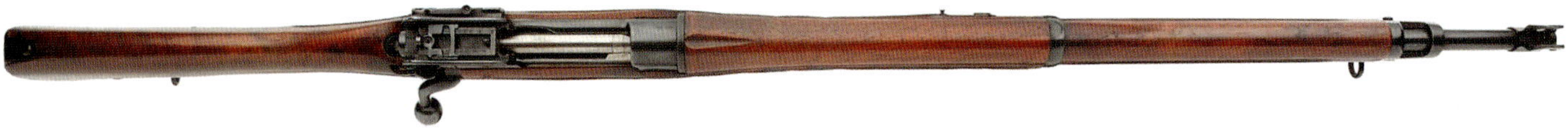

Receiver of the Winchester tool room sample Model 1917 rifle. Note the lack of markings on the receiver ring. *(National Firearms Museum)*

a sample of a Pattern 1914 rifle modified for use with the U.S. .30-06 cartridge and forwarded the guns to Springfield Armory for test and evaluation. This didn't go well and presented General Crozier and Ordnance with yet another problem. In his book *America's Munitions*, Benedict Crowell related the difficult decisions that had to be made to balance obtaining the greatest number of rifles without unduly sacrificing the performance and utility of the weapons:

"The Enfields sent to the Springfield Armory test were not standardized at all, but were largely hand fitted. Little or no attempt had been made to obtain interchangeability of parts among the rifles turned out by the three plants. Even the bolt taken from one company's rifle would not enter the receiver of another company's.

The Ordnance Department was confronted with of approving and issuing a weapon pronounced unsuitable by its own experts and thus obtaining speedy production, or of delaying until interchangeability was established. It chose the latter course.

On July 12th, a second set of rifles had been tested. These came more nearly up to our ideas of standardization, but were still not entirely satisfactory. Nevertheless, we decided to go ahead with production and improve the standardization as we went along. The Winchester and Ilion plants elected to start work on that understanding, but Eddystone preferred to wait for the final requirements. Ilion afterwards decided to postpone production until the final specifications were adopted. **It would have been well if the same course had been followed at the Winchester plant, for word came later from Europe not to send over rifles of Winchester manufacture of that period.** The final drawings of the standardized and modified Enfield did not come from the plants until August 18. Six days later the thousands of dimensions had been carefully checked and finally approved by the ordnance officers, and after that production started off in earnest.

The wisdom of adopting the Enfield rifle and modifying it to meet our requirements instead of extending the manufacture of Springfields was almost immediately apparent, for in August, almost as soon as the final drawings were approved, the first rifles were delivered to the government. This was possible because the modifications we adopted did not require any fundamental changing of machinery."[21] *[Bold print added by author]*

Crowell further elaborated on the initial production of the rifles, "Winchester delivered the first modified Enfields to us on August 18, Eddystone September 10, and Ilion (Remington) about October 28."[22]

Specifics regarding the manufacture of Model 1917 rifles by the three contractors will be discussed in more detail in a subsequent chapter of this book. Secretary's Crowell's somewhat cryptic comment above that **"It would have been well if the same course at been followed at the Winchester plant..."** will also be examined in more detail in a later chapter.

Col. John T. Thompson resigned from the U.S. Army to become works supervisor at the Eddystone plant. He subsequently went back into the Army with the rank of Brigadier General. *(U.S. Army)*

Once initial production got underway, perhaps not surprisingly, there were still problems to resolve and glitches to correct. General Crozier, who had always been something of a lightning rod for controversy, was under intense scrutiny and criticism by Congress, including the powerful Senate Military Affairs Committee. The complaints weren't restricted to rifle production but also included unsatisfactory handling of machine gun manufacture and other matters regarding America's munitions production.

The projections were that 100,000 Model 1917 rifles would be leaving the factories every month, but this wasn't happening. Although Secretary Crowell indicated that Winchester delivered its first M1917 rifles on August 18th, production soon afterward was virtually non-existent. As related by Colvin:

> "War was declared by President Wilson via Congress on April 6, 1917, and sample rifles that functioned satisfactorily were submitted early in June. Yet as late as August of that year I found, as a result of a personal inspection tour, that the Winchester Arms Company, which should certainly have been in (quantity) production by that time, was placidly marking time, awaiting the go-ahead signal from Washington. Orders for rifles had been expected every day for a period of weeks, but since no orders were being received the men were simply standing by idle and, of course, being paid full-time wages for doing it. No rifles could be produced without orders from headquarters, and the only orders received from Washington up to this time were merely 'Stand By.' I took several photographs in the plant of groups of men playing cards, reading newspapers, and otherwise killing time during regular working hours while the war was already four months old and we were still losing it.
>
> The facts and photographs I brought back from this visit to the Winchester Plant formed the basis for a scorching editorial in the *American Machinist* by managing editor Van Deventer. We did not print the photographs, but indicated that we had them on file as evidence in case anyone should think we were imagining things. My old antagonist General Crozier, having learned in advance from the grapevine that the editorial was coming out, went to considerable pains to have that issue of the *Machinist* suppressed or impounded, but freedom of speech triumphed in the end and the editorial came out on schedule."[23]
>
> Not surprisingly, the editorial was not well received. As Colvin put it, "...it made me about as popular as the proverbial skunk at a garden party. General Crozier flatly refused to have anything more to do with me, and I had great difficulty in getting to see his right-hand man, Gen. John T. Thompson. The result was that I was 'requested' to meet a committee of ordnance officers so that they might show me the error of my ways, and demonstrate how my view of things was exactly the reverse of the true situation."[24]

During this requested meeting, the various ordnance officers subtly, or sometimes not so subtly, denigrated Colvin's views and cast aspersions on his credibility. The following excerpts from the meeting illustrate the opinions of the "Ordnance establishment" along with Colvin's responses. One of the officers, a Brigadier General, commented:

General #1: "...Mr. Colvin, we mean no personal slight when we say that your knowledge of firearms cannot be as extensive as ours, for we have made it our life study so to speak. Whatever delays you may have observed, and we admit there have been some, stem from the urgent necessity of securing interchangeability of parts on these rifles — a problem, which we are working on night and day and which will be solved in the very near future. Then you will see how production will begin to roll."

Colvin: "But, General," I replied, 'What about the tens of thousands of Enfield rifles that some of these very plants we're talking about have already manufactured and shipped to British troops?"

General #1: "You're forgetting, aren't you, that those rifles had a different bore and a different chamber from those now being designed for our own troops."

Colvin: "I don't see how that detail makes any difference," I replied, forgetting to add 'sir'.

General #1: "Well, it does, and you must take our word for it."

Colvin: "Just what do you mean by interchangeability?" I asked one of the older generals.

General #2: "Interchangeability...harrumph—er, by interchangeability is meant that feature of one rifle, or 'piece,' as it is called, whereby a part of one rifle can be replaced with its corresponding part from another rifle

— a very important advantage in modern warfare. For example, — we have an infantryman in the heat of battle whose piece — rifle, that is — becomes disabled or defective for some reason — the trigger jams or the bolt freezes or the front sight gets knocked off. If these and other parts of the rifle are interchangeable with those of every other rifle, he has merely to stoop down, remove the corresponding part from the piece of the nearest fallen comrade, replace it on his own rifle, and continue firing at the enemy."

Colvin: I almost laughed out loud… "Why in the name of… for Pete's sake, General, why wouldn't the solider grab the rifle from the Fallen comrade instead of acting like a crazy fool and trying to make complicated repairs under fire? Is the enemy going to call a truce while our men sit down on the battlefield and take their guns apart?"

General #2: "The example I gave, the general answered coolly, was perhaps not too well chosen. The more complicated repairs would, of course, be made behind the lines by specially trained crews, who would thus salvage thousands of rifles that would otherwise be lost."[25]

It didn't take long for Colvin to realize that, as noble and patriotic as they may have been, many of the ordnance officers were not open to an outsider telling them what they should do.

Colvin wasn't the only critic of the Ordnance Department in general and General Crozier in particular. During a January 24, 1918, meeting of the Senate Military Affairs Committee regarding adoption and manufacture of the Model 1917 rifle chaired by Senator George E. Chamberlain and attended by General Crozier and a number of witnesses, including the officials from Remington, Winchester and Eddystone, there was much criticism of Crozier's handling of the change-over from the Pattern 1914 rifle to the Model 1917 rifle. Senator Chamberlain was clearly in favor of adopting the modified Pattern 1914 rifle but felt that Ordnance unnecessarily delayed the beginning of production by ignoring input from the companies that had made the Pattern 1914 and would produce the Model 1917. As stated in Senator Chamberlain's comments:

"Let us now consider the question of rifles. We were furnishing Lee-Enfield [sic] rifles to the British government in large numbers. The factories were prepared for them. It is true that Great Britain was trying to make an improvement upon the rifles used by her when she became involved in the war, but when the war came on Great Britain said we will not waste any time improving our rifles, but will get them out just as fast as we can, and they have been manufacturing them ever since.

What did America do? With 700,000 rifles in America and in our colonial possessions, a motley group of different kinds of guns, America was seeking, through the Ordnance Department, to improve the rifle that Great Britain was manufacturing here and which we could have put out without any trouble in the factories. We went to work, through the Ordnance Department, to improve the Enfield rifle. I am frank to say it is a great improvement. I believe it is a better gun than the English gun, but here while the house was burning America was determining through its Ordnance Department what instrumentalities ought to be adopted to put out the fire. It took weeks and months before they finally got the Lee-Enfield rifle into condition where the Ordnance Department thought it was all right. And after this was agreed upon there were further delays caused by indecision.

Here were the engineers of these great arms companies, who got together and finally agreed upon a program for the manufacture of …these guns, and concluded that they would manufacture them with seven interchangeable parts, and they started to manufacture the gauges, the jigs, and dies, and everything necessary for the manufacture of guns with seven interchangeable parts.

After the Ordnance Department had practically accepted the suggestion, it went to work through a distinguished ordnance officer and changed the plan from 7 to 40 interchangeable parts, and finally raised it to over 50 interchangeable parts, with the result that everything had to be stopped for a while that additional gauges might be made. This may have resulted in improvement, but why the delay in the midst of the smoke of battles?"[26]

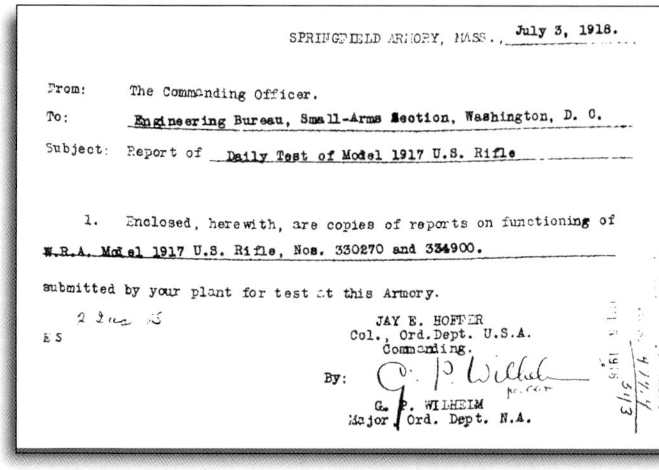
SPRINGFIELD ARMORY, MASS., July 3, 1918.

From: The Commanding Officer.
To: Engineering Bureau, Small-Arms Section, Washington, D. C.
Subject: Report of Daily Test of Model 1917 U.S. Rifle

1. Enclosed, herewith, are copies of reports on functioning of W.R.A. Model 1917 U.S. Rifle, Nos. 330270 and 334900.

submitted by your plant for test at this Armory.

ES

JAY E. HOFFER
Col., Ord.Dept. U.S.A.
Commanding.
By: G. P. Wilhelm
G. P. WILHEIM
Major, Ord. Dept. N.A.

July 3, 1918, memo from Col. Jay E. Hoffer, Commanding Officer of Springfield Armory to the Engineering Bureau, Small Arms Section, Washington, D.C., regarding the reports of daily functioning tests of Model 1917 rifles. *(Springfield Research Service. Courtesy of Frank Mallory)*

Eventually, steps were taken to correct the delays and obstacles that had unnecessarily hindered efficient manufacture. Secretary Crowell related in *America's Munitions* how the Model 1917 manufacturing program began to rapidly improve and highlighted the disparity between production of the Model 1917 and Model 1903:

"The progress in the manufacture was steadily

upward. During the week ending February 2, 1918, the daily production of military rifles in the United States was 9,257, of which 7,805 were modified Enfields produced in the three private plants, and 1,442 were Springfields built in the two arsenals. The total production for that week was 50,873 guns of both types, or nearly enough for three Army divisions. In spite of the time that went into standardization of the Enfield rifle, all troops leaving the United States were armed with American weapons at the ports of embarkation."[27]

Although total interchangeability was never reached, 95 percent was achieved, which was deemed to be sufficient. The manufacturers sent small "samples" of rifles daily to Springfield Armory to be checked for proper functioning and to make sure they were within the mandated tolerances. The commanding officer of the armory sent reports of the findings to the Engineering Bureau, Small Arms Section, in Washington, D.C.[28]

Even though reasonable numbers of Model 1917 rifles were beginning to leave the factories, there were concerns in some quarters that Ordnance's insistence on a high degree of interchangeability unduly hampered production. For example, a December 20, 1917, Winchester memo from company Vice President Henry Brewer to the War Department stated: "...up to Saturday, December 15, we had delivered to the Government 83,000 rifles but...we could have delivered in that time 170,000...made to the British degree of interchangeability."[29]

Again, the Ordnance Department wasn't interested in such conjecture, and the "company line" was that the high degree of interchangeability actually speeded up manufacture significantly. As an illustration, according to the assistant secretary of war, the fastest rate of assembly by a single mechanic for the Pattern 1914 rifle was 50 in one working day. After the standardization of components for the Model 1917, the fastest rate of assembly was 280 with an average rate of 250 per day in the same factories.[30]

There were numerous factors that went into achieving such an impressive rate of production. General William Crozier's performance, as has been discussed, wasn't conducive to optimum manufacturing efficiency. In early 1918, Crozier was transferred to the War Council and subsequently retired from the Army in 1919. In May 1918, Brigadier General Clarence C. Williams was appointed Acting Chief of Ordnance, and on July 19, 1918, was promoted to Major General and officially became Chief of Ordnance, a position he held until 1930.[31] From all accounts, General Williams did an exemplary job. Colvin remarked about the "...splendid way in which Williams handled the task, both during the war and in the years that followed up to his retirement."[32]

On November 9, 1918 (three days before the Armistice), the total combined production of Model 1917 rifles by the three manufacturers was 2,133,429.[33] Even though all outstanding production contracts were cancelled soon after the cessation of hostilities, the contracts contained a provision that the companies would receive payments for any unfinished rifles "still in process." Since it made sense to finish these rifles rather than have the government pay for partially completed guns, the three manufacturers remained in production for several months after the cancellation of the contracts. Individual totals for each of the prime contractors are shown in the next chapter of this book. Published sources vary slightly regarding the total number of rifles made,

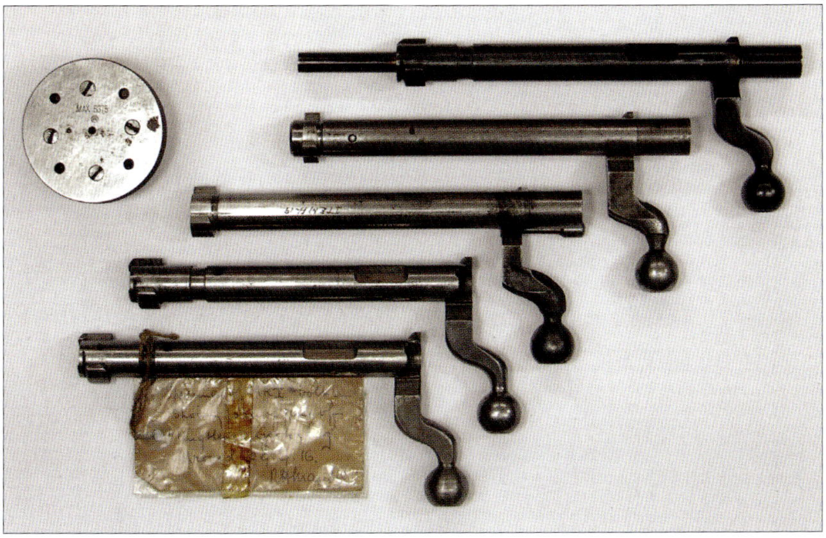

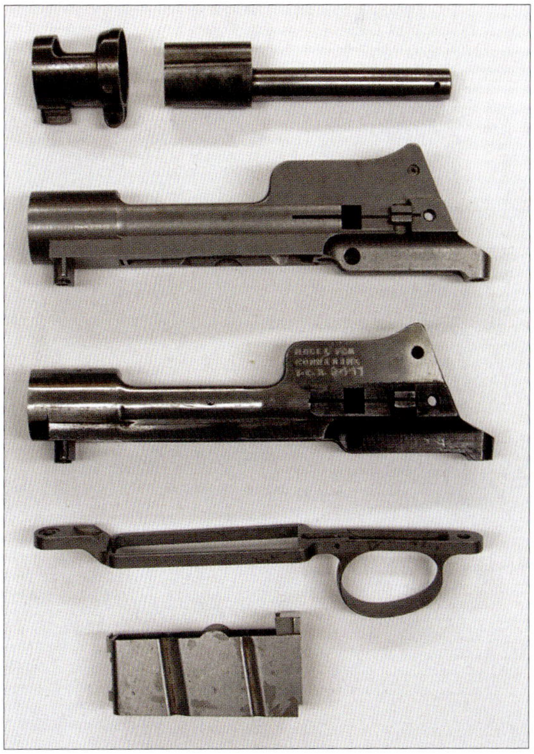

Manufacturing inspection gauges for Model 1917 receiver (shown at right). Manufacturing inspection gauges for Model 1917 bolt (shown above). These gauges were used to measure the dimesions of the various components during the manufacturing process to verify they were within specifications, which was necessary for interchangeability of parts between the manufacturers.
(Hayes Otoupalik collection. Bill O'Donnell photo)

but based on a May 3, 1919, Ordnance report, 2,556,274 rifles were produced by the time manufacture completely ceased.[34] These totals were indeed impressive for a weapon that was only in production for less than two years.

While most doughboys would likely have preferred to have been issued the trimmer, lighter Model 1903 rather than a Model 1917, in most cases, it was never a choice between the two rifles. If the Model 1917 hadn't been available, the only other option would have been between nothing or, perhaps, a hodgepodge of leftover Lebel or SMLE rifles from our French and British allies. While the "American Enfield" may have not been quite as good of a weapon as the '03 in some respects, it was equal, or better, than any of the rifles fielded by our allies or adversaries. As the Secretary Crowell put it:

> "We should have had to spend preliminary months or even years in building up an adequate manufacturing equipment for Springfields, the while our boys in France were using what odds and ends of rifle equipment the Government might be able to purchase for them, except for a condition in our small-arms industry in early 1917 that now seems to have been well-nigh providential."[35]

Noted firearms writer Edward C. Crossman penned an article titled "British Rifles for the American Army" in the November 1917 issue of *National Service Magazine* in which he thoroughly covered all aspects of the new rifle. As evidenced by the following excerpts, he didn't care for the rifle's aesthetics but deemed it better in almost all respects than the Model 1903 as a combat weapon. Crossman concluded his article with perhaps the highest accolade possible:

> "The new rifle, while better finished than the New Springfield, is not at all a pretty rifle in outline. The wide magazine, holding six shots, and the high bridge on the receiver for the protection of the rear sight, gives the rifle a very clumsy, bulky looking aft, like the built-up poop deck of a long freight steamer of the type seen so often on the Great Lakes.
>
> The stock is also very homely, having no comb and having a peculiar protuberance on the grip, and which looks as if the comb has been turned upside down.
>
> ...For the draft army soldier the rifle is much superior to the Springfield, because of its better rear sight. It is a better man-killing weapon for anybody. Then again, it is better finished, and better designed in some respects, and uses better material, such as nickel steel in the bolt and receiver.
>
> The fates were kind to us in having the machinery available to make such a splendid rifle, just as they were in giving us time to get ready, with the stout arms of the British and French between us and our enemy.
>
> If it be that some relative of yours is in the draft army, and peace does not come before he goes to the front, rest assured that in the rifle he carries, he has the finest man-killing rifle ever put into the hands of our soldiers."[36]

By the time of the Armistice, an estimated 75 percent of the soldiers of the American Expeditionary Force were armed with Model 1917 rifles. The ready availability and excellent performance of the rifle during the First World War certainly validated the proclamation that adopting the modified rifle was indeed "One of the Great Decisions of the Executive Prosecution of the War."[37]

Chapter 3

Components

U.S. Model of 1917 Rifle. *(National Firearms Museum)*

Principal Dimensions, Weights, and Miscellaneous Data from Basic Field Manual FM 23–6

Weight, w/o bayonet	9.187 pounds
Weight, w/bayonet	10.312 pounds
Length, w/o bayonet	46.3 inches
Length, w/bayonet	62.3 inches
Diameter of bore	.30 inches
Trigger pull, minimum	3 pounds
Rifling: Number of grooves	5
Rifling: Twist, uniform, left hand, one turn in	10 inches
Sight radius (distance from top of front sight to side of left, leaf raised)	31.76 inches
Sight radius (battle sight)	31.69 inches
Magazine capacity	6 rounds[1]

While it was desired that the newly adopted Model of 1917 rifle be as dimensionally close to the Pattern 1914 as possible so as to be able to use the same manufacturing machinery with minimal alterations, some changes would obviously be required. Of course, most of these modifications would be due to the switch from the British .303 cartridge to the American .30 Springfield (.30-06) round. A number of the parts that were used on the Model 1917 were identical to those of the Pattern 1914, including the buttplate, the stacking swivel along with the sling swivels, sear, front handguard, front sight blade, nose cap, safety, trigger, upper and lower bands and related parts.[2]

While the Model 1917 rifles made by Winchester, Remington-UMC, and Eddystone were, for the most part, fundamentally identical, there were some minor differences in a few parts and, of course, the various markings applied to the guns. The markings and other features will be discussed individually in Chapter 4 of this book on the three prime contractors. Also, the manufacturer markings for each part are listed on Table #3. The following are the primary components and major physical characteristics of the Model 1917 rifle:

Finish — The external metal parts were finished in what was termed "browning," which was an attractive dark tinted rust bluing. In September of 1918, Eddystone began to use Parkerizing ("Parker Rust Proof Finish") in lieu of bluing.[3] While the tint can vary, the factory Parkerizing of the late-production Eddystone "U.S. Enfields" typically had a smooth texture, as the company did not sandblast the metal prior to the finish being applied. Parkerizing by Eddystone was instituted about the same time by Springfield Armory and Rock Island Arsenal for Model 1903 rifle production and was routinely used on American military firearms from the WWII era and later. When military small arms were refurbished/overhauled, including Model 1917 rifles, they were routinely Parkerized, which is why rifles retaining the original factory bluing are rather uncommon today. Winchester and Remington continued to finish their Model 1917 rifles in bluing throughout their production runs.

Receiver — Made of strong nickel steel, the Model 1917 receiver was very similar to the Pattern 1914, but the two were not interchangeable due chiefly to minor variances in internal machining. The nomenclature "U.S. Model of 1917" and the name of the manufacturer (with one exception to be discussed later) and the serial number were stamped on top of the receiver ring.

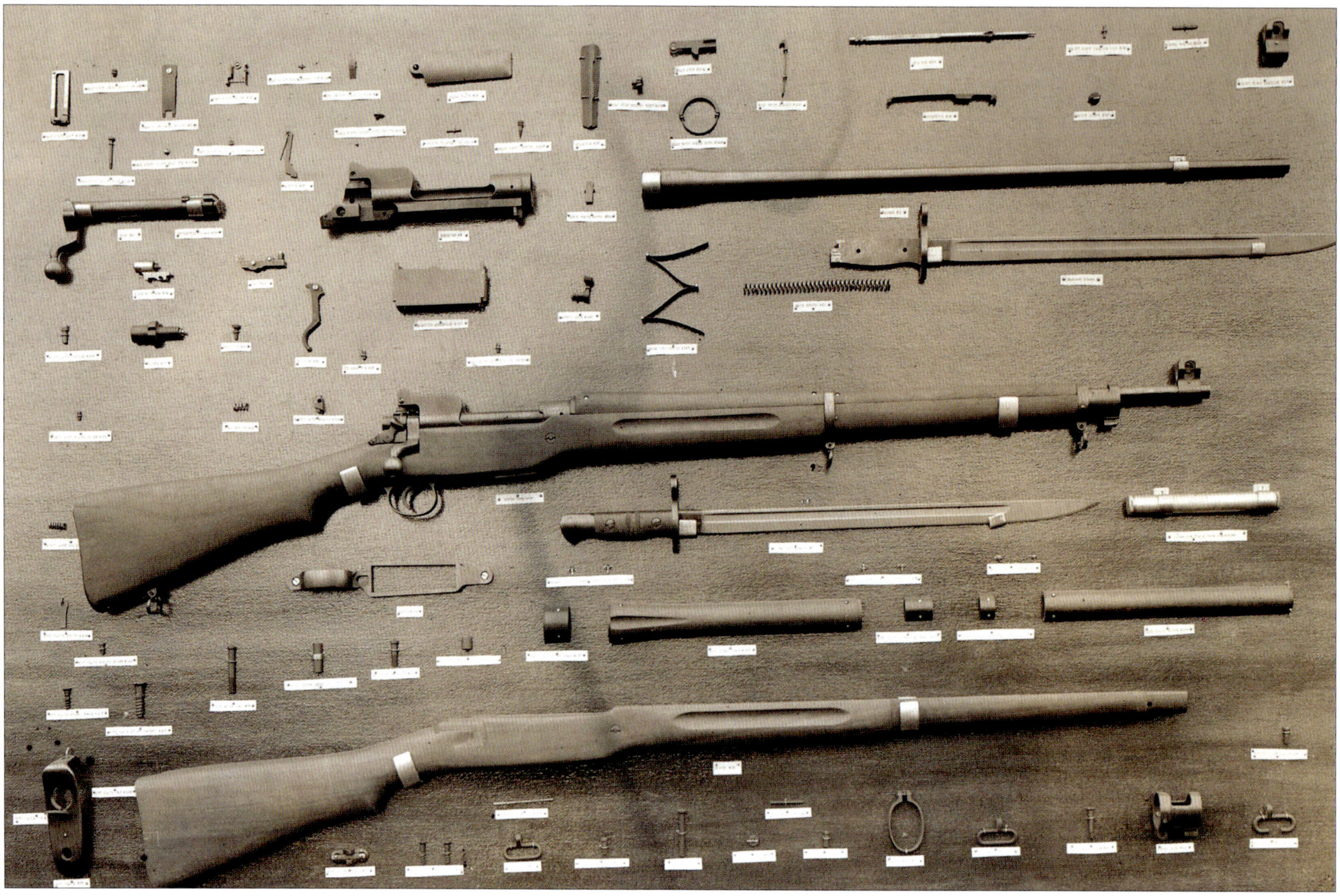

Eddystone factory display board showing all components of the Model 1917 rifle and bayonet with a finished rifle and bayonet in the center. *(Tom Laemlein)*

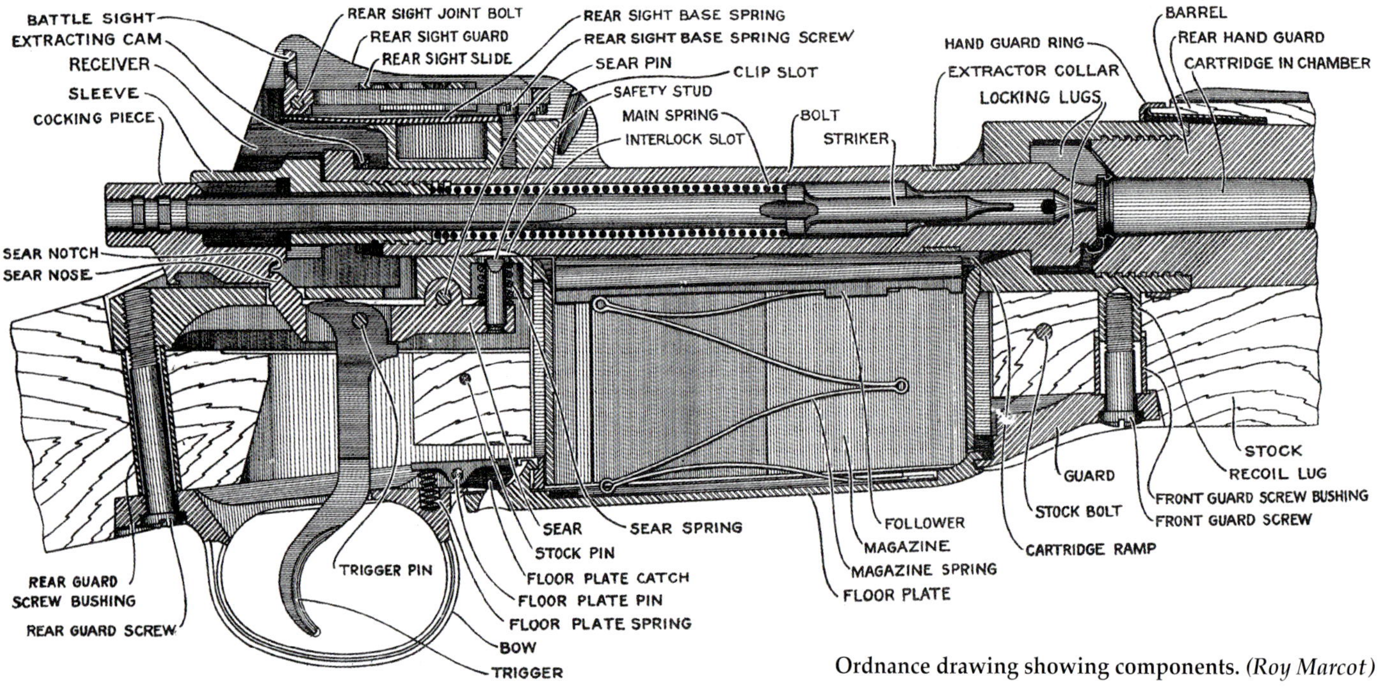

Ordnance drawing showing components. *(Roy Marcot)*

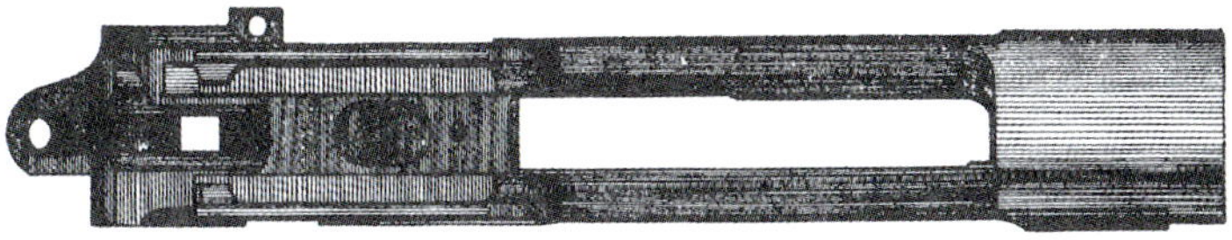

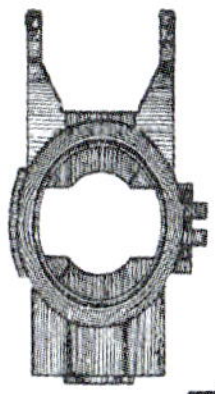

Ordnance drawing of Model 1917 receiver.

(Description and Rules for The Management of The United States Rifle, Caliber .30, Model of 1917)

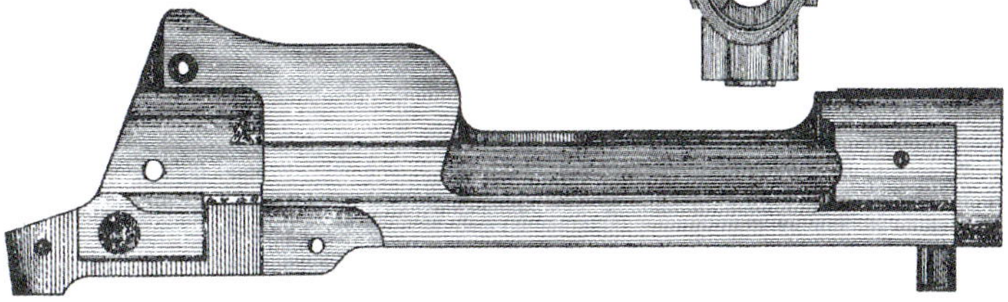

Model 1917 rifle barrel.

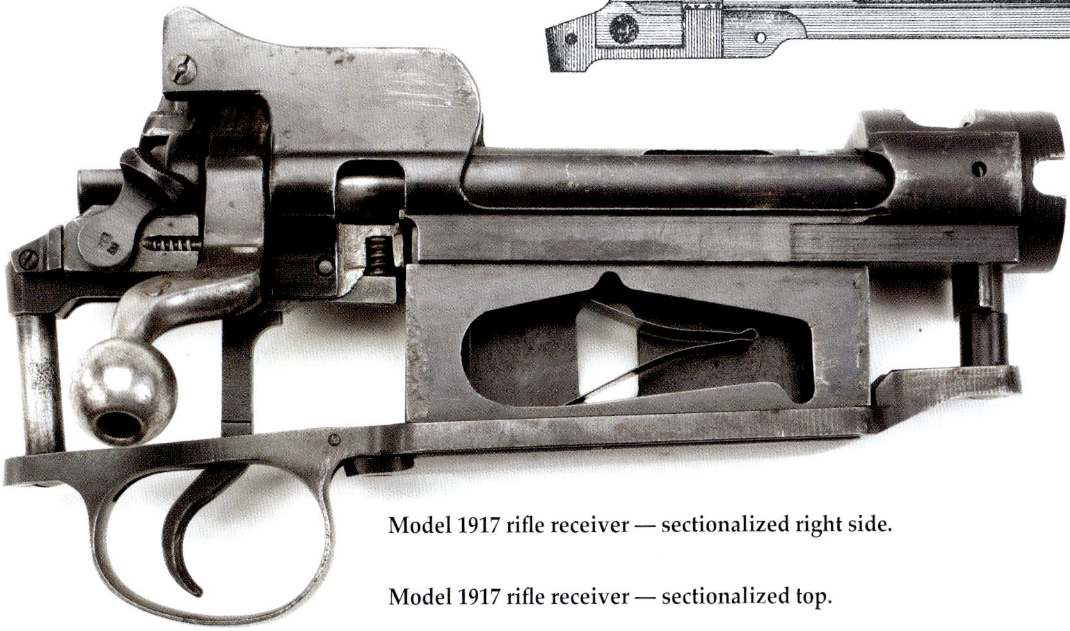

Model 1917 rifle receiver — sectionalized right side.

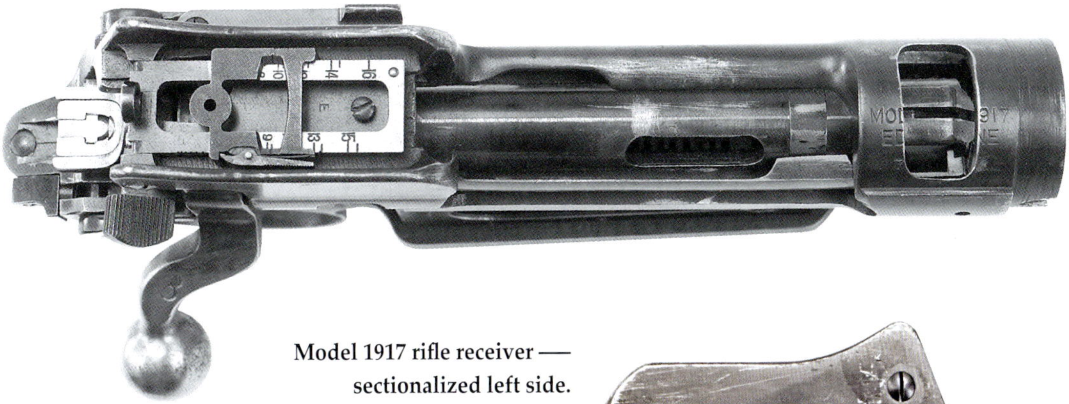

Model 1917 rifle receiver — sectionalized top.

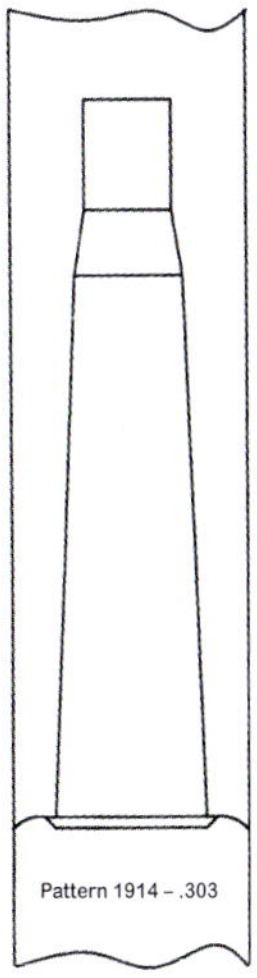

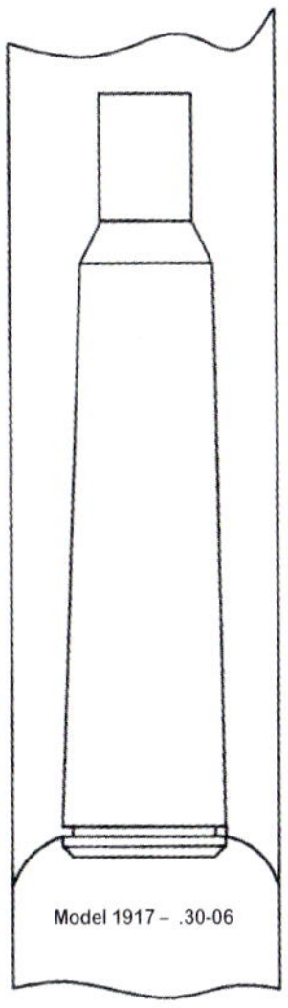

Chamber configurations: (top) Pattern 1914 for .303 cartridge; (bottom) M1917 for .30-06 cartridge. *(Pattern 1914 and U.S. Model 1917 Rifles)*

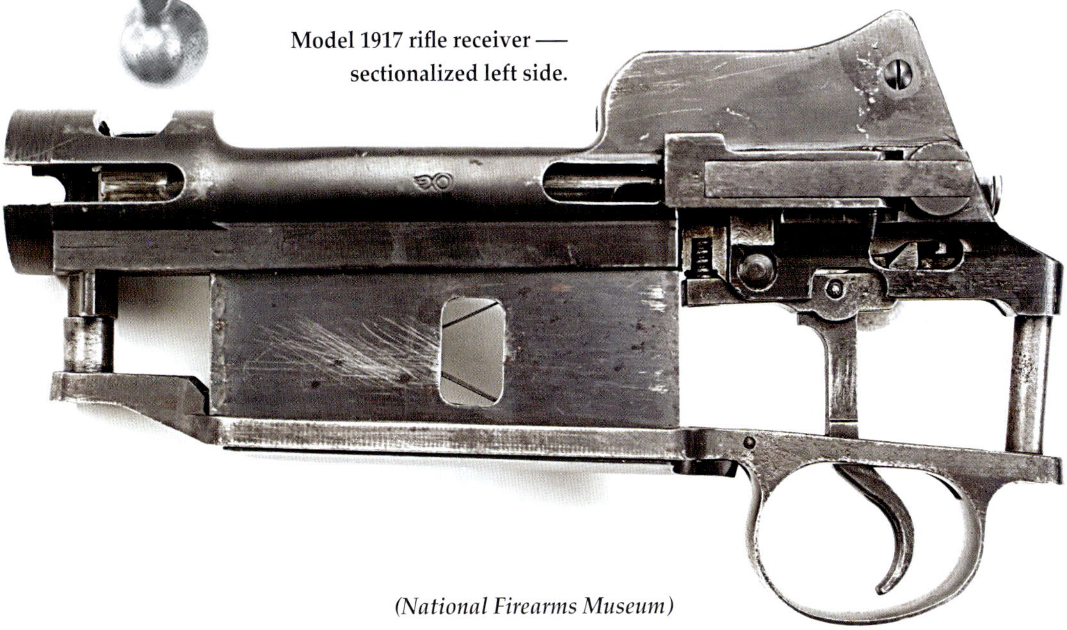

Model 1917 rifle receiver — sectionalized left side.

(National Firearms Museum)

Barrel — The 26-inch barrel was of the same general configuration as the Pattern 1914 barrel and differed chiefly in the bore diameter, chamber dimensions and rifling design due to the differences between the British .303 and the American .30-06 cartridges.[4] In his authoritative *Hatcher's Notebook*, Maj. General Julian S. Hatcher described the type of rifling used in the Model 1917 rifle and how it differed from the Pattern 1914:

> "For this rifle (the Pattern 1914), the British used the Enfield type of rifling with wide lands and deep grooves... This had a left-hand twist, with five lands and five grooves, of equal width, the grooves...
>
> We retained the Enfield form of rifling but changed the bore and groove dimensions to suit our own bullet diameter of .3086 inch. To accomplish this, we changed the bore from .303 inch to .300 and made the grooves .005 inches deep instead of .0058 as in the .303. This gave a somewhat tighter barrel than was used on the Springfield...
>
> ...we so often see in print the erroneous statement that the 1917 has a bore that is too loose for our bullet, because we used the 'British dimensions' when we took over the manufacture of the Enfield. This statement is not so...
>
> It is interesting to note...that many tests made during the years since then have shown that the 1917 barrels will always outwear the '03 barrels."[5]

The barrels were "proof tested" after manufacture to check for any potential defects before the assembled rifles received final inspection. Any defective barrels were discarded.

The initials of the manufacturer, the ordnance flaming bomb insignia, and the month and year the barrel was manufactured were stamped on top behind the front sight.

Rear Sight — The rear sight assembly had two strong protective "ears" and was fitted with a two-position folding leaf. The leaf contained a peep sight, which moved vertically on a slide and had no windage adjustment to compensate for the drift of the bullet. When the leaf was in the folded position, the rifle was sighted through a peep aperture "battle sight" set for 400 yards. When the leaf was raised, the rifle was sighted through another peep aperture graduated in 100-yard multiples from 200 to 900 yards on the left side of the leaf and in 50-yard multiples from 900 to 1,600 yards.[6]

Barrels that failed the proof-firing tests. *(National Archives)*

Rear sight leaf in raised position.

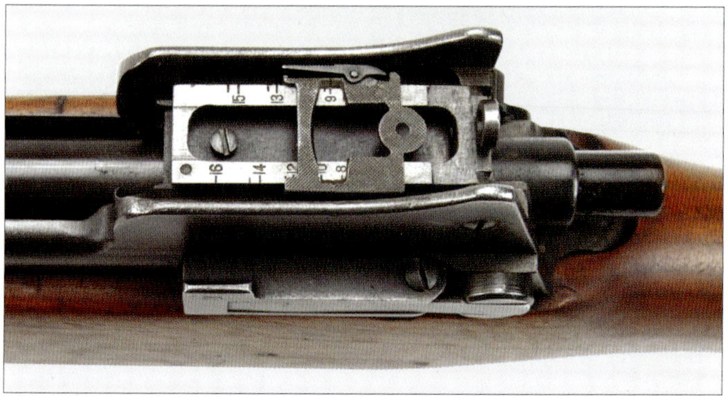

Rear sight leaf folded in "battle sight" position.

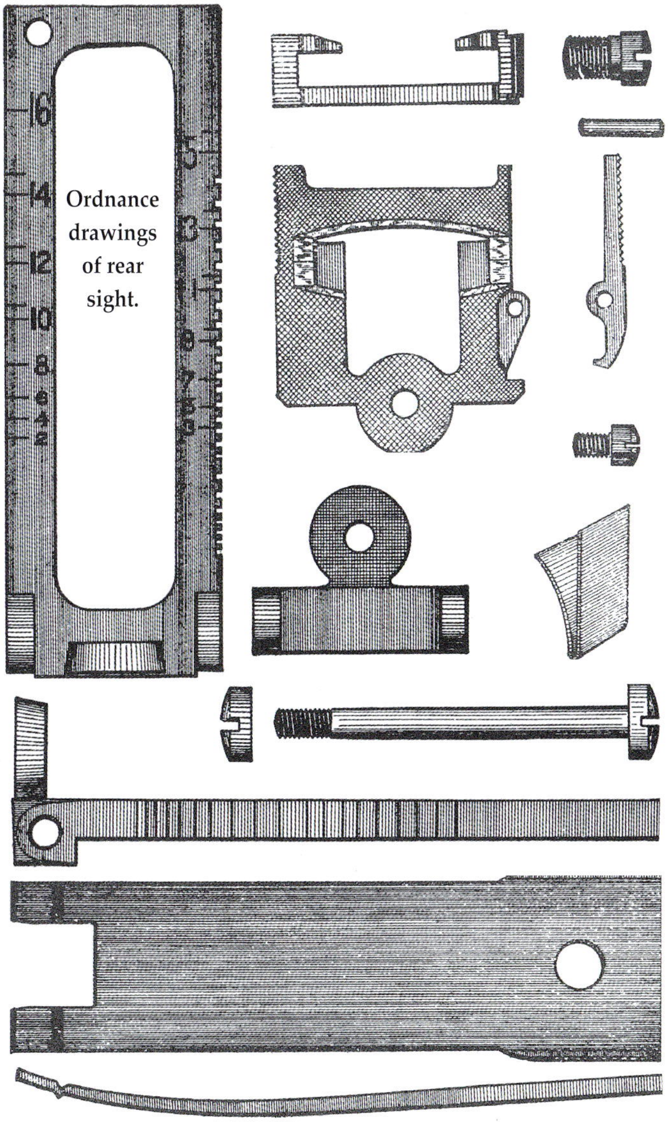

Ordnance drawings of rear sight.

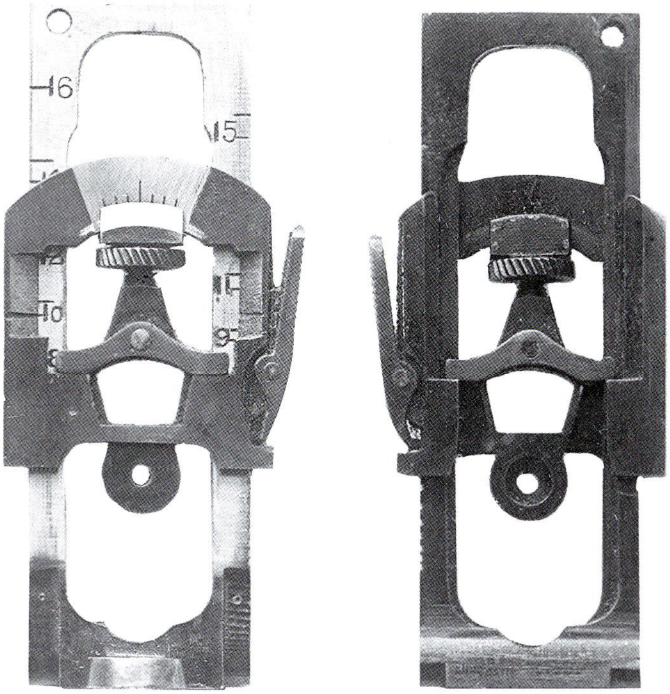

Ordnance photo of front and back of rear sight leaf. *(Tom Laemlein)*

Front Sight — The Model 1917 front sight assembly consisted of a blade with protective "ears" and was mounted on a dovetail base, held in place by a metal pin. The front sight assembly was staked in place at the factory after being "zeroed" for elevation. While the front sight was not designed to be easily adjustable for windage, corrections could be done by moving the blade by means of a brass drift and hammer or a purpose-made adjustment tool.[7] Such adjustments were not intended to be done by individual soldiers "in the field."

The Model 1917 sights were a matter of some controversy when compared to the Model 1903. The '03 sight leaf was graduated for greater distances and could be easily adjusted for windage. This made it a better sight for precision and/or competitive shooting than the Model 1917. However, the

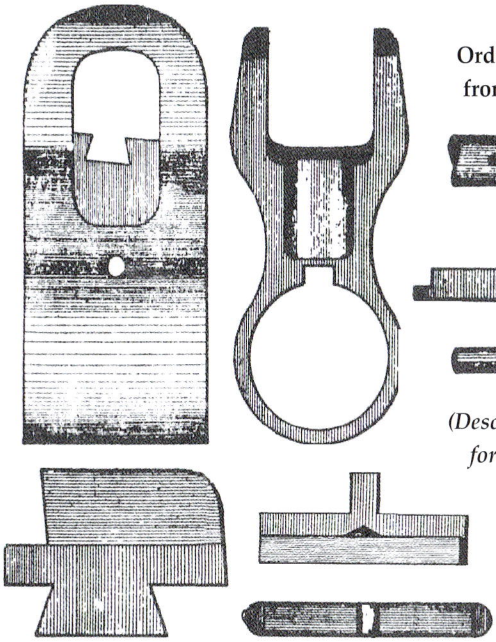

Ordnance drawing of front sight assembly.

(Description and Rules for The Management of The United States Rifle, Caliber .30, Model of 1917)

Front sight assembly.

Front view of front sight assembly.

Model 1917 sight mounted on the rear of the receiver was a better sight for the average soldier. The superiority of the Model 1917 sights was echoed in *Hatcher's Notebook*: "The well protected peep sight mounted on the receiver, close to the shooter's eye, with a front sight likewise protected by strong steel ears, gave a sighting combination that was far superior to that on the Springfield, and by a considerable margin the best and most practical of any seen in that war (the First World War)."[8]

Bolt — Like most of the other parts, the Model 1917 bolt was very similar to that of the Pattern 1914, differing chiefly in the dimensions between the rimmed .303 and the rimless .30-06. The configuration of the bolt had a distinctive "crooked" appearance designed to put the bolt handle in close proximity to the trigger to aid rapid manipulation of the action. In his book *America's Munitions 1917–1918*, Secretary Crowell mentioned this feature with more than a little hyperbole by alleging that it permitted one rifleman armed with an "American Enfield" to fire twice as fast as a German adversary armed with a Mauser, which had a straight configuration bolt handle: "...the bolt handle was practically right at the trigger the instant after it had thrown in a new cartridge...by bending back the bolt handle we had placed two men...(on the firing line)."[9]

The Model 1917 bolt handle was generally stamped on the back with the initial of the manufacturer and, normally, on the front with an ordnance flaming bomb insignia.

Some bolts may be encountered today with serial numbers applied. These were added during subsequent use by various foreign countries. One attribute of the Pattern 1914 rifle that remained on the Model 1917 was the "cock-on-closing" bolt-action mechanism. American soldiers were accustomed to the "cock-on-opening" of previous U.S. bolt-action rifles such as the Krag and the Model 1903. This resulted in a number of complaints when the rifles began to be issued, but the grievances had more to do with the type of mechanism being unfamiliar than any inherent deficiency. This was discussed by General Hatcher:

"The bolt mechanism, like that of the S.M.L.E., was designed to cock on the closing motion. This was to make extraction easier in rapid fire, when the heat always tightens things up. With the Springfield, the work of cocking the firing mechanism is added to that of extracting the cartridge, with the result that unless very hard brass is used, the bolt becomes very difficult to open after a few rounds of rapid fire. This feature, considered a great advantage by the British was in general not liked by our people who were thoroughly used to the Springfield, which cocks on the opening of the bolt."[10]

Bolt Sleeve — The bolt sleeve, known as a "bolt plug" on the Pattern 1914 rifle, functioned as a main spring stop and attached to the rear of the bolt body. There were two variations of Model 1917 bolt sleeves. The first type was essentially identical to the Pattern 1914 bolt plug, and the later pattern had a safety dismounting notch on the rear to assist in disassembling the bolt.[11]

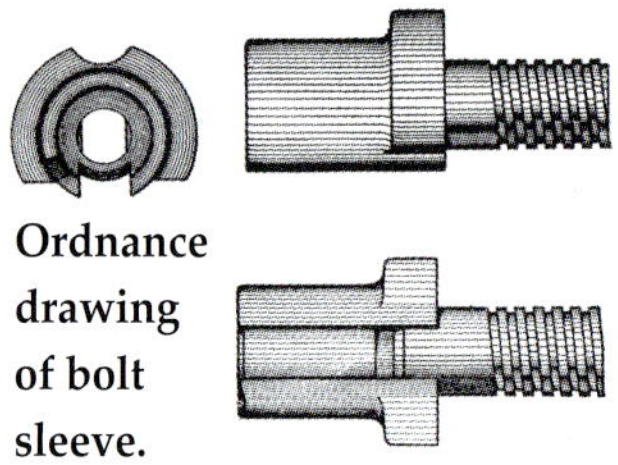

Ordnance drawing of bolt sleeve.

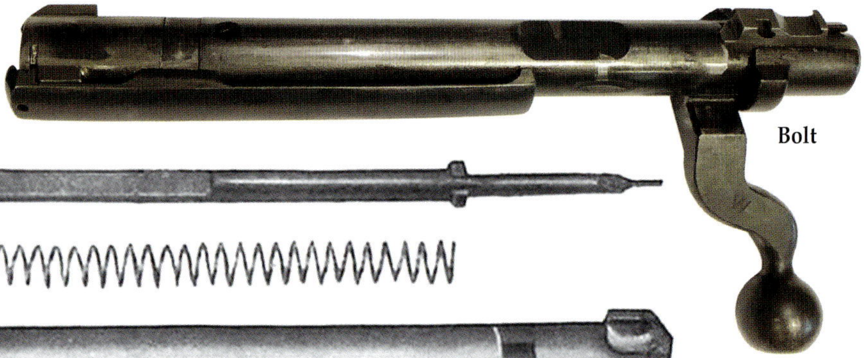

Bolt

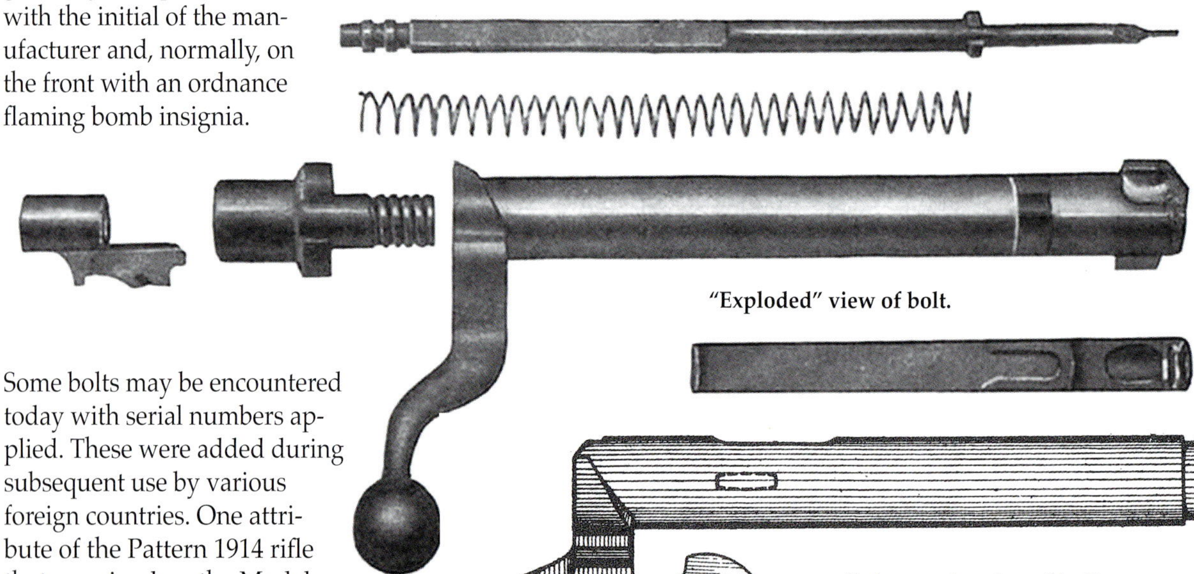

"Exploded" view of bolt.

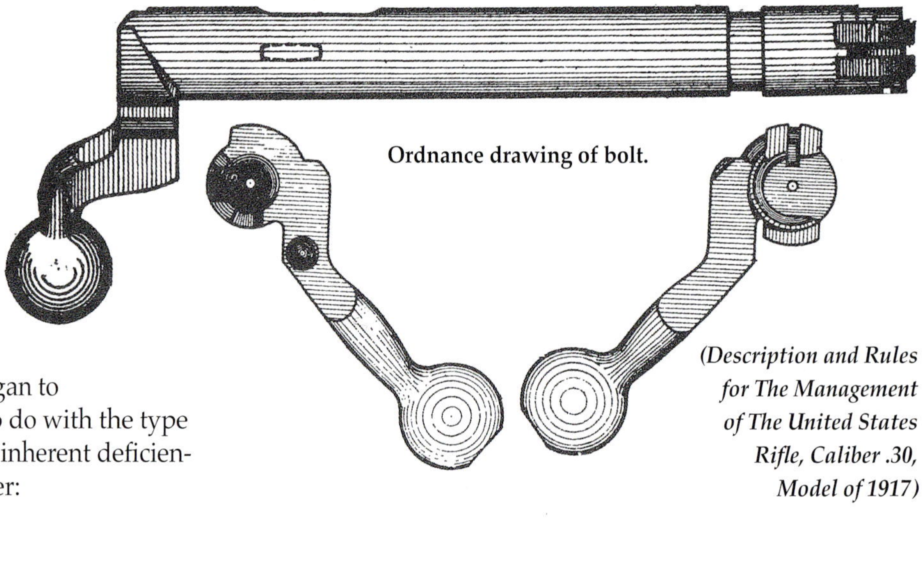

Ordnance drawing of bolt.

(Description and Rules for The Management of The United States Rifle, Caliber .30, Model of 1917)

Model 1917 bolt handle. *(National Firearms Museum)*

Firing Pin Assembly — The firing pin assembly consisted of a cocking piece and striker. The main spring surrounded the striker. It could be compressed sufficiently to permit the cocking piece to be released from the sleeve.[12]

Extractor — As the name indicates, the extractor was designed to remove a fired cartridge case and permit it to be ejected from the rifle to allow a fresh cartridge to be chambered. Since the Pattern 1914 rifle was chambered for the rimmed .303 cartridge and the Model 1917 was chambered for the rimless (cannelured) .30-06 round, the two extractors are not interchangeable. The extractor assembly included an extractor ring collar, which secured the extractor to the bolt body.

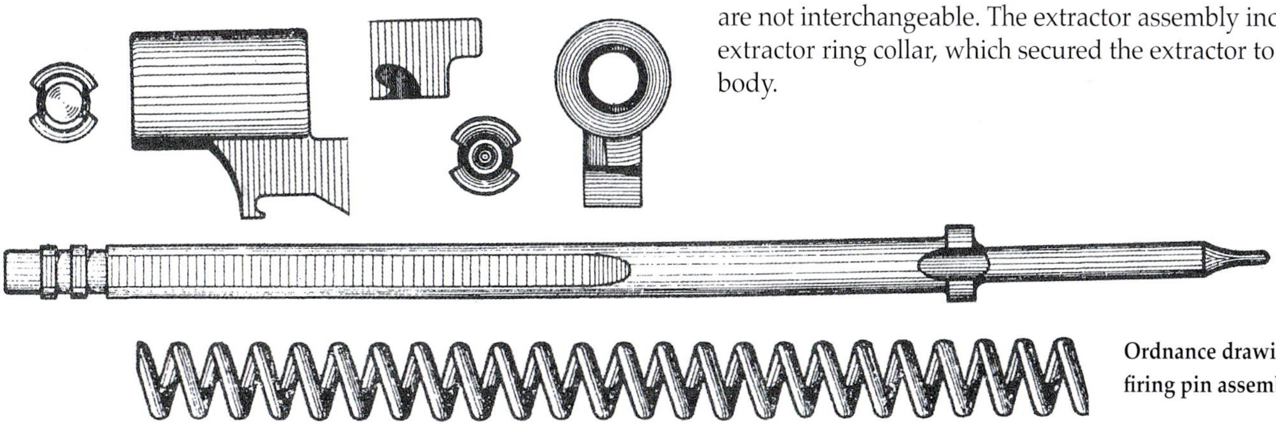

Ordnance drawing of firing pin assembly.

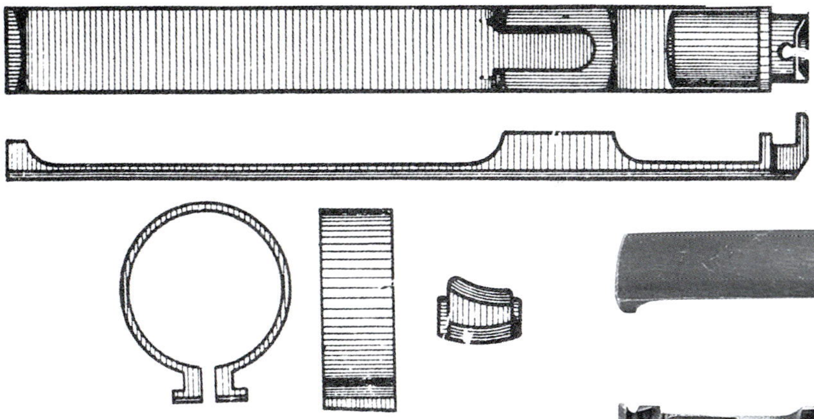

Ordnance drawing of extractor.

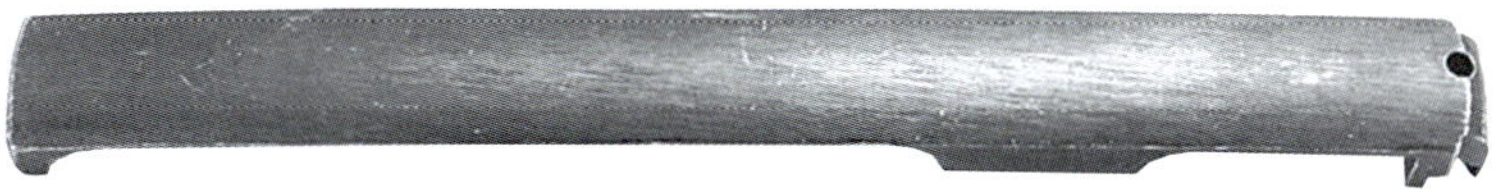

Front and rear views of extractor.

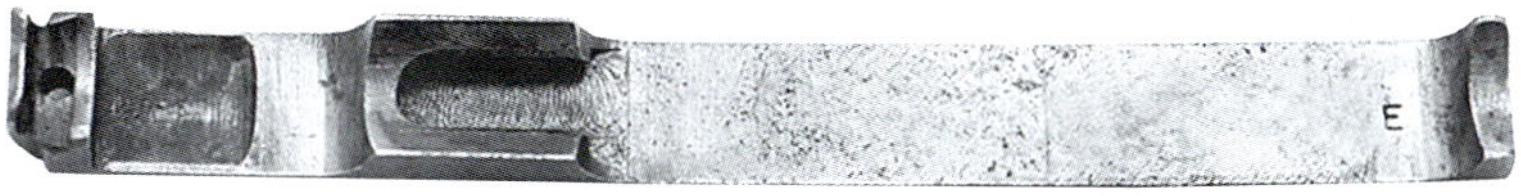

(Description and Rules for The Management of The United States Rifle, Caliber .30, Model of 1917)

Bolt Stop Assembly — The Model 1917 bolt stop assembly was functionally similar to the Pattern 1914 but had some minor configurational differences. The two are not interchangeable. The bolt stop assembly was attached to the left side of the receiver. Its main purpose was to prevent the bolt from being inadvertently withdrawn from the receiver. In order to completely remove the bolt, a thumbpiece (release lever) was moved to the left, which retracted the retaining lug.[13]

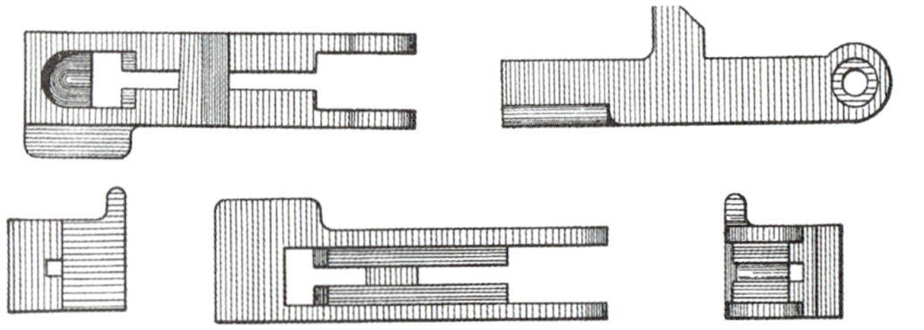

Ordnance drawing of bolt stop assembly.

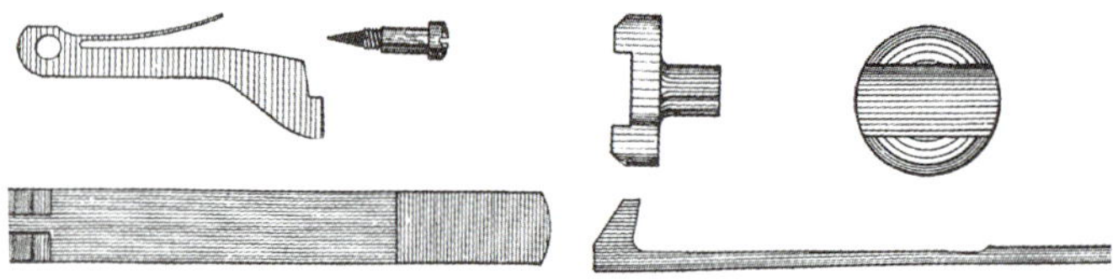

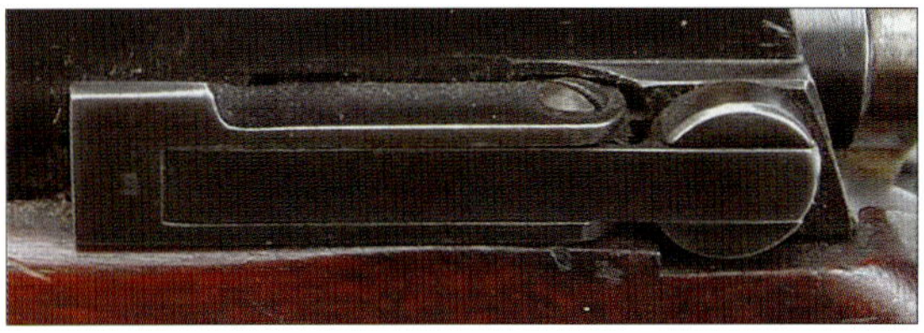

Bolt stop release lever.

Safety — One of the excellent features of the Model 1917 rifle was the safety, which could be easily manipulated by the thumb without having to move the hand. With the thumbpiece in the rear position, the safety was engaged, and when rotated forward, the safety was off and rifle could be fired.

The configuration of the safety was found to be ideal for use with telescopic sights as compared to the '03 Springfield. When a prototype '03 sniper rifle was fabricated during World War I for evaluation, it was fitted with a Model 1917-type safety. Also, the design of the safety was one of the reasons why a later World War I prototype sniper rifle was based on the Model 1917 rifle rather than the Model 1903. (See "Variations," Chapter 9 of this book).

Side and top views of the safety. Note the eagle-head inspection stamp.

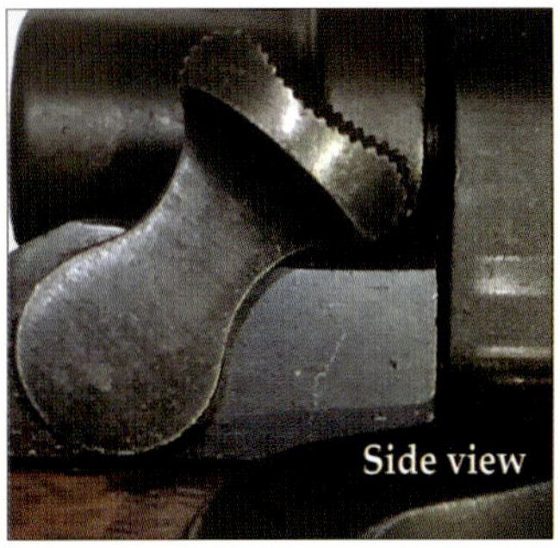

Trigger and Sear Assembly — The safety and trigger and sear assembly of the Model 1917 rifle were essentially identical to the Pattern 1914 except for the markings. There were some minor differences in the method of fabrication of the sears made by Winchester and those made by Remington and Eddystone.[14] The front of the trigger was grooved.

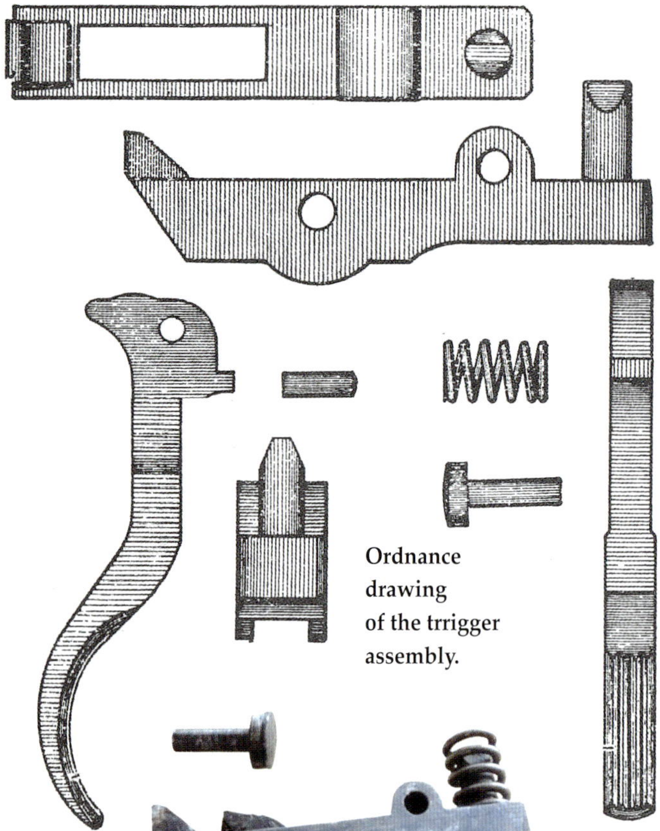

Ordnance drawing of the trrigger assembly.

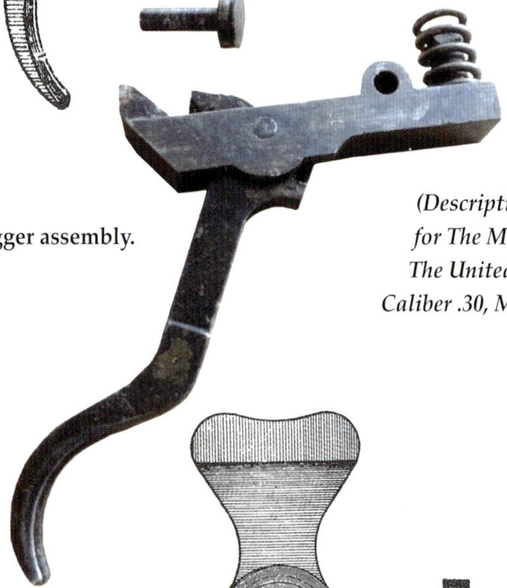

Trigger assembly.

(Description and Rules for The Management of The United States Rifle, Caliber .30, Model of 1917)

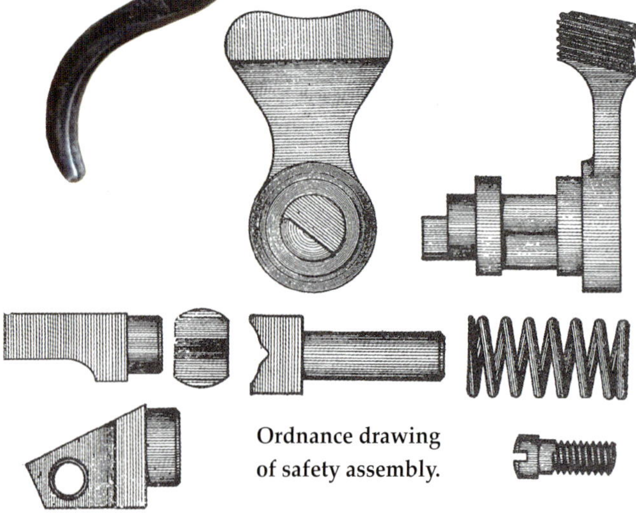

Ordnance drawing of safety assembly.

Magazine / Trigger Guard Assembly — Like the Model 1903 Springfield rifle, the Pattern 1914 and Model 1917 rifles had an internal magazine, sometimes termed a "magazine case" or "magazine box." While very similar in design and function, there were some dimensional differences between the Pattern 1914 and Model 1917 assemblies. They are not interchangeable. There were also some differences between the Eddystone and Remington Model 1917 magazine assemblies as well as those made by Winchester, and they do not interchange without some hand-fitting.[15]

Ordnance drawing of magazine assembly. *(Description and Rules for The Management of The United States Rifle, Caliber .30, Model of 1917)*

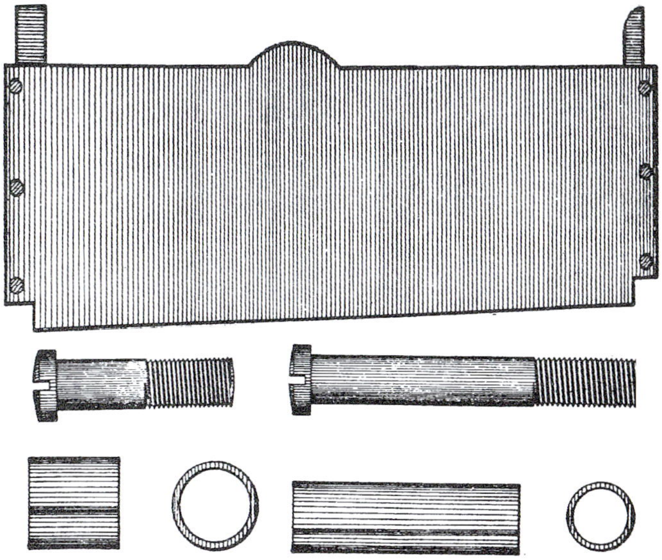

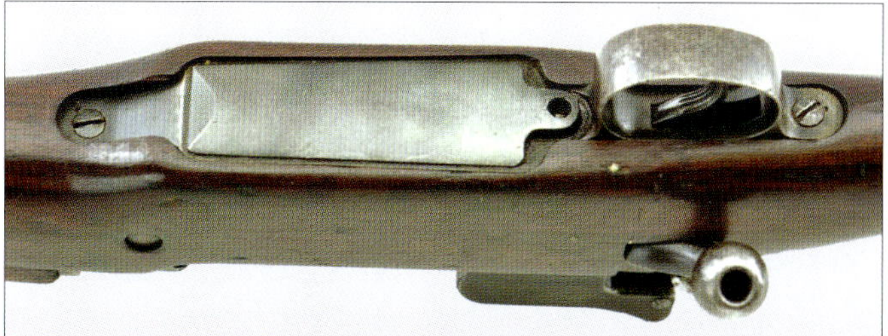

View of the magazine floorplate and trigger guard.

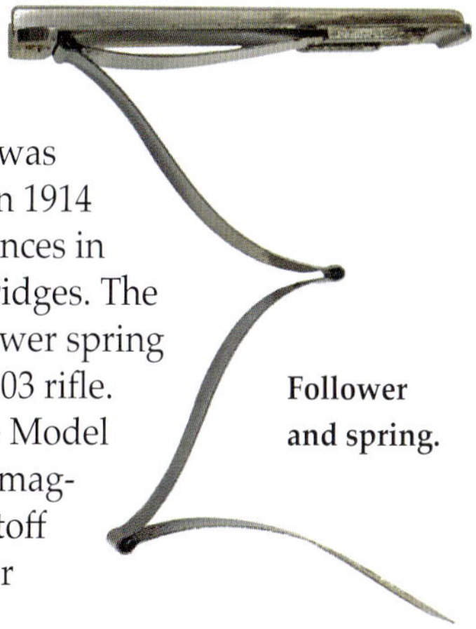

Follower and spring.

The Model 1917 follower, termed "magazine platform" in British parlance, was slightly longer than the Pattern 1914 due to the dimensional differences in the American and British cartridges. The Model 1917 follower and follower spring were identical to the Model 1903 rifle.

Unlike the Model 1903, the Model 1917 was not equipped with a magazine cutoff. The magazine cutoff for the '03 enabled the follower

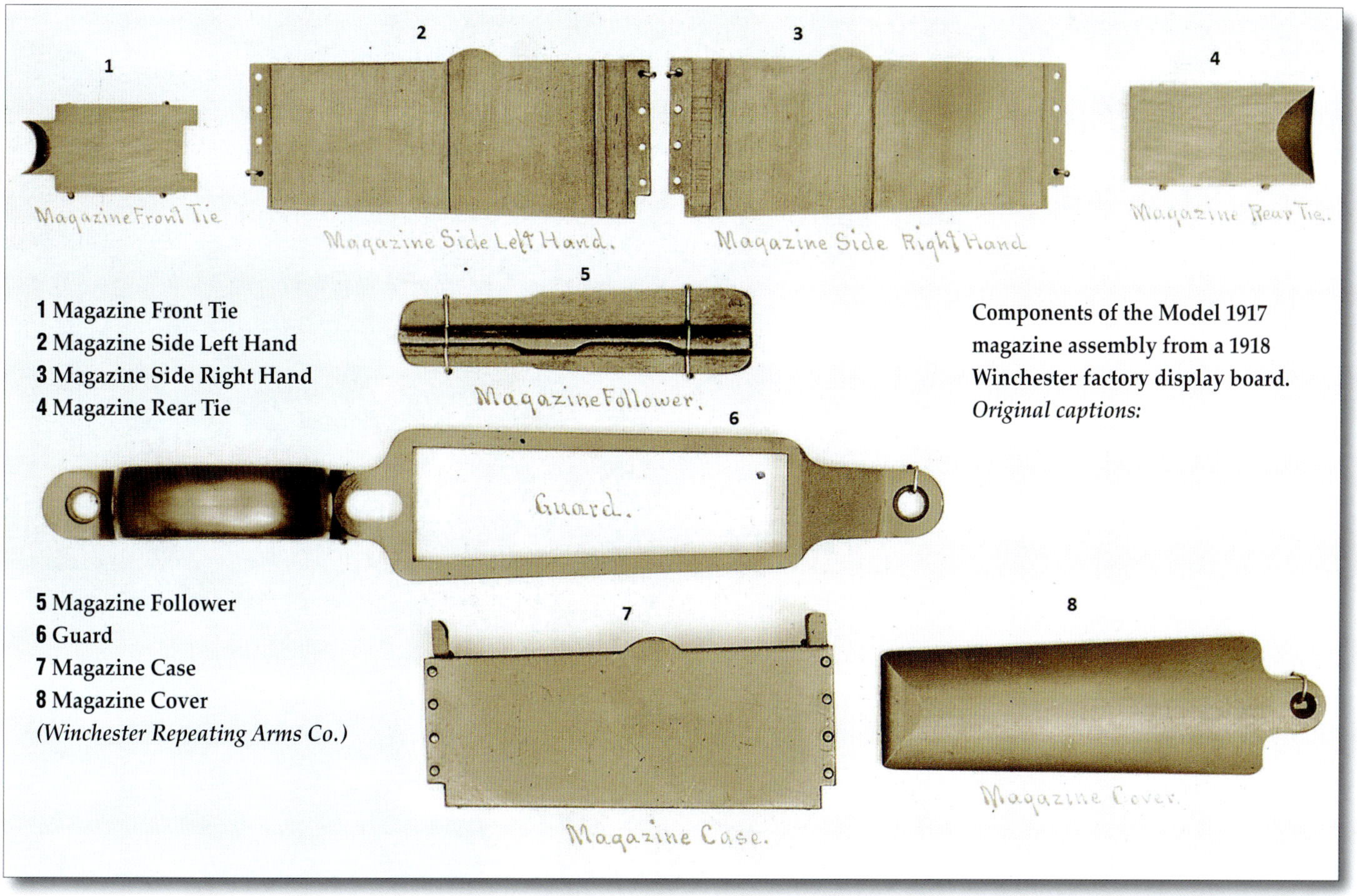

1 Magazine Front Tie
2 Magazine Side Left Hand
3 Magazine Side Right Hand
4 Magazine Rear Tie

5 Magazine Follower
6 Guard
7 Magazine Case
8 Magazine Cover
(Winchester Repeating Arms Co.)

Components of the Model 1917 magazine assembly from a 1918 Winchester factory display board. *Original captions:*

to be retracted when the magazine was empty, thus permitting the bolt to be easily closed. This was rarely, if ever, a problem in combat with the Model 1917 rifle, but it often resulted in some consternation during training because it made closing the bolt on an empty rifle a bit more cumbersome during inspection since the magazine follower had to be manually depressed. It has been reported that some soldiers would resort to "field expediency" by inserting a coin to keep the follower depressed and allow the bolt to close on an empty magazine. In order to correct this situation, a sheet metal "magazine platform depressor" was produced by all three manufacturers and issued. (See Chapter 10, Accessories, Appendages and Accouterments).

Although the magazine capacity of the Model 1917 rifle was six rounds, it was intended to be used with the same five-round "stripper clip" (magazine charger) as the Model 1903 rifle. A single cartridge could be subsequently loaded by hand to completely fill the magazine, although this was not routinely done.

The trigger guard assembly consisted of the trigger guard, magazine cover (typically referred to as "floor plate") and related hardware and screws. The magazine cover/floor plate was designed be removed by inserting the tip of a bullet to release the catch holding it in place. It had a dovetail recess into which the follower spring was attached.

A dime was inserted between the follower and the bolt to allow the bolt to close on an empty magazine without depressing the follower. This was an early field-expedient measure until magazine depressors could be manufactured and issued. It is reported that both nickels and dimes were used.

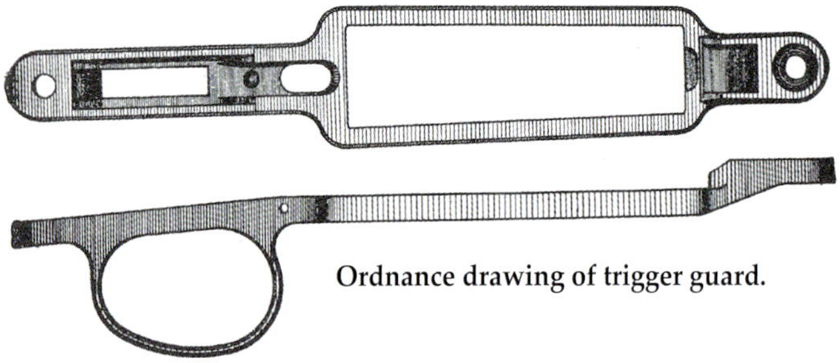

Ordnance drawing of trigger guard.

Ordnance drawing of the trigger guard floor plate and follower assembly.

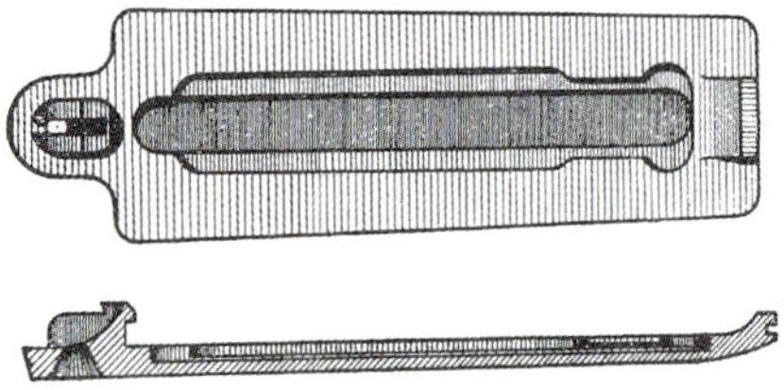

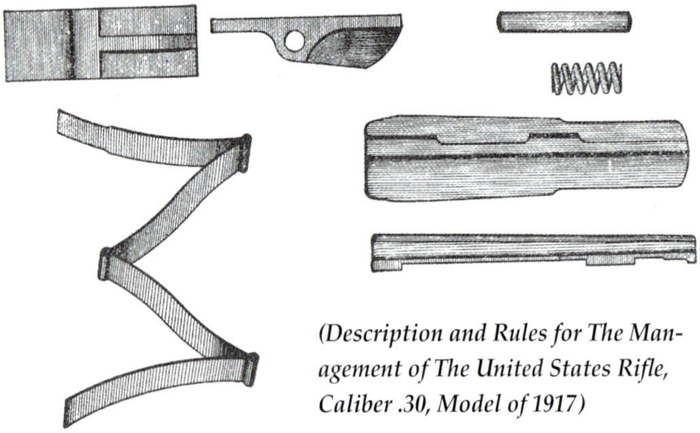

(Description and Rules for The Management of The United States Rifle, Caliber .30, Model of 1917)

Stock — While similar in appearance, there were several differences between the Pattern 1914 and Model 1917 stocks, and they were not interchangeable. The Model 1917 stock was not inletted for volley sights thus did not have a metal marking disk on the left side.

Also, the Model 1917 grasping grooves on the fore-end were longer and positioned a bit more forward than the Pattern 1914. Metal reinforcing bolts (tie bolts) to reduce the possibility of cracking during recoil were fitted transversely through the stock just forward of the magazine well and were secured by a nut. The Pattern 1914 made by Winchester and Remington and Model 1917 stock bolts were identical except that the latter had a slotted head.[16] In addition to the stock bolt, a threaded brass "stock pin" was screwed into a transverse hole between the magazine well

Stock reinforcing bolt and pin (see arrows).

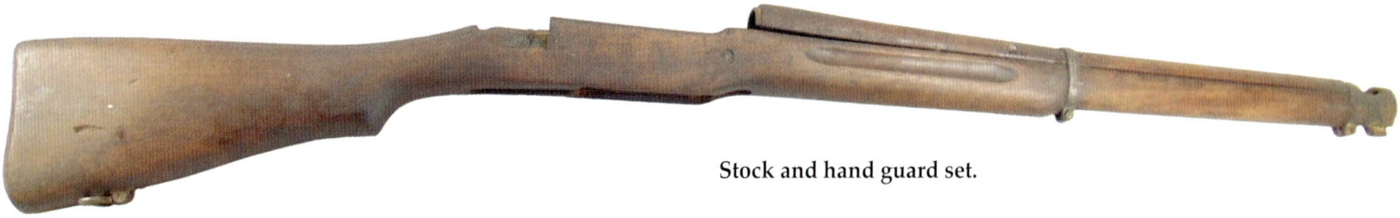

Stock and hand guard set.

Pattern 1914 rifle.

Model 1917 rifle.

(National Firearms Museum)

and trigger slot with both ends being cut off flush with the sides of the stock.[17] The inletting for the trigger guard and the magazine well was slightly different between the British and American rifles.

The Model 1917 buttstock was drilled for a 0.82-inch-diameter hole to accommodate the standard oiler and thong case as used with the Model 1903 rifle. It has been reported that a few Model 1917 stocks have been observed that are a bit shorter in butt length than the standard type and were purportedly made from salvaged Pattern 1914 stocks.[18]

Hand Guard Group — The hand guard group consisted of semi-circular front and rear hand guards and the metal fittings required to secure them in place. The purpose of the hand guards was to protect the hands of the shooter from a hot barrel and, to a lesser extent, reduce any sight distortion from heat rising from the barrel. The rear was secured under the hand guard ring, and the front rested under the barrel band. The underside of the hand guard for the Model 1917 had a longer clearance cut for the slightly longer swelled area of the barrel over the chamber but was otherwise the same as the Pattern 1914.

Model 1917 hand guard set:

Rear hand guard.

Front hand guard.

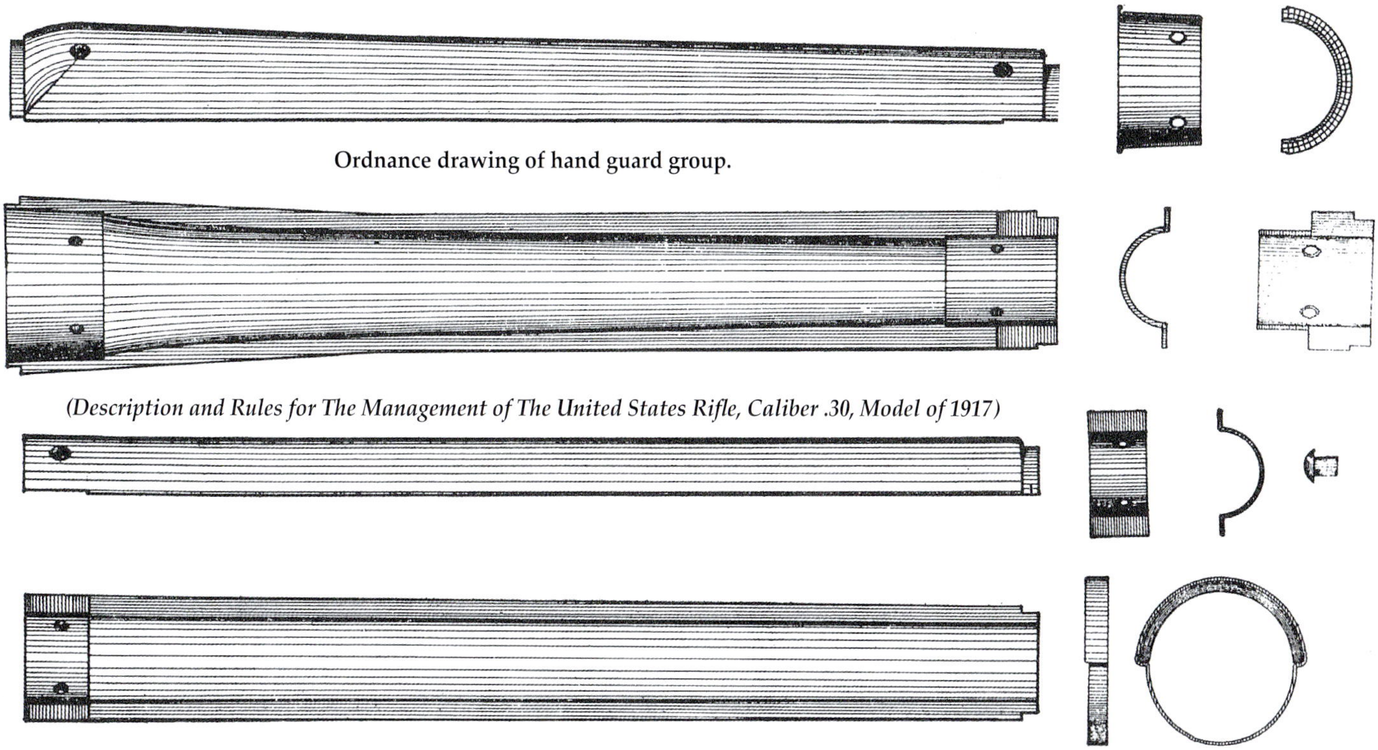

Ordnance drawing of hand guard group.

(Description and Rules for The Management of The United States Rifle, Caliber .30, Model of 1917)

The rear hand guard was 10.6 inches long and 1.75 inches wide and was attached over the rear portion of the barrel and rested on the stock fore-end. Front and rear metal liners were attached, secured by brass rivets.[19]

The front hand guard was approximately 10 inches long and 1.2 inches wide and was attached over the front portion of the barrel ahead of the lower barrel band. As with the rear hand guard, it rested on the fore-end. The front hand guard had a liner at the rear, which helped reinforce the thin wood.

As stated in the well-researched monograph *Eddystone Rifle Plant, History of the Last Remaining Building* by Kurt Sellers, Major U.S. Army (Retired):

> "Wood for (Model 1917) rifle stocks and hand guards had to be of acceptable grain and seasoning and were subjected to rigorous inspection. Springfield Armory 1916 specifications are representative. The 'best grade of black walnut' was preferred. Wood could not 'contain more than 5% moisture content after final treatment in the dry house and exposure to the atmosphere for 36 hours.' Wood had to be 'hard, straight grained and thoroughly sound.' Substitute woods such as birch were allowed, provided they met the same standard. There is no evidence any wood, but walnut was ever actually used."[20]

Although World War I production Model 1917 rifle stocks were made of walnut, some birch hand guards stained to resemble walnut were produced in 1918 and 1919.[21] However, as will be discussed in Chapter 4, the Victor Talking Machine Company received a contract in late 1918 for the manufacture of 100,000 Model 1917 birch rifle stocks, but it is unclear how many, if any, were actually made before the cancellation of all outstanding contracts soon after the Armistice. During World War II, some Model 1917 replacement hand guards, and reportedly, some stocks were made from birch wood.

Butt Plate Assembly — The buttplate assembly included the buttplate, buttplate screws, and butt trap and associated parts. The sturdy metal buttplate protected the rear of the stock from damage and contained a hinged metal door to permit access to the hole drilled into the butt of the stock to hold an oiler and thong case.

Band and Swivel Group — This group consisted of the upper band, lower band, two sling swivels, a stacking swivel and the hardware and screws necessary for attachment.

Upper Band — The upper band was mounted on the stock fore-end and secured in place by the upper band screw and held the front hand guard to the stock. There was an integral bayonet lug on the underside. The stacking swivel was also attached to the upper band. The component is sometimes referred to as the "nose cap," but "upper band" is the official designation as stated in U.S. Army Ordnance manuals.[22]

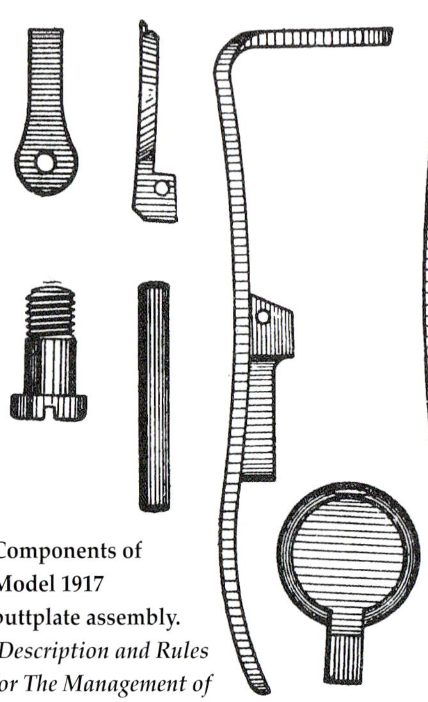

Components of Model 1917 buttplate assembly. *(Description and Rules for The Management of The United States Rifle, Caliber .30, Model of 1917)*

Model 1917 buttplate assembly with rear sling swivel and attachment screws.

Upper band with stacking swivel.

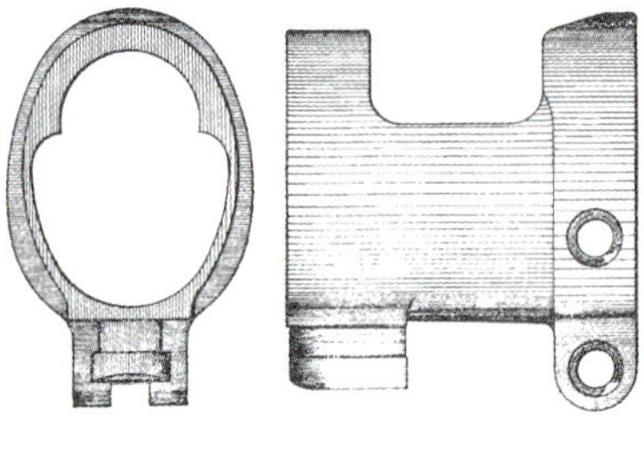

Ordnance drawing of upper band.

Lower Band — The purpose was to hold the stock and hand guards together and serve as an attachment for a sling swivel. The sling swivel was attached by means of a screw inserted from left to right.

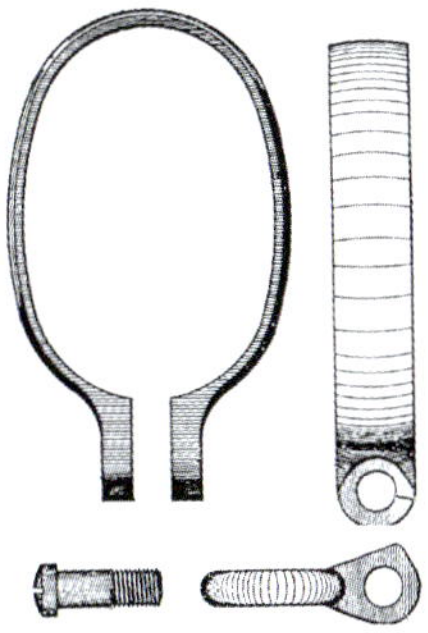

Lower band with sling swivel attached.

Ordnance drawing of lower band. *(Description and Rules for The Management of The United States Rifle, Caliber .30, Model of 1917)*

Butt Swivel — The butt swivel served as an attachment for the sling. It fitted into a recess on the bottom rear of the stock and was secured by two wood screws.

Stacking Swivel — The stacking swivel, also known as the "piling swivel," was attached to the upper band and was secured by a screw. The purpose of the stacking swivel was to permit several rifles to be hooked together in a tripod-like fashion.

It is interesting to note that the sling swivels and the stacking swivel are asymmetrical with one side longer than the other. The feature was carried over from the Pattern 1914 rifle, as it served to provide clearance for the volley sight. Since the Model 1917 rifle wasn't fitted with a volley sight, the offset configuration of the swivels served no purpose, but the design didn't affect the functionality and was one less tooling change that had to be made when production shifted from the Pattern 1914 to the Model 1917.

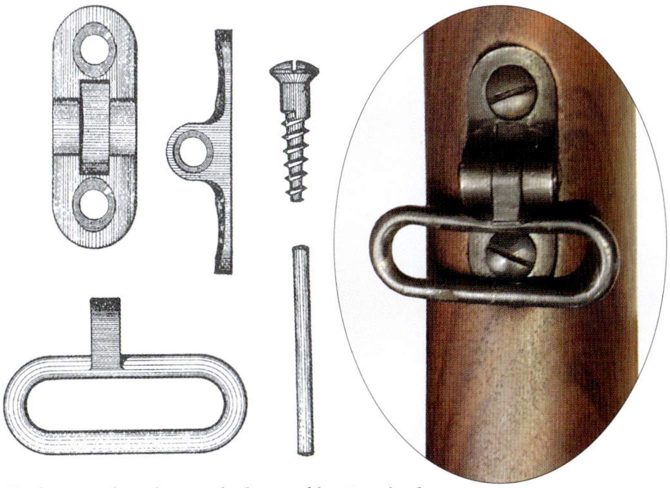

Ordnance drawing and photo of butt swivel. *(Description and Rules for The Management of The United States Rifle, Caliber .30, Model of 1917)*

Model 1917 rifles stacked in "tripod" formation using the rifles' stacking swivel — 14th Infantry, Camp Bowie, Ft. Worth, Texas, spring 1918. *(Tom Laemlein)*

The line drawings above are from the 1918 "Description and Rules for the Management of the United States Rifle, Caliber .30, Model of 1917." Originals of the manual are rather uncommon, but there are good-quality reprints available. Anyone wanting more in depth information on the various and sundry parts of the Model 1917 rifle should obtain a copy. Also, the Stratton and Ferris books cited previously have excellent information on the subject as well.

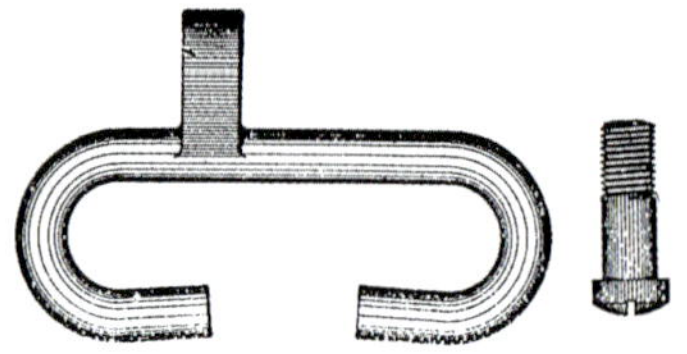

Ordnance photo and drawing of the stacking swivel. *(Description and Rules for The Management of The United States Rifle, Caliber .30, Model of 1917)*

Chapter 4

The Manufacturers

While it is seldom discussed, each manufacturer had different business philosophies, corporate structures and policies of dealing with the ordnance bureaucratic morass. The management and boards of directors of all the firms were undoubtedly patriotic citizens who wanted to help their country prevail in the brutal fight to preserve freedom in which the United States had become embroiled. However, unlike the government owned and operated Springfield Armory and Rock Island Arsenal, all three firms were private for-profit businesses. This required that each had to deal with returns on investments, make sure cash flow and capital were kept at appropriate levels, satisfy stockholders, establish management guidelines, assemble and sustain a trained and efficient workforce, pay compensation commensurate with the prevailing labor market, follow pertinent labor laws, deal with unions and union organizing attempts, maintain production when faced with material shortages and price increases from suppliers and subcontractors, keep creditors contented and deal with the often arbitrary policies of the Ordnance Department.

The U.S. Army Ordnance publication *History of Rifles, Revolvers and Pistols*, published in 1920, related the following:

> "Production was hampered in the spring of 1918 by several limiting factors, principally labor, lack of gun stocks, and poor steel for barrels. The labor troubles were chiefly caused by:
>
> 1. Lack of skilled labor
> 2. Congestion
> 3. Industrial competition
> 4. Lack of wage standardization
>
> Number 1 was one of the first and most serious difficulties. Skilled labor was at a Premium; war time industry is very different from peace time industry; we did not have time to educate an army of experts to jobs, many of which require not months but years for proficiency. For instance, there was an immediate demand for 40,000 gauge makers; there were 4,000 in the country. We needed a very large number of barrel straighteners; 7 were found in the Nation. A barrel straightener is a man with a peculiarity of eyesight, which cannot be developed; he looks down the rifle barrel held at a certain angle to the light, detects the slightest flaw and remedies it by a delicate tap of a hammer on the outside. Advertising was useless; they must be discovered by some other means.

Inspecting Model 1917 rifle barrels at the Winchester plant. *(National Archives)*

Toolmakers, gauge fitters, dozens of other kinds of skilled artisans were needed immediately in armies."

Practically all of the plants suffered from the second and third difficulties, owing to their location in the heavy New England and New York manufacturing districts. Eddystone particularly suffered severely from these because of its location in an extremely congested district and its proximity to Hog Island plants, Sun Ship Building Co., New York Ship Building Co., Westinghouse, Cramp's, Eddystone Munition Co., Baldwin Locomotive, as well as others.

There was constant competition among the government plants, such inducements being offered included higher wages, payment of transportation, provision of houses, and assurances of deferred classification. The resulting tremendous labor turnover cut down production from 10 to 20 percent. There is a letter from this plant, which is a fair example of the difficulties experienced in all three plants at that time:

"(From) Capt. R.H. Eisenbrey, Washington, D.C.

> If labor can be stabilized at Eddystone rifle plant fully 15 per cent increase in production can be obtained with the same force of 14,000 employees. The monthly turnover is about 4,200 a month, fully 75 per cent due to lack of housing facilities.
>
> The turnover means the loss of parts all the way through the plant. Only a short time ago I saw 6,000 receivers scrapped. About 140 operations are required to produce the receivers; many of those condemned were lost on the last operation. Under present living conditions, it is impossible to keep men long enough to make them first-class operators.
>
> From the time a man first enters the employ of this company and leaves at the end of 10 days, it costs the company about $150, counting loss of tools, parts, etc.
>
> The crying needs in connection with rifles production: (1) Houses, (2) power to take over houses offered for sale and ordered vacated by owners so that immediate delivery can be made. This one method has caused us to lose many men."[1]

While the above narrative pertained to the Eddystone factory, Winchester and Remington-Ilion experienced similar difficulties. There were other problems facing the manufacturers as production continued to be ramped up. As discussed in *History of Rifles, Revolvers and Pistols*:

> "In August [1918] gun stocks became one of the limiting factors. For this, as for airplane propellers and veneers, **black walnut was used exclusively**, and it was thought that the supply of this was almost exhausted prior to the war; however, a country-wide canvas for this material resulted in locating a good deal of privately owned stock. Then came the difficulty of solving the time problem in drying it. Hitherto it had been customary to permit the wood to dry out for at least two years in order to avoid splitting and checking. This emergency was finally met by the development of a new kiln-drying process capable to thoroughly drying the wood for the lathe in 60 days, but for some time the production of metal parts was far ahead of gun stocks."[2] *[Bold print added by author]*

Despite the above-enumerated problems, along with some missteps, sometimes self-inflicted, along the way, the three manufacturers did a remarkable job in supplying our armed forces with an excellent service rifle at a critical time in our history. The outcome once again proved that Uncle Sam's ultimate "secret weapon" was the expertise, resourcefulness and production prowess of the nation's commercial manufacturers and their workers. It is interesting to note that the Model 1917 rifle was manufactured for approximately $26 each as compared to $42 that the British paid the same firms for manufacturing Pattern 1914 rifles.[3]

Manufacturers' Markings — While all Model 1917 rifles were manufactured to the same basic set of specifications, there were some minor configurational variances in some of the components between the weapons made by the three prime contractors. As mentioned in the previous chapter, with the exception of late-production Eddystone-made rifles, the metal parts were finished in an attractive rust bluing, known as "browning" at the time.

Many of the components were marked with the initial of the manufacturer ("W," "R" or "E"). The exact placement of the initial of many of the parts could vary and some were unmarked. (See Table #3). Each stock had the initial of the maker stamped on the tip ("front end"). It is generally accepted that each manufacturer produced their own stocks. However, noted researcher, the late Frank Mallory, discovered that the records of the Machine Gun-Small Arms Section, Engineering Div. dated December 14, 1918, revealed a list of identification marks assigned to stocks to be made by other manufacturers. In addition to the familiar "E" (Eddystone), "W" (Winchester) and "R" (Remington-UMC/Ilion), the list also included:

New England Westinghouse Co.	"N"
Remington U.M.C. Bridgeport	"B"
Victor Talking Machine Co.	"V"
Wood-Art Machine Co.	"A"[4]

There are no known examples of M1917 stocks observed with the above initials. It is entirely possible that stock production contracts from these firms were cancelled soon after the Armistice, and few, if any, actual stocks were fabricated. However, September 18, 1918, minutes from the executive committee of the Victor Talking Machine Company stated:

> "A discussion was had on our supply of birch lumber after we have filled out present order for one hundred thousand birch gun stocks. A purchase of five hundred thousand feet at $100 per thousand feet was authorized, but it was understood that the subject would again be considered after the manufacture of the gun stock blanks is completed."[5]

These minutes clearly indicate that the company was in the process of manufacturing gun stock blanks. It is unclear if these stock blanks were made of standard black walnut to be supplied to Eddystone or one of the other manufacturers or were birch wood and intended to be used by Victor for the order for 100,000 birch stocks that had been received. Since the referenced meeting was less than two months before the Armistice, there may have been few, if any, actual birch stocks manufactured in World War I.

The top of the barrel behind the front sight was marked with the initial of the manufacturer, the ordnance flaming bomb insignia along with the month and year that the barrel was produced. While there could be some variance, there was a close correlation, typically no more than a month or two, between the barrel date and the date of the rifle assembly. This was due primarily to the rapid rate of production during wartime, as the completed barrels were used soon after they were manufactured, which made the barrel date and the date of rifle assembly in fairly close proximity.

There were assembly marks consisting of an "eagle head" and a code numbers applied to 24 of the components. Each manufacturer was assigned a specific code number sequence, which will be discussed in the subsequent section of this chapter on each maker. When a rifle was assembled and passed all requisite inspections, the stock was stamped behind and in front of the trigger guard with the "eagle head" stamp and the assembly mark. The book *U.S. Rifle Model of 1917* by C.S. Ferris has detailed information the various inspection stamps.

Winchester Repeating Arms Co., New Haven, Connecticut

One of America's most recognized and respected firearms manufacturer was the Winchester Repeating Arms Company founded by Oliver F. Winchester in 1866. The company initially manufactured its signature lever-action rifles, which became sought-after weapons by many westward bound settlers in the decades following the Civil War.[6] The firm subsequently developed and manufactured other weapons including repeating shotguns and several types of single shot and slide-action rifles for the civilian market. Winchester did produce relatively limited numbers of military weapons under government contract from the mid-1870s through 1898. However, military weapons played a rather minor role, and Winchester continued to primarily manufacture numerous types of firearms for the domestic civilian market. However, by 1914, "the times they were a-changin."

When most of Europe became embroiled in the great conflict eventually known as the First World War, the belligerent nations were scrambling to secure additional sources of all types of war material, including and especially, firearms. As discussed in Chapter 1 of this book, Great Britain made arrangements with Winchester, Remington, and Eddystone to manufacture the Pattern 1914 rifle. However, Great Britain wasn't the only foreign nation to engage the services of Winchester to manufacture military firearms during the First World War. Despite its rejection by the

Winchester's New Haven, Connecticut, plant, 1918. *(National Archives)*

Model 1895 7.6x54 Model 1895 "musket" manufactured by Winchester for the Russian government, 1914–1916. *(Rock Island Auction Co.)*

Winchester Model D rifle. Designed by Thomas C. Johnson as a civilian sporting rifle, two "sample" Model D rifles were modified for the .30-06 cartridge and the Pattern 1914 "lock work." *(Cody Firearms Museum)*

U.S. Army in 1898, Russia was interested in the company's Model 1895 lever-action "musket" modified for the standard Russian 7.62x54 cartridge. The first production contract was signed in November of 1914 with subsequent contracts to follow. By the time of the last delivery of these rifles in December 1916, Winchester had produced the not-insignificant total of 293,318 Model 1895 rifles for Russia.[7]

Winchester Model of 1917 Rifle — The late author Herbert Houze, noted Winchester scholar and former curator of the Winchester Arms Collection at the Cody Firearms Museum, related interesting and little-known information on Winchester's initial participation in the Model 1917 rifle production program, which he gleaned from factory archival documentation. As can be seen, even before the United States entered the war, Winchester was quite proactive in anticipating the likelihood of the government requesting an American version of the Pattern 1914 rifle. As related in his authoritative book, *Winchester Bolt Action Military & Sporting Rifles 1877–1937*:

> "When it became clear in early 1917 that the United States was being drawn closer and closer into the European War, the Winchester Company began making preparations. Realizing the Army would probably call upon it and Remington Company to produce small arms, experiments were carried out to determine how easily the Pattern 14 Enfield (then being manufactured by both firms) could be altered for use with the .30-06 U.S. government cartridge. To accomplish this, T.C Johnson and Frank F. Burton rebuilt two of the Model D samples (numbered 14 and 15) so that they incorporated Enfield lock work. Completed by February 22 1917, these rifles were to serve as the original prototypes for the U.S. Model 1917 Service Rifle."[8]

It is a little-known fact that in November of 1916, while Winchester was still manufacturing the Pattern 1914 rifles, the company sent a representative to Russia to demonstrate the Model D rifle. This resulted in Russian ordering 500 Model D rifles, which were sent to Russia in March of 1917 for further testing and evaluation. The subsequent

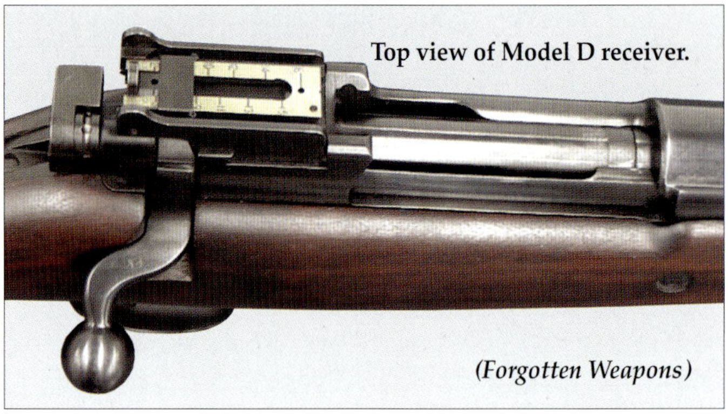

Top view of Model D receiver.

(Forgotten Weapons)

Bolshevik Revolution that November spelled an end to any further orders for Model D rifles. The American Ordnance Department had little interest in the design and opted for the development of a Pattern 1914 rifle chambered for the .30-06 cartridge.[9]

Almost immediately after the United States declared war on Germany, General William Crozier, Chief of Ordnance, made overtures to Winchester regarding the potential for the company to manufacture rifles for the government. In reply, John E. Otterson, 1st Vice President and General Manager of Winchester, sent the following letter:

> "April 3, 1917
> General W.C. Crozier
> Chief of Ordnance
> War Department, Washington, D.C.
> Dear Sir:
>
> Agreeable to your instructions, I beg to furnish you herewith a standard statement of capacity of the Winchester Repeating Arms Co. for the production of Enfield rifles, modified to take the U.S. Government cartridge, the figures presented being the production within one year after placing of contract.
>
> Assuming the full use of our facilities and favorable manufacturing conditions, we have a potential capacity of about six hundred thousand (600M) rifles in the time mentioned. If the need of the government is urgent enough, we shall be pleased to make an effort to approach this capacity, but it does not seem

Winchester "pre-production sample" Model 1917 rifle. *(National Firearms Museum)*

Winchester Model 1917 rifle, ser. no. 1. *(Springfield Armory Museum. Courtesy of Alex MacKenzie, Curator)*

wise under existing circumstances for any company to offer the Government assurances calling for the full use of its facilities under favorable conditions, and we would not obligate ourselves to this extent except on the basis of a clear understanding.

We do not recommend that we be given a contract calling for our full potential capacity, but we do recommend that we be given a contracting calling for between three and four hundred thousand rifles, and believe that we can deliver this number with reasonable assurance.

The following schedule of deliveries is proposed upon the basis of a contract for 350 M rifles –

Deliveries to commence in four months, at the rate of one thousand (1,000) per day. Such a rate to be increased to two thousand (2,000) per day in six months and to continue at this rate for the balance of the year making —

1,000 per day for 50 days, equal 50,000
2,000 " " " 150 " " 300,000
350,000"[10]

Although a production contract for Winchester was not forthcoming at this time, as mentioned in Chapter 2 of this book, the Ordnance Department did authorize the company to proceed with the development of a Pattern 1914 rifle modified for the .30-06 cartridge. The first sample was sent to the Ordnance Department on April 30, 1917. The design was found to be acceptable, and two more samples were made. All three rifles (numbered "1", "2", and "3") were sent to Springfield Armory on May 7, 1917, for testing and evaluation. The results indicated that the sample rifle was not believed to be as good as the Model 1903 Springfield, but it was equal or superior to any of the rifles fielded by allies and adversaries.[11]

Winchester was authorized to begin work on five sets of pre-production samples. These were numbered in five blocks; A to Z, 2A to 2Z, 3A to 3Z, 4A to 4Z and 5A to 5Z . The rifles were completed shortly after May 8, 1917.[12] Reportedly, these were to serve as "master reference models" by both Winchester and Remington when the firms prepared to engage in mass production of the Model 1917 rifle.[13]

The events that led to the adoption and production of the Model 1917 rifle has been covered in some detail in Chapter 2 of this book. As mentioned, the Ordnance Department was faced with the choice of approving production of a rifle that didn't conform to the level of interchangeability required or delay production until this problem would be rectified. The manufacturers were given the option of starting production or postponing it to work out the interchangeability issues. As related by Crowell:

"...we (Ordnance) decided to go ahead with production and improve the standardization as we went along. The Winchester and Ilion (Remington) plants elected to start work on that understanding, but Eddystone preferred to wait for the final requirements. Ilion afterward decided to postpone until the final specifications were adopted. It would have been well if the same course had been followed at the Winchester plant, for word came later from Europe not to send over rifles of Winchester manufacture for that period."[14]

The reason(s) why Remington abruptly reversed course and elected to wait for the final specifications, as was also the case with Eddystone, is not known. In any event, Winchester was not going to wait around for the Ordnance engineers to sort things out. As related by author Harold Williamson in his excellent book *Winchester – The Gun That Won the West*:

"Otterson took a different position. He did not wait for official action or a contract before starting production of the modified Enfield. On April 23, 1917, he ordered the start of the manufacture of 10,000 rifles...on the theory that when (Winchester) got those rifles made, and it developed they were not satisfactory...(they) would be able to sell them commercially or perhaps sell them to some other government.

By not waiting for final specifications Otterson drew criticism from some American military authorities... From other branches of the service, especially from officers in charge of training, he was congratulated because the scarcity of rifles was such that many recruits were drilling with broomsticks."[15]

Even though no contracts had yet been issued, as per Mr. Otterson's instructions, Winchester began production of the Model 1917 rifle during the third week of May 1917. As reflected in factory documentation, the first deliveries of finished components were sent to the assembly room on May 22, 1917.[16]

As mentioned previously, each manufacturer faced a number of daunting difficulties, especially systemic labor problems. As discussed in *History of Rifles, Revolvers and Pistols: 1917–1919*:

> "Winchester solved its labor problems in a novel and most effectual manner. Instead of following the usual tactics of recruiting from other industries and cities, thereby depleting those, causing a mushroom growth in their own city, which must inevitably overcrowd it during the war and leave it overdeveloped when the need for the inflated population should cease to exist, thus increasing the general dissatisfaction and labor turnover among the employees of the entire district, this company, at the suggestion of the general manager, Mr. Otterson, devised the scheme of enlisting the interest of the following classes of people in the immediate vicinity: those unemployed by reason of retirement or otherwise , those employed in non-essential occupations, those whose work might be combined with that of another, thus releasing one person for Government work; women whose circumstances did not require them to work, housewives who could give part of the day, high-school pupils who could work part time, domestic help, which could be released through the assumption of their duties by others not accustomed to work who did not wish to work in the munitions shops. Very extensive advertising through the public press, store-window exhibits, speakers in theaters and other public places, patriotic rallies addressed by women connected with war work in other places, posters, medals for those already in the plants and banners for their homes operated to give the movement a wide publicity and to arouse such enthusiasm that not only was the acute shortage relieved but it became necessary to take the names of applicants in order that notification might be sent to them when work developed to a point requiring their services. The benefits thus derived were manifold: The needed help was acquired without the usual attending disadvantages, interest in the war was kept aflame though personal service,

Chambering Model 1917 rifle barrels at the Winchester plant, 1918.

Winchester employees at the factory's main entrance gate, 1918.

and thousands of dollars extra were made available for investment in Liberty bonds. It was interesting to note that in many cases people of social position who had never worked before made average records as regards production and stability.

The campaign proved such a success in New Haven that it was copied by many other cities and produced satisfactory results wherever tried."[17]

Otterson's imaginative ways to augment the company's workforce certainly proved to be a viable model and showed "thinking outside the box." At its peak in July of 1918, those employed at Winchester numbered 19,806.[18]

Herbert Houze, provided the following data on the company's Model 1917 rifle production program beginning June 1, 1917:

"On June 1, 1917, the War Department authorized the Winchester Company to begin production of 225,000 Model 1917 Rifles plus their necessary spare parts. Under the terms of the actual contract (No. 14064) signed on July 12, delivery of all these arms was to be concluded by June 30, 1918. On December 11 and 12, the company was given authorization, respectively, to produce a further 100,000 and 50,000 Model 1917 Rifles. These authorizations were later combined under the terms of a single contract that was signed on January 1, 1918. It called for deliveries totaling 45,000 rifles in May 1918, 45,000 in June, another 45,000 in July and the final lot of 15,000 in August. On August 12, 1918, a supplemental order was attached to Contract Number 16302, calling for the manufacture of another 290,000 Model 1917 Rifles, together with their bayonets, plus the following material:

400,000 magazine platform depressors
2,900 sets of spare parts
150,000 Model 1917 bayonets
28 skeletonized Model 1917 Rifles
6 spare sets mounted on boards
1 set of taps
5,000 Model 1917 receivers."[19]

Magazine platform depressor.

The official declaration that "...word came later from Europe not to send over rifles of Winchester manufacture of that period" has led to some erroneous conclusions regarding these early Winchester Model 1917 rifles. It has been stated, or implied, that a number of these early rifles had been sent to France and proved to be defective because of the lack of parts interchangeability, thus the AEF didn't want anymore sent over. This was not the case. As discussed by author C.S. Ferris in his excellent book *United States Rifle Model of 1917*:

"The (Crowell) comment about the unacceptability of early Winchesters takes on a different color when it is learned that the Ordnance Department prohibited all Model 1917 rifles, of any manufacturer, from going overseas until March 1918, and then allowing only those made after January 1, 1918, by any manufacturer."[20]

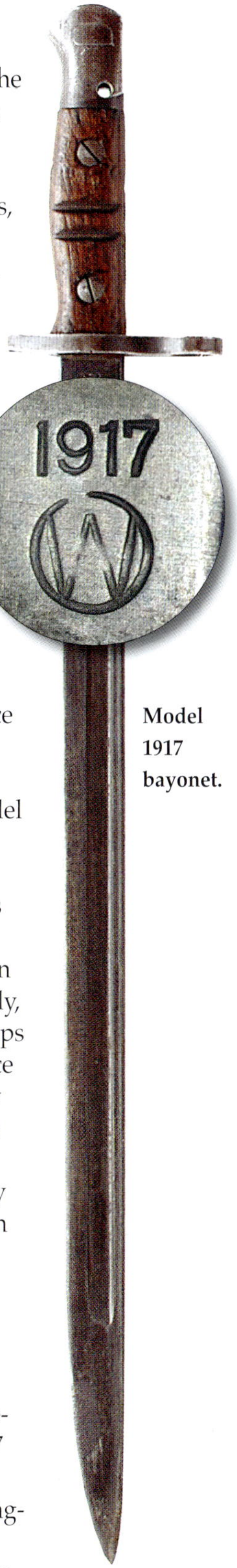

Model 1917 bayonet.

It should be pointed out that all of the rifles made by Winchester had passed functioning inspections before leaving the plant. The only "defect" the early Winchester Model 1917 rifles had was that the parts didn't have a sufficient degree of interchangeability with rifles made by Remington and Eddystone to meet the initial requirements of the Ordnance Department.

Ordnance documents indicate that Winchester manufactured 103,229 Model 1917 rifles prior to January 1, 1918.[21]

While total interchangeability with the rifles made by the three contractors was never achieved there was a sufficient level of 95 percent standardization attained by January 1, 1918. Inexplicably, despite the burgeoning number of troops being deployed to France, the Ordnance Department still would not release any of the now-conforming rifles to be sent overseas for another two months. A March 4, 1918, memorandum issued by the Chief of Ordnance, Supply Division contained the following provisions:

"1. All organizations ordered overseas and ordered to be equipped with rifles were being issued Remington Model 1917s.
2. Organizations and divisions heretofore equipped with Model 1917 rifles of Winchester manufacture were being equipped with Remington rifles.
3. Winchester Model 1917s would be issued only to troops in continental possessions of the United States."[22]

It was suggested by the Supply Division of the Office of the Chief of Ordnance to mark the stocks of the "non-conforming" Winchester rifles. As related in a June 25, 1918, memorandum:

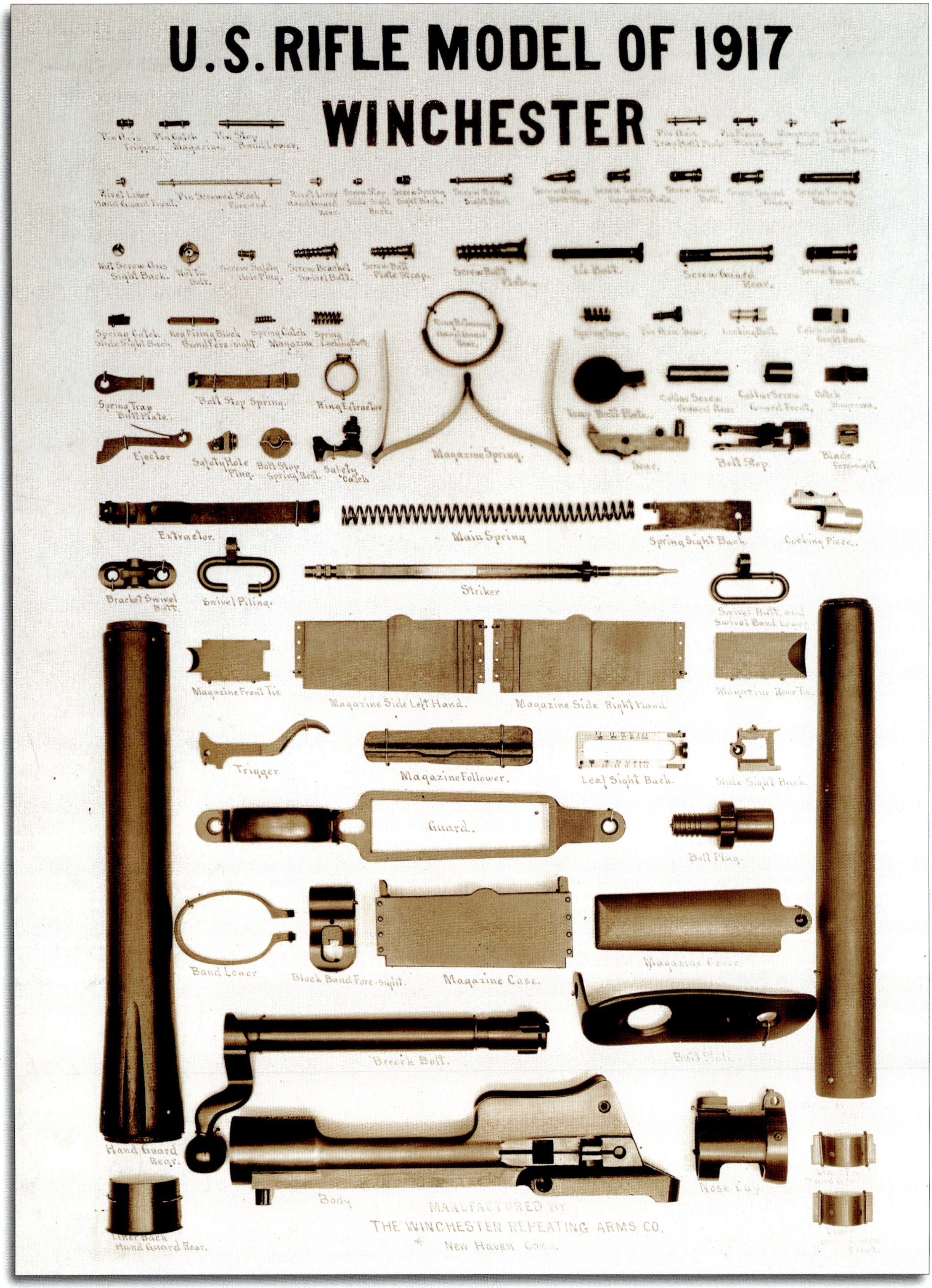

Model 1917 rifle components mounted on a display board. *(Tom Laemlein)*

Photo from 1918, captioned "Stamping out bayonets at plant of Winchester Arms Co., New Haven, Conn." *(National Archives)*

Winchester employees watching a "Suffrage" demonstration outside the factory c.1918. *(Library of Congress)*

Loading and testing newly made Model 1917 rifles at the Winchester plant. *(Tom Laemlein)*

Test-firing Model 1917 rifles at the Winchester plant's indoor firing range. Note the rack of rifles. *(National Archives)*

August 1917 photo of Winchester workman inletting a stock for the trigger guard assembly for a Model 1917 rifle. *(Tom Laemlein)*

Winchester Model 1917 rifle being fired at the New Haven plant. *(Tom Laemlein)*

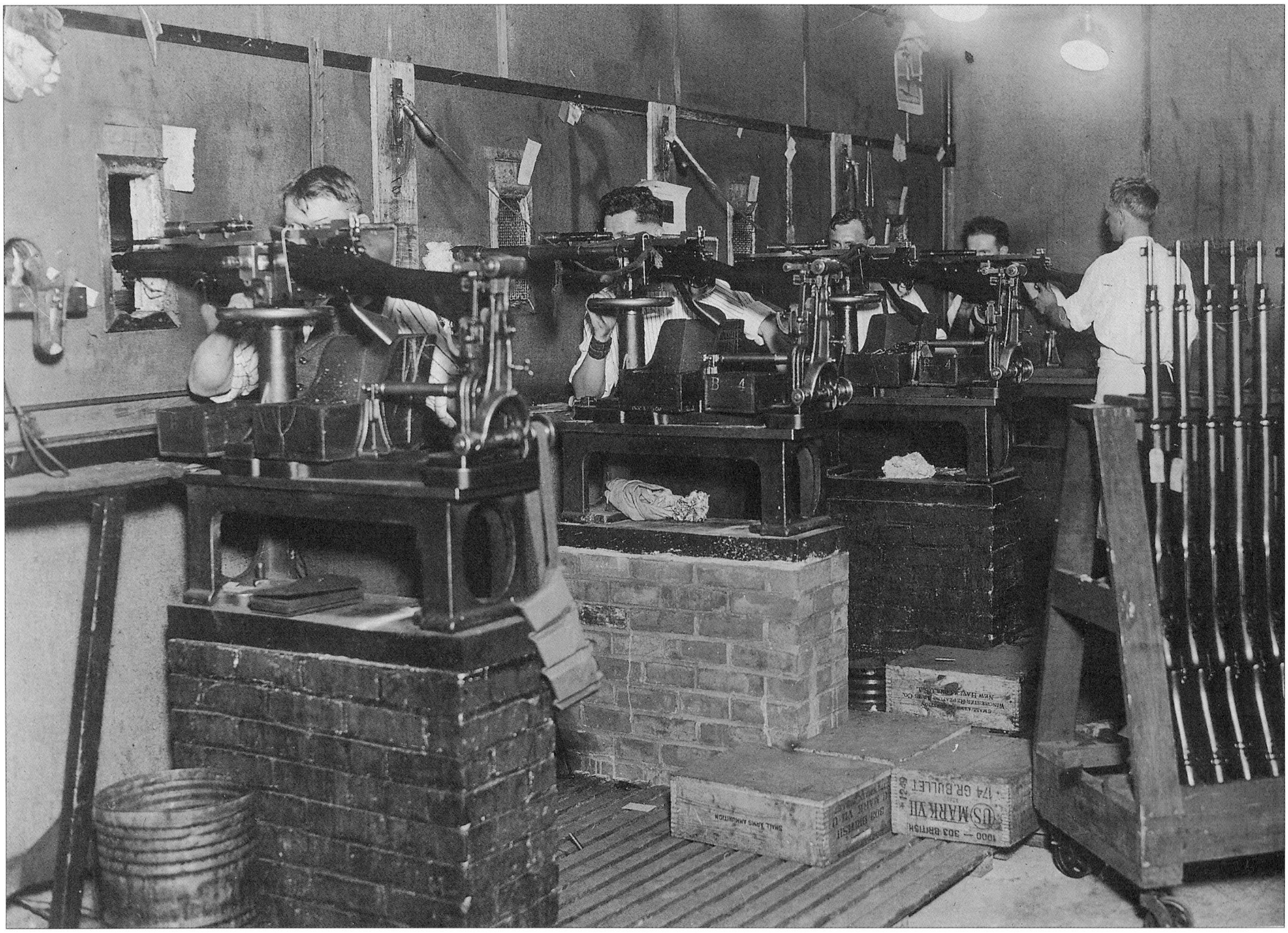

Test firing Model 1917 rifles at the Winchester plant. Note the use of telescopic sights. These telescopic sights are not actually attached to the rifles, but were temporary fixtures resting on the front and rear sight bases for use in targeting only. *(Tom Laemlein)*

"1. Orders have been received that no rifles, Winchester, manufactured prior to January 1, 1918, are to be shipped overseas.
2. Instructions are requested as to whether the Engineering Division advises any further marking be placed on these rifles. Suggest that they be marked 'T.P.O.' or 'Training Purposes Only.' "[23]

A subsequent memorandum from the Machine Gun-Small Arms Section to the Engineering Division, which was dated two days later, further proposed that "certain low grade stocks" be identified by a "star" marking:

"For your information, the Engineering Division has recently authorized the marking of certain low grade stocks, which will be accepted for temporary use only, with a star; such star to be approximately ½" in diameter and placed on the right side of the butt, about 3" from the buttplate. This will not occasion any confusion with the star formerly used to indicated non-interchangeable Winchester rifles, as this star was stamped into the metal on the left side of the receiver near the acceptance marking. This point is mentioned in order that no one be inspired to use a star in marking the stocks of training rifles."[24]

Barreled actions being transported to the assembly area at the Winchester plant. *(National Archives)*

Winchester workman fitting and adjusting M1917 rifle bolt and receiver. *(National Archives)*

Workman assembling the barreled action into the stock. *(National Archives)*

Inspecting finished M1917 rifle by inserting a clip into the receiver for proper fit. *(National Archives)*

Given the dearth of observed specimens having either of these markings applied, it can be assumed that limited numbers of rifles were stamped in this manner.

The pronouncement to consider Model 1917 rifles made by Winchester, along with Remington and Eddystone, to have an acceptable degree of interchangeability was the subject of a June 29, 1918, memorandum from General Pershing's headquarters, A.E.F.:

> "Careful tests, which have now been completed show a satisfactory degree of interchangeability of components of the caliber .30 model 1917 rifle as manufactured by all three of our rifle plants, that is, Eddystone, Ilion and Winchester. All rifles from (the) first two manufacturing plants and those rifles assembled by Winchester since January 1, 1918, as shown by dates on top of the barrels near the front sight are now being issued indiscriminately for service both in this country and overseas, and there will be no complications in issuing spare parts due to non-interchangeability. This situation makes possible the complete (equipage of) troops with rifles even with the greatly accelerated embarkation schedule."[25]

When Winchester, along with Remington, initially elected to begin production before the standardization requirements were finalized, it was directed that the receivers be marked with a "star" emblem on the left side to indicate insufficient interchangeability of components. However, since Remington subsequently chose to join Eddystone and wait for the final interchangeability requirements before making any production rifles, the "star" emblem was only applied to Winchester rifles of the period.[26] The marking is found on rifles made by Winchester through circa February of 1918 ending in the approximate 141563–145950 serial number range.[27] Some receivers stamped with the "star" marking were undoubtedly used on rifles assembled after Winchester had achieved acceptable standardization.

In addition to this star marking, there was one other anomalistic feature only found on very early Winchester rifles. When production began, the receivers were not marked with the standard "U.S. / Model of 1917 / Manufacturer Name/ser. no." format. Instead, the receivers were simply marked "U.S./ W/ ser. no.", somewhat similar to the Pattern 1914 receivers made by the firm, which were marked "W" and the serial number. This marking format continued on Winchester Model 1917 rifles until approximately ser. no. 4663 (c.Feb. 1918) when the full marking format began to be used.

There was one reported anomaly found on rifle ser. no. 4064, which had the full marking format on the receiver.[28] The reason(s) why Winchester used this early marking format is not known. While it is sheer speculation, a clue may lie in the comments of John Otterson when he stated that if the early-production rifles that didn't confirm to Ordnance specifications proved to be "...not satisfactory... (they) would be able to sell them commercially or perhaps sell them to some other government." Even though these

"Star" marking on the receiver of Winchester Model 1917 rifle manufactured prior to the standardization of interchangeable parts.

Receiver markings on early- production Winchester Model 1917 rifle.

Standard Winchester Model 1917 rifle receiver markings.

"W" marking on tip of Winchester-made Model 1917 rifle stock.

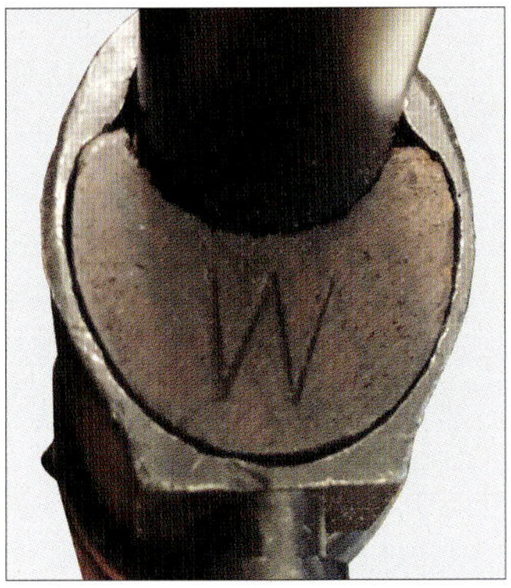

Typical markings on Winchester Model 1917 rifle barrel. This is an early example that was manufactured in July of 1917.

"W" marking on bottom of Winchester M1917 bayonet lug.

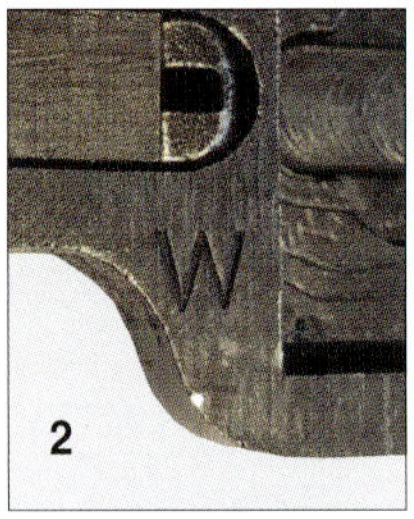

1 Marking on Winchester Model 1917 bolt handle.
2 Manufacturer's marking inside of Winchester Model 1917 rifle trigger guard.
3 Eagle-head inspection stamp behind the trigger guard on a Winchester Model 1917 rifle. Note the "500" code number. Winchester was assigned numbers in the 400–600 range.

early receivers were marked "U.S." the fact that they didn't carry the official nomenclature of a standardized American military rifle might have made them a bit easier to sell to buyers other than the United States government. In any event, as things transpired, this was a moot point.

Winchester rifles had the same marking format as the other makers on the barrel consisting of initial of the manufacturer ("W"), the date of barrel production and an ordnance flaming bomb insignia. Likewise, the manufacturer's initial was also stamped on many of the components.

Winchester was assigned a three-digit code used with the eagle-head assembly mark found on various components, including the stock. Winchester's code numbers were in the 400 through 600 range.[29]

Eventually, Winchester rifles were made to the same general degree of interchangeability as the other contractors, and all Model 1917 rifles were issued without regard to the manufacturer. However, Winchester officials believed there was a lingering bias of the Ordnance Department against the company. On June 26, 1918, Mr. Otterson requested that the Ordnance Department provide specific reasons why the company was, as he put it, "discriminated against" by not permitting rifles to be sent overseas. Major J.T. Kenyon of the Ordnance Department apparently took umbrage to Winchester's claim of discrimination and stated in typical bureaucratic fashion "...The term 'discrimination' as used by Mr. Otterson should be termed 'cautionary measures against vital results overseas' and applies to all plants manufacturing the 1917 rifles."[30]

In order assuage Winchester's concerns, General C.C. Williams approved a recommendation that Winchester's Model 1917 rifles be issued for overseas use with no restrictions. In a June 19, 1918, letter, Ordnance Department officer Colonel John Simpson wrote a letter confirming that General Williams directed that no distinction was to be made regarding Model 1917 rifles made after January 1, 1918. In his letter, Colonel Simpson stated: "This finally determines the question that rifles made by the Eddystone, Ilion, and Winchester plants since January 1, 1918, may be regarded as equally interchangeable and, therefore, equally available and serviceable for use with United States forces anywhere, either at home or overseas."[31]

Regardless of General Williams' assurances, there was seemingly still some lingering bias against Winchester rifles. As May 3, 1918, memorandum from Colonel Harry B. Jordan, Acting Chief of Ordnance, A.E.F., to Ordnance Depot No. 1 with the Subject "Winchester Rifles:"

> "This office has instructed the Regimental S.O., 17th Engineers...to turn over to you approximately 75 Winchester Enfield Rifles, which were in the possession of some new recruits who have been permanently assigned to the 17th Engineers. When these rifles come to hand, forward them to Intermediate Ordnance Depot No. 2, Gievros, Loir et Cher."[32]

The reason for singling out only Winchester rifles was not given, but it would seem apparent it was due to ongoing bias against the company's products.

Despite such periodic ostensible prejudice, Winchester continued to turn out Model 1917 rifles, and the weapons were issued to troops deploying to France as quickly as logistics permitted. Winchester Model 1917 rifles served our armed forces well, and the company's contributions to the war effort were impressive. Herbert Houze gave an excellent summary of Winchester's Model 1917 Rifle production program taken from Winchester factory archival documentation:

> "Contract Number 16302 [cited above] was expanded again on October 19, 1918, when the government attached a second supplement that ordered 200,000 Model 1917 Rifles (without bayonets), 200,000 magazine platform depressors and 59,742 Model 1918 Telescopic 'Sniper Rifles.'
>
> During the course of the war, the Winchester Company fulfilled the following contracts:
>
> - No. 14064 – 225,000 Model 1917 Rifles
> - No. 16302 – 150,000 Model 1917 Rifles
> - No. 16302 Supplement 1 – 170,511 Model 1917 Rifles (the balance cancelled 12-28-18)

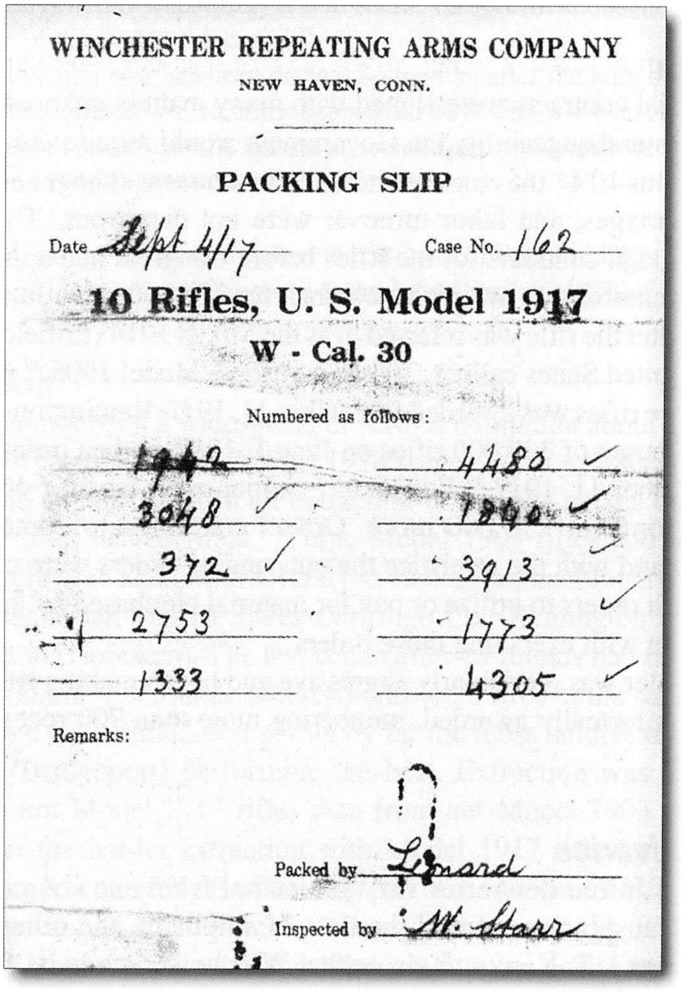

WINCHESTER REPEATING ARMS COMPANY
NEW HAVEN, CONN.

PACKING SLIP

Date Sept 4/17 Case No. 162

10 Rifles, U. S. Model 1917
W - Cal. 30

Numbered as follows:

1942	4480
3048	1890
372	3913
2753	1773
1333	4305

Remarks:

Packed by Leonard

Inspected by M. Starr

Winchester factory packing slip dated September 4, 1917, showing the serial numbers of ten Model 1917 rifles shipped from the factory in Case No. 162. Note that the high and low serial numbers of the rifles in the crate are more than 4,000 numbers apart. ***(John Spangler)***

- 28 skeletonized Model 1917 Rifles (order originally cancelled 1-19-19, reauthorized on 2-1-19 and subsequently completed)
- 6 parts boards
- No. 16302 Supplement 2 – 189 Model 1918 Telescopic Sniper Rifles (balance of order cancelled on 1-10-19)

In addition to the above, Winchester produced a total of 718,689 Model 1917 bayonets, 6,955 spare parts sets and 215,512 magazine platform depressors."[33]

It was reported in *America's Munitions 1917–1918*, that Winchester received contracts for the production of 672,500 Model 1917 bayonets but only delivered 395,894.[34] Curiously, these figures taken from U.S. Government Contracts contained in the Winchester archives are markedly different from the total of Winchester Model 1917 bayonet production cited by Crowell.

This discrepancy can perhaps be partially explained because Crowell's figures were only for the Model 1917 bayonets produced by Winchester up to the time that hostilities ceased with the Armistice and did not include bayonets made afterward until production totally came to an end in the spring of 1919. Also, Crowell's numbers do not include the sizable number of Pattern 1913 bayonets purchased from the British. While conjecture, it is possible that the Pattern 1913 bayonets made by Winchester may have been counted as part of the company's total Model 1917 bayonet production figures.

In addition to the standard Model 1917 rifles manufactured under government contract, Winchester made several "presentation" rifles. The most notable example was discussed by Houze:

"All of the U.S. Model 1917 Rifles made under government contract were plain service weapons, but shortly after their production ceased, the Winchester Company built a 'Deluxe' grade rifle for presentation to President Woodrow Wilson. Unmarked except for inspection and proof marks, this rifle had an 'extra polish' blued finish as well as a 'semi-fancy' walnut stock. For reasons now unknown, this rifle was never given to Wilson, and it remains to this day, in the firearms reference collection originally assembled by the company."[35]

There was another Model 1917 presentation rifle, formerly in the Winchester Museum collection, that has the same hallmarks as the rifle intended for President Wilson, including no markings on the metal except for a Winchester proof mark and a specially finished stock and polished blued metal. There is no record of any intended recipient of the rifle.[36]

The production totals for each manufacturer published in different reference sources may vary. For example, the Ordnance publication *America's Munitions 1917–1918*, reflects a total of 465,980 rifles made as of November 9, 1918, which is the same figure quoted by Hatcher and Skennerton.[37] The final production contract was cancelled on December 28, 1918. Even though the contract was cancelled on that date, contractual provisions allowed each manufacturer to continue assembling rifles that were "in process" at the time of cancellation. The last Model 1917 rifle serial number used by Winchester was 548255, which was stamped on the receiver on March 25, 1919.[38]

Houze indicated that a total of 545,700 Model 1917 rifles (including 189 M1918 sniper rifles) were supplied to the U.S. Government.[39] Ferris quotes a figure of 545,482, which is very close to the Winchester archival records.[40] The Winchester Model 1917 rifle assembly line closed on April 5, 1919.[41]

Winchester Model 1917 rifle (right side). *(National Firearms Museum)*

Winchester Model 1917 rifle (left side). *National Firearms Museum)*

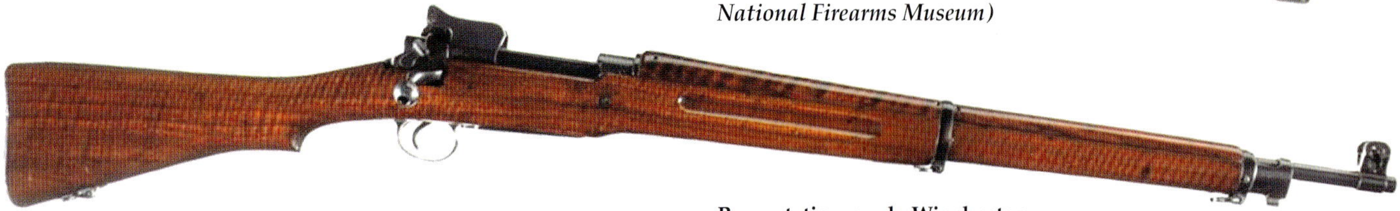

Presentation-grade Winchester Model 1917 rifle. *(Rock Island Auction Co.)*

Remington Arms Company, Ilion, New York

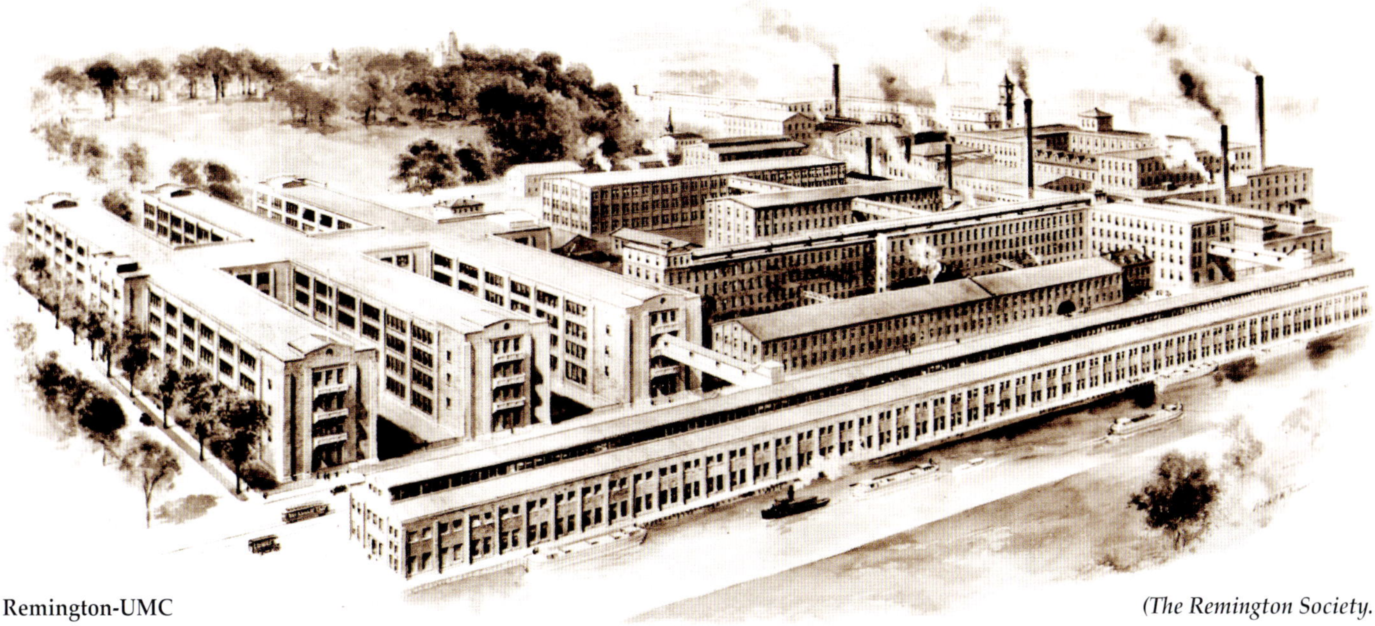

Remington-UMC plant in Ilion, New York, c.1915. *(The Remington Society. Courtesy of Roy Marcot)*

America's oldest established firearms manufacturing company, the roots of the Remington Arms Company can be traced back to 1816 when Eliphalet Remington II asked his father for money to buy a rifle. When the request was refused, the young man forged and welded a barrel blank and took it to a gunsmith in Utica, New York, to have it reamed and rifled. The gunsmith who finished the barrel was favorably impressed with the initial work that Remington had done, and the young man was soon on the path to making his own complete firearms.[42] In 1828, Remington established a firearms manufacturing operation in Ilion, New York, which remained the site for the company's operations for the next two centuries.

E. Remington & Sons manufactured a variety of rifles, shotguns, and handguns for the civilian market. The company also produced several types of firearms for the U.S. Government from the early 1840s through the late 1880s including the Model 1841 "Mississippi Rifle," the Jenks Navy carbine, the New Model Navy .36 caliber revolver and .44 caliber New Model Army revolver, several types of split breech and rolling block rifles and carbines, and a limited numbers of Remington-Keene Navy rifles and Remington-Lee Army and Navy rifles. Following the American Civil War, several foreign nations, including France, Mexico, Japan, China, Russia, Spain, Peru, purchased Remington firearms, chiefly of the rolling block design.[43] However, the domestic civilian market continued to be the company's "bread and butter" with its wide array of popular rifles, handguns, and shotguns. Despite the popularity of its firearms, E. Remington & Sons experienced severe financial issues in 1888, and the company was restructured. As related by noted Remington historian Roy Marcot:

> "An unusual partnership of Hartley & Graham, owners of Union Metallic Cartridge Co. (UMC), and Winchester Repeating Arms Co. acquired joint ownership of E. Remington & Sons and renamed it Remington Arms Co. The dual ownership continued until 1896 when Hartley & Graham purchased the other fifty percent and acquired full control."[44]

The restructured entity was often referred to as simply "Remington-UMC." The Remington Arms factory was also sometimes designated as "Remington – Ilion" to differentiate it from the Union Metallic Cartridge Company's plant in Bridgeport, Connecticut.

As was the case with the Winchester Repeating Arms Company, the outbreak of war in Europe in 1914 resulted in a flurry of orders for Remington-UMC from several allied nations for the manufacture of military firearms. The company accepted contracts from the Russian czarist

Remington-UMC plant in Bridgeport, Connecticut. *(Library of Congress. Courtesy of Roy Marcot)*

Model 1891 Mosin-Nagant manufactured by Remington-UMC for the czarist Russian government.

Model 1908-15 Berthier rifle produced for the French government by Remington-UMC.

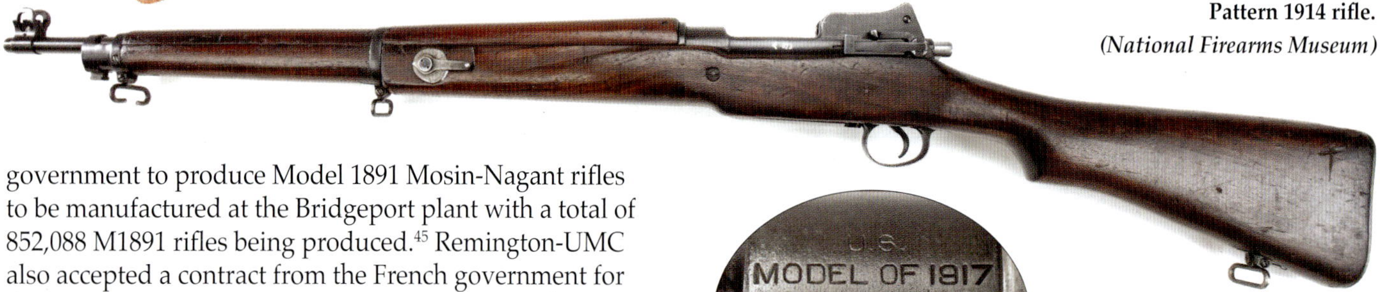

Remington-UMC British Pattern 1914 rifle. *(National Firearms Museum)*

First week of production, Remington Model 1917 rifle, serial number 137. *(National Firearms Museum)*

government to produce Model 1891 Mosin-Nagant rifles to be manufactured at the Bridgeport plant with a total of 852,088 M1891 rifles being produced.[45] Remington-UMC also accepted a contract from the French government for the manufacture of 250,000 Model 1907–1915 Berthier rifles, also to be made in Bridgeport. However, production and contractual difficulties resulted in less than 20,000 being made, and fewer than 10,000 of those were actually accepted by the French.[46]

As previously discussed in Chapter 1, Great Britain approached several American firms, including Remington, to manufacture the Pattern 1914 rifle. With the existing production commitments, the company was initially hesitant to commit to the venture without sizable advance payments and reasonable delivery schedules. Once all the obstacles were addressed, according to ordnance records, Remington-UMC manufactured a total of 403,117 Pattern 1914 rifles by the time production ceased on July 21, 1917.[47] Remington factory records indicate a slightly higher production total of 403,193.[48]

Remington Model of 1917 Rifle — A Remington Arms Company memorandum dated August 17, 1927, titled "World War Production of Remington Ilion Works" discussed the origins of the company Model 1917 rifle production program:

> "Early in January, 1917, when it became apparent that the United States would sooner or later become involved in the World War, officers of this Company were requested by Ordnance Department officials to make preparations for the tooling up and production of the U.S. Rifle, Caliber .30, Model of 1903, familiarly known as the .30-06 Springfield. This rifle, while of bolt-action type, differed in practically all particulars from that of the .303 Remington-Enfield, which we were producing. After many conferences and much correspondence, the Ordnance officials agreed with Remington Arms Company officials that the proper expedient would be to adapt the 'Remington-Enfield' to the .30-06 U.S. Government cartridge, upon which our engineering department had been working 'on our own hook' since negotiations were first begun with the U.S. Ordnance Department. Accordingly, early in April 1917, we had such a rifle ready for submitting to the Ordnance Department for their approval. A conference was arranged at which representatives of the Ordnance Department, of the Remington Arms Works at Eddystone, of the Remington Arms Works at Ilion, and of the Winchester Repeating Arms Company were present. The Ilion sample rifle was carefully studied, and with a very few minor changes – necessary because of the slight variations in manufacturing procedures at the various plants – was adopted finally.

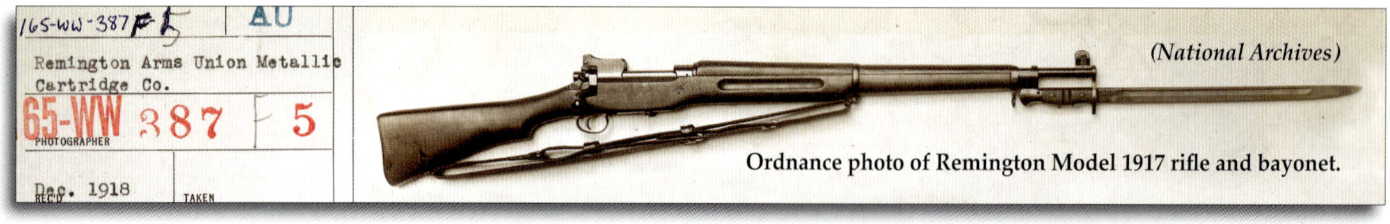

(National Archives)

Ordnance photo of Remington Model 1917 rifle and bayonet.

As the production of the British 'Remington-Enfield' rifle had approached the 400,000 mark, all of our activities were directed toward the adaptation of this rifle to the .30-06 cartridge, with a view to reaching the maximum production possible at the earliest date. As the .303-inch British cartridge was of the rim type, while the .30-06 U.S. cartridge is rimless, the task of changing jigs, fixtures, cutters, gauges, etc., to provide even for the cartridge change alone, not to consider other changes, which seemed necessary, was enormous, but within a few months deliveries of rifles had begun, and at the time the Armistice was signed in November 1918, production had reached a volume of 3,000 or more U.S. Rifles, caliber .30, Model of 1917, per day. As stated in the summary, a total quantity of 646,444 of these rifles were furnished the U.S. Government."[49]

Model 1917 bayonet and scabbard.

As mentioned in Chapter 1 of this book, the Remington factory designation for the Pattern 1914 rifle was the "Model 28," but the Model of 1917 rifle was designated as the "Model 22."[50] There was no explanation as to why the Pattern 1914 rifle had a higher factory designation number than the Model of 1917, which was produced later. Also, as discussed previously in this chapter, the Ordnance Department authorized Winchester to proceed with the development of a Pattern 1914 rifle chambered for the .30-06 cartridge, and the first sample was submitted on April 30, 1917. However, as recorded in the above-cited document, Remington apparently had such a rifle in parallel development as evidenced by the statement that the company had a rifle ready to be submitted to the Ordnance Department for approval in "...early in April 1917."

It was further stated that "The Ilion sample rifle was carefully studied, and with a very few minor changes...was adopted finally."[51] This would seem to suggest that the sample rifle submitted by Remington was actually selected by Ordnance rather than the example fabricated by Winchester.

In any event, a contract (War-Ord-14063) was executed on July 12, 1917, for the manufacture by Remington-UMC of 300,000 Model of 1917 rifles. The following are selected excerpts from the contract:

"(a) Two Hundred Fifty Thousand (250,000) U.S. Rifles, Caliber .30, Model 1917 (without bayonets)

(b) Two Hundred Fifty Thousand (25,000) Magazine Platform Depressors for U.S. Rifle, Caliber .30, Model 1917.

ARTICLE II. The delivery of the additional (50,000) rifles and magazine platform depressors shall commence immediately following the completion of the delivery of the rifles and parts as provided for in the original and first supplement contracts and shall be completed as soon thereafter as possible.

ARTICLE III. All of the additional rifles and magazine platform depressors shall herein provided for will be paid for by the United States on the same basis as provided in the original contract, to-wit: Cost plus ten (10) per cent for profit.

ARTICLE IX. The contracting officer desire and the contractor shall use his best endeavors to the end that delivery of rifles covered by this contract shall begin at the end of the fourth month and continue thereafter at a rate not less than that shown on the following schedule:

Fifth month........................10,000
Sixth month........................23,000
Seventh month......................35,000
Eight month........................45,000
Ninth month........................57,000
Tenth month........................60,000
Eleventh month.....................70,000"[52]

The manufacturer's marking on a Remington-made Model 1917 bayonet, which was produced in 1917.

Field Symbol
No. in Unit Pkg. 25
Ord. Office Symbol R24RRN03K
Magazine Spring
For U. S. RIFLE Cal. 30
MODEL OF 1917
(ILION)
Packed by W. Rockefeller
Date Packed NOV 29 1918
Nomenclature and Count Guaranteed by
REMINGTON ARMS U. M. C. CO.
WHILE SEALS ARE INTACT
DUPLICATE OF THIS SLIP PACKED INSIDE

Package of 25 Model 1917 rifle magazine springs made by Remington-UMC at the Ilion plant, packed on November 29, 1918.

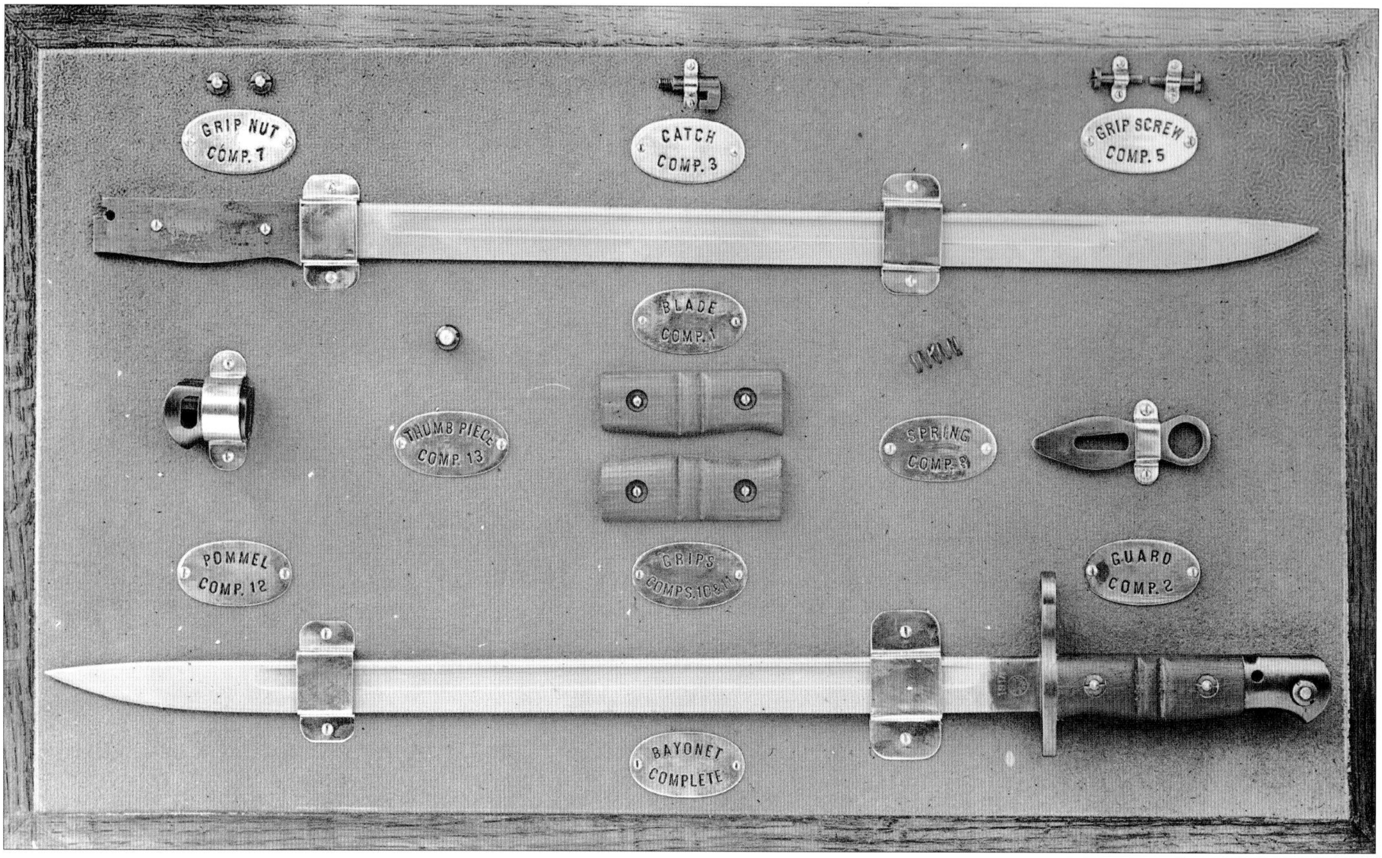

Model 1917 bayonet components on Remington factory display board. *(National Archives)*

Target used for accuracy testing at Remington. *(National Archives)*

Model 1917 rifle barrels ready for inspection at the Remington plant. *(National Archives)*

It is interesting to note that the contract called for rifles "without bayonets," and the procurement of bayonets was contracted for separately. The Pattern 1913 bayonets used with the British Pattern 1914 and subsequently the Model 1917 rifles were manufactured by Remington's plant in Bridgeport, Connecticut, and were shipped to the Ilion factory for distribution.[53] Details of Model 1917 bayonet production will be discussed in a subsequent chapter of this book.

On October 9, 1918, Remington-UMC received an additional contract for 250,000 Model 1917 rifles.[54]

As was the case with Winchester, Remington had its share of problems with the labor situation. There was also periodic difficulty in obtaining sufficient quantities of suitable steel. As related in *History of Rifles, Revolvers and Pistols, 1917–1919*:

> "Ilion (also) experienced)...difficulties in the labor turnover problems. For the first three months of 1918 their labor turnover was 160 per cent monthly; that is out of 6,000 employees, 10,000 left every month; in order to keep an average of 6,000 employees they hired 30,000 men between January 1 and March 31. Here again it was largely due to its location in a congested district, which mean labor competition and poor housing conditions. In the case of this plant, too, the low wages and methods of payment were seriously limiting factors; in a 10-hour day a man doing piecework was permitted to make only a limited sum; if he exceeded this sum on one day, he was cut down accordingly on the following day. This company also experienced considerable difficulty in regard to the poor steel for barrels, and in May, lack of steel."[55]

Like Winchester, the Remington rifles had an attractive rust bluing on most of the metal parts. Any Model 1917 rifles made by these two contractors that are Parkerized have been refinished after World War I, either as part of a postwar ordnance overhaul or by a subsequent civilian owner.

As mentioned in the Winchester section of this chapter, all three manufacturers had the option of beginning production prior to the finalization of standardization specifications required for satisfactory interchangeability of parts or waiting until the specifications were completed. Initially, along with Winchester, Remington elected to start production but reversed the decision before manufacture got underway. The rifles were marked "U.S./ Model of 1917/Remington/ser. no." on the receiver ring. Many of the components of the rifles made by Remington-UMC were marked with a "R" to denote the maker.

The initial of the maker, along with the month and year of production and the ordnance flaming bomb insignia, were stamped on top of the barrel behind the front sight. As previously mentioned, the date referred to the production of the barrel and not necessarily when the completed rifle was assembled.

Like Winchester, Remington was assigned a three-digit code number range for the eagle-head assembly mark. The company's code number range was 201 through 399.[56] The tip of the stock on Remington Model 1917 rifles was marked "R" sometimes with single digit presumed to identify the ordnance stock inspector. Numerous other components were stamped "R" to denote manufacture at the Remington-UMC Ilion plant.

Except for a few prototypes and modifications, which will be covered in a subsequent chapter of this book, the Model 1917 rifles made by all three contractors had the same features and general configuration. There was, however, one little-known variant believed to have been made solely by Remington. The Model 1917 sight was often criticized for its lack of windage adjustment and unsuitability as a precision long-range sight. Circa October 1918, Remington assembled a small number of Model 1917 rifles

Typical receiver markings on Remington Model 1917 rifle. *(Rock Island Auction Co.)*

Remington Model 1917 rifle barrel markings. This barrel was manufactured in September of 1918. On original rifles, there was typically only a couple of months variation between the date of barrel production and the assembly date of the complete rifle.

Remington eagle-head assembly mark with code number "339" on Model 1917 rifle stock. Remington code numbers were in the 201 through 399 range.

Marking on tip of Remington-made stock. The "3" is believed be an inspector code marking.

Assembly mark on bottom of floor plate, code number "325."

Remington manufacturer marking on back of Model 1917 rifle bolt handle.

Eagle-head assembly mark, code number "301," on Model 1917 rifle bolt stop assembly.

equipped with an experimental rear sight with a micro-meter adjustable peep sight. At least two genuine rifles of this type, serial numbers 411,484 and 556,826, have been observed, but others likely exist.[57] Some have claimed that this sight was also used on a few Winchester Model 1917 rifles, but the only known verified examples are Remington-made. Although the rifles were apparently never issued, the sight was found to be highly accurate and quite effective, but by the time it was tested, the First World War was over and the concept was dropped.[58]

Presumably, the sight would have permitted the Model 1917 rifle to have been better suited for competitive long-range shooting following the war, but the concept of a "match-tuned" Model 1917 rifle never materialized. There were some Model 1917 rifles used in the 1918 National Matches, but they retained the standard-issue, non-adjustable rear sights. Also, briefly mentioned in a later chapter in this book, there was some experimentation done late in WWI by the Marine Corps with another type of adjustable rear sight for a Model 1917 rifle, fitted with a telescopic sight, but it was apparently a one-off and differed in several respects from the above-mentioned sight.

Like any other manufacturer, Remington was faced with numerous challenges during the Model 1917 rifle production program. Perhaps the most endemic problem was the age-old quandary of simple supply and demand. Getting large quantities of badly needed armaments, including rifles, into the hands of our soldiers deploying to Europe resulted in shortages of trained workers as well as the scarcity of raw materials that many other arms makers were seeking. The Remington Company was fortunate that it had more than one factory in operation that could work in concert in getting the rifles manufactured as expeditiously as possible.

For example, the large number of rifles being manufactured obviously needed a like number of stocks. In order to prepare the stock blanks to be properly fabricated, it was necessary to kiln dry the wood before shaping the stocks

Experimental "micro-meter" peep adjustable rear sight, raised position. *(Former Peter Wainwright collection. Courtesy of Roy Marcot)*

Same rear sight in lowered position. *(Author's collection)*

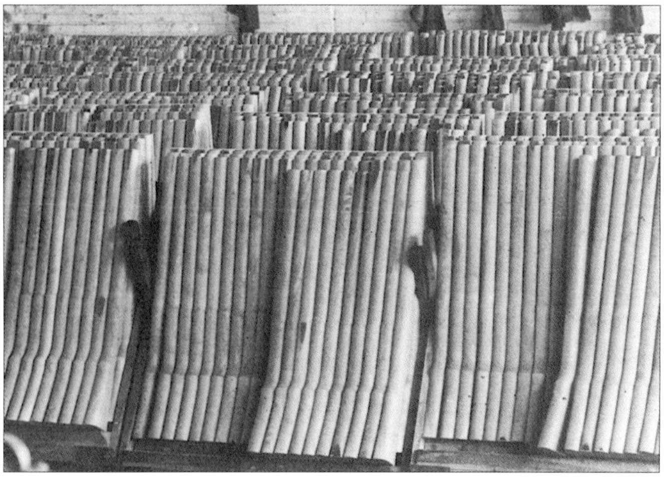

Unfinished Model 1917 stock blanks. *(The Remington Society. Courtesy of Roy Marcot)*

Final assembly of Remington Model 1917 rifles. *(The Remington Society. Courtesy of Roy Marcot)*

to the correct dimensions. To this end, kilns and shaping machinery were required. Although some of this equipment was on site at the Ilion facility, most of the stocks were made at Bridgeport, Connecticut, and that entity was under constant pressure to keep up with the required production rate.[59]

Remington was also faced with some complaints from the Army regarding the quality of their rifles shipped to various training camps. For example, on June 25, 1918, an ordnance officer at Camp Sevier, South Carolina, notified the company that some 3,000 rifles shipped to the base on April 23rd could not be issued because of excessive pitting and rusting on some of the components, particularly the bolts and rear sights, even though they were coated with cosmoline. It was determined that part of the reason was due to improperly dried wood used to make the shipping crates.[60]

Additionally, it was reported that in one battalion, a total of 194 Model 1917 rifles, all Remington, exhibited excess headspace.[61] To be fair, excessive headspace was not uncommon in a number of rifles made by all three manufacturers.

During this same period, Remington was facing difficulty in making extractors to the proper specifications. An ordnance officer provided the company with extractors that had been modified by Winchester and Eddystone. Remington worked diligently to correct the issue, but production was slowed for a while as things were sorted out. Eventually, Remington engineers solved the problem, and Ordnance was sufficiently impressed to request that the company provide 15,000 of the improved extractors. Other problems cropped up, including a rifle that had been sent to Springfield Armory for function testing but had been rejected due to difficulty in retracting the bolt.[62]

When Remington received the last production contract for Model 1917 rifles in the fall of 1918, the Rochester Ordnance District determined that some parts should be obtained from outside vendors to be sent to the Ilion factory for finishing. These firms included Ithaca Gun Company, Hunter Arms Company, and Baker Gun & Forging Company. Remington was not thrilled about using parts made by subcontractors and demanded that any parts received would be marked with the initials of the firm that made them so any rifles found to be unsatisfactory due to defective parts made by these entities could not be blamed on Remington. As events transpired, the coming of the Armistice resulted in all outstanding orders for such parts being cancelled and none were produced.[63]

Despite the inevitable glitches and setbacks, Remington-UMC garnered an enviable record in manufacturing the "American Enfield." The ordnance book *America's Munitions 1917–1918*, indicates that 545,541 Model 1917 rifles had been manufactured by November 9, 1918.[64] As of December 23, 1918, the total had reached 646,444, and by the time the production ceased in February of 1919, Remington had manufactured 654,149 Model 1917 rifles for the United States government at the Ilion plant.[65]

Remington Arms Company, Wilmington, Delaware and Midvale Steel and Ordnance Company, Eddystone, Pennsylvania

Baldwin Locomotive Plant, Eddystone, Pennsylvania. *(The Remington Society. Courtesy of Roy Marcot)*

While Winchester Repeating Arms Company and Remington-UMC were well-established firearms manufacturing firms that had been around for many decades, Eddystone was a newly formed entity that was established solely for the production of Pattern 1914 rifles for the British and, subsequently, Model 1917 rifles for the United States.

As discussed in Chapter 1, on November 24, 1914, the British government contracted with Winchester Repeating Arms Company and Remington-Union Metallic Cartridge Company for the manufacture of 200,000 Pattern 1914 rifles from each entity.

On February 10, 1915, Remington received a second contract for 200,000 more rifles. Followed on September 27, 1915, by a third contract for 600,000 additional rifles. Such a substantial commitment would require the company to enlarge its manufacturing plant in Ilion, New York. These contracts encouraged Remington management to raze many existing, antiquated factory buildings in Ilion and construct new ones. With the additional construction and renovation completed, there would be over 1,000,000 square feet of manufacturing and office space available.[66]

With construction completed, on March 25, 1916, work began in Ilion on the .303 Pattern 1914 rifles for England. An average rate of production of 1,000 rifles per day was eventually achieved, with the greatest rate of production of 2,000 rifles in one day on March 10, 1917. Pattern 1914 production in Ilion ceased on these rifles in mid-1917.[67]

Even while the company was in the process of increasing the production capacity at Ilion, it was apparent that additional sources of production would be needed. As related by Roy Marcot and Joe Poyer in *The Story of Eddystone:*

> "Even while plans were underway to enlarge Ilion's manufacturing capacity in 1915, Marcellus Hartley Dodge, Chairman of the Board at Remington, decided to organize a new company, the Remington Arms Company of Delaware. He convinced the management of the Baldwin Locomotive Works at Eddystone, Pennsylvania, to construct new buildings, and to lease them to the new arms manufacturer. Construction took place in 1915, and the completed buildings were leased to the Remington Arms Company of Delaware. This new complex was immense, with over 33 acres of floor space. Baldwin made sure that the buildings were designed so that at the end of the lease they could be reclaimed by the Baldwin Works and utilized as locomotive shops.
>
> Colonel John T. Thompson (who later became famous for developing the Thompson sub-machine gun) resigned from the U.S. Army [in November of 1914] to become Chief Engineer to oversee the construction of the new Remington facility in Eddystone. This vast manufacturing complex soon became one of the largest arms factories in the world. Colonel Thompson remained in management as Works Supervisor at Eddystone until he reentered the service [in 1917], eventually being promoted to

Early 1920s photo of John T. Thompson holding his signature submachine gun. ***(National Archives)***

Thompson retired from the army in November of 1914. He was then named Chief Engineer of the Remington Arms Company and supervised the construction of the Eddystone plant.

Remington Arms Co., Delaware main plant at Eddystone.
(The Remington Society. Courtesy of Roy Marcot)

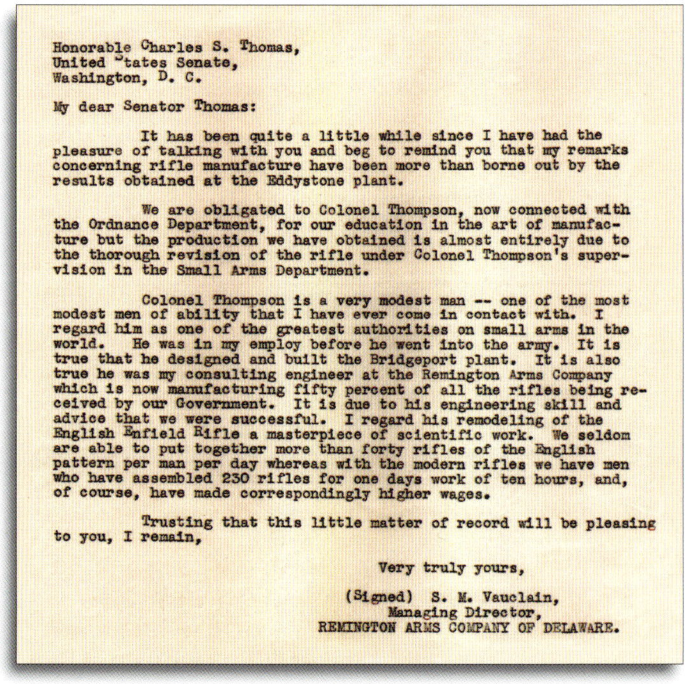

Honorable Charles S. Thomas,
United States Senate,
Washington, D. C.

My dear Senator Thomas:

It has been quite a little while since I have had the pleasure of talking with you and beg to remind you that my remarks concerning rifle manufacture have been more than borne out by the results obtained at the Eddystone plant.

We are obligated to Colonel Thompson, now connected with the Ordnance Department, for our education in the art of manufacture but the production we have obtained is almost entirely due to the thorough revision of the rifle under Colonel Thompson's supervision in the Small Arms Department.

Colonel Thompson is a very modest man -- one of the most modest men of ability that I have ever come in contact with. I regard him as one of the greatest authorities on small arms in the world. He was in my employ before he went into the army. It is true that he designed and built the Bridgeport plant. It is also true he was my consulting engineer at the Remington Arms Company which is now manufacturing fifty percent of all the rifles being received by our Government. It is due to his engineering skill and advice that we were successful. I regard his remodeling of the English Enfield Rifle a masterpiece of scientific work. We seldom are able to put together more than forty rifles of the English pattern per man per day whereas with the modern rifles we have men who have assembled 230 rifles for one days work of ten hours, and, of course, have made correspondingly higher wages.

Trusting that this little matter of record will be pleasing to you, I remain,

Very truly yours,

(Signed) S. M. Vauclain,
Managing Director,
REMINGTON ARMS COMPANY OF DELAWARE.

Undated letter from S.M. Vauclain, Managing Director of Remington Arms Co. of Delaware to Senator Charles S. Thomas, with fulsome praise for John Thompson regarding his work at the Eddystone plant. ***(Tracie Hill,* Remington Collectors Journal, *4th qtr., 2022)***

Brigadier General and Chief of the Small Arms Division of the Office of the U.S. Army Chief of Ordnance. He was appointed Director in Charge of Arsenals, and was responsible for all small arms production."[68]

It was, and still is, common practice for some entities to incorporate in the State of Delaware due to favorable tax and legal provisions as provided for under state law.

Eddystone Model of 1917 Rifle — Perhaps the most authoritative and comprehensive overview of the Eddystone Model 1917 rifle production is found in a very rare book by William Bradford Williams, *Munitions Manufacture in the Philadelphia Ordnance District for U.S. Army in the World War*, which was published in 1921. The following narrative excerpted from this book was titled *The British Enfield & U.S. Model 1917 Rifles made by the Midvale Steel & Ordnance Company*. Although a bit lengthy, the passage will be presented in its entirety due its historical importance and the myriad of interesting facts it contains. The original punctuation and writing style remain intact:

> "Actual work of manufacture of the modified Enfield rifle, known as the United States model of 1917, was started at the Eddystone Plant of the Midvale Steel & Ordnance Co., on a lot of twenty-five rifles, April 17th, just eleven days after the War was declared by the United States, and on May 27th work on all components for which no changes were contemplated was begun."[69]

Photo dated October 5, 1918, captioned, "Secretary of the Navy [Josephus] Daniels attended the ceremonies, which marked the delivery of one million rifles to the government within a year by the Midvale Steel Company of Eddystone, Penn. Secretary Daniels is shown with a few of the women workers in the plant. Brig. Gen. John Thompson, Director of Arsenals, is on the right." *(National Archives)*

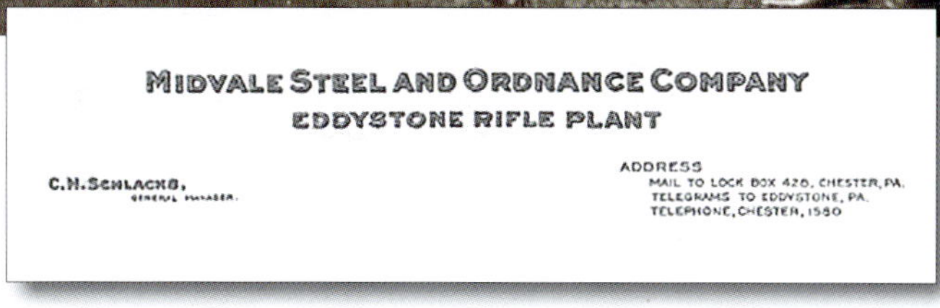

MIDVALE STEEL AND ORDNANCE COMPANY
EDDYSTONE RIFLE PLANT

C.H. SCHLACKS,
GENERAL MANAGER.

ADDRESS
MAIL TO LOCK BOX 420, CHESTER, PA.
TELEGRAMS TO EDDYSTONE, PA.
TELEPHONE, CHESTER, 1550

Eddystone letterhead. *(Remington Society. Courtesy of Roy Marcot)*

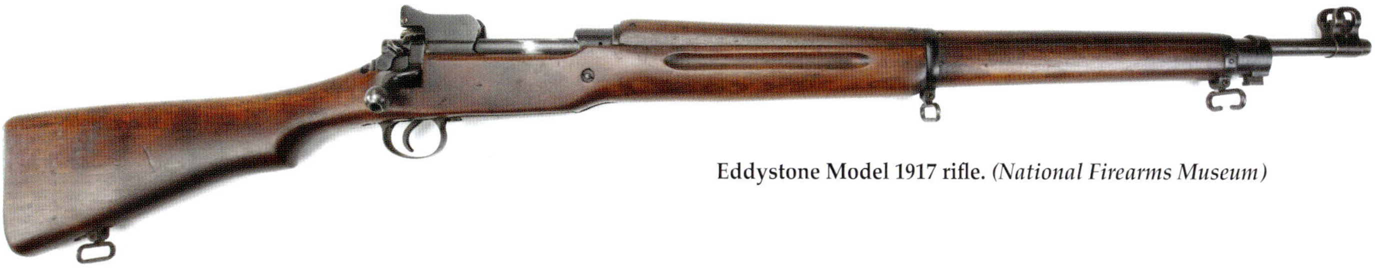

Eddystone Model 1917 rifle. *(National Firearms Museum)*

Although seldom mentioned in many reference sources, Eddystone manufactured a number of prototype Model 1917 rifles before the company was ready to go into full production for the official contract. These rifles had a receiver marking format of "US / ERA / ser. no.", which was similar to the early Winchester Model 1917 rifles. These prototype rifles did not have any barrel markings or inspection stamps. The number of prototypes fabricated is not known, but at least three have surfaced serially numbered 12, 19 and 24. Assuming sequential serial numbering, it can perhaps be inferred that at least two dozen were made.[70]

Interestingly, prototype ser. no. 19 has a brass presentation plaque on the stock, which reads:

"HON. FRANK CLARK. M.C.
Compliments
D.M. HESS
Eddystone Rifle Plant."

Frank Clark was a U.S. Congressmen for 20 years before retiring in 1925, but the reason for receiving the rifle from Eddystone is not known. To continue the narrative by Williams' *The British Enfield & U.S. Model 1917 Rifles made by the Midvale Steel & Ordnance Company:*

"The total number of employees enrolled at Eddystone, at the time of beginning work on the U.S. rifle, was 10,500, but only a small part were working on the U.S. rifles, most of them engaged in completing the Enfield rifles, which the Eddystone Plant had been making for the British Government. Because of the fact that the British work was completed and therefore diminished more rapidly than the U.S. work could progress during the early stages, the total number enrolled gradually decreased to a minimum of 3,917 on July 15, 1917, after which it again began to increase. The maximum number of employees enrolled at Eddystone was 15,409, which was reached on September 27, 1918, of which about 3,000 were women.

From the time when the Eddystone Rifle Plant was started, during the Summer of 1915, for the manufacture of rifles for the British Government, care was taken to protect the huge plant against damage at the hands of sympathizers with the Central Powers and to, as far as possible, exclude that class from their personnel. This required a well-organized Police Department and careful examination by the Employment Bureau of all applicants who were accepted for service. When it became apparent that the Country would almost certainly enter the War, and later, when War was actually declared, provisions for protection were much increased. Amongst other measures of protection, they organized their most trusted leaders and mechanics into a Vigilance

Eddystone M1917 rifle, ser. no. 1, the first rifle manufactured by the company on the U.S. Government contract. ***(Springfield Armory Museum. Courtesy of Alex MacKenzie, Curator)***

Early prototype Eddystone M1917, ser. no. "19". Note the receiver marking, which was similar to that used with the Pattern 1914 but with "US" added. ***(Hayes Otoupalik collection. Bill O'Donnell photo)***

Presentation plaque on Eddystone M1917 prototype ser. no. 19. ***(Hayes Otoupalik Collection. Bill O'Donnell photo)***

Female employee at the Eddystone plant doing broaching work on a Model 1917 rifle. ***(The Remington Society. Courtesy of Roy Marcot)***

Heat treatment of Eddystone Model 1917 receivers. *(Tom Laemlein)*

Corps, each member of which was sworn to do all in his power to safeguard the property and effects of the Government and of the company.

Because of the accelerated interest in enlisting in the military service, which began at the time War was declared, and because of the provisions of the Selective Draft Law, many men left the big rifle plant for active service, in spite of every effort to secure exemption for them.

As time went on and large numbers of troops were being moved to and from camps, transportation facilities were overtaxed and many of the employees were subjected to delays and hardships in getting to and from their homes. This contributed to the increase in the turnover and added to the difficulty in securing employees.

In April 1918, the difficulty of securing adequate help had become so great that women were employed for the first time in the manufacturing departments. Prior to this date they had been used only in the company's general offices. The number of women inspectors and machine operators gradually increased until, in September 1918, there were approximately 3,000 women employed. At this time, it became difficult to

Eddystone factory parade float featuring Model 1917 rifles. *(The Remington Society. Courtesy of Roy Marcot)*

secure sufficient women workers, as well as men.

During the major part of the time that the Midvale Steel & Ordnance Co. was making the U.S. rifle at Eddystone, it was seldom that any skilled, or even slightly experienced, machine operators could be obtained, as the supply of this kind of labor was becoming exhausted. In consequence, it was necessary to take many men and women who had never before worked on a machine or even in a factory. It was possible to train employees of this kind to become satisfactorily skillful if they would remain a sufficient length of time. Herein lay the great difficulty, due to the large turnover resulting in a large proportion of the employees always being new-comers or of short length of service, and consequently unskilled.

One class of work where skilled labor was absolutely essential, particularly as to tool and gauge makers, the demand throughout the country far exceeded the number of skilled men available.

During the first year of the operations on U.S. rifles, shortages of steel and of walnut for gun stocks and facilities for drying same occurred at times. The facilities for drying gun stock blanks at the plant were increased by the construction of sixteen new kilns, increasing the capacity for drying fifty percent, and more nearly balancing the capacity of the remainder of the plant.

In addition to the manufactured rifles, 1,352,862 spare parts were produced at Eddystone, based upon value, equivalent to ten percent of the monthly production; the total of spare parts were equivalent, it is estimated, to be 135,000 rifles.

From time to time there were many minor and some major improvements made in methods and equipment, but it was the combined result of all these and of the longer acquaintance of the organization, as a whole, with the work in hand, which was responsible for the continued increase in output in the face of ever-increasing difficulty in securing and holding labor.

The browning of components (barrel, receiver, bolt, guard, floor plate, upper band, front sight carrier, lower band, buttplate) had, from the start required a large force of men, much space and equipment, and from one to three days' time and the results were frequently not permanent and otherwise unsatisfactory, necessitating repeating the process. After careful tests made at Eddystone and elsewhere, the process of the Parker Rust Proof Co. was substituted. This was not in full operation until September, 1918. The browning formerly required 305 men, whereas better work was obtained with the new process with eighty men and the components required less than three hours for this treatment.

It has always been difficult to prevent the rifle barrel, which has no projections, from turning in the vise while the receiver was being screwed on. The largest and strongest commercial vises were used, making

Model 1917 rifles being test fired at the Eddystone plant. *(Tom Laemlein)*

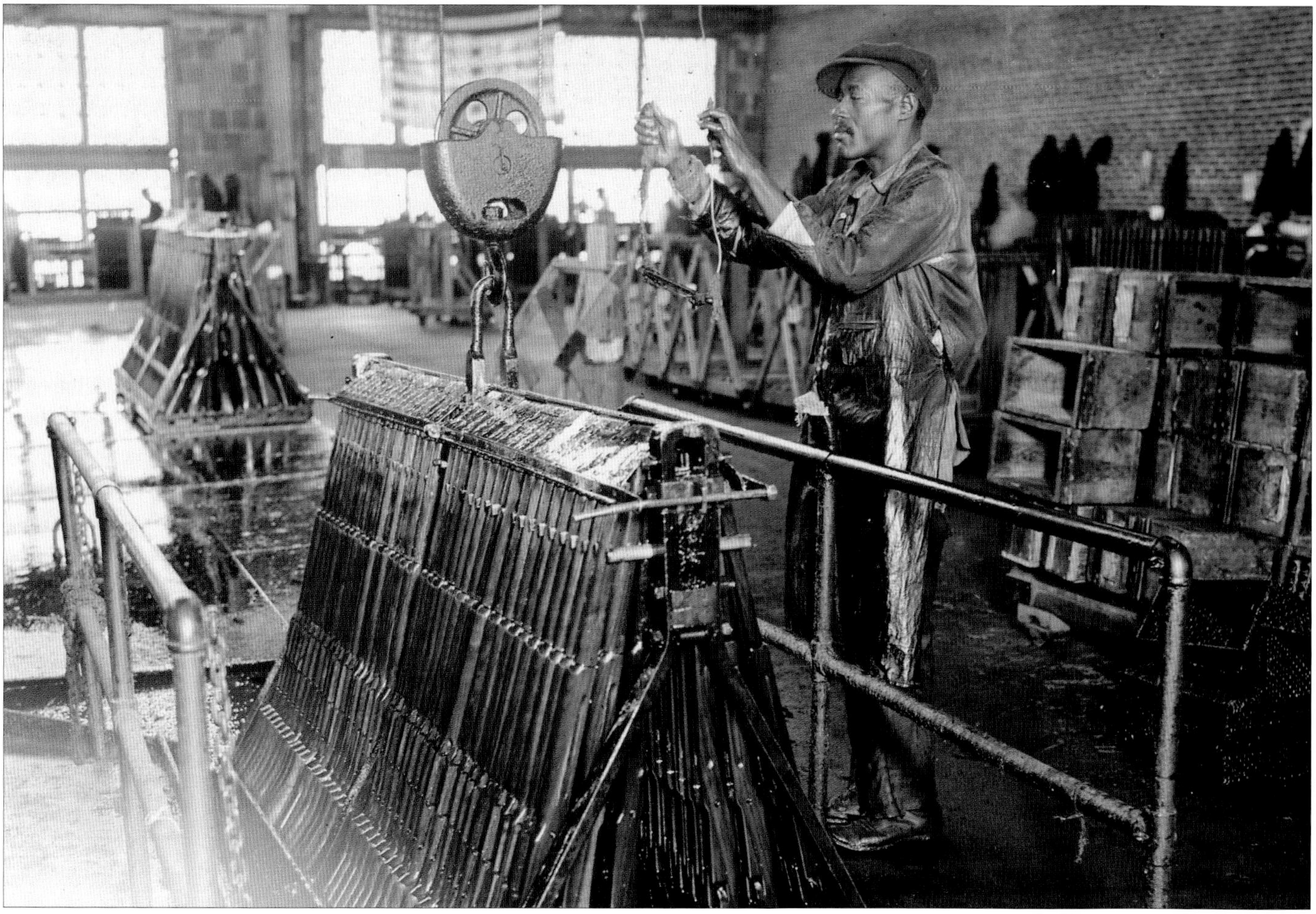

Newly completed Eddystone Model 1917 rifles being dipped in linseed oil. *(Tom Laemlein)*

Another view of Eddystone Model 1917 rifles being dipped in protective oil prior to shipment from the plant. *(National Archives)*

the operation a very heavy one for the men, as well as slow, and the vises were not infrequently broken. In August, 1918, pneumatic vises, designed and made at Eddystone, were substituted, enabling five men to do the work, which formerly required ten, and with the pneumatic vises the work was much easier and did not require men of exceptional endurance.

The turning lathes used for the rough turning of gun stocks were early replaced by side shapers. The substitution of six shapers released twenty-four lathes, which were then devoted to second turning, and also four tip turners and three centering drills. The operation of the shapers was much more rapid and required six men, where eleven had formerly been required. Many other small improvements in devices and methods, in the making of gun stocks, resulted in the production of gun stocks, at Eddystone, which were much more complete and accurate to dimensions, than were at first ever made elsewhere. The result of this was most marked in the assembly of the action to stock. Whereas, on the British rifle, which was not essentially different, it was more difficult to make, it was not possible for one man to assemble more than 150 stocks and actions in ten hours, the improvements...made it possible for a record of assembling 281 stocks to actions by one man in the same time!

During the British contract, much trouble was experienced in assembling the several components of the rear sight, although much improvement was made during the early part of the United States contract. As the result of a number of small improvements in methods, tools, and gauges, the quality of the work on the individual components involved was so improved that the assembly could be done much more rapidly and the result, as measured by the U.S. Inspection Reports, was very much better. Perhaps the greatest single improvement at Eddystone Rifle Plant was in grinding the rear sight leaf for finishing, instead of filing. During the British contract, the index line, on the slide, was cut to match the leaf graduations when assembled. At the beginning of the United States work an assembler could not assemble more than 300 sights per day, but when the work was stopped as many as 800 were assembled by one man! A record far exceeding any other rifle plant visited by the author.

The final reaming operation was added to the barrel after rifling. This was with a scrape reamer and gave a very high finish, cutting down, very greatly, the number of barrels to be lapped.

The short days of winter and other dark days, materially interfered with the straightening of barrels and their inspection for straightness and interior defects. After trying many forms of artificial lights, one was devised, which gave excellent results, enabling this work to continue regardless of the sunlight condition, and at night. It consisted of a mercury vapor lamp enclosed in a large box with plane mirror reflector and ground glass to diffuse the light. This was hung before the windows normally used in daylight. When using the artificial light, shades were drawn over the windows and large canvas curtains lowered back on the operator, so that all natural light and artificial light, other than from the box, was excluded. Much loss of time was saved by these devices.

The United States contract with the Midvale Steel & Ordnance Co. required that the rifle parts should be interchangeable, not only as regards to all rifles from the same plant, but also as to rifles made from the other two great rifle plants: Winchester and Ilion. This was not a requirement under British contract, and it necessitated a much greater accuracy in Workmanship, and a greater number of gauges to secure it.

The total orders for rifles entrusted to the Midvale Steel & Ordnance Co.'s big Eddystone Plant, aggregated 2,175,000, together with an immense number of spare parts and, at the time of the Armistice, this plant was producing over 5,500 rifles per day, with a substantial proportion of spare parts in addition!

On the whole and despite difficulties, due to abnormal conditions existing, the output of rifles and spare parts by the Eddystone Rifle Plant was not only fully up to, but in advance of the schedule of output provided for in the contract, the first shipment being made on September 1, 1917, two months in advance of contract schedule, and within one year, or by September 17, 1918, 1,000,000 rifles had been manufactured and delivered! The total output, 1,362,862 rifles and spare parts, is estimated to have been about sixty percent of all rifles manufactured in the United States during the period of the War!"[71]

While not explained in the above narrative, on January 2, 1918, the Remington Arms Company of Delaware was absorbed by the Midvale Steel and Ordnance Company. Thereafter, the official owner of the Eddystone plant was Midvale, not Remington, and as stated above, the facility was thereafter referred to as the Eddystone Rifle Plant. The final delivery of Model 1917 rifles by Eddystone occurred on March 8, 1919.[72]

These innovations cited above resulted in Eddystone achieving some truly remarkable results in manufacturing a tremendous number of Model 1917 rifles, which resulted in the following impressive statistics:

- The highest average daily production at the Eddystone Arsenal was 5,000 rifles, and the highest maximum daily production was 7,201 rifles.
- At the peak of production, there were 15,097 employees with a payroll of $402,817 per week. In addition, there were 1,317 government inspectors assigned to the plant.
- A rifle stock blank has a peculiar shape and requires eight board feet of lumber for cutting.

- It requires one walnut tree for every 28 stocks, and to cut the above quantity of stocks, 53,333 trees would be required, or 3,137 acres of walnut trees. While this plant was in operation, 320 wal nut trees were cut into gunstocks for shipment here each working day.
- The compilation of the inventory in the Eddy stone Plant comprised 45,500 written pages.
- 5,175 china plates were used in the company restaurants, which would feed 2,500 per day."[73]

Noted researcher the late Frank Mallory reported that Eddystone made some laminated stocks in June of 1918, but apparently, these were done on a quasi-experimental basis.[74]

As mentioned, Eddystone began Parkerizing their Model 1917 rifles in September of 1918 and was the only prime contractor to do so, as Winchester and Remington continued to use rust bluing through the end of production.

Eddystone was also the only contractor that did not manufacture Model 1917 bayonets. As related by Skennerton:

> "One particular facet of production, which Eddystone had problems was the bayonet. All the contract details indicate that Eddystone did produce the Pattern 1913 bayonet, yet no examples have been noted.

Close-up of Eddystone Model 1917 rifle with telescopic sight attachment being test fired for accuracy at the factory. *(Tom Laemlein)*

U.S. Army Ordnance officers inspecting Eddystone Model 1917 rifles prior to being shipped from the factory. *(National Archives)*

British assessment reports soon after the war, which studied wartime production in depth, note that, regarding bayonet production, 'one U.S. contractor failed so badly as to have to give up attempting the production altogether,' and this can only refer to Eddystone. Whether their contracts for bayonets were taken over by the other Remington plan, or sub-contracted, is uncertain, but they appear to have been made at Bridgeport…even though on paper they were divided up between Ilion and Eddystone."[75]

Actually, the lack of bayonet production at Eddystone was not a failure but a deliberate corporate decision by Remington to make the most effective use of the various production facilities. Bayonet production will be covered at length in Chapter 10.

Except for some minor configurational differences in a few of the components, the Model 1917 rifles made by Eddystone were essentially identical to those made by Winchester and Remington. Many components were marked "E" to denote the maker, including the barrels, which also had the date of barrel production and the ordnance flaming bomb insignia. The Eddystone assembly mark format was slightly different from that used by Winchester and Remington. Rather than a three-digit code number, Eddystone's assembly markings consisted of an eagle head and a code generally consisting of a letter and one or two digits.[76]

In addition to the standard Model 1917 service rifle, there were several Eddystone rifles fabricated as specially finished rifles for presentation to corporate officials and other notables. One such early-production rifle, serial number 38425, had a highly figured walnut stock and polished blued metal. A brass plague was affixed to the buttstock, which read: "To / Mr. Samuel F. Pryor / Compliments / of / C.H. Schlacks". Mr. Schlacks was General Manager of the Midvale Steel and Ordnance plant – Eddystone Rifle Factory, and Mr. Pryor was President of Remington Arms Company.[77]

Another notable Eddystone presentation rifle was made to commemorate the production of one million Model 1917 rifles at the plant. A celebration on September 12, 1918, was attended by more than 14,000 plant employees and numerous army, navy and industry officials. The serial number of the rifle was, of course, "1000000". It had a highly figured walnut stock with a high-gloss oil finish and polished blued metal but otherwise had the same features as a standard Eddystone service rifle but lacked inspection markings. The barrel was dated "8-18" with standard markings on the receiver. A scroll-shaped silver plaque was affixed to the right side of the buttstock and was marked:[78]

"1000000th Rifle
September 17th 1917 to September 17th 1918
To
C.H. Schlacks
General Manager
From the Employees of the
Eddystone Rifle Plant"

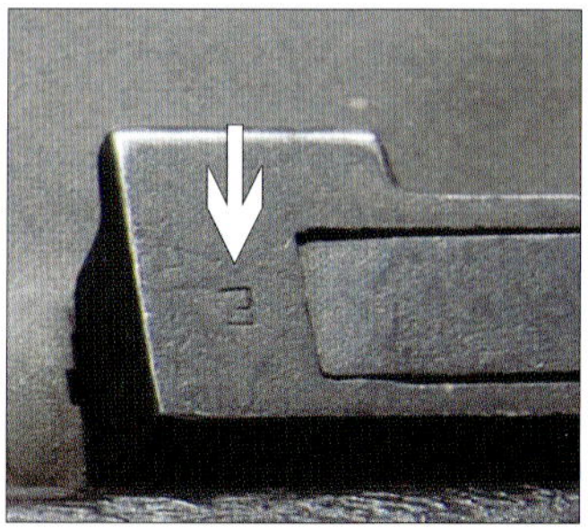

Eddystone bolt release. Note the "E" marking.

Eddystone Model 1917 rifle barrel manufactured in September of 1917.

Close-up of the rear handguard ring with an "E" marking. *(Roy Marcot)*

"E" marking on the tip of the Eddystone Model 1917 rifle stock.

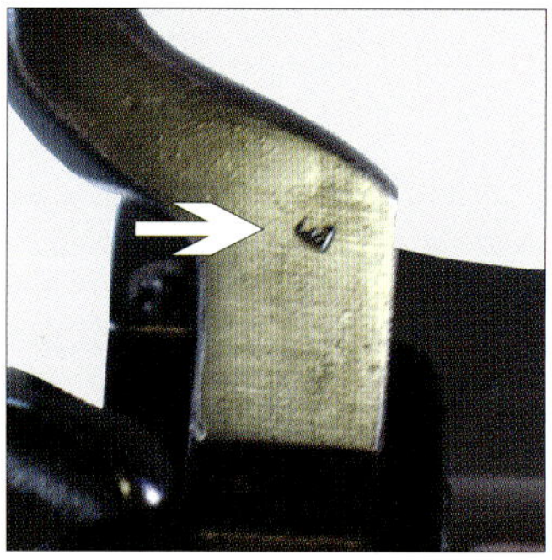

"E" marking on the underside of the bolt handle.

"E" marking on the safety.

Eagle-head assembly stamp. Note the "C9" number.

Eddystone M1917 presentation rifle, ser. no. 1000000. This rifle was presented to C.H. Schlacks from the employees of the Eddystone Rifle Plant. *(Rock Island Auction Co.)*

There was a Model 1917 Eddystone with an interesting plaque attached to the stock that is the subject of some debate. The rifle is a standard Model 1917 with a brass plaque attached to the left side of the stock, which reads:

"MODIFIED LEE ENFIELD RIFLE
ADOPTED BY THE UNITED STATES GOVERNMENT
FOR USE IN THE WORLD'S WAR
STOCK AND FITTINGS
MADE BY
VICTOR TALKING MACHINE CO.
CAMDEN, N.J., U.S.A. / 1918"

It has been stated that a former Victor Talking Machine Co. executive indicated that the rifle was on display at the company's headquarters for many years.[79] The fact that the Victor factory was located almost directly across the Delaware River from the Eddystone plant, and given the company's experience in crafting fine wooden cabinets for its products, would suggest that the rifle was intended to commemorate Victor's role in making Model 1917 rifle stocks. As previously stated in this chapter, Victor Talking Machine Co. was one of several commercial firms given contracts late in WWI for the manufacture of Model 1917 stocks.

Eddystone Model 1917 presentation rifle given by C.H. Schlacks, General Manager of Midvale Steel and Ordnance to Samuel F. Pryor, President of Remington Arms Company. *(Rock Island Auction Co.)*

Close-up of the C.H. Schlacks plaque on the Eddystone Model 1917 rifle that was presented to Samuel Pryor. *(Rock Island Auction Co.)*

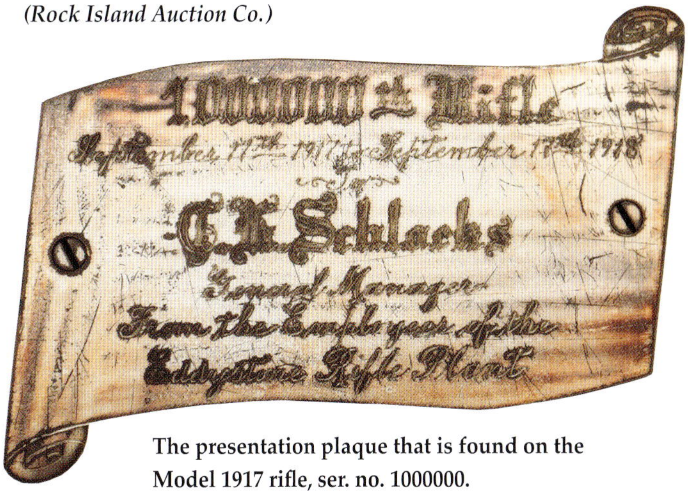

The presentation plaque that is found on the Model 1917 rifle, ser. no. 1000000.

Eddystone Model 1917 rifle, ser. no. 1000000 in a custom-made presentation case. *(Rock Island Auction Co.)*

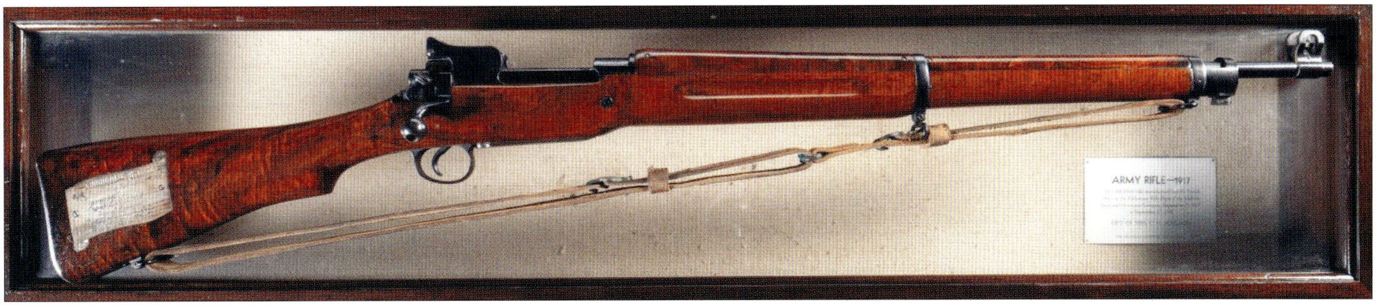

Eddystone Model 1917 rifle, with stock attributed to Victor Talking Machine Co. *(Rock Island Auction Co.)*

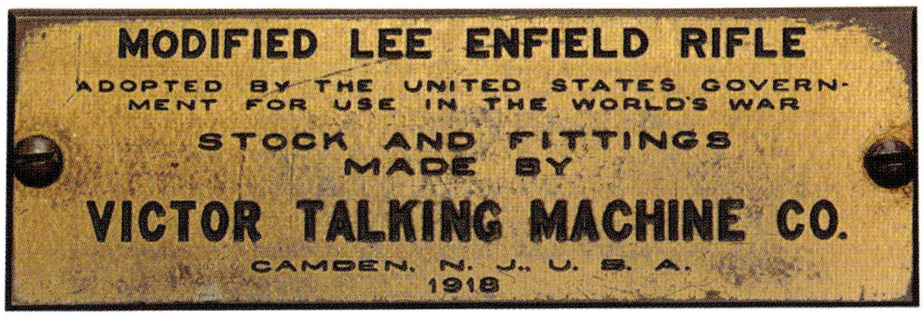

Close-up of the Victor Talking Machine Co. plaque.

Victor did receive a contract for 100,000 birch M1917 rifles stocks but it not known how many, if any, were made before the cancellation of outstanding contracts at the end of the war. Perhaps this was a sample stock made by Victor after receiving the substantial order for stocks, which was likely cancelled before quantity production got underway.

There is some discrepancy in the total Eddystone production numbers. *America's Munitions 1917–1918*, reflects that 1,181,908 Eddystone Model 1917 rifles had been manufactured as of November 9, 1918, and Skennerton quotes the same figure.[80] The well-researched monograph *Story of Eddystone* co-authored by Roy Marcot, states that when production ceased on March 8, 1919, Eddystone had manufactured a total of 1,332,477 Model 1917 rifles.[81] Ferris quotes a figure of 1,328,260 Model 1917 rifles made by Eddystone as of January 9, 1919, and a total of 1,356,643 by the time production ceased on March 8, 1919.[82]

Regardless of which figure is used, Eddystone had manufactured significantly more Model 1917 rifles than the combined production of Remington-UMC and Winchester.[83] For a corporate entity that didn't exist prior to the outbreak of the First World War in 1914, in the span of less than four years they had manufactured over 600,000 Pattern 1914 rifles for the British and well over 1,000,000 Model 1917 rifles for the United States. It was a remarkable achievement.

The Eddystone story vividly illustrates how the industrial might of the United States was a major factor in helping secure the hard-fought victory in WWI and in subsequent conflicts.

The commemorative sign that was erected in 2019 at the site of the Eddystone Rifle Plant. *(Pennsylvania Historical and Museum Commission)*

Chapter 5

World War I

The First World War was the *raison d'être* for the Model 1917 rifle. Due to the previously discussed shortage of weapons at the start of the war, the adoption of the Model 1917 rifle was vital to America's war effort. Prior to entering military service, some of the new enlistees and draftees may have had prior shooting experience with sporting arms or even with the Model 1903 service rifle, but they knew nothing about the "American Enfield." Their introduction to the rifle was when it was issued to them at one of the training camps that sprang up around the country to accommodate the burgeoning number of new troops.

Anticipating and meeting the demand for rifles was a top priority of the War Department. The ordnance publication *History of Rifles, Revolvers and Pistols, 1917–1919* gave the following information regarding the number of rifles to be issued and their allocations:

"Rifle Requirements are estimated as follows:

For every 1/24 of an army:

Initial equipment	33,125
Monthly wastage and consumption in France	1,089
Working reserve in France	4,355
Monthly wastage and consumption in U.S.	331
Total per 1/24 army	38,900
Total per army	933,600

Organization tables giving equipment based on divisions call for 17,731 rifles per division (27,152 men). This means that approximately 60 percent of all troops are armed with rifles, and that 85 percent of the weapons are used as initial equipment, while the remaining 15 percent are used for reserve purposes. A working reserve of 300,000 rifles was desired, considered as necessary at the camps and cantonments in this country to provide for any sudden speeding up of mobilization and the prompt equipping of new drafts.

U.S. Army recruits in close order drill at Camp Dix, New Jersey, using a mixture of Krag and Model 1917 rifles. *(National Archives)*

The total requirements to January 1, 1919, as issued by the Requirements Division were 4,213,476 rifles of both models (M1917 and M1903)."[1]

An interesting April 9, 1918, report from the Ordnance Department, Inspection Division to Brig. General Charles B. Wheeler gave the following average daily rifle production as of April 6, 1918:

"Model of 1917	5,856
Model of 1903	116
Russian Rifles*	1,797

The total production of all models for the week was 42,743; from April 6th, 1917 to April 3, 1918, inclusive was 1,180,593."[2]

It is interesting to note that there was some consideration given to equipping the American Expeditionary Force with Model 1917 rifles and relegating the Model 1903 for

**The "Russian Rifles" were the M1891 Mosin-Nagant rifles originally made by Remington and New England Westinghouse for the czarist Russian government, a number of which were later acquired by the U.S. primarily for stateside training purposes. Also, American troops in the North Russian campaign of 1918–1918 were armed Mosin-Nagant rifles that were never delivered to Russia and remained in storage in England.*

stateside training purposes. As stated in a February 11, 1918, Ordnance memorandum:

"This section learns unofficially of the prospective substitution of Model 1917 Rifles for Model 1903 for the A.E.F. and the withdrawal of the latter to the United States for training purposes. No official information has been received. Such a decision, particularly in view of the marked differences in spare parts for training and spare parts for Europe, would probably require a very marked modification in our manufacturing program, which in turn would require very early action if confusion, delay and loss are not to result."[3]

Needless to say, this proposed policy was not forthcoming, and the A.E.F. was equipped with both Model 1903 and Model 1917 rifles.

Indoctrination and Training — Training the rapidly growing number of inductees at dozens of training camps across the country to properly use these rifles would be a major endeavor.

As discussed in *History of Military Mobilization in the United States Army, 1775–1945*:

"World War I marked the Army's first attempt at systematic individual Soldier training, but the requirement to organize divisions while building training bases impeded the units' ability to train. When the U.S. declared war on Germany in April 1917, the U.S. Army consisted of 127,588 troops scattered in small garrisons across the country. In order to raise a mass Army to fight a war on a European scale, the Army created its first system of centralized individual basic combat training. There was, however, no centralized training headquarters or organization. The system based all training at the regimental level, with the division commander responsible for all combat training preparation. The training program for divisions preparing to go to France in 1918 was more mobilization than training.Drafted men went directly to regiments, which provided rudimentary training to familiarize Soldiers with military discipline and operations. The Army provided overall training guidance for the divisions, which included training topics and prescribed hours for each.

The 16-week basic training plan directed a total of 624 hours divided into four, four-week phases. The first three phases focused on individual Soldier skills and weapons training, while the final phase focused on collective combined arms and maneuver training."[4]

On November 24, 1917, Chief of Ordnance Major General William Crozier directed that training schools be established for selected officers to attend and become familiar with the new rifle. Drawn primarily from the Ordnance Reserve Corps and National Guard, the officers assigned as "rifle demonstrators" would be dispatched to various training camps around the country to demonstrate the weapon and assist the commanders of the camps in establishing courses for training the men of the rapidly growing army on the Model 1917 rifle.[5] The selected training officers reported directly to Col. John T. Thompson, who had resigned from the Army to oversee the construction of the massive Eddystone rifle plant but later re-entered the service and was promoted to Brigadier General.

Part of the initial training consisted of a detailed curriculum titled *A Brief History of the U.S. Rifle, Model 1917*, written by a Major Tilton who was affiliated with the Eddystone plant, that was provided to the Ord-

(U.S. Army photo. Courtesy of Tom Laemlein)

U.S. Army troops training during WWI. The soldier on the far left has a Model 1917 rifle. The soldier to his right has a U.S. .30-40 Krag rifle. The Krag saw a surprising amount of use in stateside training during the war.

nance "demonstrators" assigned to the various training camps. Some of the technical aspects in the document were compiled by an Eddystone engineer, H.E. Clearman. The document was cited in a Weekly Report of Captain James Kidder to Col. John T. Thompson on August 13, 1918.[6] Selected portions of the training syllabus follow:

"(a) 1. The new rifle being put into the hands of the U.S. troops is an adaptation of the British rifle, known as the Enfield, 1914.

2. Its official name is the U.S. Rifle, Model 1917.
3. It should not be called the Springfield, or the new Springfield, as it is distinctly a different rifle than the Model 1903 or Springfield.
4. The Enfield Rifle was in the process of development by the British Government when war was declared, and it was originally designed to shoot a .276 cartridge. As the British Army was armed with the Lee-Enfield rifle, which shoots a .303

American infantrymen with M1917 rifles and bayonets practicing "going over the top'" at a simulated trench position, Fort Foote, Maryland., June 10, 1918. *(National Archives)*

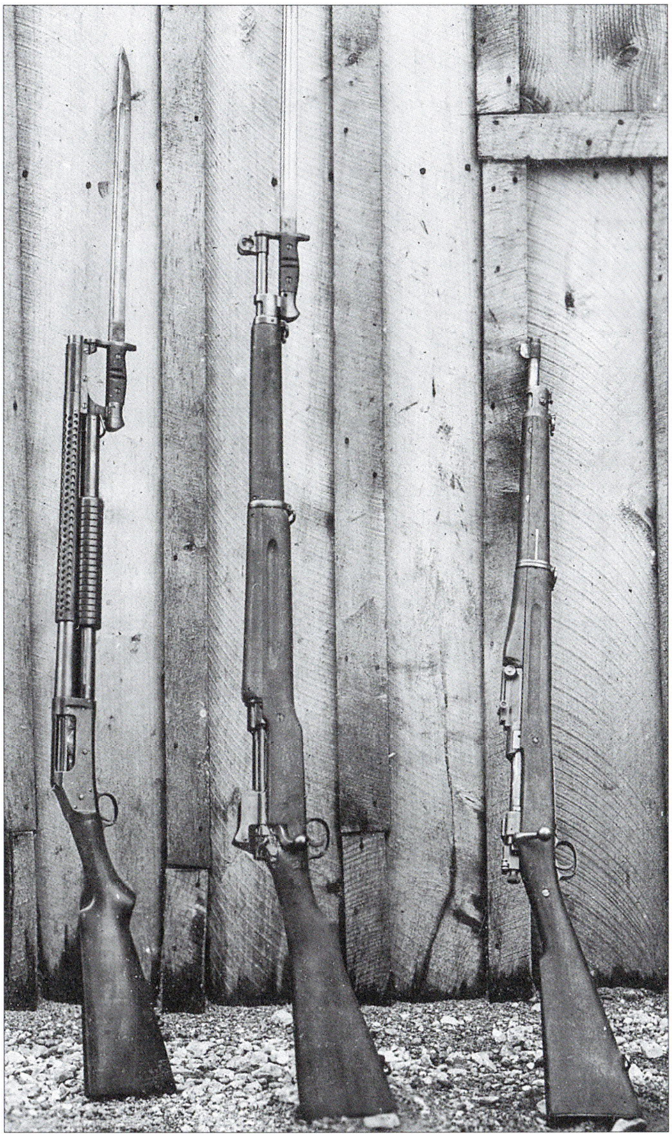

Three weapons of the AEF: the M1897 Winchester 12 ga. trench gun, the M1917 rifle, and the M1903 rifle. *(U.S. Army Signal Corps photo)*

Photo from 1918 captioned, "Infantry Drills, Camp Dix, N.J." *(National Archives)*

U.S. Infantryman at stateside training camp during WWI with a Model 1917 rifle. *(Tom Laemlein)*

Cartridge, the British Government decided to modify the Enfield to such an extent that would enable it to be used with a .303 cartridge.

5. This modification was accomplished by two commercial rifle manufacturing companies in the United States, and a certain number of the rifles as modified were delivered to the British Government.
6. Shortly after declaration of war by the United States, the U.S. Government decided to arm its troops with the Enfield, 1914, but in a form so modified from the Enfield 1914 as to permit it to be used with the U.S. Caliber .30, Model 1906 cartridge.
7. The Enfield Rifle as made by the commercial rifle manufacturing companies was not interchangeable, that is, the components as made by one firm would not interchange with the components made by other firms.
8. While modifying the Enfield 1914 to take the Caliber .30 (rimless) Model 1906 cartridge, the dimensions were altered and tolerances established so that the components parts of a rifle made by any one firm would interchange with the parts as made by any other firm.
9. The U.S. Rifle, Model 1917, is now being manufactured by the Remington Arms, U.M.C. Co., Ilion. N.Y.; the Winchester Repeating Arms Co. of New Haven, Conn.; and the Midvale Steel and Ordnance Co./Eddystone Rifle Plant, Eddystone, Pa.

(b) Brief description of the making, assembling, and inspection of the 1917 rifle, showing that it proba bly is the best rifle with which any large army has ever been equipped.

(c) To familiarize the officers and noncommissioned officers, and mechanics with the various parts, their names, and how they function, of the rifle in detail and all its workings.

(d) Emphasizing the ease in keeping the rifle in order.

(e) Difficulty of getting a rifle back in condition, after once neglected, or allowing it to get out of order.

(f) In referring to the care of the rifle, special importance was laid on the barrel, showing how necessary it is for accurate shooting, that the barrel must be protected from rust, metal fouling, etc.

(g) How the last three inches near the muzzle must receive special care, this being one of the important reasons for cleaning from the breech; never to use emory, or any substance likely to reduce, injure, the five-thousandths of an inch lands;

U.S. Army WWI doughboys with Model 1917 rifles. *(U.S. Army Signal Corps)*

30th Infantry Division troops posing with Model 1917 rifles, Camp Sevier, S.C., 1918. *(Mills Goodlet. Courtesy of Stuart Mowbray)*

Montage of a Model 1917 rifle atop a WWI U.S. Army Reserve Captain's tunic. *(American Rifleman. Courtesy of Jonathan Krisko)*

how necessary cosmic is to protecting the rifle in transport or storage; that when a rifle is once issued, it should become the enlisted man's greatest pride. The barrel and bolt should be carefully cleaned from this heavy oil, and a lighter oil should be used in daily care. Especially should the barrel and main-spring may be free from cosmic, showing that to clog the barrel with a heavy grease, or any substance, as an oily rag, does not permit free passage of the air ahead of the bullet, which may cause a burst or bulged barrel; showing that to clog the main-spring may retard the striker in its functioning, and cause a misfire. A series of photographs, showing burst and bulged barrels and what caused them, was exhibited.

(h) The rifle must always be ready for instant inspection or action. Description of the flight of the bullet with the battle sight, from the time the primer ignites the powder charge and the bullet takes the rifling, until it reaches its objective at slightly over 400 yards, which is the battle sight range. It was shown that the ideal relationship between the primer, powder charge, bullet, and length of the barrel is obtained in the 1917 rifle as demonstrated by the slightly added velocity of [over] the Springfield.

THE GREAT ADVANTAGE OF THE REAR SIGHT OF THE 1917 RIFLE OVER SPRINGFIELD

1. Nearness to the eye, giving greater sight radius.
2. Advantage of the aperture sight in obtaining a quick aim.
3. Great necessity of centering the front sight in the center of the rear sight.

It was especially brought to the attention of all classes, that the aperture in the rear sight should be watched to prevent mud, snow, or rain clogging it. Special notice should be taken of the rear sight in stormy weather. It is easy to clear the aperture if it once becomes chocked, with the use of a small object, such as a straw.

Great care must be taken to avoid flinching, showing that if the muscles of the thumb and three lower fingers of the fore-arm are exerted in gripping the stock firmly, and the forefinger is used only in pressing the trigger, one cannot flinch unless the thumb and three lower fingers are allowed to release the firm grip.

To sum up, in all the talks the aim has been to show how worthy of trust this 1917; to instill pride in the enlisted man in his special rifle; to show that carelessness in small matters may lead to dangerous results; that within 600 yards, no wind gauge is

New inductees engaged in the routine chore of cleaning their Model 1917 rifles at Camp Devens, Mass., 1918. *(National Archives)*

WWI marksmanship training with Model 1917 rifles. *(Tom Laemlein)*

Army troops at Camp Dix, New Jersey, in April of 1918 training with Model 1917 rifles in improvised trench positions. *(National Archives)*

needed in battle; to show that one shot well placed is worth hundreds fired at random.

If every American enlisted man in France gets in one well-placed shot, there will be no enemy left. It is all up to the enlisted man to see that every shot in preparation on the range or in actual battle is well placed, and this can only be done by knowing well, and taking special care, of his rifle."[7]

While this sort of classroom work was important in helping new soldiers become acquainted with the background and basic functioning of the Model 1917 rifle, as well as learning technical aspects, real familiarity with the weapon occurred during close-order drills, marksmanship training and practice on the range. As the Model 1917 rifles began to make their way to stateside training camps in late 1917 and early 1918, many of the soldiers who were issued the weapons were less than enthusiastic. Those who had experience with the svelte Model 1903 Springfield often found fault with the heavier and bulkier "American Enfield."

The cock-on-closing bolt action of the Model 1917 as compared to the cock-on-opening of the Springfield was frequently seen as an undesirable feature, although in many cases that was due more to unfamiliarity than any inherent deficiency. There were some opinions that the design of the Model 1917's bolt was actually advantageous. As related in an *American Rifleman* article:

"Captain Mortimer Munn, an Ordnance instructor at Camp Lee, Virginia, adopted a more positive strategy regarding the M1917 bolt by hailing it (and the whole rifle) as having been designed especially for modern warfare and instructed his students that, 'Even the crook in the Bolt Handle, means valuable seconds, for as an improvement over other models, it brings the index finger naturally into position, for pulling the trigger without lost motion. We must get in two shots to the Boche's one, when he sticks his head Over the Top.'"[8]

Another frequently cited criticism heard by Ordnance instructors was that the Model 1917's sights did not have windage adjustments. As discussed in the same *American Rifleman* piece:

"The peep-style aperture sights were also a marked departure from the M1903. While the M1903 does have a tiny aperture sight on the ladder, suitable for precision use at long range, the battle sight remained a standard open U-notch affair. The M1917 by comparison has a generously sized opening that allows for quick target acquisition, even in situations of low light. Like anything new, this caused some controversy amongst both the troops of the American Expeditionary Force and the rifle demonstrators themselves. Besides the nature of the bolt, Capt. Morris noted that two of the most frequent complaints he heard were the lack of a windage gauge, and 'the open field around the aperture in the rear sight.' "

For the men under his instruction, Capt. Munn had a

Men were trained at Camp Meade, Md., in transporting wounded soldiers without stretchers while under fire with their Model 1917 rifles at the ready. *(National Archives)*

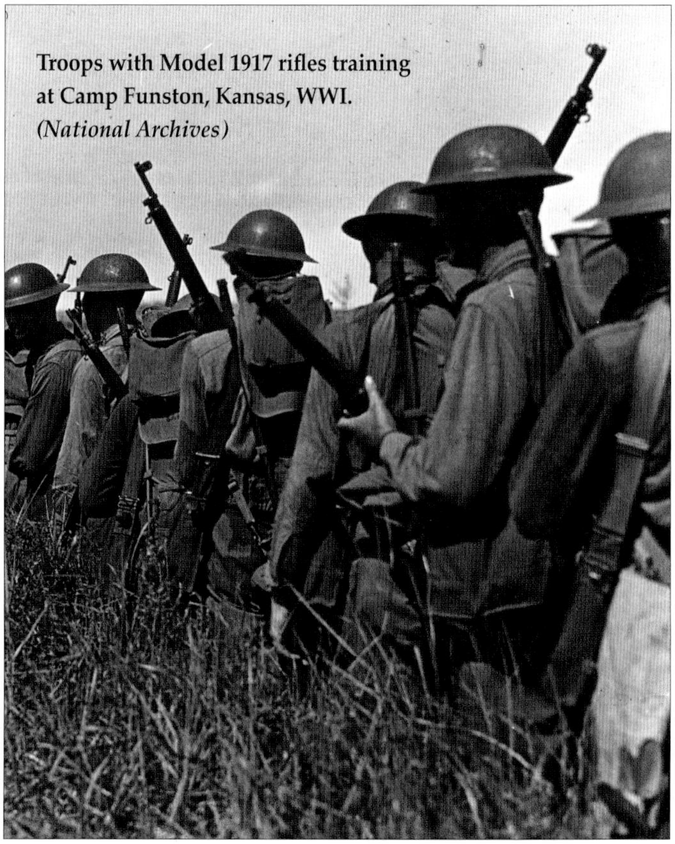

Troops with Model 1917 rifles training at Camp Funston, Kansas, WWI. *(National Archives)*

ready answer for the former complaint and stated that,

"It was readily realized that windage adjustment would be useless [and] small peep sights a hindrance. The 'Hun' will not wait for the Yankee to prepare for his execution," and that 'Expert riflemen who have been engaged by the Government to test the accuracy and rapidity of the 1903 and 1917 rifles, have become equally proficient in the latter model.' The long-range champion Capt. Preussner likewise included in his instruction that, 'The advantages of an aperture sight over an open sight is no longer questioned by the best marksmen' – an insight he was uniquely qualified to give."[9]

As mentioned previously, the lack of a magazine cut-off made the Model 1917 a bit more awkward in close-order drills and inspections, which took up an inordinate amount of the average recruit's day. Until sufficient numbers of the sheet metal magazines depressors became available, there were some informal improvisational measures employed such as inserting a coin to depress the follower in an empty magazine in order to keep it from blocking the bolt. The magazine follower depressors will be profiled in Chapter 10 of this book.

The lack of a cocking knob for the Model 1917 rifle as found on the '03, also came in for criticism since a misfire could not be corrected without opening the bolt, which could lead to the possibility of a dangerous hang-fire. The prescribed procedure for soldiers to follow in the event of a misfire was related in the standard manual of the time, *Description and Rules for the Management of the United States Rifle, Caliber .30, Model of 1917*:

"If the rifle misses fire, the bolt should not be opened or unlocked until sufficient time has elapsed to assure that the cartridge is not hanging fire.

Photo is captioned "Capt. P.T. Carey, Commanding U.S. Rifle Range, Camp Gordon, Georgia, scores 24 points out of 25 at 1,000 yards with M1917 rifle, holding off for windage." Captain Carey's performance vividly illustrated that, with proper training, the M1917 rifle's lack of windage adjustment did not unduly hamper its long-range performance. *(National Archives)*

This photo is captioned, "Rifle Range, Camp Devens, Mass., 1918." *(National Archives)*

> Inasmuch as the rifle cannot be cocked except by opening the bolt, there will be a temptation to open the bolt too soon, and it is wise to wait even a full minute to be sure."[10]

Hang-fires were not the only potential hazard involved on the rifle range. There were a number of accidents reported that resulted in an array of injuries to the shooters. A memorandum from the "Eye Consultant, Army Base Hospital No. 45 to the Commanding Officer, Justice Hospital Group," related a number of specific instances of eye injuries, with varying degrees of seriousness, incurred during target practice with the Model 1917 rifle. The report was "...forwarded to the Chief Surgeon as well as for any action that may be taken by the Commander in Chief and the Ordnance Department."[11]

Medical Corps officer Col. Walter D. McCaw responded to the report: "It would appear to be a serious indictment of the modified Enfield. It is recommended that investigation be made at the place where the accident occurred."[12] As requested, the Ordnance Department investigated the issue as related in the following memorandum from the Chief of Ordnance, Second Army to the Deputy Chief of Ordnance, G.H.Q.,A.E.F.:

> "Attached hereto is correspondence recently received from your office relating to eye injuries from use of the Enfield rifle, together with other reports from the several units concerning these accidents.
>
> ...It may be noted that in but one case is the ammunition used definitely stated to have been U.S. Rifle ammunition, while in two cases it was known to have been of German manufacture. It is quite likely that in other cases the ammunition was German although nothing definite was said.
>
> Captain Roberts, of this office, experimented with an Enfield, Eddystone, trying to blow it up with the bolt partly unlocked. The results of this test are negative as the rifle functioned perfectly. This test was performed as a result of an idea that possibly the accidents may have may have occurred by firing with the bolt partially unlocked. Attempts were made to fire the rifle with German ammunition but this was not possible as the bolt could not be locked."[13]

Further investigation revealed that virtually all of these accidents were indeed due to defective ammunition, which was not an uncommon occurrence, especially early in the war. Similar problems were encountered during this period, particularly with the low-numbered Model 1903 Springfield rifles. It was found that one particular brand of ammunition was to blame for many of the reported incidents. As stated in "Administrative Memorandum No, 71" from Headquarters, Second Division, American Expeditionary Force:

> "On account of caliber .30 rifles having been blown up in target practice while used in firing Winchester Rifle ammunition, NO MORE Winchester Rifle ammunition will be used by the organizations of this Division until investigation now being made has been completed and until it has been determined that this ammunition can be used with safety. Organizations will be advised further in this matter."[14]

These investigations confirmed that the Model 1917 rifle was not the cause of the various accidents and placed the blame squarely on substandard quality ammunition. Suspending use of the faulty ammunition and more stringent quality control at the ammunition factories resulted in fewer such unfortunate incidents. Due to the huge inventory of .30 caliber cartridges on hand at the time of the Armistice, very little additional ammunition was made until the eve of the Second World War.

Complaints Upon Receipt of Rifles — In addition to the problems with ammunition failures, there were myriad other complaints against the rifles including improperly machined parts, broken ejector springs and extractors as well as many reports of excessive headspace. Most of these were corrected by more thorough inspection procedures at the factories and the redesign of some components.

A frequently cited complaint was that a number of rifles arrived in training camps showing substantial rust accumulation. As mentioned in an *American Rifleman* article:

Soldiers from the 29th Division engaging in target practice in France with their Model 1917 rifles. *(Tom Laemlein)*

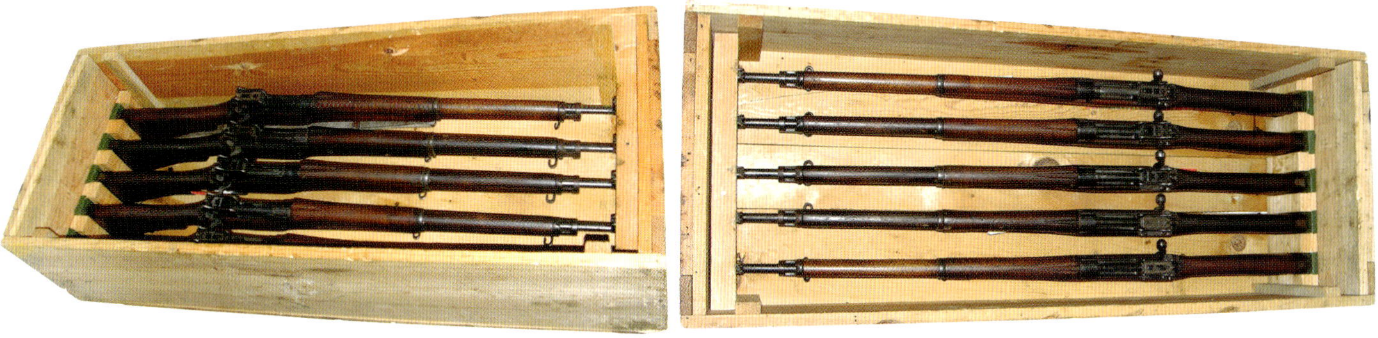

Two views of the shipping crate for the M1917 rifles. The crates were made at the Ogden Ordnance Depot and had a capacity of eight rifles. The WWI-era factory crates were constructed differently. The crates held ten rifles each. Note the label (shown below), which reads "UNIT PACK FOR OVERSEAS SHIPMENT / Do Not Open UNTIL READY FOR USE". *(Hayes Otoupalik collection. Photos by William R. O'Donnell)*

"Unfortunately for the rifle demonstrators, they had to address some things beyond just the unique features of the new rifle. One such problem, discovered upon uncrating brand-new rifles direct from the factory, were that some were arriving in a rusted state. Capt. Munn wrote that, 'A large number of Eddystone rifles (examined) in January and February 1918… issued to this Division, have been received in a frightfully rusted condition within the bore and receiver, likewise on the leaf sights.' He also noted that the rust actually appeared under the applied shipping grease, and that there was 'considerable difficulty' in removing the rust on the rifles. Capt. W.B. Porter also reported similar issues with Eddystone rifles in March of 1918, although the problem does not appear to have been as severe as that experienced by Captain Munn.

Other rifles also arrived at Camp Travis, Texas, either not cosmolined or improperly cosmolined, which resulted in 346 Eddystone rifles of the 165th Depot Brigade to be classified as either 'rusty' or 'very rusty.' Fortunately, identifying and tackling such issues was one of the stated aims of the rifle demonstrator program, and by April of 1918, the Assistant Chief Inspector of the Eddystone Rifle Plant was assuring the Ordnance Department that the greasing process was sufficiently modified and refined to ensure that 'absolutely no difficulty will be encountered in the future from rust accumulation.' "[15]

MARKSMANSHIP MEDALS — Row 1: 14kt gold Distinguished Marksman; 14kt gold Chemical Warfare Trophy. **Row 2:** WWI American Expeditionary Forces and American Forces Germany Department Shooting Prizes. Note the top bars are with Letters "AEF" or "AFG". These Department shooting badges were issued in gold, sterling and bronze and are very scarce. **Row 3:** Expert Rifleman, Sharpshooter and Marksman bar all in sterling silver. **Row 4:** Marksman Course A, Expert Rifleman Short Range, and Sharpshooter Short Range. **Row 5:** AEF Table medal awarded to AEF shooters in France in 1919. *(Hayes Otoupalik collection. Photo by William R. O'Donnell)*

Row 1

Row 2

Row 3

Row 4

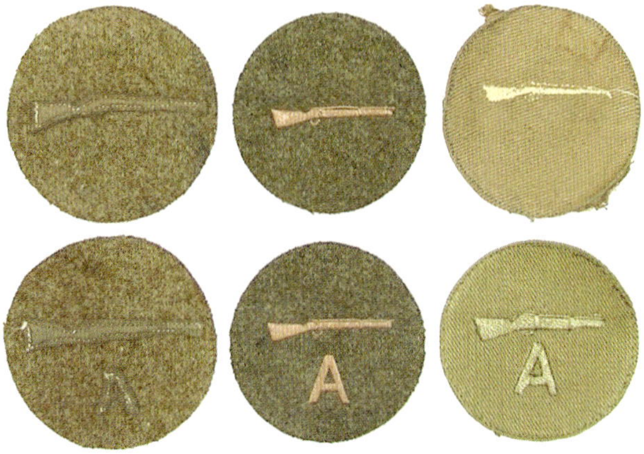

MARKSMANSHIP CLOTH PATCHES — Between April of 1918 and August of 1919, in order to save metal for the war effort, the U.S. Army was given these cloth discs to replace the silver and bronze marksmanship badges. Never popular with the troops, the soldiers continued to flaunt the regulations and wear silver and bronze shooting badges. Today, these cloth marksmanship discs are rare. **Row 1**: Expert Rifleman. **Row 2**: Sharpshooter. **Row 3**: Marksman. **Row 4**: Marksman Course A for Coast Artillery Corps. *(Hayes Otoupalik collection. Photo by William R. O'Donnell*

Some Remington-made rifles were also found to have been shipped in packing crates made of green wood, which retained moisture and caused rusting.[16] Better quality control measures helped alleviate that problem, but it was never totally solved. Most soldiers were very proud of their accomplishments in marksmanship training with their Model 1917 rifles, and various types of proficiency badges and medals were awarded. The badges and medals were a real source of honor amongst soldiers in recognition of their abilities on the rifle range.

Bayonet Training — In addition to marksmanship training, inductees received extensive instruction in bayonet fighting, which was considered to be a vital component of warfare even though it may not have been as ubiquitous as is sometimes believed. During the 16 weeks of Basic Training, 32 hours were dedicated to Bayonet proficiency.[17] As stated in a 1918, U.S. War Department Training publication:

> "The division Chief of Staff, COL Christian A. Bach, summed up the problem: Many of the men drafted had never struck a blow in anger in their lives. The bayonet instilled a fighting spirit and gave them individual aggressiveness, but it was never really popular. The rifle was the national arm of the American people, and they do not take kindly to the use of cutting or thrusting weapons. But, although the men of the 4th Division had few occasions to use their bayonets in hand-to-hand fighting, the training received was of real value and had a distinct psychological effect."[18]

U.S. doughboy in winter overcoat with a Model 1917 rifle and fixed bayonet. *(Tom Laemlein)*

Pvt. Harry W. Decker, Co. A, 116 Infantry, Reg., 296th Div. AEF, colorized photo. *(Charlie Williams, Harry Decker's great-grandson)*

Photo, dated 1918, captioned, "Mike Gibbons and Lieut. Fainey, bayonet instructors, Camp Dodge, Iowa." *(National Archives)*

Bayonet training with Krag rifle and bayonet early in WWI. *(U.S. Army photo)*

This photo, dated 1918, is captioned, "Drills with bayonet, Camp Jackson, S.C." *(National Archives)*

This photo is captioned "Infantry drill, Camp Devens, Mass., 1918." *(National Archives)*

Photo dated November 21, 1918 (ten days after the Armistice), showing soldiers at Camp Dix, New Jersey, being mustered out of service and turning in their equipment. This photo confirms that Krag rifles and bayonets continued to be used for training purposes in the United States throughout the war. *(National Archives)*

Due to the dearth of Model 1903 and Model 1917 rifles and bayonets, some U.S. Krag rifles and bayonets were pressed into service for training until more of the standardized rifles and bayonets became available. However, perhaps surprisingly, a number of the Krag rifles and bayonets continued to be used for stateside training purposes until the end of the war.

When the doughboys arrived in France, many of them received even more thorough training in the use of bayonets. When employed in combat, the sturdiness and overall dimensions of the Model 1917 rifle, coupled with the length of Model 1917 bayonet, made for an ideal combination for bayonet fighting.

Deployment to France — The first contingent of 14,000 doughboys arrived on June 26, 1917 and disembarked at the French port of Saint-Nazaire. American troops arriving early in the war were armed with Model 1903 Springfield rifles but as the inventory of Model 1917 rifles began to grow, more and more were issued. According to AEF cable dated February 28, 1918:

> "…it is now evident that in the near future, all combatant troops beyond second phase and service of rear troops in all phases must be equipped with Enfield rifles, which are now available in sufficient quantities with spare parts to make these plans practicable. Information requested as to whether you will

American troops engaged in bayonet training with Model 1917 and Model 1903 rifles. *(U.S. Army Signal Corps photo)*

Photo dated August of 1918 of doughboys training at Camp Funston, Fort Riley, Kansas. The exercise involved removing a "wounded" soldier from a barbed wire entanglement. Note the M1917 rifle with fixed bayonet stuck into the ground at left. *(NARA photo. Courtesy of Tom Laemlein)*

recall Springfield rifles from service of rear troops in 1st and 2nd phase upon receipt in France of Enfields of equal number with spare parts therefor, thus confining Springfield rifles to use of combatant troops in first and second phases. This matter (is) urgent and action is awaiting reply."[19]

On March 2, 1918, just a few days after the above-cited cable was sent, the Adjutant General sent a letter to the Chief of Ordnance clarifying the issuance of Model 1917 rifles *vis-à-vis* the Model 1903:

"1. You will take immediate steps to substitute model 1917 rifles for model 1903 in all organizations, except Cavalry and Mounted Engineers, which are not included in the 1st and 2nd phases of the A.E.F. Project and further excepting troops serving in insular possessions of the United States. This exchange will include all appendages, spare parts and pertaining equipment not interchangeable between the two arms.

2. The exchange will be made in the following order: 28th Division, 27th, 7th (Regular), 37th, 29th, remaining National Guard Divisions in numerical order, to all remaining troops in the order which circumstances may indicate to be the proper one.

3. This change of arms will also apply to Service of Rear Troops, which are part of either the 1st or 2nd phases, not yet overseas, providing the change can be completed in time without delaying their departures.

Two U.S. Army infantry sergeants with M1917 rifles. The soldier at left is also holding a M1917 .45-caliber revolver. *(U.S. Army)*

U.S. Army infantrymen advancing through a French village armed with M1917 rifles. *(US. Army Signal Corps photo)*

Three doughboys of the 36th Division chatting with a young French boy in August of 1918, at Bar-Sur-Aube, France. All of the soldiers have Model 1917 rifles. The rifle on the left has a canvas breech cover attached, and the soldier on the right has his rifle in a full-length case. *(Tom Laemlein)*

4. You will give the necessary instructions to division and Post Ordnance Officers for disposal of rifles turned in."[20]

Reportedly, the reason for not issuing Model 1917 rifle to the cavalry and mounted engineer units was primarily because the weapon would not fit the saddle scabbard for the Model 1903 Springfield rifle.[21]

As the Model 1917 rifles arrived in France and began to see active service, the reception of the weapon by the doughboys was decidedly mixed. However, once the soldiers armed with the Model 1917 rifle faced the crucible of combat, many found the weapon to have a number of redeeming features. As related in an *American Rifleman* article:

"In March 1918, 1st Lt. James Wallace, reporting from Camp Upton, N.Y., stated that while the men of the 77th Division were initially skeptical upon be re-equipped with M1917s, he found that 'as soon as the Officers and men are shown the care of manufacture, and inspection methods, and the really excellent arm they have, they too become enthusiastic about it.' The men of the 77th Division would put their M1917s to hard use at the Battle of Château-Thierry just a few months later, as well as throughout the rest of the war.

After training with the men of the 367th Infantry Regiment, 92nd Infantry Division, one of their experienced regular service field-grade officers reported to Lt. Wallace that 'in his judgment the [M1917] met every requirement' and 'spoke of the splendid success the regiment had in training green troops to shoot with it.' In addition, while he felt it was not as well-balanced as the M1903, the M1917 'compared favorably with [the M1903] in every respect and in the matter of the location of rear sight and size of aperture, and battle sight that it was much to be preferred to the Springfield.' "[22]

Ordnance and Rifles in France — Ordnance files contain numerous documents ordering the replacement of Model 1903 rifles by Model 1917s. A typical example was a memo from the American Expeditionary Force, Office of the Chief of Ordnance, to the Commanding Officers of four Ordnance depots in France:

"1. It is requested that you ascertain the number of rifles, M1903, now in the hands of organizations about your depot and a report rendered to this office.

2. It is the intention to have all M1903 rifles in the hands of such organizations, replaced by M1917 rifles.

3. Supplementing this report, it is requested that you advise this office the exact number of M1917 rifles, which your depot will require. Shipments of M1917 rifles will be authorized and upon their receipt, you will turn in the M1903 rifles in accordance with instructions, which will be furnished at this time."[23]

An April 10, 1918, American Expeditionary Force "Report of Small Arms Situation" elaborated on the status of the Army's small arms procurement program for the A.E.F. The report, was authored by Major C.C. Sheppard, A.E.F., and sent to the Ordnance Department in the United States. The report covered Model 1903 and Model 1917 rifles, Model 1911 pistols, Model 1917 revolvers and shotguns. Selected excerpts pertaining primarily to the Model 1917 rifle follow:

"General Situation. The most salient features of this general discussion are:

(1) Necessary rifles for reserve of the A.E.F. Forces have not been supplied as per our cables.

(2) All troops were supposed to come completely equipped with rifles and small arms. There have been many exceptions to this.

(3) Spare parts for maintenance of these arms have not been maintained in quantities necessary.

(4) There is a very great shortage of pistols and revolvers for the arming of troops in France.

(5) Considerable confusion exists with reference to the distribution of the various models assigned to the A.E.F. Requirements."

The requirements of arms for the A.E.F. in this support are computed on the following assumptions:

American soldiers in a French village in 1918. Some of their Model 1917 rifles (with protective canvas breech covers attached) are arranged in a tripod-like configuration using the rifles' stacking swivels. *(U.S. Army Signal Corps)*

(1) Schedule of arriving troops as shown in memorandum taken by Lt. Henry.
(2) Requirements of 13,000 rifles per 25,000 men regardless of the arm of service.
(3) Replacements and spare parts for U.S. Rifle, Model 1903 as shown in our automatic supply tables dated March 30, 1918, and for rifle Model 1917 as shown in our automatic supply tables dated April 1, 1918, herewith.
(4) The requirements with reference to replacement and spare parts are based on the best information we could obtain from British and French sources...

U.S. Rifle Model of 1917. The situation of spare parts for the U.S. Rifle Model 1917 is not clear at the present time. It is requested that you furnish us with definite lists of the spare parts being provided for this arm and that these spare parts be shipped with the arms, and that further supplies of spare parts based on the number of arms in service here be furnished automatically.

With reference to Model 1917 rifles it is noted in your memorandum dated January 4, 1918, that spare parts for these rifles manufactured by three different firms engaged in this work are not strictly interchangeable, and that you have deemed it necessary that spare parts manufactured by each concern is shipped to us separately. It is believed that it would be possible to select the product of the largest producer of these rifles and reserve them exclusively for use of the A.E.F. and thus avoid the possibility of having non-interchangeable parts in store here. Under present distribution of troops and depots the separation of spare parts for different factories is impractical. Our cable 775 dated March 25 requests that this distribution be made."[24]

This report contained a number of interesting items:

- The projected requirement for rifles was 13,000 per 25,000 men regardless of the branch of service.
- There were clearly insufficient numbers of small arms and spare parts arriving in France.
- The estimated requirements for replacement rifles and spare parts were based on the experience of our British and French allies.
- The most surprising comment was the concern about the lack of interchangeability of parts between the three manufacturers and the request to only send rifles made by the "largest producer" to the A.E.F. In order words, only Eddystone rifles were to be sent to France since Eddystone produced the largest number of rifles. Apparently, the intent was to keep the A.E.F. from being burdened with the "inconvenience" of non-interchangeable parts. Presumably, the rifles made by the other two companies would be used for training and other non-combatant purposes. This, of course, didn't happen. Other than being a cumbersome logistical imposition, the primary reason for this request not taking effect was

U.S. infantryman taking a much-needed drink from his canteen while on his way back from a field hospital, August 29, 1918, at Ville-Sur-Tierre, France. His Model 1917 rifle is on the grass beside him. Note the Model 1917 bayonet and scabbard on his cartridge belt. ***(U.S. Army photo. Courtesy of Tom Laemlein)***

A weary, mud-encrusted doughboy with his Model 1917 rifle. ***(U.S. Army Military History Institute)***

likely the directive from General C.C. Williams, which clearly stated, "This finally determines the question that rifles made by the Eddystone, Ilion and Winchester plants since January 1, 1918, may be regarded as equally interchangeable and, therefore, equally available and serviceable for use with United States forces anywhere, either at home or overseas."[25]

Throughout most of America's involvement in the war, there was a constant struggle to supply sufficient quantities of weapons, spare parts and other items necessary for our troops arriving overseas to be properly equipped.

Spare Parts Shortages — Ordnance continuously grappled with pervasive shortages of spare parts for both the Model 1903 and Model 1917 rifles. There were numerous dispatches, reports and communiques from units in France regarding the need for spare parts. A.E.F. Headquarters in France was inundated with messages from units in the field advising as to the shortage of parts. A typical example was this terse August 9, 1918, memo from Ordnance officer of the 42nd Division: "This organization has no spare parts for the Model 1917 rifle on hand."[26]

Another lengthier memorandum dated August 8, 1918, from the Ordnance Officer, 79th Division, A.E.F. to the Office of the Chief of Ordnance, A.E.F. stated, in part:

> "...You are advised that units of the 79th Division were not equipped with sets of spare parts for the M1917 Rifles before leaving the United States. The undersigned joined this Division May 25, 1918, and to his knowledge, no sets of parts were issued prior to departure of the division from Camp Meade. Inquiry has been made of each supply officer of the organizations at present in Europe with this division and the statement I have made is borne out by their statements. Further investigation shows that none of the records of this office show an issue of the spare parts in question."[27]

An August 10, 1918, memorandum was sent from the Chief Ordnance Officer, A.E.F. to the Chief of Ordnance in Washington pertaining to the on-going problem of spare parts shortages for the Model 1903 and Model 1917 rifles. The following are excerpts from the Model 1917 rifle section of the memorandum:

> "U.S. Rifle, Model 1917- It appears...that if the necessary complete rifles are to be manufactured, it will be impossible to supply the required parts without the development of additional plant facilities and that such additional facilities would not result in any material production during the current year...
>
> It would not seem to be in accordance with the probable conditions, therefore, to 'wipe out' the requirements for parts for repairs merely because such parts were not available at the time when the repairs became necessary. Furthermore, if additional plant facilities must be developed it is probable that the portion of requirements, which could be eliminated would not be any large proportion of the production for which such plants would be designed. The above comments regarding the elimination of requirements for factory deliveries in the United States up to August 1st for use in France 60 days or 90 days later, assume that it will be impossible to have the necessary deliveries of such parts in France before the first part of 1919...
>
> The situation appears to be that if our troops in France are to be kept properly armed, it will be absolutely necessary to have either the necessary parts or completed rifles representing the quantities as shown in the automatic supply table. This might well make it necessary to take a portion of the rifles from the troops in training in the United States until sufficient additional facilities were developed to meet the requirements in both France and the United States. It is understood that you have available approximately 250,000 Russian rifles and their use in training is suggested."[28]

The memo gave some rather audacious comments and advice as what steps should be taken to help rectify the problem including:

- Build additional factories to make more rifles and/or spare parts.
- If our troops in France are to be "kept properly armed," they must be supplied with "necessary parts or completed rifles."
- The problem won't go away simply by simply stating that parts aren't available.
- Even if the factories were built, the problem wouldn't be resolved until early 1919.
- It might be necessary to take service rifles currently used for training troops in the United States and send them to units in France until more rifles became available.
- Consider using more of the 250,000 Model 1891 Mosin-Nagant rifles supposedly available for training purposes to replace the standard service rifles diverted to France.

It seems the suggestions proffered in the above memo may not have been particularly well-received by the Ordnance "brass" in Washington. This apparent displeasure perhaps resulted in the Ordnance Officer of the A.E.F. sending the following August 24, 1918, memorandum to the Commanding Officer, Advance Unit, Army Ordnance in France:

> "1. Be advised that temporarily there will be a shortage of spare parts for rifle, model 1917. This apparent shortage was caused by a condition of circumstances over which we had no control.
> 2. There is no more on hand with the divisions than are required.
> 3. A very sizable shipment is in transit from the port to I.O.D. #2. It is suggested that you forward a requisition to that Depot for a reasonable amount.
> 4. **Should there be any shortage of spare parts you may rest assured that the proper authorities in the U.S. are well aware of this fact, and it is**

entirely unnecessary for any advice to be given from this side regarding the operating of any of their plans."[29] *[Bold print added by author]*

In order to help alleviate the endemic shortage of spare parts, rifles in poor condition were to be cannibalized so as to utilize salvageable components. As outlined in an October 18, 1918, memorandum from the "Office of the Construction and Maintenance Division, Ordnance Department" to the Supply Division:

"It is believed that rifles...beyond repair will be of little value for return to the United States. If their condition is such that repair is impossible, it is recommended that these rifles be torn down and the parts salvaged for use in repair work. It is understood from the Supply Division that **there are practically no spare parts for Model 1917 rifles on hand**, and that spare parts for Model 1903 rifles are limited in number."[30] *[Bold print added by author]*

An October 22, 1918, memorandum from the Commanding Officer of the Advance Ordnance Depot of the A.E.F. highlighted how a lack of spare parts can keep large numbers of badly needed rifles out of commission:

"1. There are now on hand about 40,000 caliber .30 Rifles, model 1917 to be repaired. The main troubles found in those rifles are unserviceable ejectors and extractors. Of those that have gone over 40% require new ejectors and extractors.

2. Work on 1917 Rifles was begun Friday, October 18, and there are not enough ejectors and extractors on hand to continue their repair. It will be necessary to hold some of the extractors for issue to advance units for similar repairs. Without these parts it will be necessary to lay aside all rifles with those parts missing and repair the others. This will lessen the expected daily production of 2200 somewhat.

3. Advice is requested whether any ejectors and extractors are available, which can be supplied for these repairs."[31]

Rifle Shortages — The importance keeping as many rifles in serviceable condition as possible was vital to meet the constant demand for weapons. To cite one example of such demand, an October 12, 1918, memorandum from the Camp Ordnance Officer, Camp Hunt (Le Courneau, France) to the Chief Ordnance Officer, Base Section detailed the equipment needed to outfit over 10,000 replacement troops scheduled to arrive the following month:

"10,000 Rifles Model 1917
10,000 Gun Slings, 1917
10,000 Bayonets, Model 1917
10,000 Bayonet Scabbards, model 1917
5,000 Helmets"[32]

Even though production of the Model 1917 rifle continued to accelerate at an impressive rate, there were still periodic shortages of rifles throughout the war, exacerbated by the above-cited shortage of spare parts. Noted researcher, the late Frank Mallory, discovered a 1918 document that revealed:

"In August, 1918, the shortage of rifles for the AEF caused Coast Artillery troops to be directed to turn in their (Model 1903) rifles, to be replaced by riot guns. The limited supply of Model 1903 rifles and the desire to continue to furnish them to the Navy was a factor in the decision to replace '03s with M1917 rifles in the AEF."[33]

Ordnance continued to seek effective ways to increase the production rate to help address the troubling shortages of rifles. One bottleneck that occurred from time to time was the scarcity of suitable black walnut wood for stocks. As briefly mentioned in Chapter 4, in June 1918, Eddystone fabricated some laminated stocks on a quasi-experimental basis. Government records indicate that on August 13, 1918, laminated stocks were tested on several other M1917 rifles including #114415 and #115157.[34] However, laminated stocks were not put into quantity production and walnut stocks remained standard.

An interesting document drafted late in the war (October 1918) detailed the shortages of Model 1917 rifles and bayonets. Quite surprisingly, it also indicated there were U.S. Army units armed with British rifles. The memorandum from the Office of the Chief of Ordnance dated October 5, 1918, to the Chief Ordnance Officer of the A.E.F. stated:

"Because of the present shortage of U.S. Rifles, Cal. .30, Model 1917, Bayonets Model 1917 and Scabbards, Model 1917 and the prospect that this shortage will not be relieved for at least six weeks, it is recommended that these three articles be added to those included in Class IV within the meaning of General Orders No. 44, G.H.Q.

You are informed that the present status of these articles is as follows:

A.O.D. #1	I.O.D. #2	In Transit	Total	Dues Out	Balance
U.S. Rifle, Cal. .30, Model 1917					
17,247	804	5,640	23,691	13,830	9,861
Bayonets, Model 1917					
25,645	9,074		34,719	21,430	13,289
Bayonet Scabbards, Model 1917					
31,048	76,620		109,668	20,870	88,798

During the past two weeks the issues of U.S. Rifle, Cal. .30, Model 1917 has averaged 1,345 per day. The promised shipments indicate that the shortage will be overcome in about six weeks. In the meantime there are 12,000 rifles of this model now held in reserve at Le Havre for the 30th Division of the 2nd Corps, **which is at present using British rifles.**

These may be used for issue elsewhere provided the military situation is such that their issue to the Division for which reserved will not be required before November 1, 1918."[35] *[Bold print added by author]*

Another noteworthy document pertaining to the shortage of rifles in France was a memorandum from the Ordnance Department to the "Director of Storage" dated October 25, 1918, which began: "Referring to your inquiry as to what steps have been taken by the Supply Division of the Ordnance Department to do everything possible to meet the acute situation existing in France..."[36]

The document went on to detail, among other things, the number of Model 1903 and Model 1917 rifles on hand, in transit and the estimated current and future needs. The section pertaining to the Model 1917 follows:

"U.S. Rifles, Caliber .30, Model 1917:

(a) Cablegram S-210-2C copy attached herewith, states in part, must receive in addition model 1917, 235,000 through October, 125,000 November. 140,000 December.

(b) Report of October 15th indicated the following in regard to shipment of 1917 rifles: floated, Oct. 1st to Oct. 15th – 45,810; docked on October 15th – 28,556; freighted 11,000 – making a total of 85,366 in transit overseas on October 15th.

(c) Approximately 30,000 of those manufactured from October 15th to date (Oct. 25) are now freighted in addition to the above.

(d) It is estimated that there are over 80,000 model 1917 rifles in the hands of troops at Replacement training camps and approximately 160,000 in the hands of troops at Divisional Camps. It is understood that a paper may come through from the Office of the Chief of Staff directing that only sufficient rifles be retained at the Training Camps for the equipment of 1 out of every 3 men for when the rifle is prescribed. It is believed that this order would probably release about 50 percent of these rifles in the Camps.

(e) The production of 1917 rifles at the manufacturing plants is estimated at about 160,000 per month during November and December.

All rifles manufactured are now being directed overseas and every effort has been made by Ordnance Department to expedite production, inspecting and shipment. The Port Ordnance Supply Officers at the New York and Philadelphia Depots from which all rifles are forwarded, have been requested to load promptly all rifles as docked."[37]

As indicated by an Ordnance Department memorandum dated October 26, 1918, the perceived shortage of rifles was deemed to be so acute that plans were formulated to significantly increase the production schedule for Model 1917 rifles for calendar year 1919 and beyond. These plans

A doughboy at trench dugout entrance with his M1917 rifle close by. *(U.S. Army Signal Corps photo)*

were based primarily on expanding the production capacity of the three rifle plants:

"...with the notice to resume the production of rifles at Winchester's, we were in receipt of preliminary notice of an increase in rifle requirements from about 3,000M up to the 1st of July, 1919 to about 5000M total up to that date. This change of program required an increased capacity during the remaining intervening period of about 10%. By a slight rearrangement the floor space was available at all three plants for such increased installation as might be necessary to so balance their plants as to produce the added 10%. This added equipment has not yet become effective, principally for lack of available labor, due to influenza and other causes

As soon as it is possible to utilize this equipment, and we have reasonable hopes of being able to do so in the course of the next sixty days, the capacities of the three plants on a basis of 60% labor efficiency and double-shift, will then be:

Midvale	5,500 per day
Remington	3,300 " "
Winchester	2,200 " "

This morning we received preliminary notification of the probability that the total Model 1917 rifle requirements for the calendar year 1919 would be increased by 2,000M rifles. Figures are now being prepared to determine precisely what this will require. Inasmuch as this increased capacity is not required until about the 1st of July, it is not impossible that it can be met by extensions of existing plants. These extensions will probably entail the construction of buildings at Eddystone and Ilion and probably the utilization of some building now occupied by a non-essential plant for Winchester, New Haven, and the necessary (acquisition) of machinery. Such extensions are not likely to be in production, at the very earliest, before the summer of 1919.

It seems hardly necessary to say that no completely new rifle plant, even with the buildings in existence, could be equipped, tooled up and organized, to meet this added requirement and be in appreciable production, prior to the 1st of January 1920, and probably not then, so that the solution must lie along the lines previously indicated, and pending final official action, steps are already being taken to that end."[38]

As events transpired, the cessation of hostilities due to the Armistice being declared just 16 days after the above memo was drafted, rendered moot the plans for the increased rifle production.

Fortunately, there is little indication that the American war effort in France was unduly hampered by these shortages of rifles but it was an on-going concern up through the time of the Armistice.

Marine Corps and Navy Use of the Model 1917 Rifle — The Model 1903 was the primary service rifle of the U.S. Marine Corps during the First World War, and is sometimes claimed that the Marines did not utilize any Model 1917 rifles during the war. However, this assertion is disproven by government documentation. In his excellent book, Equipping the Corps, Vol. 1, author Alec Tulkoff referenced a number of Marine Corps documents regarding the Model 1917 rifle including this report:

"Due to the large number of M1903 rifles required by all of the armed forces during WWI, they were in short supply. Marine Corps Order No. 20 (Series 1918) stated the following about M1903 rifles: 'Owing to the shortage of the model 1903 rifle and the uncertainty in procuring and [sic] additional quantities from the Army in the near future, it has become necessary to procure a supply of model 1917 rifles…It is the intention that all troops ordered overseas duty be equipped with the model 1903 rifle, and that this rifle be issued at Recruit Depots as long as possible, and all Springfield rifles be withdrawn from permanent posts of the Marine Corps in the United States and replaced immediately by the model 1917 rifle."[39]

A memorandum dated January 24, 1918, from the U.S. Navy's Director of Gunnery Exercises and Engineering Performances to the Bureau of Ordnance related the following:

"It is requested that 1,000 rifles, caliber .30, Model 1917, be procured as soon as possible from the Ordnance Department, U.S. Army, for immediate distribution to Navy Rifle Ranges, and upon receipt, be distributed as follows:

U.S. Marine with Model 1917 rifle, c.1918. He is wearing a "Sharpshooter Marksmanship Medal." *(Alec Tulkoff)*

Qty.	Ship and Mark
200	Commanding Officer, Camp Logan, Ill.
50	Range Officer, Great Lakes, Ill.
100	Commanding Officer, Wakefield, Mass.
180	Commanding Officer, Peekskill, N.Y.
20	Range Officer, Navy Yard, Phila., Pa.
50	Range Officer, Cape May, N.J.
100	Commanding Officer, Glen Burnie, Md.
100	Range Officer, Naval Acad. Rifle Range, Annapolis, Md.
100	Commanding Officer, Virginia Beach, Va."[40]

(Author's note: The quantities cited total 900 rifles. The disposition of the remaining 100 rifles requested is not known.)

This was followed by a memorandum dated March 7, 1918, from the Office of the Chief of Ordnance, Supply Division, citing an "urgent need" for Model 1917 rifles by the Navy Department for use at Marine Corps rifle ranges. It is interesting to note that the Navy didn't want to pay for the rifles:

"1. Reference is made to the 1st Ind. of this office of date of March 5, 1918, (OOS.045/413) to the Military Information Branch concerning 1,000 caliber .30, Model 1917 rifles desired substituted by the Navy Department in lieu of 1,000 Springfield rifles Model 1903…

2. This office has received a telephone message from the Bureau of Ordnance, Navy Department, today stating that 1917 rifles are not being used by the Navy Department and that 1,000 1917 rifles urgently desired, are for the use of the Marine Corps at its rifle ranges. These ranges are jointly used whenever desired by the Army and accordingly it is the feeling of the Navy, in view of the facts as herein stated, that they should not be compelled to pay for the rifles but rather they should be invoiced to a representative of the Marine Corps who would be accountable to the Ordnance Department for same.

3. It suggested that the papers accordingly be referred to the Control Bureau with the request that it will be permissible for the Supply Division to issue and not sell 1,000 Model 1917 rifles requested. If this course is approved for the Military Information Branch to arrange for a modification of the present requisition of the Navy Department for 1,000 Model 1917 rifles."[41]

A summary chart in a June of 1919 Ordnance report titled "Sold to Navy and Marine Corps" gives the following information regarding the total number of Model 1917 rifles procured by both branches of service contained in a June 7, 1919, Office of the Chief of Ordnance document:

"Rifle, cal. .30, model 1917:
Marine Corps – 76,840
Navy – 2,050"[42]

Early in the war, there was some use of Krag rifles by Navy personnel, primarily for stateside sentry duty and related applications, but most were replaced by Model 1917 or Model 1903 rifles as they became available.

An April 18, 1918, memorandum from the War Department, Office of the Chief of Ordnance to the Engineering Bureau contained some interesting information as well as concerns about the Model 1917 rifles to be procured by the Marine Corps:

"Subject: Spare parts Winchester U.S. Model 1917 rifles

1. This office is given to understand by Major Clark of the Supply Division that your office expected to get shortly complete requirements on the above-named rifle.

2. As it is quite necessary to place orders very quickly with the Winchester Repeating Arms Co. for all spare parts needed due to the fact that these works are expected to be utilized for another purpose, this office will appreciate further advices on the subject.

Photos of two WWI U.S. Marines with Model 1917 rifles. The photographs were likely taken while training at Parris Island, S.C. *(Cheairs Aguilar)*

3. There are to be some seventy odd thousand rifles of the Winchester manufacture subsequent to January 1, 1918, issued to the U.S. Marine Corps. It is learned unofficially that their spare parts requirements would be the same as our home requirements of spares, plus 20% to 30%.

4. There is one other feature on these spare parts for the Winchester rifles on which this office would appreciate advice, and that is, how many spare parts should be ordered for the 103,229 Winchester U.S. Model 1917 rifles, which were manufactured prior to January 1, 1918, and which rifles it has been decided to retain in this country for training purposes. There are the rifles on which spare parts would not be interchangeable with those produced by Remington, Ilion and Eddystone. The only spares ordered for this last lot of riles are 4150 sets, and it seems desirable that enough spares should be ordered to last until these rifles are discarded."[43]

Obviously, Ordnance remained concerned about the lack of interchangeability of the Winchester rifles made before January 1, 1918, and wanted to make sure the rifles supplied to the Marines would have a sufficient supply of spare parts. It is interesting to note that the vast majority of the 76,840 Model 1917 rifles that were procured by the Marine Corps were Winchester.

Initially, the primary reason for the Marines acquiring the Model 1917 rifles was for training purposes in the stateside camps in order to have a sufficient number of Model 1903 rifles available for combat in France. However, there is anecdotal evidence from reliable sources that indicate some U.S. Marine units landed in France in late 1918 with M1917 rifles, but the war was over before they could be employed.[44]

Even though the Marine Corps decided to replace their Model 1917 rifles with Model 1903s soon after the Armistice, some of the "American Enfields" remained in the USMC's inventory for a number of years. Table #1 in this book lists several Marine Corps Model 1917 rifles that were inventoried between March of 1923 and November of 1925. Serial numbers of these rifles ranged between 222692 and 273335.[45] Their ultimate disposition is not known.

In order to attempt to give an edge to the doughboys engaged in the unfamiliar arena of trench warfare, there were several categories of weapons very useful in such fighting but were lacking in Uncle Sam's arsenal. One such weapon that had proved its worth in the war was the rifle grenade launcher. The launchers and grenades will be discussed in

Group of WWI-era U.S. Marines with Model 1917 rifles. This photo was probably taken when they had finished boot camp and completed rifle training with the Model 1917s. *(Alec Tulkoff)*

U.S. Navy sailor on sentry duty with a U.S. Krag rifle and bayonet. As was the case with the Army, due to a shortage of Model 1903 and Model 1917 rifles, some of the obsolescent Krags were pressed into service by the Navy for such tasks. *(National Archives)*

Photo dated October 5, 1918, of a U.S. Navy sailor holding a Model 1917 rifle. He is conversing with a member of the Women's Motor Corps. They are both wearing masks due to the influenza epidemic. They are assisting patients injured during an explosion at a shell-loading plant in New Jersey. *(National Archives. Courtesy of John Spangler)*

U.S. Navy sailor firing a M1917 rifle at the Naval Rifle Range at Caldwell, N.J., 1918. At right is a close-up view of the rifle. *(National Archives)*

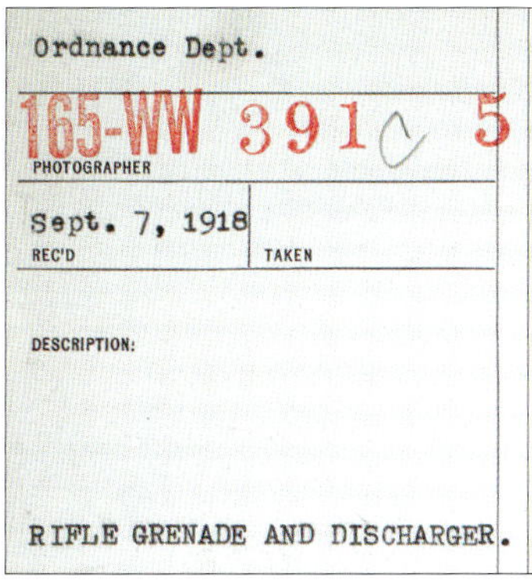

1918 Ordnance Department photo of V-B rifle grenade and launcher. *(National Archives)*

U.S. troops training with M1917 rifles and V-B rifle grenade launchers. *(Tom Laemlein)*

detail in Chapter 10 of this book. Other potentially valuable weapons such as an improved sniper rifle, "trench periscope rifles" and the innovative "Pedersen Device" were considered for development but most never advanced beyond the prototype stage and were not issued during the war. These will also be discussed at some length in later chapters.

As the war progressed in the summer and fall of 1918, more and more U.S. Army front-line infantry combat units were armed with "American Enfields," and by the time of the Armistice, an estimated 75 percent of the AEF were armed with Model 1917 rifles.

The authoritative book *History of Rifles, Revolvers, and Pistols: Army Ordnance 1917–1918* contained the following information regarding disposition of the rifles on hand at the end of the war:

> "The total production of model 1917 (rifles) during the war (i.e., actually to Nov. 11, 1918) was 2,202,429; up to and including week ending December 14, 1918 — 2,305,414; and including week ending December 28, 1918 — 2,422,529.
>
> Of these there were sent overseas 1,1123,259 — 800,987 with troops and 322,292 floated in bulk. None of this model was ever sold to our allies, but 61,000 were delivered to the marines and 604 to the Navy."[46]

It should be noted that the figure of 61,000 quoted here and the 76,840 number shown above likely reflects additional Model 1917 rifles transferred to the Marine Corps and Navy between December 1918 and June 1919.

1918 National Matches — Although the Model 1917's rear sight did not have the same functionality as the Model 1903 rifle's sight, it is interesting to note that a number of the "American Enfields" were used in National Match competition. With the outbreak of the First World War, the War Department decided to postpone the National Matches due to concerns that the availability of material and manpower might be compromised. As reported in the December 1917 issue of *Arms and the Man*: "The coming of war, with its consequent demand for all available weapons to be used by the fighting forces of the nation, brought these plans to a sudden halt, resulted in the suspension of the free issue and purchase

Interesting photo dated September 24, 1918, captioned, "Guards with full set of equipment, Gas Defense, Division." The soldiers are armed with M1917 rifles and have M1917 bayonets and scabbards attached to their packs. *(National Archives)*

A unique feature of military weddings, is the "rifle archway." The ceremonial event pictured here was conducted using M1917 rifles and bayonets at Camp Lewis, Wash., on April 22, 1918. *(National Archives)*

Troops of the U.S. Army's 77th Division, Camp Upton, N.Y., 1918. *(U.S. Army)*

privilege, and in calling off the National Matches of 1917."[47]

However, it soon became apparent that many trained marksmanship instructors were needed for proper training of the growing number of soldiers entering service. To this end, the Small Arms Training School was established on May 15, 1918, to be directed by Lt. Col. Morton Mumma, a noted marksman and a member of the executive committee of the National Rifle Association and the National Board for the Promotion of Rifle Practice. As discussed in the NRA publication *Enfield Rifles at the National Matches*:

> "The Small Arms Firing School (SAFS), in 1918, was different from the school as it operates today. In 1918, the SAFS trained personnel to act as rifle marksmanship instructors. The course lasted a month, with a new class beginning every two weeks. Most students were military officers learning how to run a marksmanship program and conduct range operations as well as learning how to shoot and how to teach the fundamentals of marksmanship. Classes numbered 500, more or less.
>
> Over and above the regular schedule of classes, a special 10-day SAFS session was held at Camp Perry. This special class was attended by 54 16-person teams—representing 36 of the then-current 48 states and the District of Columbia; the Arkansas National Guard, the Army (2), Navy (2), Marines (2), and the Navy Rifle Range Service (10). In addition, 400 Reserve Officer Training Corps (ROTC) students attended, representing various U.S. colleges and universities. Instead of Springfields, most of the students in this special class drew Model 1917 U.S. Enfields and reported directly from graduation to the range, where they participated in the NRA National Championships, and the National Trophy Individual and Team Rifle matches. The ROTC students were divided into four groups and were issued Model 1917 rifles, Model 1903 rifles fitted with one of two types of experimental sights, or standard issue Model 1903 rifles as a control.
>
> Shooters' reactions to the Model 1917 rifles began as 'cool,' but warmed up as shooters began to learn the rifle. The chief objections were to the rifle's 'cock-on-closing' bolt and the lack of a windage adjustment on the rifle's rear sight. The former was a matter of familiarity. Since the adoption of the Krag-Jorgensen in 1892, American military rifle shooters had a rifle that cocked the firing mechanism as the bolt handle was lifted to unlock the action. British designers of the precursor to the Model 1917 rifle elected to use a mechanism that cocked as the bolt was pushed forward to chamber a cartridge and close the action. Either system had its strong and weak points. The advantages of a cock-on-closing mechanism do not begin to become obvious until firing a rapid-fire string. And, in the rapid-fire matches, Enfields won unreserved praise.

A competitor at the 1918 National Matches firing a Model 1917 rifle. *(National Rifle Association)*

> The sights were another matter, though, and caused something of a division among the student marksmen. Shooters generally from east of the Mississippi River (and presumably familiar with the windage adjustable sights of the Krag and the Springfield) denounced the lack of a windage adjustment on the 1917 rear sight and had trouble holding off, or 'favoring' to compensate for changes in the wind. Shooters from out West (and presumably familiar with the non-adjustable sights found on Winchester and other sporting rifles) had little difficulty adapting to the generally excellent aperture rear sight used on the Enfield. They simply applied what is generally referred to as 'Kentucky windage' and fired away accurately and effectively."[48]

The Model 1917 acquitted itself quite well on the one occasion when it went head-to-head against the Model 1903 Springfield:

> "Only in one match was Springfield pitted against Enfield. That was the NRA National Individual Rifle Championship, 10 shots sitting rapid, 10 shots prone rapid, 20 shots prone slow from 500 yards, and 20 shots prone slow from 600 yards—60 shots for a possible score of 300 points. H.J. Mueller, a civilian from Terra Haute, Indiana, who attended the SAFS as part of the Indiana team, won the contest with a score of 289, using a Model 1917 rifle."[49]

As can be seen, the Model 1917 performed quite well in the matches even though they were standard-issue rifles with no special modifications. After the Armistice, the Model 1903 was selected to remain the standard American military service rifle and was used in subsequent National Matches until after WWII when it was replaced by the M1 Garand and, later, the M14.

Sergeant York — No discussion of the Model 1917 rifle's use in World War I would be complete without some mention of the most famous doughboy of the war, Alvin York. It has been conventional wisdom for some time that York was armed with a Model 1917 rifle during his extraordinary feat of arms on October 8, 1918. Despite Hollywood images to the contrary, it is beyond dispute that the authorized service rifle listed on the TO&E for his unit, Co. G, 328th Infantry, 82nd Division, was the Model 1917. While the 82nd Division did have some Model 1903 rifles early in the unit's deployment to France, a July 1, 1918, memorandum (three months before York's intrepid deed) from the Chief Inspector of Small Arms to the Chief Ordnance Officer, A.E.F., ordered that these rifles be purged from the division's inventory:

"1. The 82nd Division has 2,100 Springfield rifles, '03 and 10,000 U.S. Rifles, Model 1917.

2. In view of (the) decision of the General Staff to maintain one type of rifle in a division, recommendation is made that Enfield rifles be supplied to this Division and that Springfields be turned in to our Advanced Army Shop at Void-Vacon."[50]

An October 6, 1918, memo from the Division Ordnance Officer, Fourth Depot Division to the Chief Ordnance Officer, AEF, requested that information regarding which divisions were using Model 1917 rifles as well as Hotchkiss, Browning and Vickers machine guns.[51] The response dated October 14, 1918, indicated that the 82nd Division was equipped with Model 1917 rifles, Chauchat Automatic Rifles, and Hotchkiss machine guns, thus confirming that the '03 rifles were no longer part of the division's inventory.[52]

Sgt. Alvin C. York. *(National Archives)*

Soldiers of the 328th Infantry Regiment, Camp Gordon, Georgia, with M1917 rifles and fixed M1917 bayonets, including Pvt. Alvin York (circled). The soldier sitting in front of him was William Cutting (actual name Otis Merrithew) who was also involved in the October 8, 1918, engagement. Cutting claimed that York's role was exaggerated and that the other members of the patrol were just as responsible for the outcome. *(Courtesy of James P. Gregory Jr.)*

Many books have alluded to York's prowess with the "American Enfield." One example is the well-regarded book *The Doughboys: The Story of the AEF 1917–1918* authored by Laurence Stallings who served as an officer with the U.S. Marine Corps during WWI and was badly wounded during the Battle of Belleau Wood, which resulted in the loss of a leg. The following excerpt is from his above-cited book:

> "York now holstered his pistol and thumbed a fresh clip into his Enfield… It was difficult for York's Germans to discern where came this murderous fire. – what ear could distinguish the well-aimed shots of an American-modified Enfield rifle amidst all the chatter."[53]

Another quote from the well-researched book *American Rifle: A Biography* by Alexander Rose stated:

> "(York) soon made 'good friends' with his Model 1917 Enfield, declaring it just as accurate as his old Tennessee muzzle-loader, at least to a hundred yards.
>
> He proved it on October 8, 1918, when, during an American raid into the German lines, York's Enfield shot at least a score of Germans in the head, with nary a miss…capturing no fewer than 132 prisoners while armed only with an Enfield rifle and spare pistol."[54]

There are myriad other references to York's use of a Model 1917 rifle. However, the subject elicits much controversy today, as a number of people are of the opinion that York actually utilized a Model 1903 Springfield rather than a Model 1917 Enfield. Those who assert he used an '03 cite several reasons for their belief. One is that York's son, Andrew, supposedly said that his father used a Model 1903 rifle. However, it has been related that another son disputed this claim.

In any event, family history can be problematic. To cite just a few examples: a son of a WWII family was absolutely convinced that his father carried an M1 Carbine ashore on D-Day made by the Universal Firearms Company despite the fact that the company only made commercial facsimile copies of M1 Carbines for the civilian market in the postwar period. Another individual stated that his Korean War veteran uncle had an M3 submachine gun made by the Continental Can Company. There are also numerous claims attributed to Vietnam veterans that they had M16 rifles made by the Mattel Toy Co. Many other anecdotes can be added to the list confirming that not every claim made regarding the use of a particular weapon by a family member is accurate.

It has been pointed out that there is a statue of Sgt. York that depicts him with a Model 1903 rifle. There are numerous statues, paintings, and other works of art depicting famous servicemen and other notable martial personages with the absolutely incorrect weapons. These include .45-70 Trapdoor Springfields in the Civil War and Winchester M1894 carbines at the Battle of Little Big Horn, just to name two. A sculptor or artist generally is interested only in capturing an accurate likeness of the subject and typically little thought is given to the type of prop weapon on hand.

Another frequently cited reason for the '03 claim is that Alvin York was a technical advisor for the 1941 Academy Award-winning motion picture *Sergeant York*, which depicted the star, Gary Cooper, armed with Model 1903 rifle (actually it was a Bannerman-type ersatz rifle), and York would never have agreed to the use of an "incorrect" rifle in the movie. If York was a stickler for authenticity, he would have nixed the use of a German Luger in the movie and insisted on a Model 1911 .45 pistol with which he was actually armed. Also, York had a mustache during the war but actor Cooper was clean shaven.

Another claim often proffered is that York didn't like the Model 1917 sights and preferred the '03 sights. With all due respect to York, at the time of his heroic exploits, he was a nondescript corporal, and his personal preference for certain weapons would have been moot. An infantryman would be armed with the rifle issued to him. There were definitely instances in wartime when a soldier picked up a discarded weapon and used it, but it was much more common for him to retain the issued weapon for which he was accountable. It is certainly not beyond the realm of possibility that York did find and then use a discarded Springfield '03, but the preponderance of anecdotal evidence points to his use of the Model 1917. The debate will go on, but ultimately it doesn't matter. York's actions on October 8, 1918, were notable regardless of what type of rifle he used.

With the coming of the Armistice on November 11, 1918, the focus of the United States military immediately shifted from training as many troops as possible and manufacturing ever-increasing numbers of weapons to planning for the postwar future. As will be seen, while the war was over, there were many decisions that needed to be made, including what would happen to the large number of Model 1917 rifles that were suddenly no longer needed.

Chapter 6

Between the Wars

By the fall of 1918, the American Expeditionary Force was making preparations for a massive offensive planned for the spring of 1919, with the aim of knocking Germany out of the war. All manner of weaponry was to be used, including tanks, aircraft, artillery, and numerous additional small arms, such as the newly introduced Model 1917 Browning Machine Gun and Model 1918 Browning Automatic Rifle. Greater numbers of Model 1903 and Model 1917 service rifles, Model 1911 pistols, and Model 1917 revolvers would also be required. It was planned for the top-secret "Pedersen Device" to play a key role in the campaign as well.

The somewhat unexpected declaration of the Armistice on November 11, 1918, resulted in an immediate reassessment of the need for the increased quantity of weapons. The authoritative book *History of Rifles, Revolvers, and Pistols: Army Ordnance 1917–1918* contained the following information regarding disposition of the rifles on hand at the end of the war:

> "The total production of model 1917 (rifles) during the war (i.e., actually to Nov. 11, 1918) was 2,202,429; up to and including week ending December 14, 1918 — 2,305,414; and including week ending December 28, 1918 — 2,422,529.
>
> Of these there were sent overseas 1,123,259 — 800,987 with troops and 322,292 floated in bulk.
>
> On dock, Nov. 30 81,950
> In transit, Nov. 30. 30,770
> At camps, depots, land arsenals in the United States, Nov. 28 70,940
> At American Expeditionary Forces depots, Oct. 1, 1918. 910
> In hands of troops in the United States, Nov. 30 127,000
> At plants, Nov. 21. . . . . 6,326"[1]

A December 24, 1918, memorandum from the Chief of Supply Division to the Office of the Chief of Ordnance addressed the fact that large numbers of Model 1917 rifles were still being manufactured after the Armistice, even though further demand for the weapons had ceased:

> "It is desired to call attention to the Small Arms Division to the fact that no reduction in the number of rifles, U.S. cal. .30, M1917, being manufactured, has at present producing as great a number, if not a greater number, than ever before. The Expeditionary Forces have stated that no more cal. .30 rifles of any model were desired overseas. The Supply Division has no obligations upon which they can apply the present production of rifles and these are now being shipped to storage. The same condition applies to the pertaining equipment issued with the rifles, such as bayonets and bayonet scabbards.
>
> It is presumed that the office of the Chief of Ordnance is cognizant of the above facts but same is brought to his attention as the Supply Division believes that reduction should be made in view of the number of small arms being produced and the monetary value involved."[2]

In order to address the issue, production contracts for the Model 1917 rifle were cancelled, although as mentioned in Chapter 4, contractual provisions allowed each manufacturer to continue assembling rifles that were "in process" at the time of cancellation. Rifles continued to be delivered, albeit at a much reduced rate, into the spring of 1919.

A memo dated June 21, 1919, from the Ordnance Office, Administration Division to the General Staff, Statistics Branch, gave a detailed tabulation of the consumption of rifles, revolvers and pistols during the war. The section pertaining to the Model 1917 rifle contained the following:

> "<u>1917 – U.S. Rifle</u>
> Actual production to Nov. 30, 1918. . . . 2,295,390
> Est. available on Nov. 30, 1918. 1,561,575
> Presumably lost, scrapped & consumed . 734,017
> Est. value of loss, scrap/consumption . . $18,181,601.09"[3]

The fact that the memo also gave an estimated available total of Model 1903 rifles of 617,230 as compared to the over 1.5 million Model 1917 rifles during the same period confirms that the "American Enfield" was indeed the primary U.S. service rifle of the war.

Making the Decision — This situation left the War Department with a conundrum. The Model 1903 rifle had firmly secured itself as the standard American military service rifle for the past decade and a half and was a popular weapon with most of the servicemen who used it. On the other hand, while the Model 1917 rifle was heavier and a bit more cumbersome, it had proven to be a reliable and rugged weapon and gave more-than-adequate service to our troops fighting in France. It was also available in much greater numbers than the '03. However, there could only be one standard service rifle. One option was to retain the Model 1903 as standard and relegate the Model 1917 to the war reserve stockpile. The other choice was to anoint the Model 1917 as standard and downgrade the Model 1903 to "Limited Standard" or "Substitute Standard." There were cogent arguments on both sides of the issue.

The decision was formalized in an April 19, 1919, memorandum from Major General Henry Jervey, Assistant Chief of Staff to the Adjutant General stating that the Secretary of War gave the following directive:

> "For the future the policy will be to place the U.S. Rifle, Model 1917 in war reserve and to issue the U.S. Rifle, Model 1903 to troops in the United States and Insular possessions, including Panama, and Alaska. This change shall be made as rapidly as model 1903 rifles are available."[4]

There were a number of factors that went into the decision to retain the '03. The Model 1903 rifle was manufactured by the government-owned-and-operated Springfield Armory and Rock Island Arsenal, while the Model 1917 rifles were made by commercial contractors. Domestic labor unrest and strife was growing in the country, which meant that future production of military rifles by civilian firms could be hampered. Production machinery and tooling was already installed at Springfield and Rock Island. If the Model 1917 should become the standard rifle, it would require that the '03 manufacturing equipment be discarded or placed in storage and at least some of the Model 1917 machinery at Winchester, Remington, and Eddystone would have to be moved to Massachusetts, and possibly Illinois, in the event production of the Model 1917 needed to be resumed in the future.

Another factor that is seldom discussed is the Model 1903's finely adjustable rear sight as compared to the Model 1917 non-adjustable sight. While fine for combat use, the Model 1917's sight, despite some success at the 1918 National Matches, was not well suited for all-around marksmanship competition at which the '03 excelled. Since many army officers, including some of very high-rank, were avid marksmen, this likely played at least a minor role in the decision. In any event, there is little doubt that retaining the Model 1903 as standard was the correct decision.

Demobilization — The end of the war resulted in plans being made for the return of a large number of servicemen of the AEF. As related in the *History of Personnel Demobilization in the United States Army*:

> "With roughly 4.4 million men in uniform at the end of the war, and about 2 million of these overseas in the American Expeditionary Force, the United States government faced a logistical nightmare in organizing their return home in the face of the ravaging influenza pandemic and the requirements for ship transport. The first and comparatively easy part of demobilization had been discharging the 1.5 million men in training camps by February 1919. Relatively

American troops with their M1917 rifles boarding the SS *Panaman*, in Bordeaux, France, for return to the United States, April 10, 1919. *(Naval History and Heritage Command)*

H.A. Lewis, Company K, 113th Infantry, returning home on the SS *Floridian* May 17, 1919. Lewis was awarded the Distinguished Service Cross and the *Croix de Guerre* with palm. Note the M1917 rifles. *(National Archives)*

unprepared for demobilization, the US army adopted a policy of demobilization by unit beginning with 'casuals' not attached to a particular post. Troops in England and Italy headed across the Atlantic before the main body of infantry from France did. In the meantime, men were held at hastily constructed debarkation centers near French ports like Brest.

At first, there was a strict regime of drills and target practice while medical checks and paperwork filled spare time. The miseries of crowded debarkation camps prompted protests and petitions from soldiers still in France such that from February 1919, General John J. Pershing followed advice to increase recreation and had special honors laid on for American units by French towns and cities. Unruly U.S. servicemen wore out their welcome in cities like Paris, however. On 1 September 1919 the last American combat unit left France followed by the last American troops on 3 January 1920, with those from the U.S. intervention in Northern Russia following in April 1920."[5]

Occupation Duty — With the war over in Europe, some 250,000 American servicemen were posted to occupation duty in Germany, as part of the allied occupation of the Rhineland, beginning in late 1918 and continuing until 1923.[6] A March 31, 1919, memorandum from the American Expeditionary Forces to the Ordnance Depot in Tours, France, anticipated the potential demand for rifles to be used during the occupation:

"...Confirming telephone authority this date will ship unserviceable rifles and pistols to U.S. via Rhine. To meet possible requirements, request ten thousand 1903 and two thousand 1917 rifles to be held in reserve in France."[7]

U.S. doughboy in the immediate post-WWI era posing with a Model 1917 rifle with fixed bayonet. He is wearing a Belgian *Croix de Guerre* medal and was likely one of the soldiers assigned to postwar occupation duty. *(Photo credited to and colorized by Johnny Sirlande)*

Photo taken in Moselle Valley, Germany, January 18, 1919. Private Charles H. Purviance of the 310th Radio Field Signal Battalion on occupation duty. Men in the background are members of 301st Engineers, Co. D. Pvt. Purviance is armed with a Model 1917 rifle and fixed bayonet. Note the stacked Model 1903 rifles (with canvas breech covers) on the left. *(U.S. Army Signal Corps photo. Courtesy of Tom Laemlein)*

Inspection, Cleaning, Repair, Overhaul, and Storage — Even before the decision was made to retain the '03 as standard was finalized, arrangements were being made to return the large number of "American Enfields" that were in France but were no longer needed due to the rapid demobilization of the AEF. A November 25, 1918, memorandum from the Chief of Ordnance gave detailed instructions on the procedures to prepare the rifles for shipment back to the United States. These instructions included the following:

> "It is desired that during the return voyage to insure (these rifles) will reach the United States in the best possible condition.
>
> Rifles should be cleaned…and thoroughly cosmolined in accordance with Ordnance Pamphlets #1903 and #1917.
>
> In packing the U.S. Rifle, Model 1917, care will be taken to segregate these rifles by manufacturer as far as possible. The outside of the boxes will be marked 'Winchester,' 'Eddystone' or 'Ilion' according to the factories in which the rifles has been manufactured."[8]

It is rather curious that Ordnance mandated that the Model 1917 rifles be identified as to the manufacturer rather than simply having the rifles cosmolined, packed and shipped without regard to the maker. Perhaps there was some lingering uncertainty as to actual degree of interchangeability of the components between the manufacturers despite the previous directive from Ordnance Department that,

> "…rifles made by the Eddystone, Ilion and Winchester plants since January 1, 1918, may be regarded as equally interchangeable and, therefore, equally available and serviceable for use with United States forces anywhere, either at home or overseas."[9]

Plans had to be made as to the method of shipping the

U.S. Army soldiers on their way back to the United States, boarding the SS *Belgic* with their M1917 rifles, January 20, 1919. Most of of rifles brought home by the large number of returning veterans were sent to various ordnance facilities to be inspected, refurbished, and prepared for long-term storage. *(National Archives)*

rifles that were to be returned. A November 1918 memorandum from the Chief Ordnance Officer, A.E.F., to the Commanding Officer, Ordnance General Supply Depot No. 5, outlined the procedure for packing rifles in the absence of standard wooden shipping crates:

"Method of packing 1903 and 1917 U.S. Rifles, Caliber .30.

1. Memorandum has been forwarded by this Office to Construction & Maintenance Division suggesting that rifles be packed three in a bundle, each wrapped in burlap, the whole thoroughly tied, and the sights protected by blocks when necessary.
2. This is to be done only when there are no arms chests on hand.
3. The method is similar to that used by the British in like emergencies.
4. A few rifles of the 1st shipments made in this manner are to be returned to shipping depot for examination.
5. Please advise this office of the first few shipments made in this manner, so proper instructions to return them can be sent to the consignee."[10]

Many of the over 1.5 million rifles that were soon to arrive back in the United States from France, or still remained stateside, had seen hard use in overseas combat and in training camps and needed some measure of refurbishment and repair. Many of the other rifles that had never been issued remained in their packing crates for an extended period of time, often in less-than-optimal storage conditions.

In order to coordinate the future disposition of the rifles, it was necessary to determine how many of the weapons were on hand. A January 9, 1919, report from Chief of Small Arms Division, to the Chief of Supply detailed the following regarding the Model 1903 and Model 1917 rifles:

	Total	Total Delvd.	Bal. to be Delvd.	Cncld.
M1917	3,908,460	2,402,514	1,505,946	1,343,498
M1903	1,220,000	340,772	879,228	532,888

"The Eddystone Plant was instructed to stop production of rifles after January 11, 1919. It is also believed that all the rifles to be completed on orders as given above will be finished by February 28, 1919."[11]

It was decided to cycle the available Model 1917 rifles through several ordnance armories and arsenals for inspection, evaluation and for necessary cleaning, repairs and refurbishment. These facilities included Springfield Armory, Rock Island Arsenal, Augusta Arsenal, and Raritan Arsenal. Not surprisingly, the bulk of this work was performed by Springfield Armory and Rock Island Arsenal.

Ordnance directed that various military installations around the country report the number of rifles in inventory and specified to which arsenal they were to be shipped. The following are typical examples of the reports submitted by various camps:

Facility	Date	Model 1917 Rifles on Hand
Camp MacArthur, Tex.	March 12, 1919	18,000 to Rock Island Arsenal 10,728 to San Antonio Arsenal
Camp Wheeler, Ga.	March 12, 1919	4,815 to Augusta Arsenal
Camp Hancock, Ga.	March 15, 1919	11,697 to Augusta Arsenal
Camp Sevier, S.C.	March 17, 1919	15,644 to Rock Island Arsenal[12]

A May 15, 1919, Report from Camp Kearney, Calif., to the Army Ordnance Field Services related the following regarding the rifles on hand at that facility:

"...there are 17,579 model of 1917 and 394 model of 1903 rifles at this camp. 2,210 model of 1917 rifles on hand in warehouse ready for shipment to arsenal to be repaired. There are also in storage 4,720 model of 1917 rifles, which have been cleaned and repaired and classified as class one. There are 6900 new model of 1917 rifles in storage. It is estimated that about 2,500 model of 1917 rifles will be turned in to depot by organizations."[13]

Springfield Armory — A Springfield Armory "Report on the Status of Small Arms and Machine Guns" gave the following information on the Model 1917 rifles at the armory as of the week ending May 17, 1919:

Packed ready for issue or permanent storage
(New) 195
(Cleaned and repaired) 8,300
Awaiting or undergoing repair . . . . 7,265[14]

Springfield Armory Annual Reports from Fiscal Years 1919 to 1921 indicate the following regarding Model 1917 rifles:

Fiscal Year 1919 (July 1, 1918 to June 30, 1919)

Model 1917 rifle rebuilt at Rock Island Arsenal in the 1920s. *(Rock Island Auction Co.)*

Springfield Armory Clean and Repair Room, early 1920s, refurbishing Model 1917 rifles. *(Springfield Armory National Historic Site Museum. Courtesy of Alex Mackenzie, Curator)*

"Demobilization at camps and overseas has resulted in the turning in to stores of a very large number of U.S. Rifles, Model 1917. In accordance with the announced policy of the War Department these rifles are not to be issued to troops for the present (the U.S. Rifle, Model of 1903, being designated exclusively for such issue). Model 1917 rifles, which have been and will be turned in are to be divided into two classes:

(a) One million to be put in new condition. This work has been distributed between the Springfield Armory and the Rock Island Arsenal.

(b) The remainder be put into serviceable condition as far as possible. This work can be done in many different places.

The first quantity shipments of Model 1917 rifles were received in Springfield on February 25, 1919, the shipment numbering 5,000 arms. A new organization was built up in the plant for handling cleaning and repair work on these guns and this organization employs at the present time about 175 men. As the rifles are received, they are disassembled, all parts are inspected and defective or broken components are replaced. All metal parts are then refinished, reassembled, function fired, bore sighted, inspected and packed.

Approximately one thousand U.S. Rifles, Model of 1917, were cleaned and repaired each eight-hour day during the month of June, 1919. It is expected to increase the capacity of output to 1800 rifles or more per day by running a partial night shift.

Repairs – 26,720 U.S. Rifles, Model of 1917.[15]

Fiscal Year 1920

"There have been cleaned and repaired to date at this Armory since June 30, 1919, 470,410 rifles, Model of 1917. These rifles have been put out in condition equivalent to new. At one time during the year, we had reached a delivery of 2,500 of these guns per day, with 261 employees. Having reached the 2,500 schedule our production was decreased and men transferred from this department to assist in departments where work was more urgently needed. The work in this department has been delayed several times during the year due to irregular arrival of rifles to be repaired, and when restarted the organization had to be rebuilt to meet conditions.

In addition to cleaning and repairing the Model 1917 rifles, we received 551,785 pieces of the 1917 Rifle to be salvaged. From barrel and receiver assembled with sub-assemblies attached, to single components, we have salvaged and returned 198,000 pieces leaving 352,785 pieces in process of repairs. There are many thousands of parts in store yet to be salvaged. Orders were given for this work on April 29, 1920.

Repairs — 481,760 U.S. Rifles, Cal. .30, Model of 1917."[16]

Fiscal Year 1921

"...27,221 rifles, Model of 1917, were cleaned and repaired in the last fiscal year, the rifles being put in condition equivalent to new.

There have been cleaned and repaired about 75,260 Model of 1917 Bayonets.

In addition to cleaning and repairing Model of 1917 rifles, this Armory has salvaged 529,960 components for the Model 1917 rifle...All parts were packed and delivered to store in condition equivalent to new."[17]

Springfield Armory cleaned and/or repaired a total of 535,701 Model 1917 rifles between fiscal years 1919 and 1921. There are no records showing any Model 1917 rifles being processed at Springfield Armory after FY 1921, since the armory's focus returned to manufacture, and the overhaul work was left to other ordnance facilities. The only subsequent rebuild work on the M1917 rifle was not until FY 1942 when the armory rebuilt an additional 859 rifles for use during the Second World War.[18] As will be mentioned in Chapter 8, some M1917 rifles were inspected and rebuilt by Springfield and Rock Island in the immediate post-WWII era as well.

Rock Island Arsenal — As related in *A History of Rock Island and Rock Island Arsenal, Volume II*:

"When World War I ended, Rock Island Arsenal was at work on orders for 190,000 Model 1903 rifles, but on May 31, 1919, all production of new rifles was discontinued. Repair of Model 1903 and 1917 rifles was undertaken on a large scale and a schedule of a thousand rifles of each model repaired daily was laid out.

In 1920, the Armory (Shops B, D, and E) cleaned and repaired more than 1,250,000 parts received from Raritan Arsenal and overhauled more than one million rifles, pistols, revolvers, machine guns, and bayonets turned in from the field. The maximum number of weapons reconditioned in one eight-hour day was 6,800.

June 7, 1920, photo of 185,180 Model 1917 rifles in storage at Rock Island Arsenal. *(Kevin Braafladt, Army Sustainment Command Historian, Rock Island Arsenal)*

Springfield Armory Experimental Machine Shop, 1920s. Doing some unspecified work on a couple of Model 1917 rifles. *(Springfield Armory National Historic Site Museum. Courtesy of Alex Mackenzie, Curator)*

As troops returned from overseas, they were sent to various camps, where their equipment was turned in to the Camp Ordnance Officer, who prepared it for storage. There was no opportunity at the camps for making repairs, and most of the arms were cleaned as well as possible under the circumstances, they were greased and placed in packing chests. It was soon found necessary to process all this equipment through the arsenals and depots, and as a result, more than a million rifles, including U.S. rifle, cal. .30, Model 1903 and U.S. rifle, cal. .30, Model 1917, were processed at Rock Island."[19]

Unlike the Springfield Armory Annual Fiscal Year Reports, the above-cited publication gave detailed information on the inspection, cleaning and repair work done at the arsenal in the years following World War I. While some of the excerpts are a bit lengthy, it is important to outline the myriad steps taken to fully appreciate the work that went into preserving these rifles for potential future utilization.

"The first successful rifle cleaning machine ever used in the United States was built and installed at the arsenal in WWI. Until that time, all rifle and machine gun bores were cleaned by hand.

Rock Island Arsenal had been criticized by the Ordnance Office for excessive cost of repairing rifles as compared with the Springfield Armory. Investigation showed that Rock Island Arsenal was doing a more thorough job of repairing than Springfield Armory, and approximately 25 percent more was paid for labor. Based on the investigation, methods were adopted at Rock Island that increased production by 45 percent and reduced personnel by 25 percent. This naturally lowered the quality of work, but the rifles thus repaired were in a satisfactory condition to issue.

In removing cosmic from rifles, it had been the practice to wash them in gasoline. Instead, ovens were installed in which approximately 2,000 rifles could be placed on regular racks containing 50 rifles each. They were heated to 200 degrees fahrenheit for one hour, causing all grease to melt and run into a container. The grease was filtered and used again. An average of 1,506 gallons of petroleum was reclaimed each month.

After removal from the ovens, the rifles were washed in a tank of oaklite composition, linseed oil and water. The new method netted a savings of $2 per 100 rifles and increased production 100 percent.

Tanks with steam coils were installed for coating rifles with cosmic after they had been cleaned and repaired. The cosmic was kept at 200 degrees fahrenheit, and the entire rifle was submerged to insure all parts being coated. One man could coat 1,600 rifles in an eight-hour day.

The bore-sighting method of targeting rifles was adopted in February of 1920, in order to maintain a production of 4,500 per day. The cost of repairing was reduced also by browning the barrel and receiver assembly. Browning equipment with a capacity of 3,000 rifle barrels a day was installed.

With the emphasis on economy, the Small Arms Department again made progress in that respect in 1927, when the department was consolidated on one floor to save heat and the handling of material.

...Another rearrangement of working areas brought additional economies in 1929. The total of employees engaged in cleaning and repairing small arms was reduced from 38 to 31 in fiscal 1930, and the direct labor cost of reconditioning Model 1917 rifles was lowered to 5.3 cents per rifle in 1929 to 4.6 cents in 1930, The following year the cost was given as 4.5 cents.

...In 1919, the Rock Island armory had 792 machines, acquired during the war, and 465 other machines, most of them new, selected after the Armistice from private plants to replace worn or obsolete equipment. To avoid a repetition of the slow building up experienced in the late war, Rock Island Arsenal resolved to prepare for a future emergency at once.

Much of the work in 1920 was devoted to that task. Every machine was set up and adjusted, sealed and coated with rust-preventing grease. When the job was done, the small arms shops had an estimated capacity production of 1,800 rifles each 24-hour working day.

After the heavy overhaul orders of 1919 and the caring for machinery of 1920, work in the department declined. The number of employees decreased year by year until a low of 14 civilian employees was reached in 1924."[20]

Springfield Armory and Rock Island Arsenal did exemplary work in cleaning and repairing the huge number of Model 1917 rifles during this period. However, as the guns were sent to various depots for warehousing, it later became painfully obvious that the procedures employed to keep the weapons in serviceable condition while in long-term storage were lacking. As early as 1927, there were myriad problems encountered when some of these guns were unpacked, including rusted metal and badly pitted bores. As will be subsequently discussed, this became critical when large numbers of sorely needed Model 1917 rifles were uncrated at the beginning of the Second World War and were found to be in seriously deteriorated condition after languishing in storage for a couple of decades. As discussed in *A History of Rock Island and Rock Island Arsenal, Volume II*:

"...Reports to the Chief of Ordnance by Rock Island Arsenal showed deterioration of arms and machines to be the result of economy in appropriations:

'In the reconditioning of 200,000 rifles, on which work this section has been engaged since October 1927, it was found that considerable more work had to be done than was contemplated, owing to the fact that a very large number had been placed in storage without having been slushed with grease; on others a good coat of grease had been applied to the outside but none in the bore of the barrel and, for that reason, many were badly rusted and pitted, and the barrels could not be saved,

The usual procedure of cleaning rifles three times, the third cleaning by hand, with a fine steel brush, was found insufficient. It was therefore, necessary to continue the hand process as long as there was a possibility of saving the barrel. Careful inspection after each successive cleaning was made. As many as seven cleanings were sometimes restored to, which resulted ultimately reducing the rejections from 20 per cent to 7½ per cent.

As work progressed it became advisable to disassemble rifles found to be unserviceable and to dip serviceable parts in group assemblies in grease, repack and transfer to store. The work of reconditioning was further facilitated by the manufacture and installation of a new barrel cleaning machine at a cost of $2,225.00 and by the introduction of an improved method of oiling the bore and chamber.

...At the close of the fiscal year an average maximum of 1,250 accepted rifles per day was attained."[21]

Rock Island continued cleaning, repairing, rebuilding and packing Model 1917 rifles at least as late as 1932, over a decade after Springfield ceased the practice. Unlike Springfield, which was primarily devoted to manufacturing, Rock Island was not making small arms, so the arsenal relied on overhaul work, even during the lean years of The Great Depression.

Columbus General Depot — In addition to the well-known government facilities mentioned above (Springfield Armory, Rock Island Arsenal, Augusta Arsenal, and Raritan Arsenal) the Ordnance Department also utilized Columbus General Depot in Columbus, Ohio, to degrease and recondition 473,000 Model 1917 rifles beginning in 1933 and intended to continue until 1938. A December 19, 1933, memorandum from the War Department, Ordnance Office in Washington, D.C., to the Ordnance Officer, Columbus General Depot, stated the following:

"1. Degreasing rifles using hot recoil oil and soda solution is approved.
2. It is desired that the reconditioning of four hundred seventy-three thousand (473,000) Rifles, cal. .30, M1917, be completed within a five-year period.
3. The use of Dearboline was suggested by this office as a substitute for gasoline only because it is not a fire hazard. For your information, Dearboline is a liquid similar to gasoline and is used the same way, preferably in a galvanized iron pan, long enough to take a rifle and 12-in. wide and 10-in. deep. Enough solvent is put in the pan t cover one rifle, when it is scrubbed with a stiff bristle brush.
4. If the method of degreasing rifles by dipping in hot recoil oil and cleaning with soda solution proves satisfactory, there will be no need for further experiments."[22]

It is not known what markings, if any, were stamped on the rifles refurbished at Columbus General Depot during this period.

Even with the improved methods of preparing the rifles for long-term storage, there were tens of thousands of Model 1917 rifles that had not been properly prepped, and continued to deteriorate in their packing crates at numerous ordnance facilities.

Rock Island Arsenal "Sporting Rifle" Conversion Proposal — With the tremendous number of Model 1917 rifles in storage and with little likelihood of their utilization in the foreseeable future, there was a novel and potentially very useful idea conceived at Rock Island Arsenal.

In the early 1930s, the nation was in the grip of The Great Depression, and funding for all sorts of government programs became increasingly problematic. Rock Island Arsenal proposed a unique way to provide much-needed funds to help retain the work force. Major W.F. Vander Hyden, Chief Supply Officer of Rock Island Arsenal, sent a letter to the Chief of Ordnance on January 13, 1933, containing an interesting proposal. Major Vander Hyden explained that the arsenal had received numerous inquiries from civilians as to the possibility of purchasing a "sporterized" Model 1917 rifle. It was pointed out that Springfield Armory, for almost a decade, had been making various types of sporter models for the Model 1903 Springfield, such as the "M1903 N.R.A. Sporter rifles," that could be available to civilians under certain conditions.

The letter to the Chief of Ordnance included two photos of a Model 1917 rifle that the arsenal had modified "to sporter" configuration. It was stated that the approximate cost of such modification would be $5.00, which would result in a price of $12.50 plus an additional $1.35 for packaging and shipping. Major Vander Hyden emphasized that such a program would likely enable Rock Island to retain many of the skilled workers employed at the arsenal. He also requested that the Chief of Ordnance contact the Director of Civilian Marksmanship for assistance in mar-

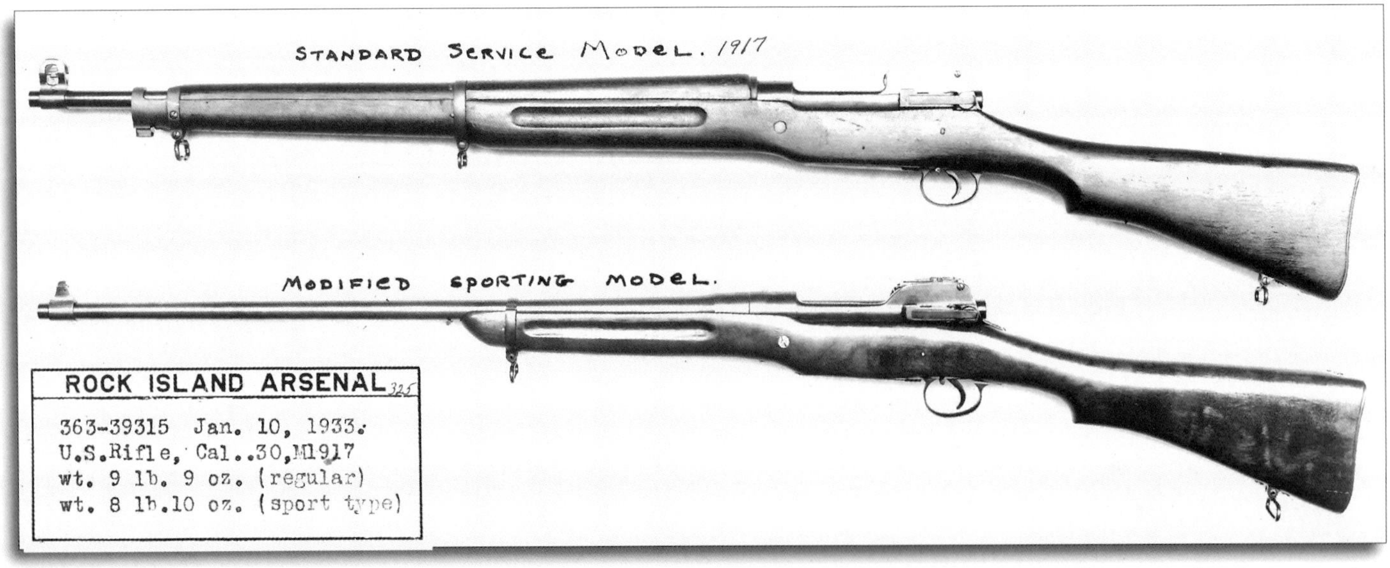

Rock Island Arsenal photograph dated January 10, 1933, showing a standard Model 1917 rifle above an example converted to the proposed "sporter" configuration. *(Kevin Braafladt, Army Sustainment Command Historian, Rock Island Arsenal)*

keting the modified guns should the proposed program be approved. Vander Hyden further suggested that no additional features, such as commercial sights, be offered. His stated rationale was: "This is necessary in order to avoid a great many nuisance orders and a vast amount of correspondence that would otherwise result."[23]

As word of the proposed program began to leak out, civilian arms-making firms were understandably against the idea. The prevailing economic conditions of the Depression negatively impacted the sale of commercial rifles. The concept of the government, in effect, competing with private companies by offering a low-cost sporting-type rifle to the public was not viewed favorably. It didn't take long for this displeasure to be registered with the respective congressmen and senators who then prompted a congressional committee meeting on the subject. The result was the Ordnance Department being advised that the proposal be denied.

The Chief of Ordnance was well aware of the positive impact of such a program providing much-needed funds to Rock Island. However, the powers that be nixed the idea, and he responded to Major Vander Hyden, "...there has been a considerable amount of agitation recently against the participation of Government in business, and undoubtedly the arms companies would consider that this provision of a large number of sporting rifles for sale at a very cheap price would be interference with private business."[24]

This controversy also resulted in the production of Model 1903-based sporter rifles at Springfield being stopped as well, as evidenced by a February 16, 1933, memorandum from the National Board for the Promotion of Rifle Practice to the Assistant Secretary of War, which stated, in part:

> "In 1923 this office was authorized to sell to the members of the National Rifle Association , the U.S. Rifle, caliber .30, M1903, Style NRA, commonly called the Springfield Sporter. This rifle is a modified Springfield M1903, the modification being made by the Ordnance Department intended to produce a Sporting Rifle of unusual quality for sale to members of the National Rifle Association and to stimulate the production of similar rifles by commercial manufacturers... In some cases the commercial manufacturers now consider that the Government, by the further sale of these rifles, is entering into competition with them and that its sale should be discontinued... The National Rifle Association and the Ordnance Department have been consulted and, in both cases, have stated that they have no objection to its discontinuance."[25]

Model 1917 Rifles with Benicia Arsenal Modifications — Noted researcher Frank Mallory discovered an interesting document dated October 12, 1934, instructing Benicia Arsenal to make certain modifications to standard Model 1917 rifles for sale primarily to army officers. This was done despite the unequivocal rejection, just a year earlier, of the concept of armory conversions of Model 1917 rifles to sporter configuration for sale on the commercial market. As stated in the memorandum:

> "HDQRS. Benicia arsenal, Benicia, California, October 12, 1934. To: Chief of Ordnance, War Department, Field Service, Munitions Bldg., Washington, D.C.
>
> 1. From time to time this office receives requests from Officers and others authorized to purchase arms, to modify the M1917 rifle, to take either the Pacific or Lyman sights together with certain modifications of the stock.
> 2. There are three separate operations, the estimated cost of which as follows:
>   (1) Modify stock, fit Krag lower band, with original sights, cost $5.07.
>   (2) Modify receiver for Pacific sight, modify stock and fit Krag lower band, estimated cost $6.56.
>   (3) Modify receiver for Lyman sight, modify stock, fit Krag lower band, estimated cost $8.57.
>
> The price of each of these operations, covers the cost of the Krag lower band at 7 (cents).
>
> 3. Should the Ordnance Officer be unable or unwilling to furnish a revolving fund for modifications of these rifles, authority it requested to perform this work when approval is considered desirable, charging the cost to our contingent allotment, and re-crediting the contingent allotment with the actual cost of the operation."[26]

Apparently, this program didn't raise the hackles of the civilian arms industry since it pertained primarily to army officers and didn't negatively impact potential commercial buyers to any great extent. The number of rifles converted by Benicia is not known but was presumably rather limited given the dearth of verified extant specimens, although some are likely to have been erroneously identified as simply civilian "sporterized" rifles.

Interwar Commercial Modifications — When the Model 1917 rifle production contracts were cancelled soon after the Armistice, two of the manufacturers, Winchester and Remington, sought to utilize their manufacturing experience and utilize some production machinery still on hand to make sporting rifles for the civilian market, based to some extent, on the Model 1917. Eddystone's venture into arms making ended with the cessation of Model 1917 rifle production.

While not intended for the civilian market, Remington also made two military rifles for foreign governments in the mid-1930s, incorporating the same type of receiver as used on the some of the M1917-influenced sporting rifles. These rifles will be profiled in "Variations," Chapter 9 of this book.

Postwar Distribution — While the Model 1903 replaced the Model 1917 as the standard U.S. military service rifle, there was still some utilization of the latter during the 1920s and 1930s. As discussed in the ordnance publication *Project Supporting Paper, Rifle, U, S, Caliber .30, M1917*:

"As these guns (M1917 rifles) were cleaned, they were placed in war reserve. Thereafter, they were made available and issued to various organizations such as the Civilian Military Training Corps and the Reserve Officer Training Corps and to State Guards. To these organizations they were made available as a Federal loan. In 1936, members of the National Rifle Association were permitted to purchase the rifle.

All of the rifles issued or sold were subject to replacement by the War Department if found defective. As a result, during the years that followed many of the rifles were returned for correction of defects, which appeared to have developed while the rifles were in storage. It became necessary to manufacture new parts to replace defective ones."[27]

An interesting ordnance document reflected the number of Model 1917 rifles sold by the government in fiscal years 1921 through 1940. As mentioned above, beginning in 1936, Model 1917 rifles were sold via the Director of Civilian Marksmanship (DCM) to NRA members:

"1921	1923	1924	1925	1927	1929	1930	1931	1932
2	6	7	2	3	81	87	86	874
1933	**1934**	**1935**	**1936**	**1937**	**1938**	**1939**	**1940**	**Total**
2,281	2,739	4,744	645	6,703	6,624	6,596	7,668	44,148"[28]

Number of sale	Date of contract	Articles sold	Value
First	Jan. 7, 1924 [1]	5,000 Enfield rifles	$165,850.00
		5,000,000 caliber .30 cartridges	104,520.00
	Jan. 12, 1924 [2]	8 DH4 airplanes	
		24 machine guns	128,816.00
Total			419,186.00
Second	Jan. 14, 1924 [1]	2,500 pistols, caliber .45	68,775.00
		2,500 pistol holsters	4,325.00
		20,000 pistol magazines	16,200.00
		10,000 pistol ammunition	1,523.00
	Mar.—, 1924 [3]	2,500 pistol belts	2,950.00
Total			93,773.00
Third	Jan. 18, 1924 [1]	5,000 Enfield rifles	185,850.00
Fourth	Jan. 19, 1924 [1]	Bombs, airplanes, machine guns and engine spare parts, and accessories.	
	Jan. 22, 1924 [2]		104,202.79
Fifth	Jan. 25, 1924 [1]	1,000 pistol holsters	1,757.50
	Mar.—, 1924 [3]	1,000 belts, pistol	1,180.00
Sixth	Feb. 8, 1924 [1]	5,000 Enfield rifles, caliber .30	
		5,000 Russian rifles, caliber 7.62 mm	
	Feb. 6, 1924 [2]	9 machine guns and belts	
		3 DH4 airplanes	255,938.35
Seventh	Feb. 7, 1924 [1]	50,000 tracer ammunition	6,675.34
Eighth	Feb. 8, 1924 [1]	100 Enfield rifles	
		250 bombs, demolition	
		10,000 caliber .30 ammunition	15,508.50
Ninth	Feb. 27, 1924 [1]	1,000,000 pistol ammunition	15,230.00
Tenth	Feb. 28, 1924 [1]	500 holsters, pistol	880.00
	Mar.—, 1924 [2]	5,000 pistol belts	590.00
Total			1,470.00
Eleventh	Mar. 12, 1924 [1]	5,000 Enfield rifles	185,850.00
Twelfth	Not consummated.	15,000 Russian rifles stored in Manila. To be sold at $3.60 plus cost of transportation from Manila to El Paso.	
Total value of munitions sold.			1,286,611.48

[1] Ordnance. [2] Air Service. [3] Quartermaster.

April 24, 1924, chart from U.S. Senate Document No. 104 showing the military equipment purchased by Mexico in 1924, including Model 1917 rifles. (Man at Arms *magazine. Courtesy of Stuart Mowbray)*

Rare photo of a Shanghai Volunteer Corps unit armed with Model 1917 rifles. The caption on the photo reads, "A unit falls in for inspection, with a U.S. Marine Corps Sergeant in the front. Note the armored-car in the background, c.1937, Shanghai, China." *(Naval History and Cultural Command. Courtesy of Rear Admiral J.P. Walker, USN, Ret.)*

As can be seen, there were no sales listed for a couple of years. When the sale of rifles through the DCM was suspended after FY 1940, there were a number of complaints registered by NRA members. One such example was a letter from an irate member sent to President Franklin Roosevelt complaining about the situation. Not surprisingly, it was forwarded to the War Department for a response. A February 21, 1941, letter from the War Department, Director of Civilian Marksmanship, was sent to the unhappy individual and contained the following explanation:

> "The reasons for the decision to restrict the sale of ordnance supplies (including rifles) to civilians was given very full consideration by the War Department before it was made. Because of the necessity for conserving our resources of arms and ammunition for the proper training of the active services and to relieve the arsenals of much extra work during the time when all their energies were being devoted to essential items of National Defense, it was considered necessary to order this restriction. Conditions have not improved to warrant resumption of sales of these items."[29]

Sales to Mexico — To digress a bit, in 1924, the War Department approved the purchase of a significant amount of military equipment by the government of Mexico, including 20,100 Model 1917 rifles. These rifles were designated as "constructive surplus," indicating that they were not considered as necessary to meet mandated requirements.[30]

In 1929, Mexico received an additional 3,000 Model 1917 rifles along with 3,000 Model 1917 bayonets and scabbards, 3,000 metal oil and thong cases and 1,500,000 rounds of .30-06 ammunition shipped from the San Antonio Arsenal.[31]

Shanghai Volunteer Corps — The Shanghai Volunteer Corps (SVC) was a multinational volunteer force organized for protection of the Shanghai International Settlement and was in existence from 1853 to 1942.

It was mobilized in 1900 for the Boxer Rebellion and again in 1914 for the First World War, although the German and Austro-Hungarian contingents were disbanded when China declared war on Germany.[32] The unit's final mobilization was in 1937 for the Sino-Japanese War. In 1940, it took over occupation duties for the International Settlement when the British withdrew. It was disbanded in 1942 with the Japanese takeover.[33]

In 1934 and 1935, some 794 Model 1917 rifles were sent to the Shanghai Volunteer Corps, with some additional rifles purportedly sent later. Table #1 in this book has a large number of Model 1917 rifles listed, including a roster of the known SVC Model 1917 rifles. The serial number of the SVC rifles ranged from 23,838 to 135,0802.[34] Presumably, the Japanese confiscated the Model 1917 rifles when they occupied Shanghai, but their subsequent disposition is not known.

Snabb Semiautomatic Conversion Prototype — Although the M1 Garand rifle was adopted in 1936, there was some experimentation with modifying the Model 1917 rifle to be capable of semiautomatic operation. The rationale behind this was to use at least a portion of the large number of Model 1917 rifles in storage for conversion into weapons capable of semiautomatic operation that would be better suited for modern warfare. A similar idea had been concocted just prior to World War I by converting Model 1903 rifles to semiautomatic operation. Circa 1916, Springfield Armory and Rock Island Arsenal fabricated prototypes that were tested with less-than-spectacular results.

In 1938, a Swedish firearms engineer developed a mechanism, dubbed the "Snabb conversion," which would purportedly permit any bolt-action rifle to be converted to a semiautomatic. In addition to the Model 1917, examples were made for the Model 1903 Springfield, Mauser, and Mannlicher rifles. A Model 1917 rifle altered for the Snabb mechanism was tested at Springfield Armory. As stated in the *Project Supporting Paper, Rifle, U.S. Caliber .30 M1917 1917–1945*:

> "In January 1940, one Snabb Company offered a description and supporting documents covering the conversion of any model bolt-action rifle into semi or full automatic. Several tests were made (at Springfield Armory) of a M1917 Rifle, which had been so modified to permit semi-automatic fire. As the tests were not satisfactory, the Ordnance Committee in May 1940, disapproved the proposed modification. No further action was taken."[35]

Interestingly, there were other accounts that didn't correspond with the conclusions of the Ordnance Committee:

> "The Snabb was well balanced, simple and rugged design that was intended as a low-cost alternative to replacing large stocks of bolt action rifles with new semi-automatics. Though mechanically successful, the higher than anticipated price, coupled with the outbreak of the Second World War, ruined the Snabb's chances"[36]

Perhaps the Ordnance Department wasn't interested in any other type of semiautomatic rifle since it was firmly committed to the M1 Garand, and the ordnance test results may not have been indicative of the Snabb's actual performance, but this is conjecture. On the other hand, similar allegations were raised during this same period regarding the semiautomatic rifle designed by Melvin Johnson Jr., which was tested several times by Ordnance and eventually rejected. As events transpired, the Model 1917/Snabb conversion is now but an obscure footnote to the story of the "American Enfield."

It is interesting to note that when the M1 Garand rifle was adopted in 1936, many considered it to be a badly flawed weapon for a variety of reasons, some valid and some not. An article in the October 1940 issue of *American Rifleman* magazine, titled "Garand Critique," written by an NRA member and WWI veteran, took the new M1 rifle to task and clearly held the Model 1917 in the highest regard. The article started out:

> "Lord knows something MUST be done. I spent an afternoon on the range with some choice copies of Mr. Garand's great mistake recently. And with some 25 years' experience with military firearms, two years as a sniper and some combat experience... it is my prophecy that unless this rifle is altered...any unit with it will be defeated by a trained determined foe that has ordinary bolt-action rifles.
>
> I have been writing and encouraging others to write their Congressmen and Senators to require the

Right and left-side views of the Model 1917 rifle altered for the Snabb semiautomatic conversion mechanism.

(Rock Island Auction Co.)

Filipino soldiers armed with Model 1917 rifles. (LIFE *photo archives. Carl Mydans, photographer)*

Infantry Board to make a…test of this rifle or to recommend the manufacture of the Model 1917, which is, at the present time, our best war rifle (absolutely fool proof and reliable!). I threw away a scope-sighted Springfield as a sniper, for its superior reliability for military purposes. Please do not give up the fight for a reliable autoloading rifle (or the model 1917)."[37]

Gen. MacArthur, the Philippines, and the M1917 Rifle — In 1935, General Douglas MacArthur retired from active service with the U.S. Army and accepted a position as military advisor to the President of the Commonwealth of the Philippines. One of MacArthur's first actions was to obtain better weaponry for that nation's armed forces, including service rifles. General MacArthur was obviously well acquainted with the large number of Model 1917 rifles sitting in storage in several ordnance facilities in the United States. MacArthur requested that the Philippine Government be permitted to purchase a substantial number of Model 1917 rifles. The Ordnance Department was not averse to the request since the number of these rifles on hand greatly exceeded the projected requirements for war reserve material. The fact that the request came from Douglas MacArthur, with his many high-level contacts in the U.S. Army, undoubtedly aided his cause.

As related in the ordnance publication *Project Supporting Paper – Rifle, U.S. Caliber .30, M1917*:

"Shortly thereafter, among other things, requisitions for M1917's were received by the Ordnance Department through the Bureau of Insular Affairs." The first call was for the sale of 25,000 M1917 rifles complete with bayonets, scabbards, gun slings and basic spare parts at a cost of $16.65 each. Shipment was made 8 April 1936. On 8 July 1936, another order was filled for 50,000 rifles only, with spare parts. The charge was $9.05 per rifle. Other requisitions followed. Field Service Stock Control records list a quantity of 220,495 M1917 rifles in the category of known losses, of which transfers to the Philippine Commonwealth were a substantial part. In some cases, material was sold, in other cases it was loaned. These rifles were well received in the Philippines except for some trouble, which arose from extractors."[38]

The problem noted with the extractors was not new, as there were numerous reports from France regarding the problem. Eventually, some 70,000 extractors of an improved design and more durable steel were sent to the Philippines.[39]

A July 1, 1937, memorandum from Lt. Col. Charles A. Walker, Ordnance Department, to Lt. Col. Kenneth Buchanan, Office of the Assistant Chief of Staff, C-4 gave the following information regarding war material sent to the Philippines as of July of 1937:

"Material Sold to the Philippine Commonwealth
75,000 U.S. Rifles, Cal. .30, M1917
25,000 Bayonets, M1917
25,000 Scabbards. Bayonet, M1917
25,000 Gun slings, M1907
Basic Spare Parts for 75,000 Rifles
100 Carts. Machine Gun, M1917
Material Loaned to the Philippine Commonwealth
25,000 U.S. Rifles, Cal. .30, M1917
25 Sets Basic Spare Parts for above
886 Guns, Machine, Browning, Cal. .30 M1917 with pertaining accessories
21,000 Cases, oiler and thong
12 Guns, 37mm, with accessories
4,500 Revolvers, Colt Cl. .45, M1917
2 3" Field Guns and Carriages, M1905, with accessories."[40]

A subsequent Ordnance Department memorandum dated October 26, 1938, gave approval for the shipment of additional large numbers Model 1917 rifles to the Philippines:

"Subject: M1917 Rifles for Philippine Army
To: Commanding Officer, Rock Island Arsenal, Rock Island, Illinois

1. It is anticipated that sufficient space can be made available on the Ludington to sale [sic] November 26th from New York to accommodate 3,000 to 4,000 cases of rifles, M1917 for the Philippines.
2. The exact information as to availability of space will not be known until November 7th, on which date you will be informed by radio if shipment is to go forward. In the meantime, it is requested that you make such plans as may be necessary to expedite such a shipment in order that it may reach New York at the earliest practical date."[41]

Actually, in addition to wanting to assist General MacArthur and the Philippine Government, the Ordnance Department had something of an ulterior motive in facilitating the sale of the M1917 rifles. A July 15, 1936, memo from the Office of the Chief of Ordnance to the Adjutant General stated:

"The recent sale of rifles, caliber .30, Model 1917 to the Philippine Government has resulted in the crediting to 'Replacing Ordnance and Ordnance Stores' the sum of $452,000. It is recommended that this sum be authorized for use in the manufacture of semi-automatic rifles, caliber. 30."[42]

The M1 Garand rifle had been adopted just a few months prior to the above memo, but very little money had been appropriated for production of the weapon. Chief of Ordnance Major General W.H. Tschappat was looking in every nook and cranny for funding to set up the production line for the new rifle, and the sale of the "American Enfields" seemed like an excellent place to start. Unfortunately, the Adjutant General responded to Tschappat's missive with the following memo:

"The recommendation that the sum of $452,000 received from the Philippine Government be authorized for use in the manufacture of semi-automatic rifles is not favorably considered. The available funds will be employed to procure…43 MG's H.B., Cal .50, $191,104 and, $257,896 to apply on aircraft armament."[43]

By the time the last shipment was made in September of 1939, a total of 220,495 Model 1917 rifles had been sent to the Philippines either by purchase or loan. As will be seen in the next chapter, some of these rifles saw extensive combat use in the Philippines just a couple of years later in the hands of Filipino soldiers and guerillas battling the Japanese invaders.

Chapter 7

World War II

Germany's invasion of Poland on September 1, 1939, started another massive conflict little more than 20 years after the conclusion of the woefully misnamed "War to End All Wars." As more of the dominoes began to fall, the already prominent isolationist sentiment in the United States grew stronger, and there was a pervasive mood that American young men should not be sent once more to fight "Europe's war." Nevertheless, the U.S. Government was aware of the potential of all, or most, of Europe turning into a totalitarian fascist state. President Franklin Roosevelt was determined to help our allies, especially Great Britain, without jeopardizing the United States' policy of neutrality. As outlined in the National Archives "Milestone Series – The Lend-Lease Act:"

> "In July 1940, after Britain had sustained the loss of 11 destroyers to the German Navy over a 10-day period, newly elected British Prime Minister Winston Churchill requested help from President Roosevelt. Roosevelt responded by exchanging 50 destroyers for 99-year leases on British bases in the Caribbean and Newfoundland. As a result, a major foreign policy debate erupted over whether the United States should aid Great Britain or maintain strict neutrality.
>
> In the 1940 presidential election campaign, Roosevelt promised to keep America out of the war. He stated, 'I have said this before, but I shall say it again and again and again; your boys are not going to be sent into any foreign wars.' Nevertheless, FDR wanted to support Britain and believed the United States should serve as a 'great arsenal of democracy.' Churchill pleaded, 'Give us the tools and we'll finish the job.' "[1]

Great Britain — One of Britain's most pressing needs at the time was small arms, especially rifles, to replace the weapons lost during the Dunkirk evacuation. The large number of Model 1917 rifles sitting in ordnance depot warehouses in the United States could be extremely helpful to our English allies. The British Purchasing Commission had been established in the United States on November 7, 1939, to facilitate acquisition of war material. However, the United States was unable to simply sell the rifles to the British Government due to provisions of the Neutrality Act, so a "workaround" was devised to facilitate sales through a non-governmental entity. The U.S. Steel Export Corporation would purchase the rifles from the American Government, and the rifles would then be resold to Great Britain.

This arrangement initially resulted in approximately 500,000 Model 1917 rifles and 130 million rounds of .30-06 ammunition being sold to the United Kingdom, with the first shipment on June 13, 1940. Another 80,000 went to Canada. The estimated value of the 500,000 rifles sent to England was $3,750,000.[2] Within the next two months, an additional 615,000 Model 1917 rifles were subsequently shipped to Great Britain. According to the Ordnance Department book *Planning Munitions for War*: "Between June 1940 and February 1941, about 1,135,000 Enfields and 188,000,000 rounds of .30-caliber ammunition were sent to Great Britain."[3] Other weapons provided to the British during this period were 25,000 Model 1918 Browning Automatic Rifles as well as some Lewis and Model 1917 Browning Machine Guns.

Most of the Model 1917 rifles that arrived in England were slated for issue to the British Home Guard beginning on July 8, 1940.[4] The Home Guard was an armed citizen militia supporting the British Army from 1940 to 1944. There were some 1.5 million local volunteers otherwise ineligible for military service, such as those who were too young or too old to join the regular armed forces.[5]

Since the Model 1917 and other rifles sent were of "non-standard caliber" compared to other British weapons, a red band was generally painted on the fore-end, often with ".30/06" in black, to distinguish them from standard British weapons chambered for the .303 cartridge. This was doubly important with the Model 1917 since it so closely resembled the .303 caliber Pattern 1914 rifle.

It was indeed very fortunate that the United States had such a large inventory of Model 1917 rifles to assist our

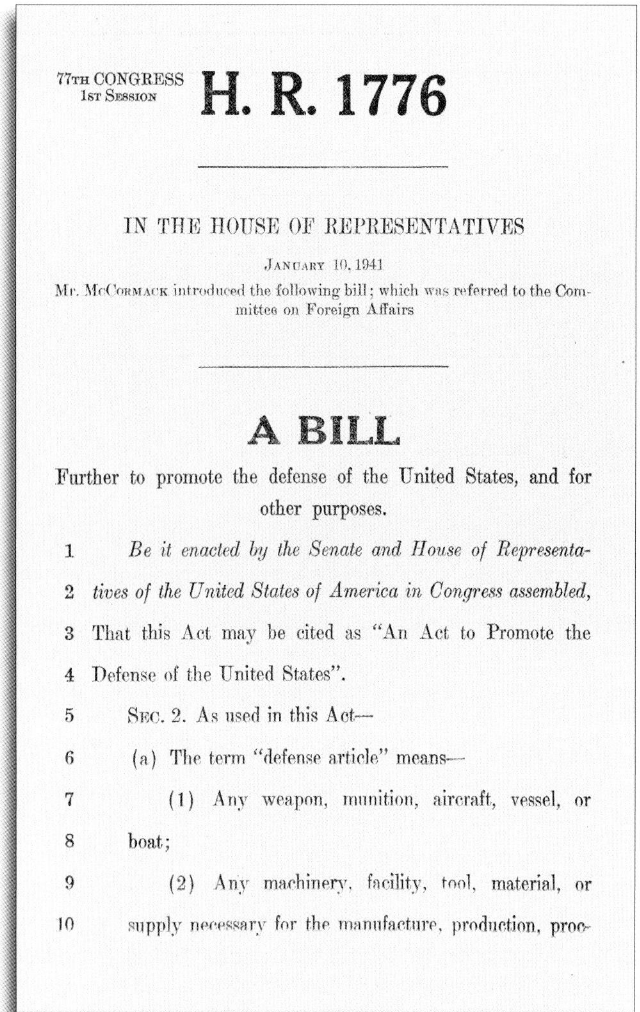

77TH CONGRESS
1ST SESSION

H. R. 1776

IN THE HOUSE OF REPRESENTATIVES

JANUARY 10, 1941

Mr. McCORMACK introduced the following bill; which was referred to the Committee on Foreign Affairs

A BILL

Further to promote the defense of the United States, and for other purposes.

Be it enacted by the Senate and House of Representatives of the United States of America in Congress assembled, That this Act may be cited as "An Act to Promote the Defense of the United States".

SEC. 2. As used in this Act—

(a) The term "defense article" means—

(1) Any weapon, munition, aircraft, vessel, or boat;

(2) Any machinery, facility, tool, material, or supply necessary for the manufacture, production, proc-

Bill H.R. 1776 dated January 10, 1941, introducing the Lend-Lease Act legislation, which was passed on March 11, 1941.

British Home Guardsmen in inspection formation with Model 1917 rifles, September of 1942. *(ForcesWarRecords.co.uk)*

British Home Guardsman undergoing bayonet training with a M1917 rifle. *(www.historicalfirearms.info)*

Royal Army Ordnance Corps recruits with drill instructor, Woolwich Barracks, London, August 1940, with M1917 and SMLE rifles. *(www.historicalfirearms.info)*

allies as well as our own armed forces. However, as correctly stated in the Ordnance Department book *Procurement and Supply*, "Transfer of this material took a big bite out of the U.S. Army's small arms stockpile."[6] Unfortunately, an undetermined number of the rifles never made it to Great Britain and ended up on the bottom of the Atlantic Ocean due to the prevalent German U-Boat attacks during this period.

Lend-Lease — The subterfuge of using a civilian firm to arrange for the transfer of weapons to allied nations was a rather cumbersome endeavor. In order to streamline the process and still retain a fig leaf of neutrality, the Lend-Lease Act was passed by Congress on March 11, 1941, which permitted the United States Government to directly supply weapons to allied nations. Under the terms of the act, the American Government was permitted to "...lend-lease or otherwise dispose of arms" as well as other supplies needed by any country whose security was vital to the defense of the United States.[7] The passage of the act was especially important for the Brits, as they were running low on money to purchase arms. Since the terms of the legislation stated that the weapons were being "lent" to the recipient nation, no payment was required.

Following passage of the act, an additional 119,000 Model 1917 rifles and 50,001 Model 1917 bayonets were supplied to the British. Other weapons provided to the British Army and/or Home Guard via Lend-Lease from 1941 to 1945 included:

- 38,001 M1 Garand rifles
- 25,362 M1 Carbines
- 1,196,706 .303 caliber No. 4 Lee-Enfield rifles (made by Savage Arms)
- 38,424 .55-caliber Boys anti-tank rifles (made by Inglis in Canada)
- 39,592 .45 caliber pistols
- 650,551 .38 caliber S&W revolvers
- 651,086 Submachine guns[8]

Dr. D.M. Clarke's well-researched and interesting PhD thesis at Cranfield University in Cranfield, England, titled "Arming the British Home Guard, 1940–1944" makes these observations pertaining to the U.S. Model 1917 rifle:

> "The first weapons to reach the Kent Home Guard in large numbers were consignments of American (M1917) rifles, known as Springfields [sic], Remingtons or Eddystones, depending on which company had manufactured them. All these rifles fired the standard American .300 rimless cartridge, which was in very short supply in Britain at this time.
>
> Clifford Shore, the Home Guard turned sniping instructor, took every opportunity to examine and shoot any small arms that came his way. Of the M1903, he recorded, 'This Springfield rifle is seldom seen in England, and my personal knowledge of it is not extensive. But I have used it enough to know it is a first-rate rifle.' He says of the M1917, 'The M17s which I used in England in the years 1940–43 [his Home Guard service] were really splendid weapons; I never came across a bad one. In certain quarters, they were not popular, but that can be primarily and summarily dismissed with the one word 'ignorance.' Many of the Home Guard in 1940 did not like the M17; usually these fellows were veterans of the First World War and the word rifle to them meant only the S.M.L.E. ... The higher velocity .300 cartridge gave slightly improved ballistics than the .303 cartridge in the P14, and I should say that the M17 was probably the most accurate rifle I have ever used.
>
> The tenth edition of former RSM, now Captain, A. Southworth's *The Home Guard Pocket Manual* was published in January of 1944. The only bolt-action rifle featured was the M1917, indicating the extent to which it had by that stage become the standard personal weapon of the Home Guard."[9]
>
> As we have seen, the U.S. M1917 rifle represented an improvement on the already quite respectable P14 rifle, and it was ten years, two design evolutions and one major war newer than the British service rifle, the Rifle No. 1 or SMLE. The irony that the excellent M1917 military rifle should be superseded by "gimcrack" (flimsy or poorly made) wartime manufactured No. 4 Lee-Enfields was entirely lost on the Home Guard."

March 1943 photo of Nottingham's Home Guard deployed for action during "invasion" maneuvers. The photo depicts a STEN submachine gun, a stripped-down Lewis machine gun and four M1917 rifles that have painted identification stripes on the fore-ends. *(Militaryhistorynow.com)*

A fairly sizable number of M1917 rifles earmarked for the British Home Guard were subsequently given to Free French Forces troops.[10]

Although the M1917 rifle never saw combat service with the British military, it served an important role in equipping the Home Guard for the expected German invasion in 1940–41, which thankfully, did not happen.

Members of the British Home Guard in the rubble of a bombed-out building aiming their Model 1917 rifles. Note the painted bands on the fore-ends of the rifles in the photo. The men shown at the top of the photo have older Lee-Enfield rifles. *(Warhistoryonline.com)*
(INSET) Close-up of the typical red band, which was painted on the fore-end of Model 1917 rifles sent to the British Home Guard in WWII. Note the ".30 / 06" designation. *(Joesalter.com)*

December 7, 1941 to 1945 — While the United States was arranging shipments of war material to allied nations via Lend-Lease, the deck was reshuffled on December 7, 1941. The following day, war was declared against Japan. The Axis nations of Germany and Italy responded by declaring war against the United States, which resulted in Congress doing likewise on December 11, 1941. The gloves were off, and the war was on.

In the hectic weeks and months following the attack on Pearl Harbor, the United States accelerated work in obtaining sufficient war material, obviously including weapons, with which to equip our rapidly expanding military. Production of small arms was immediately ramped up, but there was a lag time between the letting of contracts and the delivery of weapons to the using services. The M1917 rifle played an important role in helping bridge that gap. As stated in Project Supporting Paper, *Rifle, U.S. Caliber .30, M1917, 1917–1945*:

> "The supply situation was continually growing tighter. Production of carbines during 1942 had not been up to expectations and authority was granted to designate the M1917 as substitute to fill immediate requirements. At various times during the war, quantities were issued for various operations under the jurisdiction of the Office of Strategic Services, as well as for the use of the Navy Department and Foreign Nations."[11]

The following table details how the M1917 rifles were distributed:

"Zone of the Interior	213,410
Navy	16,521
Office of Strategic Services	54,030
International Aid	1,503,485
Shipped Overseas (for use of the American Forces)	14,855
Known Losses (turned over to the Philippines)	220,495"[12]

- The Zone of the Interior rifles were primarily those that remained stateside for training purposes and for supplemental weapons to arm sentries, etc.
- Navy, or course, refers to the M1917 rifles acquired by the U.S. Navy. The Navy's use of these rifles will be discussed later in this chapter.
- The rifles obtained by the Office of Strategic Services were provided to various irregular units funded and operated by the OSS, primarily in the CBI (China-Burma-India) Theater.
- The rifles in the "International Aid" category will be discussed next.
- The Shipped Overseas (for use of American Forces) is a very interesting category, as it clearly refutes the claim that is sometimes made that no Model 1917 rifles were utilized in combat theaters by American troops. While a number of the almost 15,000 rifles that were sent overseas for American troops were utilized in ostensibly non-combat roles such as issuance to Military Police units, as will subsequently be discussed, some were employed in combat roles as well.
- The known losses "turned over to the Philippines" were those rifles procured by the government of the Philippines in the late 1930s under the aegis of General Douglas MacArthur."[13]

International Aid — As stated in the oft-referenced Ordnance Department Project Supporting Paper, *Rifle, U.S., Caliber .30, M1917, 1917–1945*:

> "With the advent of World War II this country once again was faced with the need for large quantities of rifles for troops. Acting in the capacity of (the) 'Arsenal of Democracy' the United States prepared to make this tremendous effort of supplying not only its own Armies but also those of the United Nations. In this latter capacity the stocks of M1917 rifles became (an) invaluable source of material."[14]

A large portion of the M1917 rifles that were classified as "International Aid" in the above table were those supplied to Great Britain in June of 1940, prior to the passage of the Lend-Lease Act, as well as additional rifles subsequently sent under Lend-Lease. Since these have already been discussed, the other nations receiving M1917 rifles in WWII will be profiled.

The Philippines — Just hours after the attack on Pearl Harbor, the Japanese launched a full-scale invasion of the Philippines. The American military units posted to the Philippines and their Filipino allies were woefully outnumbered but fought valiantly.

Most of the regular U.S. Army troops were armed with Model 1903 rifles and a few M1 Garands, but the men of the Philippine Commonwealth Army and Philippine Constabulary were armed chiefly with M1917 rifles procured in the late 1930s. It is reported that some U.S. Navy sailors who participated in the defense of Corregidor and Bataan were armed with M1917 rifles obtained from the Filipino allies.[15]

The following observations were related in the book, *Bataan, Our Last Ditch*, by author John W. Whitman who served in Bataan with the 31st Division:

> "The Filipino infantryman carried World War I .30 caliber Enfield and Springfield rifles with a stock too long for their short arms and shoulders. 'That rifle stock we had was about three-quarters of an inch too long for a Filipino to get up and reach around and get hold of the trigger,' remembered Lieutenant Colonel James V. Collier. 'It was just murder when he fired that thing because it had a kick anyway.
>
> When I saw them on the range, their shoulders were black and blue in no time.' Filipinos also had trouble getting their eyes close enough to the rear sight to use it effectively. The Enfield rifles were plagued by a weak extractor made brittle by age, and although the Ordnance Corps tried to re-temper the

extractors so they would not break, success in this venture was not great."[16]

The island of Corregidor was the last allied holdout, but it eventually succumbed as well. On May 6, 1942, Lt. General Jonathan Wainwright surrendered all U.S. troops in the Philippines. When the Japanese began their occupation, they reorganized the Philippine Constabulary and renamed it "Bureau of the Constabulary." It was comprised of discharged POWs from the army of the Philippines and the former Philippine Constabulary who pledged loyalty to Japan. The M1917, rifle was the primary service arm of the new organization.[17]

In his book, *Ordnance Went Up Front,* author Roy Dunlap stated, "I had a good 1917 Enfield, taken from the Japanese — their 16th Division was armed with as much American equipment as their own. They had taken the Enfields from the Filipino forces after the Philippine surrender and had armed their own troops with them."[18]

However over 200,000 Filipinos and several hundred American holdouts, many armed with M1917 rifles, con-

Model 1917 rifles with fixed bayonets, Philippines, World War II. *(Martin K.A. Morgan)*

Japanese soldiers carrying captured M1917 rifles, Philippines, c.1942. *(Anonymous private collection)*

Filipino guerilla with an M1917 rifle. *(National Archives)*

Filipino guerillas and American army Rangers assembling for the January 1945 raid on at Cabanatuan to free POWs. Note the Model 1917 and M1 Garand rifles. (LIFE *photo archives. Carl Mydans, photographer)*

tinued to fight the Japanese in guerilla-style engagements throughout the war. Dunlap had the following to say about the rifles used by the guerillas:

> "The 1917 Enfields, furnished to prewar Philippine forces, cropped up as guerilla guns, most of them having been buried to escape Jap searches. In the average Filipino shack there was no hiding place, so the guns were concealed outside. As a rule they were badly rusted, with worm-eaten stocks. I saw two or three replacement stocks, made of beautiful and heavy red mahogany. The Filipinos had faithfully copied the old English-style 1917 stock in every detail. Most rifles had the spring part of the ejector broken off and replaced with a bit of rubber inserted under the bolt stop spring (it worked). Springfields did not wear quite so well as Enfields, although a few of them saw service through the years as native rifles."[19]

Some of the Filipino guerillas and their M1917 rifles, participated in a number of combat actions in conjunction

Filipino guerillas armed with an M1 Garand rifle and a M1917 rifle. (LIFE ***photo archives. Carl Mydans, photographer)***

Weapons captured from the Japanese in the Philippines. Note the number of M1917 rifles in the pile. ***(National Archives)***

A Filipino soldier and a Filipino geurilla (at right), both carrying Model 1917 rifles. (LIFE ***photo archives. Carl Mydans, photographer)***

with American troops, including the famed rescue of allied POWs at Cabanatuan in January of 1945.[20]

As can be seen, the arrangements to obtain M1917 rifles for the Philippines orchestrated by Douglas MacArthur, beginning back in 1936, certainly paid good dividends just a few years later.

Use of the M1917 rifle in the Philippines did not end with V-J Day. The Communist-inspired Hukbalahap Rebellion, a.k.a. "Huk Rebellion," in central Luzon (c.1946–54) saw use of the M1917 rifle by both sides.

China — The passage of the Lend-Lease Act permitted the United States to supply China with weapons to aid in their war with Japan that had been going on since 1937. This aid included 152,241 M1917 rifles and 154,336 Model 1917 bayonets.[21]

A large number of these rifles were sent to the troops of the "X-Force," which was a component of the Chinese National Revolutionary Army. Trained by American advisors beginning in 1942, it consisted of the 30th, 22nd, 38th, 34th, 14th, and 50th Divisions comprising a total force of some 75,000 men and saw extensive action in Burma, armed in large measure, with M1917 rifles.[22]

There was an interesting theater-modified "carbine" variant of the M1917 rifle, which had a shortened barrel and stock and was intended to be better suited for the smaller-statured Chinese soldiers to use. This weapon will be profiled in "Variations," Chapter 9 of this book.

American army advisor training Chinese soldiers in the use of the Model 1917 rifle. *(National Archives)*

Chinese soldiers training in bayonet techniques with a Model 1917 rifle under the watchful eye of an instructor. *(National Archives)*

Chinese troops at marksmanship training with Model 1917 rifles. *(U.S. Army Military History Institute)*

Chinese soldiers during inspection with their M1917 rifles. *(National Archives)*

The M1917 rifle was modified by shortening the barrel and stock for easier manipulation by Chinese soldiers. *(C&Rsenal)*

"X-Force" soldiers in training with Model 1917 rifles. Note the sandals and the German-style helmets. *(National Archives)*

Soldier of the "X-Force" 30th Division in Burma with a Model 1917 rifle. *(Tom Laemlein)*

A number of standard Model 1917 rifles were subsequently used in the Chinese Civil War (1945–1949) by Nationalist troops, and some were captured by U.S. troops in Korea and Vietnam years later.

France — The most authoritative reference source on this subject, *Rearming the French, Special Studies, United States Army in World War II,* provides the following excerpts regarding French use of the M1917 rifle in the war:

> "The first small arms which the French obtained from Allied sources consisted, it will be recalled, of the 8,000 rifles assigned in November 1942 and delivered to them from the United Kingdom in mid-January 1943. These were .30 caliber M1917 (Enfield) rifles.
>
> … In addition to the M1917 rifle, which they received throughout the year 1943, the French were also given large quantities of the M1903 (Springfield), likewise .30 caliber. The continued issue to them of these two weapons was being made at a time when the M1903 rifles in the hands of U.S. troops were gradually being replaced by newer and more efficient arms, the .30 caliber M1 rifle and M1 carbine. The French who were not being issued these weapons feared that their units would be less fit than the U.S. troops fighting along with them. American officials, on the other hand, acted on the principle that the M1 rifles and carbines, being scarce at the time, could be made available only at the expense of U.S. troops. They felt justified, as a result, in prescribing acceptable substitute weapons of standard issue during the current war. As both the M1903 and M1917 fell in that category, the former being still used by U.S. troops and the latter by the British, the two weapons became standard issue for the North African program units.
>
> The French would have preferred to equip all of their combat units with the M1903. This was not possible since the larger part of the rifles issued to them, approximately two thirds, consisted of M1917s. At any rate, they attempted, in the interests of economy and simplicity in the distribution of spare parts and maintenance, to standardize the type of rifle used in each unit. This they were not able to do until January 1944 because they had not been issued grenade launchers for the M1917 rifle and therefore had been forced to retain a number of M1903's in units equipped with the other rifle. Further aggravating maintenance problems was the fact that a consider-

able proportion of the rifles (estimated by the French as 10 percent, especially the M1917s) were found to be of a low order of serviceability. This was not surprising for the weapons involved were old. In fact they were part of the stocks shipped by the United States to the British after the evacuation of Dunkerque in 1940 and had served the Home Guard for over two years before being turned over to the French. The latter were forced to effect repairs made possible only by the receipt, in July of 1943, of a substantial allocation of spare parts and maintenance materials. By the end of January 1944, program units engaged on the Italian front were equipped with pistols in lieu of carbines, and with M1903 and M1917 magazine rifles in lieu of semiautomatic rifles such as the M1.

Infantryman of the French 4th Moroccan Division armed with a M1917 rifle near Mulhouse, France, November 22, 1944.
(Tom Laemlein)

French troops with M1917 rifles during Operation Dragoon, allied invasion of Southern France, August / September 1944.
(Tom Laemlein)

U.S. Army General Harold F. Loomis...estimated that the French had received a total of approximately 167,000 M1917 rifles prior to Operation Anvil (allied invasion of South France).

As U.S. stocks of M1917s were rapidly becoming negligible, it was urgent to determine some policy in anticipation of French demands for replacement rifles. NATOUSA (Headquarters North African Theater of Operations, United States Army) ruled, as it had a month earlier, that M1903 rifles would be used as battle loss replacements for either M1903 or M1917 rifles, but that there would be no wholesale exchange of M1903 with M1917 rifles.

The last discussion concerning rifles took place in mid-September. French military authorities having estimated their over-all requirements at 17,500 rifles and carbines more than the figure established by the JRC, General Loomis asked them to furnish detailed justification for their estimate. He seized this opportunity to request them to withdraw all M1903 and M1917 rifles from units not on the approved troop list. Finally, he informed them that recent requisitions submitted by them for M1 rifles were disapproved, adding a reminder that it was the War Department policy to supply the M1903 and M1917 rifles as substitutes for M1 rifles. He then enjoined them to make no further demands on U.S. supply agencies for M1 rifles, a request which was duly acknowledged, on 27 October, by Brig. Gen. Antoine Poydenot. To sum up, the French North African forces engaged in the campaign of Italy and later in the campaign of France fought with the M1917 rifle and to a lesser extent with the M1903 rifle and the M1 carbine, in addition to some old rifles of French manufacture."[23]

The Netherlands — When the Netherlands fell to the Germans in May of 1940, the Dutch were able to get most of their navy to England. Some pilots managed to make their way across the channel, and a number joined the RAF and participated in the Battle of Britain. Many Dutch remained in their country, and numerous resistance groups were organized, which spread chaos among the German occupiers.

In 1941, the Royal Netherlands Independent Brigade was formed from a group of approximately 1,500 Dutch soldiers who had been evacuated from the Netherlands following the German invasion. Renamed the "Princess Irene Brigade" in February 1941, the unit's first combat operation was on August 6, 1944, when it landed in Normandy. The Brigade later participated in Operation Market Garden. The Dutch forces were supplied with a variety of small arms by the Americans and the British, including some 20,000 M1917 rifles.[24] Dutch troops also participated in the ill-fated defense of the Dutch East Indies and New Guinea in 1942 until they fell to the Japanese onslaught, but it is unlikely that any M1917 rifles were used in those campaigns.[25]

Canada — As mentioned previously, in July of 1940, Canada received 80,000 M1917 rifles and 5,000,000 .30-06 cartridges, with another 20,000 rifles acquired in 1941. These were issued to several entities such as the Canadian Army Basic and Advanced Training Centers, the Royal Military College of Canada, and the Royal Canadian Army Service Corps as well as other Canadian Army Reserve units. In addition, some were also utilized by the Royal Canadian Air Force, Pacific Coast Militia Rangers and the Veterans Guards of Canada. The Canadians typically painted a red band around the fore-end as was the custom with the rifles issued to the British Home Guard. Most were marked on the stock and/or rear of the receiver with the Canadian "broad arrow" or other form of government ownership marking.

Although these rifles were not used by Canadian troops in combat, they did make a substantial number of standard SMLE .303 caliber rifles available to troops deploying overseas.[26]

Other recipients of M1917 rifles in WWII included the Irish Army's "Local Defence Force," which was somewhat analogous to the British Home Guard.[27] Reportedly, these rifles provided to the Irish were from the substantial number of M1917 rifles initially supplied to the British.

A number of M1917 rifles were subsequently acquired by several other nations in the post-WWII period. These will be listed in the following chapter of this book.

M1917 rifle acquired by Canada during WWII. Note the painted red band on the fore-end and the close-up of "Royal Canadian Army Service Corps" markings on the stock. *(Joesalter.com)*

Red painted band with "30-06"marking on the stock of a Canadian M1917 rifle to denote "non-standard" caliber as was the practice the American rifles acquired by the British Home Guard. *(Joesalter.com)*

Typical broad arrow in "C" Canadian ownership marking on M1917 Lend-Lease rifle and a rifle stock with "Royal Canadian Air Force" ownership markings.

United States' Use of the Model 1917 Rifle During World War II — As detailed above, the M1917 rifle played an important role in assisting allied nations to defend themselves and take the fight to the enemy. However, it also proved to be invaluable to the United States' war efforts. The rifles on hand were a quarter century old, and many required refurnishing. As just one example, the Naval Supply Depot at Norfolk, Virginia, on February 14, 1942, reported that the facility had seven M1917 rifles on hand in serviceable condition and 34 requiring overhaul.[28] There were myriad other reports from ordnance facilities across the country citing similar complaints regarding the poor condition of their M1917 rifles, some of which had been in storage since the early 1920s. As stated in Supplement I to the *History of U.S. Rifle, Caliber .30, M1917*:

> "On 18 November 1941, concern was expressed over the possible need of spare parts (for the M1917 rifle), including bayonet, and, on 30 January 1942, Chief of Supply Section, Office of Chief of Ordnance, advised that the rifle had been designated as a substitute for the Carbine and was being used to meet a shortage of the M1903 Rifle. It was estimated that there were over 300,000 now in the U.S. Army and approximately 150,000 were in depot stocks, to be issued in the near future — consequently, urgent demands were being received for M1917 spare parts, and the distribution factor was considerable."[29]

World War II Overhaul — The most common defect was the condition of the barrels, many of which were badly pitted and/or rusted, which was due primarily to bore corrosion caused by the corrosive-primed ammunition of the era. This was from the barrels having not been promptly and properly cleaned after use. Also, the often unsatisfactory storage conditions in which the rifles were subjected to for a couple of decades were not conducive to keeping the components in pristine condition. An Ordnance Department inspection of overhauled rifles indicated that about 25 percent required replacement of barrels. Obviously, rifles in such condition had to be refurbished before they could be issued to our troops or sent to our allies. The importance of getting these M1917 rifles into serviceable condition was the subject of a lengthy memorandum dated September 8, 1942, from Brig. Gen. Lucius D. Clay, U.S. Army Asst. Chief of Staff or Material. Selected passages from the memo are as follows:

> "It is evident that there will be a serious shortage of Cal. .30 Rifles during the next five to six months. It has been reported that there are available in field service sufficient stocks of all parts except barrels for the assembly of 50,000 M1917 rifles, and that the assembly of these rifles could be effected without depleting the maintenance stock required for any item other than barrels. By using such stocks of spare parts, 50,000 serviceable M1917 Rifles could be made available in about two months' time.
>
> Since the current stock of rifle barrels for this weapon is already below a safe level, the above procedure should be followed with additional procurement of rifle barrels for this weapon. Field Service has recommended that if 50,000 M1917 Rifles are assembled at this time, authority be granted for the procurement of 137,000 spare barrels in order to replace those used in the assembly of the above weapons and to restore to a proper level spare barrels for maintaining those in service.
>
> ...In view of the urgent demand for rifles, Cal. .30, you are authorized to take the necessary action to cover the procurement of 137,000 spare barrels for the M1917 Rifle and to initiate at once the assembly of 50,000 M1917 Rifles from currently available stocks with a view to making this number of rifles available not later than November 15, 1942."[30]

In order to comply with General Clay's directive, the following efforts were detailed in the Ordnance Project Supporting Paper pertaining to the M1917 rifle:

> "Rock Island Arsenal was called upon to recondition every M1917 rifle possible of reconditioning and, in addition, to assemble new rifles from Field Service parts. This brought about concern over possible need for spare parts and also for bayonets. Prompted by the urgent demand for caliber .30 rifles, on 8 September 1942, Headquarters, Services of Supply, authorized the procurement of 137,000 spare barrels for this rifle. It further requested Ordnance to initiate at once the assembly of 50,000 complete weapons from currently available stocks of spare parts; this number of rifles was to be available not later than 15 November 1942.
>
> Locating a facility was difficult. An effort was made to place an order with the Remington Arms Company at Ilion, New York, which concern was at the time tooled up and producing M1903 rifles. The company finally agreed to accept an order for 2100 barrel blanks to be made from steel on hand. These would then be turned over to Savage Arms Company at Chicopee Falls, Mass., where they would be sized as prescribed by the Rock Island Arsenal. The Arsenal was to finish the barrels prior to assembly in the guns.
>
> A contract was finally made with Johnson Automatics, Inc., Boston, Mass., to complete the total quantity of 135,000 barrels. The first supplement to the contract called for delivery of the total quantity by the end of 1943. Final delivery was actually completed in September 1944 for a total of 123,850 barrels. Just prior to the signing of this contract, Ordnance Committee action established two grooves as standard for all caliber .30 rifles thereafter to be manufactured. Drawings were revised. Almost immediately it became evident to manufacturers already in production that it would not be advantageous to change their broaching and hook rifling equipment. There would not be any reduction in cost nor increased production possible because of the cost and time lost on conversion. At a meeting of the Rifle Industry Integration Committee held in New York City on 11 December 1942, it was recommended that four grooves be made standard and two grooves be made optional. On 29 January, a recommendation to this effect was made by Springfield Armory and was approved."[31]

In addition to the replacement barrels made during WWII by Rock Island Arsenal and Johnson Automatics, the High Standard Manufacturing Company of New Haven, Conn., also made 61,250 M1917 four-groove barrels.[32] Of the 123,850 M1917 barrels made by Johnson, 81,571 were of the two-groove variety.[33] Relatively few M1917 barrels were manufactured by Rock Island.

The replacement barrels were marked on top, behind the front sight, with the initials of the maker ("RIA", "HS", or "JA") along with the Ordnance Department "flaming bomb" insignia. The barrels were not dated but were typically marked with a "P" to indicate the barrels had been successfully proof-fired. After the rebuild process was completed,

This M1917 rifle was arsenal rebuilt in World War II.

Markings on a replacement M1917 rifle barrel made by Rock Island Arsenal. Relatively few Rock Island barrels were made. Unlike the original WWI barrels, the WWII replacement barrels were not dated.

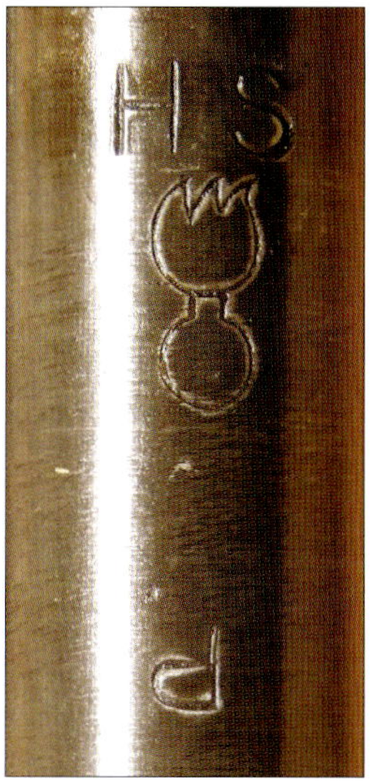

Markings on a High Standard M1917 rifle replacement barrel.

Markings on a M1917 rifle replacement barrel made by Johnson Automatics, Inc.

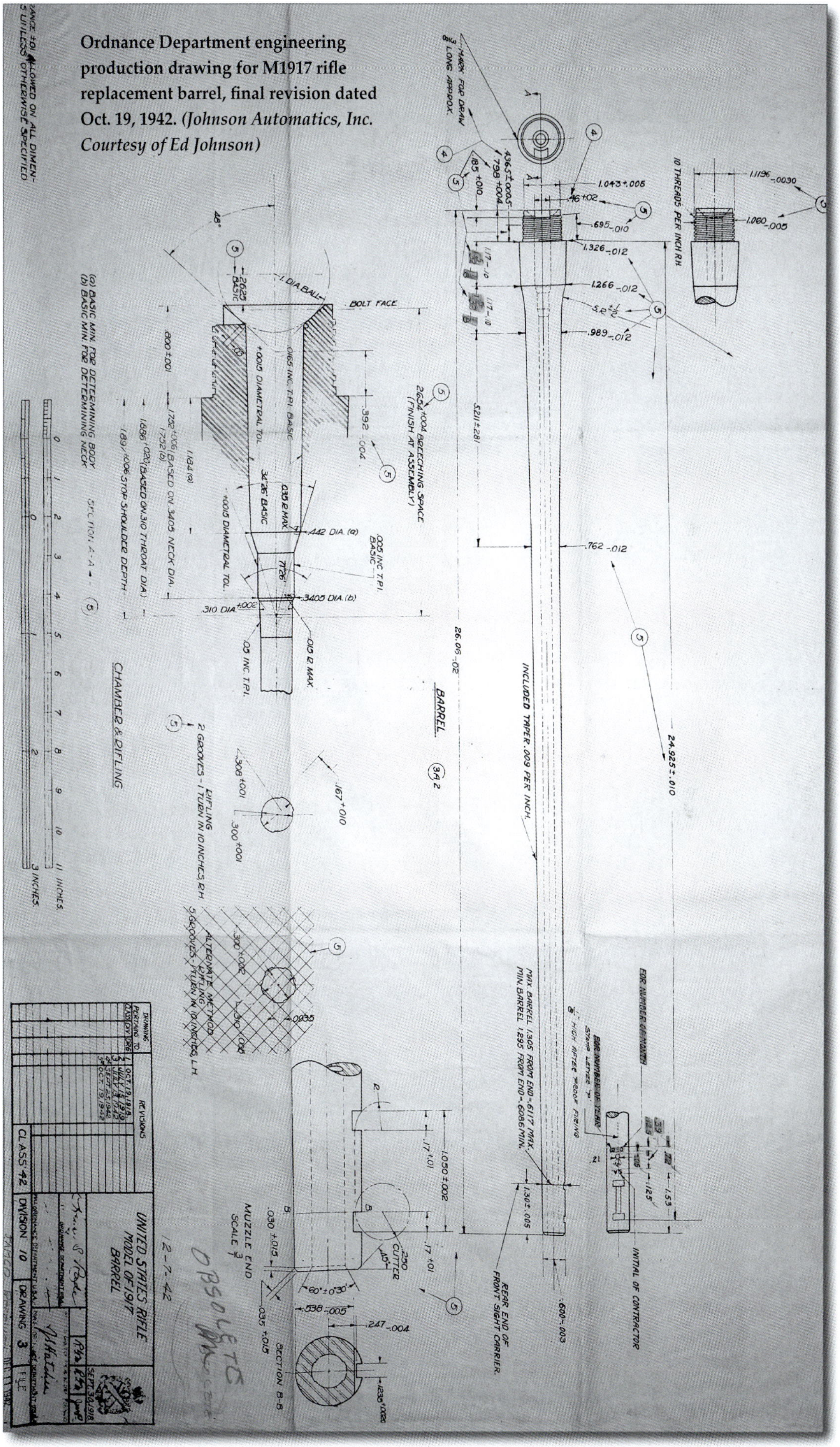

Ordnance Department engineering production drawing for M1917 rifle replacement barrel, final revision dated Oct. 19, 1942. *(Johnson Automatics, Inc. Courtesy of Ed Johnson)*

the stock was also generally stamped with a "P" on the bottom of the grip to denote the rifle had been proof-fired.

Barrels were not the only parts that often needed to be replaced when the rifles were pulled from storage to be rebuilt. In addition to barrels, two of the more common replacement parts required were stocks and bolts. It is interesting to note that Rock Island Arsenal documents reflect that, beginning in June of 1942, the Arsenal "…shipped 592,998 gunstock blanks for U.S. Rifle, Cal. .30 M1917, together with 180,000 handguard blanks for that rifle to Springfield Armory, where they are being used in the manufacture of U.S. Rifle, Cal. .30, M1 (Garand)."[34] In other words, Rock Island Arsenal was shipping stock and handguard blanks for the M1917 rifle to Springfield Armory to produce M1 Garand rifles, which resulted in ordnance having to find other sources for new M1917 stocks and handguards. To anyone familiar with government procurement practices, this sort of thing really wasn't a particularly unusual occurrence.

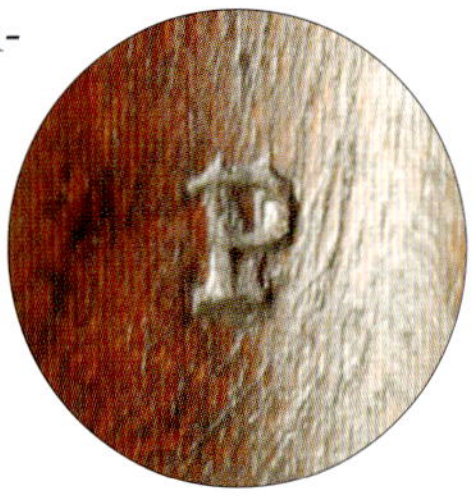
The "P" on the bottom of the grip denotes that the rifle had been successfully proof fired as part of the rebuild procedure.

Another M1917 rifle part that was produced in WWII for maintenance and repair purposes was bolt assemblies. At least two firms received contracts for replacement bolts. One was United Shoe Machinery Corp. of Beverly, Massachusetts. These bolts were marked with the initials of the maker "USMC". This marking has resulted in a number of subsequent civilian owners of surplus M1917 rifles assuming they have a rare and desirable "United States Marine Corps" M1917 rifle when what they actually possess is a rebuilt rifle with a replacement bolt that had no connection whatsoever with the Marine Corps. The United Shoe Machinery Corp. received a contract in July of 1943 for the manufacture of 75,000 M1917 bolts at a per-unit price of $10.76. During this same period, the Kingsbury Machine Tool Company of Keene, New Hampshire, was also granted a contract for 225,000 M1917 bolts at a price of $7.67 each.[35]

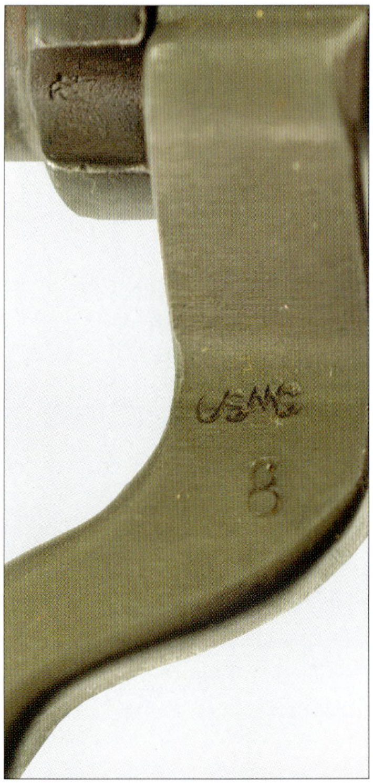
Replacement M1917 rifle bolt made by United Shoe Machinery Corp. Note the "USMC" marking, which is sometimes mistakenly believed to indicate a U.S. Marine Corps connection.

M1917 rifle barrel manufacturing equipment at the Johnson Automatics, Inc. factory in Cranston, R.I. *(Johnson Rifles & Machine Guns)*

Rock Island Arsenal WWII Model 1917 Rebuilds — As outlined in "Small Arms Overhauled and Modified at Rock Island Arsenal – March 1940 – June 1958," the arsenal overhauled the following substantial numbers of M1917 rifles between May of 1942 and August of 1945:

May 1942	Sept. 1942	Oct. 1942	Nov. 1942
2	1,130	6,130	9,810
Dec. 1942	**Feb. 1943**	**March 1943**	**April 1943**
1,519	400	2,450	1,596
May 1943	**June 1943**	**July 1943**	**August 1943**
2,290	1,190	200	990
Oct. 1943	**January 1, 1944–August 15, 1945**[36]		
110	24,131		

The M1917 rifles and many other types of weapons that were rebuilt at Rock Island during WWII were stamped "RIA / FK" on left side of the stock, generally between the grip and buttplate, to denote that the work was done under the supervision of Frank Krack, Assistant Foreman of the Small Arms Inspection Division at the Arsenal from September 27, 1941, to July 19, 1946. This does not necessarily mean he actually inspected every rebuilt weapon but rather that they were inspected and approved by other inspectors acting under his authority. Weapons rebuilt from 1947 to 1958 were stamped "RIA / EB", the initials of Krack's successor, Elmer J. Bjerke. Since his tenure didn't begin until 1947, it is unlikely any "RIA / EB" stamps will be found on M1917 rifles, since RIA stopped rebuilding the rifles in 1945.

Inspection stamp of Frank Krack, Assistant Foreman of the Small Arms Inspection Division at the Arsenal from September 27, 1941 to July 19, 1946.

Inspection stamp of Elmer Bjerke who became the Assistant Foreman after Frank Krack. Bjerke served in that capacity from 1947 to 1958.

A Model 1917 stock showing rebuild markings from Raritan Arsenal. The markings also include the unidentified "3GM-K" stamp.

An inspection stamp on a Model 1917 rifle, which was rebuilt two different times, once at San Antonio Arsenal and then again at Augusta Arsenal.

If a weapon was subsequently rebuilt at Rock Island, new inspection stamps were not added, but if later rebuilt at another ordnance facility, new stamps were applied.[37]

Springfield Armory WWII Model 1917 Rebuilds — Springfield ceased rebuilding M1917 rifles after FY 1921. Armory records do not indicate that any such work was done by the armory until FY 1940 when five M1917 rifles were "targeted" for Benicia Arsenal.[38] As the United States became actively involved in the war, Springfield Armory resumed the overhaul of some M1917 rifles but not nearly to the same extent as did Rock Island Arsenal. SA Annual Reports state that a total of 859 M1917 rifles were overhauled in FY 1942.[39]

It should not be inferred that Springfield Armory and Rock Island Arsenal were the only entities that overhauled M1917 rifles during the war. In 1944, a civilian firm was engaged to rebuild a sizable number of M1917 rifles as well. As discussed in Project Supporting Paper, *Rifle, U.S. Caliber .30, M1917, 1917–1945*:

> "In February 1944, negotiations were initiated with the Remington Arms Company covering the recondition and repair of from 12,000 to 15,000 [M1917] rifles per month for Field Service. This never materialized. In September 1944 at the request of Field Service, negotiations were opened with the Wright Engineering Co., Lisbon, Ohio, for the overhaul of 25,000 rifles. In January 1945, this was increased to 50,000 rifles and in May 1945 there was a further increase of 6,000 to 56,000. The company was paid between $5.00 and $4.67 each to overhaul, clean and repack of 5,036 and 19,1964 rifles respectively."[40]

The type of overhaul inspection markings, if any, applied to M1917 rifles rebuilt by Wright Engineering has not been established.

There were several other ordnance facilities that also rebuilt M1917 rifles, as well as other weapons, in WWII and the early postwar period. Little documentation to cite the specific numbers of rifles overhauled by these entities during this time period has been discovered, but they processed smaller numbers than did Rock Island Arsenal or Wright Engineering. Ordnance arsenals and depots known to have overhauled M1917 rifles during and after WWII included Raritan Arsenal, Augusta Arsenal, San Antonio Arsenal, Ogden Arsenal and perhaps one or two others. There was one facility that overhauled a number of M1917 rifles that were stamped on the stock "3GM-K", and although a lot of theories have been bandied about, there has not been positive confirmation of the identity of the entity. It is not uncommon to find a M1917 rifle bearing multiple inspection stamps indicating rebuild and/or inspection at different ordnance facilities. There will be some discussion of the various rebuild inspection markings in Table #4 of this book.

There is an interesting anecdote in a book about legendary gunsmith P.O. Ackley, who served as Foreman of the Small Arms Shop at Ogden Arsenal during WWII.

Also employed at the arsenal during the same period was Elmer Keith, a well-known firearms author who was instrumental in the development of the first "magnum" handgun cartridges. Keith referenced his tenure as an inspector at Ogden in his autobiographical book titled *Hell, I was There.* As related by the author of the book *P.O. Ackley – America's Gunsmith*:

> "Ackley tells about several hundred Enfield rifles being damaged by an inspector. I was able to document that it was in fact Keith who damaged the rifles by forcing the bolt closed on a gauge. Military gauges are normally, 'Minimum' and 'Maximum' with minimum being equal to a 'Go' gauge and maximum is equal to a 'Field' gauge. It is possible that the Small Arms Shop and/or the Inspectors at the arsenal may have had gauges that fall in between. This fact is not specified in the story.
>
> What is known is that Keith used whatever the Inspectors were using as a gauge to reject rifles for excessive headspace, likely a Maximum, to check Enfield 1917 type rifles. He apparently put the gauge in the chamber and then forced the gun closed with his considerable strength. Remember, Keith was a rancher long before he was a writer. You will have to read the book to get all the details of the story, and there are many more details.
>
> When I wrote about this event in the book, I had not tested any military gauges for hardness. Recently I acquired a set of US Military 30-06 headspace gauges, so it seemed like a good idea to follow up on this story by hardness testing the gauges. Testing in multiple locations front and rear of the gauge we found 62.5 Rockwell C average. Barrel steel in WWI when most of these Enfield guns were produced was softer than modern barrel steels. Modern Chromoly (blue steel) barrels are normally in the low to mid 30s on the C scale. The actions were case hardened, so most of the damage that happened when Keith forced the bolt closed was to the barrels.
>
> Keith was so stubborn that he would not admit he was wrong until General Hatcher visited the arsenal after the rejection of the rifles. Hatcher asked Keith to demonstrate how he used the gauges. Keith was forcing the bolt closed, so Hatcher explained that you should us a light touch and showed Keith the correct method of gauging. Keith was willing to accept instruction from Hatcher, because they were personal friends and he trusted Hatcher."[41]

In addition to the replacement of barrels, bolts and any other broken or worn parts, the external metal of the M1917 rifles, including bolts, were Parkerized, which is the reason why relatively few examples extant still have the original First World War rust-blue finish. The tint of the Parkerizing will vary and can range from gray to matte black. After reassembly, the rifles were fired from a special targeting fixture for accuracy and functioning.

It was indeed fortuitous that the large-scale overhaul programs were established, as the demand for M1917 rifles proved to be even greater than initially anticipated. An example of the increased demand for these rifles was contained in a January 10, 1942, requisition from the U.S. Navy for a number of weapons, including a request for "75,000 Rifles, Caliber .30, M1917 complete with cartridges, belts and bayonets, M1917 and scabbards M1917."[42] As previously related, the response from Norfolk Naval Supply Depot dated a month later (February 14, 1942) revealed the number of M1917 rifles in inventory were "…Seven (7) on Hand and in Serviceable Condition and Thirty-four (34) Requiring Overhaul."[43] Obviously, the Navy wasn't going to get 75,000 M1917 rifles from its Norfolk Supply Depot, or for that matter, anywhere else, since the total allocation to the Navy was just 16,521 M1917 rifles. It is interesting to note that at least a few M1917 rifles were donated to the Navy during the war, presumably by civilians who had previously purchased the weapons via the Director of Civilian Marksmanship. Two such donations were found in government documentation dated January 9, 1943, reflecting that M1917 rifles ser. no. 221803 and ser. no. 285669 had been donated to the U.S. Navy.[44]

While the M1903 rifle was the most common type found in the arms lockers of U.S. Navy warships during WWII, ordnance records indicate that some M1917 rifles were in

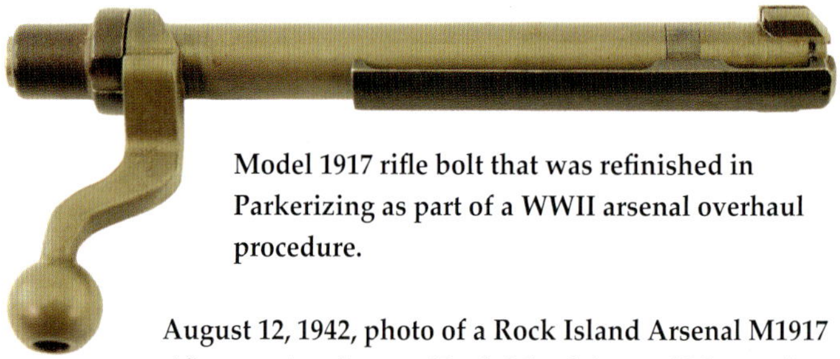

Model 1917 rifle bolt that was refinished in Parkerizing as part of a WWII arsenal overhaul procedure.

August 12, 1942, photo of a Rock Island Arsenal M1917 rifle targeting fixture. ***(Rock Island Arsenal Museum)***

use as well. For example, a report dated March 3, 1944, reflected that M1917 rifle, ser. no. 365879, was on the destroyer, USS *Carmick* (DD-493). Other reports showed M1917 rifle ser. no. 478027 was aboard the sub-chaser USS *SC-1009* on July 22, 1943, and ser. no. 1266628 was on the sub-chaser USS *SC-725* on July 4, 1943.[45]

In addition to use on some warships, the Navy needed the M1917 rifles for issuance to their construction battalions (SEABEES) until sufficient numbers of M1 Carbines or other standardized weapons could be obtained. Author and researcher Gary Cunningham reported that, "My father, a Navy SEABEE, said his unit was issued Model 1917s when they were sent to the Aleutian Islands in early 1943."[46] Also, some "American Enfield" rifles were used by U.S. Coast Guard personnel, which were likely obtained from the Navy's allocation of rifles.

Photo (obviously posed) of a U.S. Coast Guardsman with a M1917 rifle and fixed bayonet. ***(Tom Laemlein)***

Training Use — The M1917 rifles were also in demand by the U.S. Army early in the war for training the ever-growing number of draftees and enlistees arriving at numerous training camps across the country.

There was a War Department Field Manual FM 23-6, dated August 3, 1943, titled *Basic Field Manual – U.S. Rifle, Caliber .30 M1917 (Enfield)*. As has been mentioned, although not part of its official nomenclature, before, during, and after WWII, the M1917 rifle was often referred to as the "Enfield" or "U.S. Enfield." Anyone who maintains that the term "Enfield" is incorrect and should not be used in reference to this rifle should consult this official government manual and other documentation, which used the "Enfield" nomenclature.

FM 23-6

WAR DEPARTMENT

BASIC FIELD MANUAL

U. S. RIFLE, CALIBER .30
M1917 (ENFIELD)

August 3, 1942

WWII Basic Field Manual – U.S. RIFLE, CALIBER .30 M1917 (ENFIELD) dated Aug. 3, 1942. Note usage of the non-standard "Enfield" designation.

U.S. Navy "SEABEES" training with M1917 rifles, WWII. ***(Paul Scarlata)***

Early WWII photo of a U.S. soldier aiming his M1917 rifle. *(U.S. Army Signal Corps)*

Since few M1 Garands were initially available, the troops being trained were usually issued M1903 or M1917 rifles. Neal Burdette, who had his basic training at Fort Benning, Georgia, in July of 1943 commented:

> "After I enlisted, I was sent to the Infantry School at Fort Benning for basic training. With the Infantry School… we got quite a bit of weapons training. We first trained with Enfield rifles, they didn't have enough M1 rifles for everyone. So we did drill and bayonet training with the Enfields, but we never fired them."[47]

Don Dencker was drafted in June of 1943 and took basic training at Fort Hood, Texas. He eventually saw overseas combat duty as a mortar crewman on Leyte and Okinawa. He remarked, "When I was at Fort Hood, we trained and qualified with the bolt action Enfield rifle. They were getting a lot of M1s in, but they went to the organized units. The basic training guys trained with the older rifles."[48]

Jim Kendall entered the army in January of 1943 and was later assigned to 103rd Infantry Division. He also experienced weapons training with the M1917:

> "I entered the army in January 1943 at Fort Devens, Massachusetts. A group of us were put on a train across the country to Camp Hahn, California. We were assigned to a 90mm anti-aircraft battalion. I didn't have a regular basic training; I just started training with the unit. We didn't do a lot of weapons training; the concentration was on the operation of the guns. We were out training in the California desert when I experienced my first weapons training. We were issued M1917 Enfield rifles and conducted some live

U.S. Infantryman with gas mask and heavy woolen overcoat kneeling in the snow while sighting M1917 rifle during WWII training. Original color photo. *(Tom Laemlein)*

Bayonet practice, Camp Robinson, Arkansas, July 1942. *(U.S. Army)*

Excellent view of the "business end" of a M1917 rifle showing the front and rear sights. *(Tom Laemlein)*

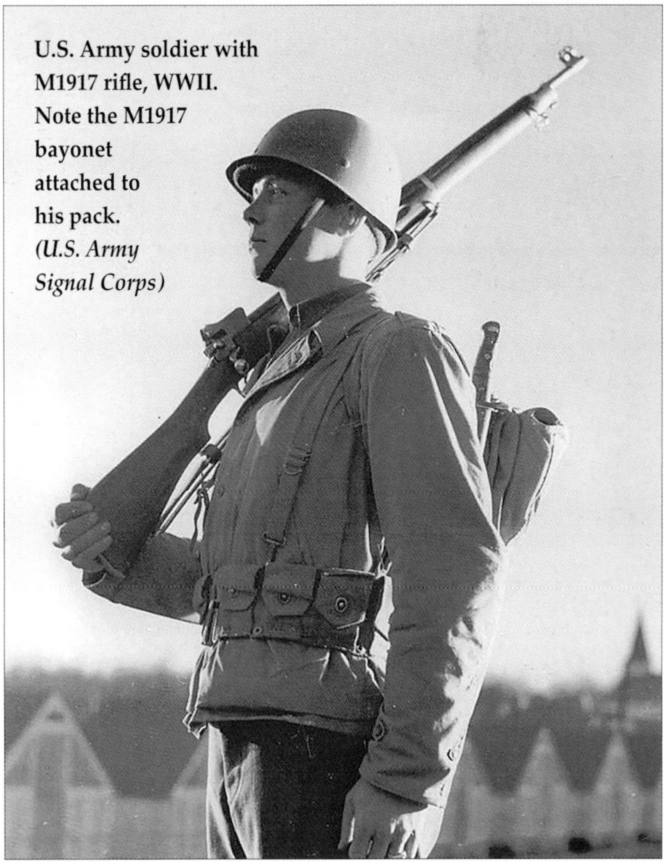

U.S. Army soldier with M1917 rifle, WWII. Note the M1917 bayonet attached to his pack. *(U.S. Army Signal Corps)*

firing maneuver drills... we only had the rifles for about a week before they took them away."[49]

Another U.S. Army draftee, who was posted to Fort Hood in the summer of 1943, recalled his experiences during basic training:

"I don't think any of us were aware of what kind of rifle to expect. Most of us hadn't even heard of the 'M1,' let alone seen one. The issued (rifle) proved to be the 1917 Enfield with long bayonet and gooped beyond reason with cosmoline...

As we lugged the big Enfields in drills, bayonet work, and interior guard duty, we heard rumors of a new rifle issue; Springfields, no less, and lighter and shorter...but (as it turned out), no Springfields and the mystic M1 'Grand' as the cadre referred to it. Finally, I got to see one...

Though I never saw combat and eventually was issued several M1s for duty, I'll always appreciate the old '17 Enfield. It helped convince me what a wonderful rifle the M1 was in comparison. Not much lighter but what a pleasure to shoot..."[50]

A draftee who was inducted into the service in January of 1943 received his basic training at an Air Corps facility in Kearns, Utah. He had a rather interesting take on the M1917 rifle he was issued:

"Those training at Kearns weren't going to be fighter pilots, we were going to be MPs or truck drivers, some sort of backup job. I was scheduled to become an x-ray technician! We did very little weapons training

Marksmanship practice with M1917 rifles with fixed bayonets, WWII. *(National Archives)*

U.S. Army soldiers wearing camouflaged gear and carrying M1917 rifles taking a break while training in WWII. *(U.S. Army)*

U.S. Army ski trooper training with M1917 rifle. Original color photo. *(Tom Laemlein)*

U.S. Army Combat Engineer training with land mines, WWII. Original color photo. *(U.S. Army Military History Institute)*

there. I qualified with an Enfield rifle, which was the damnedest rifle I'd ever got my hands on. There was something a little different about the stock, when you fired that thing, it would spin you around on your belt buckle."[51]

In addition to being used by the regular Army, some state and territorial guard units were also armed with "American Enfields." One of the most interesting was the Alaska Territorial Guard, sometimes referred to as the "Eskimo Scouts." This was a U.S. Army reserve force formed in 1942 under the authority of Territorial Governor Ernest Gruening as a response to the Japanese occupation of the Aleutians.[52] The unit was armed with M1917 rifles. As mentioned in the book *Men of the Tundra: Eskimos at War*:

"When they were finally ordered to provide weapons — Gruening had connections — they went out of their way to find old 1917 Enfields, stored since 1918. The Eskimo guardsmen didn't care. They cherished and loved those Enfields, and learned to shoot them. It was an excellent weapon."[53]

Combat Use — It is reported that M1917 rifles were issued to U.S. Army Chemical Mortarmen at the beginning of World War II. Anecdotal evidence indicates that a number of M1917 rifles accompanied U.S. Army artillerymen and mortar crewmen during "Operation Torch," the landings in North Africa in November of 1942. This is one of the earliest campaigns of the U.S. Army, and sufficient numbers of M1 Garand rifles and M1 Carbines were not available to arm all combat troops, thus the utilization of the M1917 rifles in this capacity was logical.

Poster of the Alaska Territorial Guard depicting "Eskimo Scouts" armed with M1917 rifles. *(National Museum of the American Indian, Smithsonian Institution)*

Commemorative sign for the Alaska Territorial Guard.

U.S. Army infantryman firing a M1917 rifle during fighting in the Sicily campaign July / August 1943. Although the M1917 was seldom used in combat by American troops during WWII, some were so employed on occasion. *(Tom Laemlein)*

Unidentified early WWII-era unit undergoing inspection with M1917 rifles. Note the woolen plaid civilian-style hunting coats worn by several members of the unit. This was likely due to a shortage of the issue M1938 or M1941 jackets. *(Original color photo)*

Early WWII photo of M1917 rifles stacked in a tripod-like configuration with a couple of M50 Reising submachine guns. Location unknown. While the olive drab uniforms with round brass collar discs identify the unit as U.S. Army or National Guard, the presence of the Reising submachine guns is quite unusual since the Army tested but did not adopt the weapon. *(Original color photo)*

Some U.S. Army Military Police units, particularly in the ETO, were equipped with M1917 rifles, as were some army support personnel. There is anecdotal evidence of M1917 rifles being used in combat roles during the Sicily campaign and a Major in the 101st Airborne purportedly stated M1917 rifles were used by rear-echelon American troops in France during the war.

As was the case in the Philippines, some M1917 rifles captured by the Japanese were used against American forces. Lt. Colonel John George discussed the use of captured weapons in his book *Shots Fired in Anger*:

> "(The) Japs frequently captured certain models of rifles and machine guns in such quantities that they thought it wise to load ammunition in their own factories to fit the captured weapons. So, we found large numbers of British Enfields, Dutch Steyrs, American M1917 Enfields, Mausers and Krags in the possession of the Japs on Guadalcanal, with ample evidence that their use was not the idea of individuals, but rather the carrying out of a directive from higher authority. The M1917 Enfields I saw had been modified a bit to take the Jap bayonet…"[54]

The Beginning of the End — Even though the M1917 Enfield rifle may not have seen a great deal of combat by American troops during the war, as has been discussed, the weapon saw significant use for training purposes and also as supplemental service rifles until M1 Garand rifles and M1 carbines could be obtained in sufficient numbers. However, once these standardized weapons became more readily available, the necessity of repairing the M1917 rifles began to lessen. As stated in an ordnance report:

> "In October 1943 it was suggested, in the interest of conservation of money, manpower and materials, 'that no more components for subject rifle [M1917] be made and that if spare parts are required, cannibalism be resorted to,' At that time it was believed that the quantities of other rifles and carbines then in manufacture in large volume would justify the obsolescence of the M1917 rifle and its components. Following this proposal, Field Service advised that it would be their policy to procure no more parts for the rifle."[55]

Although the policy of not procuring any new replacement components for repair and refurbishment of the M1917

Eddystone M1917 rifle and bayonet presented to Lt. Col. Frank J. Atwood, Head of the Rochester Ordnance District in WWII from "Friends and Admirers at Springfield Armory." It is dated "Christmas 1945." *(Rock Island Auction Co.)*

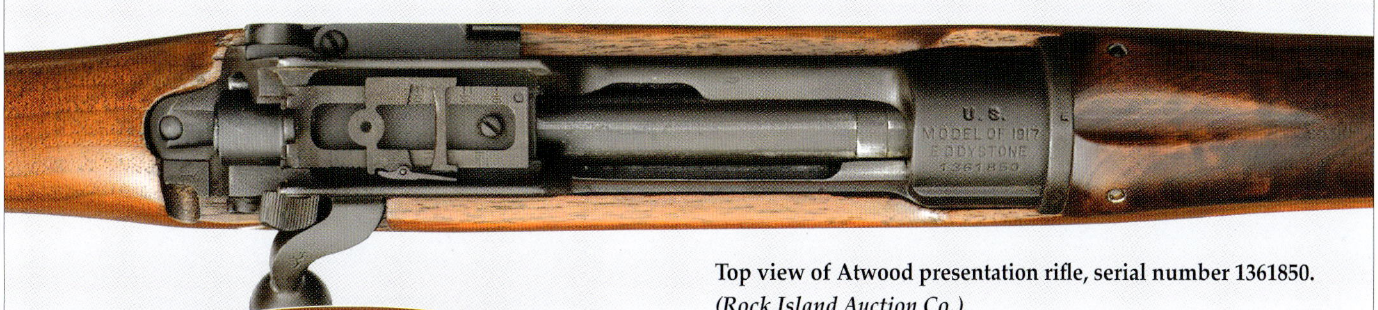

Top view of Atwood presentation rifle, serial number 1361850. *(Rock Island Auction Co.)*

Magazine floorplate with presentation engraving to Lt. Col. Atwood: "CHRISTMAS 1945 / TO LT. COL. FRANK J. ATWOOD / FROM YOUR FRIENDS AND ADMIRERS / AT SPRINGFIELD ARMORY. GOD SPEED." *(Rock Island Auction Co.)*

rifles was instituted, the rifles still in inventory were being used and required periodic repair. These efforts were discussed in the above-referenced report:

> "On 16 June 1945 Field Service advised that the repair of M1917 rifles had been ordered to be discontinued, but that since the contract at Wright Engineering Company was almost completed, all weapons at the facility should be remanufactured. Parts were supplied from Government Free Issue from Field Service Stocks. This action by Field Service was the result of a memorandum from the Chief of Ordnance, which stated that no further procurement of spare parts for this model would be made except those parts already under contract; unserviceable weapons in depot stocks will be repaired as required to meet approved demands to the extent of spare parts now available."[56]

As World War II was winding down and inventories of M1 Garands, M1 Carbines and other standardized weapons were sufficient to meet any additional perceived demand, it would seem that the old M1917 work horse had served its purpose and would be put out to pasture by formally declaring the weapon "obsolete." However, there was still some life left in the gun, as it was believed it might be useful to help fill future military foreign aid requirements. As related by an Ordnance Department publication:

> "As of June 1945, consideration was still being given to the obsoletion of the M1917 rifle. Consideration of the matter by the Ordnance Committee had been given on 23 November 1944. The Committee was advised by Research and Development Service that 'in view of the possibility that the rifle will be the subject of assignments by the Munition Assignment Committee (ground) to meet International Aid requirements, the recommendations [to declare the rifle obsolete] cannot be favorable [sic] considered at this time. The present classification, namely Limited Standard, remains in force."[57]

While always in the shadow of the M1 Garand, and even its contemporary, the M1903, the M1917 "American Enfield" rifle provided valuable service in various roles to our armed forces during two World Wars. Its role in the Second World War is often unheralded, if not forgotten, today, but the M1917 rifle deserves its rightful place among the weapons that defended democracy from the foes of freedom in 1941–1945.

Chapter 8

Postwar – 1946 to Present

After V-J Day, the United States military was faced with a vast array of war material that suddenly became superfluous. Outstanding production contracts were cancelled, and plans were formulated regarding what to do with the various weapons on hand. As mentioned in the previous chapter, recommendations were made that the M1917 rifle should be declared "obsolete" and purged from the government's inventory. However, it was determined that there may still be a need for the weapons for international aid requirements, and the "Limited Standard" remained in effect. As events transpired, with the war ending, the perceived need for continued large-scale assistance to allied nations became much less important, and most of the "American Enfields" were viewed as expendable. General Breton Schervell, Commanding General, Army Service Forces, reiterated the rationale for getting rid of the venerable old rifle:

> "The M1917 Rifle is considered to be too long and cumbersome for use as a modern combat weapon. It has no windage adjustment for the rear sight and it's not fitted with the M1 bayonet, thus requiring maintenance for the Bayonet, M1917, which is peculiar to this weapon. Breakage of bolt stops and extractors are frequent causes of failure."[1]

Danish-made "VAR" replacement barrel for M1917 rifle. *(wwiiafterwwii-WordPress)*

Postwar Inspection and Rebuild — These efforts resulted in the Secretary of War approving the request for the M1917 rifle to be reclassified from "Limited Standard" to "Obsolete" in September of 1945.[2] The decision had to be made as to what to do with the now-obsolete weapons. There were still sizable numbers in the United States, as some 213,410 remained in the "Zone of the Interior" (United States) during the war. While considering the various options for disposition of the M1917 rifles remaining in Uncle Sam's inventory, plans were made to refurbish those rifles that had seen hard use for training and as supplemental service rifles.

The Springfield Armory Annual Report for FY 1947 indicates that 5,559 M1917 rifles were inspected and overhauled.[3] Rock Island Arsenal also rebuilt substantial numbers of M1917 rifles in the postwar period. Arsenal records indicate that 31,804 M1917 rifles were overhauled from July 1, 1952 to June 30, 1953.[4]

Post-WWII Utilization by Foreign Nations — In addition to the rifles that were still on hand in the U.S, there were well over a million and a half M1917 rifles scattered all over the globe that were no longer under the control of the American Government due to the the various Lend-Lease and foreign aid programs instituted during the war. As was the case with the United States, many of these rifles that were now in the hands of the various nations were deemed expendable.

Denmark — For example, Canada began to dispose of the M1917 rifles that they had received, and reportedly sent approximately 38,000 to Denmark in 1953 to augment the rifles previously supplied by the United States. These were designated by the Danes as the "7.62mm Gevaer M/53-17," and some were used by the Danish Navy.

Many of these rifles received several types of modifications by the Danes, including replacement of parts as necessary including excellent quality barrels marked "VAR." Most of these M1917 rifles were later replaced by M1 Garand rifles provided by the United States.

Sirius Sled Patrol: Greenland — It may be surprising that there is an active military unit still using the M1917 rifle, the Sirius Sled Patrol (Slædepatruljen Sirius) of the Danish possession of Greenland. The following narrative regarding these rifles is a bit lengthy, but the fact that the venerable old "American Enfield" is still in active service over a hundred years after production ceased is worthy of note:

Emblem of the Sirius Sled Patrol. *(Grey Dynamics)*

"In April 1940, Denmark was occupied by Germany during WWII. Across the Atlantic, this left the tiny Danish settlements on Greenland isolated. The Germans allowed a puppet government to remain in power in Denmark, but this was not recognized by any of the Allies.

The highest level of authority in Greenland was two sheriffs, who both refused to obey instructions from occupied Denmark. In the U.S.A. (still then neutral) a Danish diplomat informed President Roosevelt, who…dispatched the cutter USCGC *Campbell* (WPG-32) to Godthaab, Greenland, where representatives of the State Department established what amounted to an embassy with the sheriffs. Besides the diplomats, the Coast Guard also quietly 'left behind' 50 M1917 rifles for the Greenlanders to protect themselves. One of the sheriffs, Eske Brun, was particularly interested in the weapons and established a fifteen-man force known as the Northeast Greenland Sled Patrol.

Training was provided by the U.S. Coast Guard and U.S. Army, which infuriated Germany, as President Roosevelt was all but ignoring neutrality in the matter.

Considering its size, the Sled Patrol was quite effective in capturing weather stations and radio posts that the Germans tried to set up in Greenland. The U.S. Coast Guard noted that the Patrol members were extremely good marksmen with the M1917. The force suffered one KIA during WWII, and in turn, it took German POWs in numbers quadruple its own size. When Denmark was liberated in 1945… The wartime force was renamed Sirius Sled Patrol and remained a permanent fixture, as did the M1917 rifles.

Origins of the Sirius Patrol's Model 1917s

The Sirius Patrol obtained its M1917s from a number of sources, with 'newer' acquisitions of the old rifle replacing worn-out examples. None of the original fifty provided by the U.S. Coast Guard are still in use. After WWII, the first large lot was a donation from the Canadian military, which had phased out the M1917 out. Some came from a lot transferred to Denmark from the United States in 1945; these were mostly used by ashore watchmen of the Danish navy and other second-line roles before being phased out there in the 1950s. Still others were M1917s which both Britain and the U.S.A. had provided to Norway, which in turn gave them to Denmark when no longer needed.

Throughout the latter half of the 20th century and first decades of the 21st, the Danish military repeatedly offered to reequip the Sirius Patrol with more modern assault rifles. Every time, the force refused. The Greenlanders feel that no cartridge is better than the .30-06 Springfield for stopping a polar bear. The cartridge has a legendary reputation on the island. Additionally the M1917 rifle is, for whatever reason, impervious to cold and snow and functions perfectly in the worst conditions.

…The Sirius Patrol uses the .30-06 standard military round and also civilian hollow-points. The patrolmen feel that the full metal jacket bullet on the military round is best against polar bears at long range, but, that the hollow-points are better against an enraged musk ox. Typically, the patrolmen arrange their stripper clips so every third round is a hollow-point.

Current Status and Future

In the 21st century, the Sirius Patrol has six two-man dogsled teams, armed with M1917s. Selection and training for the Sirius Patrol is amongst the most grueling of any NATO unit. Applicants have to be Danish, must have honorably completed prior service in the Danish military, and the hitch is a 26-month tour with two-man sled patrols operating in the arctic wastelands for weeks or even months at a time. Although the Sirius Patrol retains a wartime combat function, it normally does not operate as a part of the Danish army. Overall, the Danish military gives the Sirius Patrol a wide berth on how to operate itself, which is understandable given the extreme conditions. The Sirius Patrol backs up the M1917s with 10mm semi-automatic Glock 20 handguns, generations removed from the old rifles.

Sirius Sled Patrol members with M1917 rifles. *(Arctic Business Network)*

Sirius Sled Patrol members near Mestervig, Greenland, armed with M1917 rifles and 10mm Glock 20 handguns. *(Nunatslag News)*

> The Sirius Patrol has refused offers of motor vehicles, as it feels that dog sleds remain the best transportation in Greenland. The dogs also offer early warning of dangerous animals and provide companionship to the patrolmen. As previously mentioned, the Sirius Patrol has rejected all offered replacements for the M1917, and has plainly stated that it will not even consider weapons firing something other than .30-06 Springfield. …There are no plans to even begin studies on a replacement so in all likelihood… the Sirius Patrol will (continue to) be armed with weapons one hundred years old."[5]

Norway — Norway made arrangements for some 23,713 .303 caliber Enfield rifles on hand to be sent to England in exchange for 24,992 M1917 .30-06 rifles from the batch supplied via the WWII Lend-Lease program.[6] Presumably, this was done to have commonality of ammunition for all of their service rifles.

Norwegian Model 1917 Clipper Adapter — Circa 1953–1954, Norway decided to convert around 250,000 captured German Mauser 8mm Kar98k rifles, which they had been using since the end of WWII, for use with .30-06 ammunition in order to be compatible with the U.S. foreign aid rifles and machine guns they were receiving. The .30-06 cartridges (7.62 x 63mm) are longer than the 7.92 x 57mm Mauser cartridges, which required a notch to be cut in the Mauser receiver ring. These modified Mausers were designated the "K98kF1." Norwegian .30-06 ammunition was to be provided loaded in 8mm Mauser stripper clips, requiring no modification to the Mauser rifle stripper clip slots.[7]

Although Mauser stripper clips are visually very similar to the U.S. clips for the M1917 rifles, they did not work in the M1917 rifle, forcing the Norwegians to choose between providing .30-06 ammunition in two different types of five-round stripper clips (which obviously created a logistical problem) or altering the Model 1917s to work with Mauser stripper clips. An inexpensive and simple stamped sheet metal adapter plate was devised to fit into the M1917 clip slot, with new guide lips adapted for the Mauser clips. However, this took up just enough space in the old clip slot that the bullet tips would catch on the receiver ring, so the M1917 rifles also needed a notch cut in them.

The cost of the conversion was nominal. If the adapter plate was removed, the U.S.-style stripper clips could be used, leaving only the notch on the receiver ring as evidence of Norwegian use and modification. It is not known if the Norwegians had a specific designation for the modified M1917 rifles.[8]

Norway sold approximately 13,000 M1917 rifles to Denmark in 1959 for use by the Danish Home Guard. Most of them had the notched receiver modification, including those subsequently acquired by the Sirius Sled Patrol.[9]

An 8mm Kar98k rifle converted to K98kF1 configuration in Norway, which required a notch to be milled into the receiver ring to provide clearance for the longer .30-06 cartridge. *(icollector.com)*

A M1917 rifle with a notched receiver as modified in Norway to permit use with Mauser stripper clips. *(reddit.com/r/milsurp)*

China — Unfortunately, some of the M1917 rifles distributed to allied nations such as China during WWII subsequently ended up in the hands of countries or factions hostile to the United States or our allies. For example, when the Civil War in China resulted in the People's Republic of China gaining control of the country in 1949, they "inherited" the significant number of M1917 rifles that had been previously supplied to Nationalist China during WWII. Perhaps not surprisingly, a number of these were subsequently given to North Korea and saw use against American and UN troops during the Korean War. Some later made their way to the Communist North Vietnam troops and Viet Cong guerillas where they saw some combat use against American and South Vietnamese forces.

A M1917 rifle captured from North Korean troops in the summer of 1950. *(Tom Laemlein)*

Photo of a French sentry armed with a M1917 rifle during the early 1950s. *(NARA photo. Courtesy of Tom Laemlein)*

Afghan Taliban members with an array of weapons including P1914s and M1917 rifles. *(Jane's Guns)*

Other Countries — As mentioned in the previous chapter of this book, M1917 rifles were used by both sides during the Communist-led Hukbalahap Rebellion, a.k.a. "Huk Rebellion," in central Luzon, Philippines (c.1946–54). Some M1917 rifles were captured in Afghanistan by American troops as late at 2020 and were still operable despite being 100 years old!

Some of the M1917 rifles sent for military assistance purposes in 1940–1945 continued to be used by other allies, such as France, well after the end of WWII. Undetermined numbers of M1917 rifles that were not part of any WWII foreign aid programs found their way to other countries after the war. There have been unverified reports that other nations received M1917 rifles as late as c.1960 including Spain, Ethiopia, Pakistan, and India but were not necessarily utilized by the armed forces of those countries.

Director of Civilian Marksmanship Sales — To digress a bit, with the conclusion of the Second World War, the U.S. Government determined there was little likelihood of significant future demand for the M1917 rifles by either the U.S. military or for foreign aid requirements. It was decided to offer substantial numbers of the rifles for sale to civilians via the Director of Civilian Marksmanship (DCM) to adult members of the National Rifle Association. Advertisements began to appear in *American Rifleman* magazine in late 1945 for the sale of M1903, M1903A3, and M1917 rifles. All rifles were initially graded either "Serviceable" or "Unserviceable" and were priced accordingly. The supply of "Serviceable" M1917 rifles apparently sold out quickly. Most of them were categorized as "Unserviceable" and were priced at (even back then) the attractive price of $5.00 each. Buyers were limited to one rifle of each type. An announcement in the October 1947 issue of *American Rifleman* titled "Good News From the DCM" contained the following information about the M1917 rifles:

> "The U.S. Rifle, Cal. .30 Model 1917 classified as 'Unserviceable.' No more serviceable rifles of this model are available except to complete the accepted orders now on file in this office. The price of this 'Unserviceable' grade of the M1917 rifle is $5.00. The price of spare parts for this rifle listed in the 1947 are now cut 50 percent. This includes a large supply of barrels for this rifle at $1.97 each."[10]

By comparison, "Unserviceable" M1903 rifles sold for $15.00. As stated in a DCM sales brochure:

> "The Enfield (U.S. cal. .30 Model 1917) Rifles described above are all UNSERVICEABLE due chiefly to excessive headspace. That is why these Enfield rifles are unsafe to fire in present condition. Enfield rifle barrels are available for this rifle at $1.97 each, one of which barrels can be ordered with the rifle, if desired. Packing and handling for the Enfield (with or without extra barrel) will be $1.85. No tax.
>
> RIFLES ARE SOLD 'AS IS.' No replacements or adjustments will be made on any rifle of either model or grade described above. Furthermore, no U.S. Government facilities are available to aid in headspace checking, barrel fitting or other repairs.
>
> RIFLES OF UNSERVICEABLE GRADE should not be fired until examination and correction is made by a competent person such as a qualified gunsmith, to insure that all safety factors exist. This examination should include a test for headspace not in excess of 1.946". The headspace is equally important when a new barrel is fitted."[11]

The M1917 rifle's headspace was the subject of a June 11, 1946, memorandum from the Office of the Chief of Ordnance titled "Headspacing U.S. Rifle, Cal. .30, M1917":

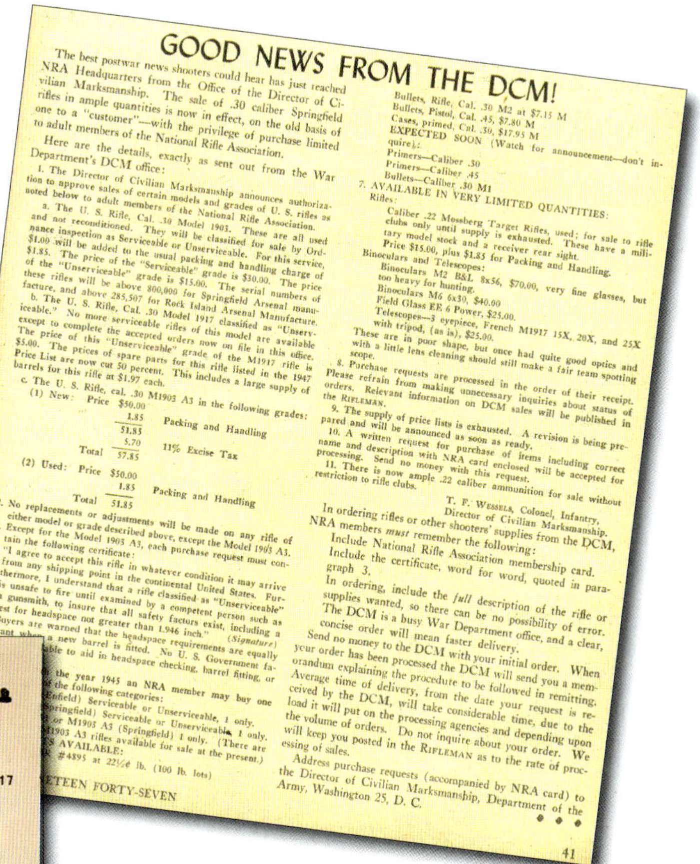

GOOD NEWS FROM THE DCM!

The best postwar news shooters could hear has just reached NRA Headquarters from the Office of the Director of Civilian Marksmanship. The sale of .30 caliber Springfield rifles in ample quantities is now in effect, on the old basis of one to a "customer"—with the privilege of purchase limited to adult members of the National Rifle Association.

Here are the details, exactly as sent out from the War Department's DCM office:

1. The Director of Civilian Marksmanship announces authorization to approve sales of certain models and grades of U. S. rifles as noted below to adult members of the National Rifle Association.

a. The U. S. Rifle, Cal. .30 Model 1903. These are all used and not reconditioned. They will be classified for sale by Ordnance inspection as Serviceable or Unserviceable. For this service, $1.00 will be added to the usual packing and handling charge of $1.85. The price of the "Serviceable" grade is $30.00. The price of the "Unserviceable" grade is $15.00. The serial numbers of these rifles will be above 800,000 for Springfield Arsenal manufacture, and above 285,507 for Rock Island Arsenal Manufacture.

b. The U. S. Rifle, Cal. .30 Model 1917 classified as "Unserviceable." No more serviceable rifles of this model are available except to complete the accepted orders now on file in this office. The price of this "Unserviceable" grade of the M1917 rifle is $5.00. The prices of spare parts for this rifle listed in the 1947 Price List are now cut 50 percent. This includes a large supply of barrels for this rifle at $1.97 each.

c. The U. S. Rifle, cal. .30 M1903 A3 in the following grades:

(1) New: Price $50.00
1.85 Packing and Handling
51.85
5.70 11% Excise Tax
Total 57.85

(2) Used: Price $50.00
1.85 Packing and Handling
Total 51.85

2. No replacements or adjustments will be made on any rifle of either model or grade described above, except the Model 1903 A3.

3. Except for the Model 1903 A3, each purchase request must contain the following certificate:
"I agree to accept this rifle in whatever condition it may arrive from any shipping point in the continental United States. Furthermore, I understand that a rifle classified as "Unserviceable" is unsafe to fire until examined by a competent person such as a gunsmith, to insure that all safety factors exist, including a test for headspace not greater than 1.946 inch." (*Signature*)

4. Buyers are warned that the headspace requirements are equally important wh… a new barrel is fitted. No U. S. Government fa…ble to aid in headspace checking, barrel fitting, or

… the year 1945 an NRA member may buy one … the following categories:
…Enfield) Serviceable or Unserviceable, 1 only.
…pringfield) Serviceable or Unserviceable, 1 only.
… or M1903 A3 (Springfield) 1 only. (There are …1903 A3 rifles available for sale at the present.)
…S AVAILABLE:
… #4895 at 22½¢ lb. (100 lb. lots)

Bullets, Rifle, Cal. .30 M2 at $7.15 M
Bullets, Pistol, Cal. .45, $7.80 M
Cases, primed, Cal. .30, $17.95 M
EXPECTED SOON (Watch for announcement—don't inquire):
Primers—Caliber .30
Primers—Caliber .45
Bullets—Caliber .30 M1

7. AVAILABLE IN VERY LIMITED QUANTITIES:
Rifles:
Caliber .22 Mossberg Target Rifles, used; for sale to rifle clubs only until supply is exhausted. These have a military model stock and a receiver rear sight. Price $15.00, plus $1.85 for Packing and Handling.
Binoculars and Telescopes:
Binoculars M2 B&L 8x56, $70.00, very fine glasses, but too heavy for hunting.
Binoculars M6 6x30, $40.00
Field Glass EE 6 Power, $25.00.
Telescopes—3 eyepiece, French M1917 15X, 20X, and 25X with tripod, (as is), $25.00.
These are in poor shape, but once had quite good optics and with a little lens cleaning should still make a fair team spotting scope.

8. Purchase requests are processed in the order of their receipt. Please refrain from making unnecessary inquiries about status of orders. Relevant information on DCM sales will be published in the RIFLEMAN.

9. The supply of price lists is exhausted. A revision is being prepared and will be announced as soon as ready.

10. A written request for purchase of items including correct name and description with NRA card enclosed will be accepted for processing. Send no money with this request.

11. There is now ample .22 caliber ammunition for sale without restriction to rifle clubs.

T. F. WESSELS, Colonel, Infantry,
Director of Civilian Marksmanship.

In ordering rifles or other shooters' supplies from the DCM, NRA members *must* remember the following:

Include National Rifle Association membership card.

Include the certificate, word for word, quoted in paragraph 3.

In ordering, include the *full* description of the rifle or supplies wanted, so there can be no possibility of error. The DCM is a busy War Department office, and a clear, concise order will mean faster delivery.

Send no money to the DCM with your initial order. When your order has been processed the DCM will send you a memorandum explaining the procedure to be followed in remitting. Average time of delivery, from the date your request is received by the DCM, will take considerable time, due to the load it will put on the processing agencies and depending upon the volume of orders. Do not inquire about your order. We will keep you posted in the RIFLEMAN as to the rate of processing of sales.

Address purchase requests (accompanied by NRA card) to the Director of Civilian Marksmanship, Department of the Army, Washington 25, D. C.

…ETEEN FORTY-SEVEN 41

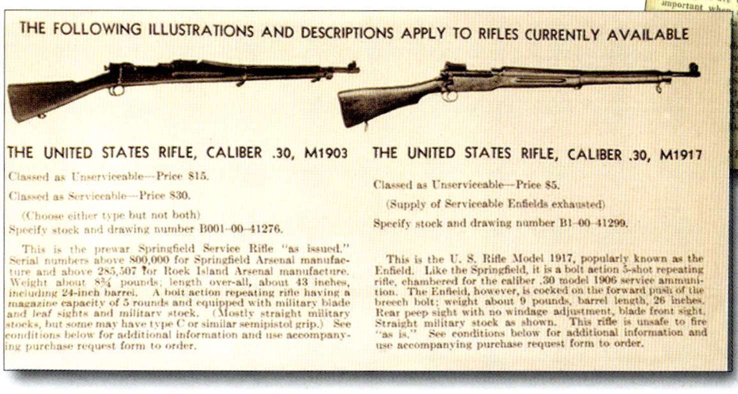

THE FOLLOWING ILLUSTRATIONS AND DESCRIPTIONS APPLY TO RIFLES CURRENTLY AVAILABLE

THE UNITED STATES RIFLE, CALIBER .30, M1903

Classed as Unserviceable—Price $15.

Classed as Serviceable—Price $30.

(Choose either type but not both)

Specify stock and drawing number B001-00-41276.

This is the prewar Springfield Service Rifle "as issued." Serial numbers above 800,000 for Springfield Arsenal manufacture and above 285,507 for Rock Island Arsenal manufacture. Weight about 8¾ pounds; length over-all, about 43 inches, including 24-inch barrel. A bolt action repeating rifle having a magazine capacity of 5 rounds and equipped with military blade and leaf sights and military stock. (Mostly straight military stocks, but some may have type C or similar semipistol grip.) See conditions below for additional information and use accompanying purchase request form to order.

THE UNITED STATES RIFLE, CALIBER .30, M1917

Classed as Unserviceable—Price $5.

(Supply of Serviceable Enfields exhausted)

Specify stock and drawing number B1-00-41299.

This is the U. S. Rifle Model 1917, popularly known as the Enfield. Like the Springfield, it is a bolt action 5-shot repeating rifle, chambered for the caliber .30 model 1906 service ammunition. The Enfield, however, is cocked on the forward push of the breech bolt; weight about 9 pounds, barrel length, 26 inches. Rear peep sight with no windage adjustment, blade front sight. Straight military stock as shown. This rifle is unsafe to fire "as is." See conditions below for additional information and use accompanying purchase request form to order.

The announcement in the October 1947 issue of *American Rifleman* along with the DCM advertisement for the sale of M1903 and M1917 rifles. *(American Rifleman)*

1. The sale of the U.S. Rifle, cal. .30, M1917 is authorized to members of the National Rifle Association. After approval of the sale by the Director of Civilian Marksmanship, the necessary papers are forwarded to an ordnance establishment for completion of the sale.
2. All of the U.S. Rifles, Cal. .30, M1917 that are available for sale are classified as used and in order to insure that they are entirely safe special precautions must be taken with regard to the headspace of these weapons. No U.S. Rifle, Cal. .30, M1917 will be sold to a member of the National Rifle Association that has headspace in excess of 1.947".
3. Where 1.947" headspace gauge are available they will be used. The bolt should not close on this gauge. If a 1.947" headspace gauge is not available, the 1.946" gauge may be used. Where this gauge is used the bolt may completely close by the exertion of a slight pressure on the bolt handle. Due consideration should be given to the fact that this rifle cocks on the forward stroke of the bolt. Selection of components to obtain required headspace is permissible.
4. The above information is furnished pending its inclusion in a War Department Supply Bulletin."[12]

In his classic *Hatcher's Notebook*, Julian S. Hatcher commented on some misconceptions elicited by the DCM's policy regarding the necessity for headspace measurements of the M1917 rifles being sold at the time:

"When the M1917 rifles were placed on sale right after World War II, a lot of misinformation about headspace in these rifles was circulated. Numbers of ex-service personnel, casting about for employment, decided to capitalize on the knowledge of weapons they had gained in the service and went into business as gunsmiths. Many of these men, unfortunately, did not have enough of the right kind of experience to make them experts in this line, and a lot of very bad advice was given by some of them.

The purchaser of an Enfield would often take it to a gunsmith to be checked and would be told it had excess headspace and was dangerous to shoot. Some of these owners wrote about that information to various gun editors, and it was printed, and soon the rumor grew that it was dangerous to fire a 1917 rifle purchased from the D.C.M. This in spite of the fact that these rifles had all been in use off and on since World War I, and had an exceptionally clean record."[13]

Problems with Eddystone Rifles — As early as 1946, it was discovered that many of the M1917 rifles made by Eddystone experienced problems with cracked receivers. The cracks were often not visible unless the rifles were commercially refinished.

On May 14, 1947, the Ordnance supply officer at the Columbus Army Depot recommended to the Chief of Ordnance that all sales of Eddystone M1917 rifles be discontinued.[14] Another memo dated May 28, 1947, from Rock Island Arsenal to the Chief of Ordnance also referenced the problems being encountered with the rifles at the Columbus Depot:

"1. During the recent meeting held in Washington in conjunction with supply from General Depots, it was pointed out by the Columbus representative that out of approximately 28,000 U.S. Rifles, Cal. .30, Model 1917, issued by the Columbus General Depot for sales, that about half were returned due to cracked receivers. This condition occurred in rifles of Eddystone manufacture.

2. In view of the above, Model 1917 Rifles set aside for sales purposes, at this Arsenal, were inspected for cracks by using magnaflux, and out of the first 30, 4 were found with cracked receivers. In order to eliminate a large return of rifles sold to members of the National Rifle Association, all model 1917 Rifles of Eddystone manufacture will be checked for cracked receivers prior to issue.

3. Information is required as to whether Model 1917 Rifles with cracked receivers should be set aside as suitable for firing blank ammunition or if they should be placed in the category for drill purposes."[15]

A subsequent memo from Rock Island to the Chief of Ordnance dated June 24, 1947, stated:

"In order to check these rifles for cracks, it is necessary to disassemble the rifle, removing all wooden components and the bolt. This is an expensive process, and the cost of disassembling, inspecting, and reassembling the entire stock of rifles of Eddystone manufacture would be prohibitive.

In view of the results of the above inspection, it is recommended that all M1917 Rifles of Eddystone manufacture be issued *only* for the firing of blank ammunition or drill purposes, whichever your office should decide."[16]

Obviously, there was an endemic problem with many of the Eddystone receivers. The problem generally occurred after WWII when the rifles were being overhauled, and the worn or damaged original 1917–1918 vintage barrels were removed in order to install new replacement barrels. Since there were few reports of similar problems with Model 1917 rifles made by Winchester and Remington, it was apparent that something different had happened at Eddystone.

It is possible that at least some batches of steel used by Eddystone were slightly over-hardened. It also appears that the problem was exacerbated due to the original barrels being excessively tightened when installed at the plant. During subsequent rebuilds, the extreme torque required to remove the barrels resulted in some of the receivers cracking.

An ad in the September 1962 issue of *Guns* magazine for surplus military rifles, including M1917s, selling for $29.95. *(Guns & Ammo)*

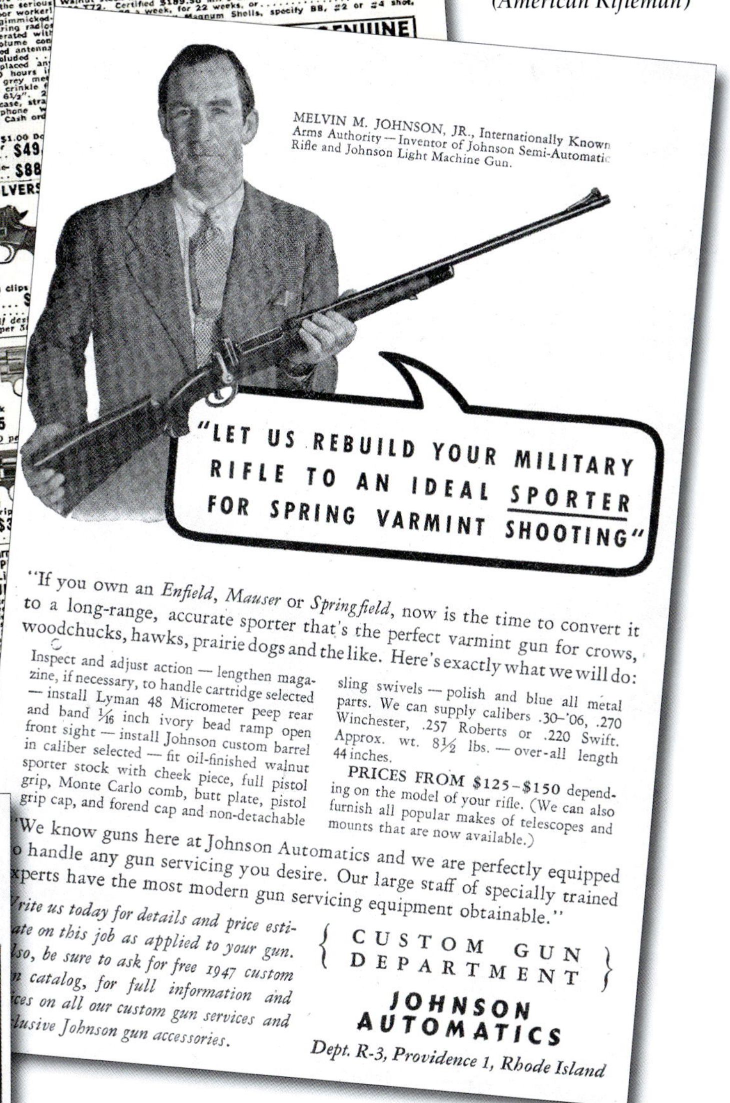

An ad in the March 1947 issue of *American Rifleman* from Johnson Automatics, Inc. for its "Custom Gun Department," which, among other things, would sporterize surplus military rifles, including the M1917. *(American Rifleman)*

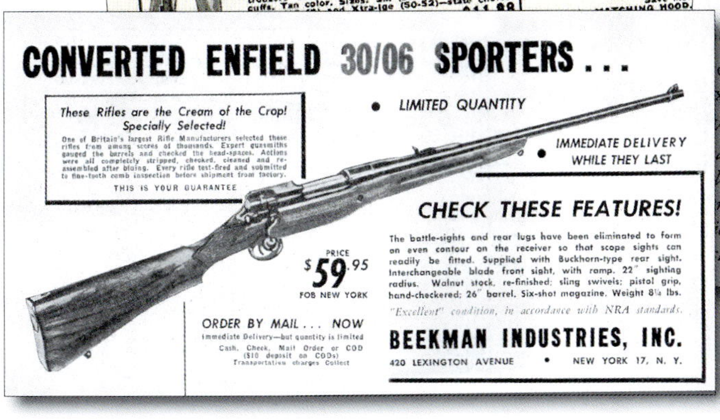

February 1949 *American Rifleman* ad for "Converted Enfield 30/06 Sporters." *(American Rifleman)*

While the rifles with cracked receivers may have been satisfactory for firing blank ammunition or for drill purposes, it is highly unlikely that subsequent buyers of such rifles through the DCM would be inclined to always adhere to such restrictions. Clearly, the easiest, safest and most expedient course of action for Ordnance was to simply stop selling Eddystone-made rifles. A November 7, 1947, memo from the War Department to the Commanding Officer of Springfield Armory contained the following instructions:

> "Dispose of as scrap…all Rifles, U.S. Cal. .30, M1917 with known cracked Eddystone receivers **as well as the entire stock of Eddystone Receivers and Eddystone Receiver Assemblies for this Rifle.** [17]
>
> *[Bold print added by author]*

Regardless, the surplus M1917 rifles continued to be sold until the supply was exhausted. There was an announcement in the September 1949 issue of *American Rifleman,* which stated: "The supply of M1917 rifles has become exhausted and it is not expected that any more will become available."[18]

Unlike today, in the 1940s and 1950s (and even later), surplus military rifles were not considered particularly collectible, and many individuals bought surplus M1917 rifles with the intent of converting them into economical sporting/hunting rifles. The very low purchase price of $5.00 (approx. $62, adjusted for inflation)[19] and the strong nickel-steel action made them excellent candidates for conversions.

It didn't take long for the M1917 rifles sold to NRA members via the DCM to start changing hands, and a number were subsequently purchased by commercial dealers. Sporting and firearms-related magazines in the late 1940s and into the 1960s had numerous advertisements for firms specializing in "sporterizing" M1917s and other types of surplus military rifles.

Some owners, wishing to have their rifles sporterized, used the services offered by these established firms or had the work done by individual gunsmiths. Others took the "do-it-yourself" route. Results ranged from extremely professional and attractive renderings to amateurish hack jobs that consisted of little more than cutting off the stock fore-end.

A M1917 rifle "sporterized" by simply cutting off a portion of the stock fore-end. ***(Rock Island Auctions)***

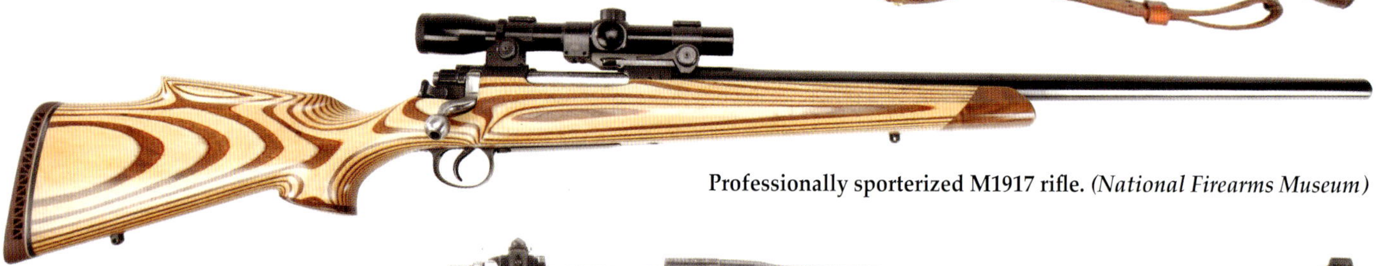

Professionally sporterized M1917 rifle. ***(National Firearms Museum)***

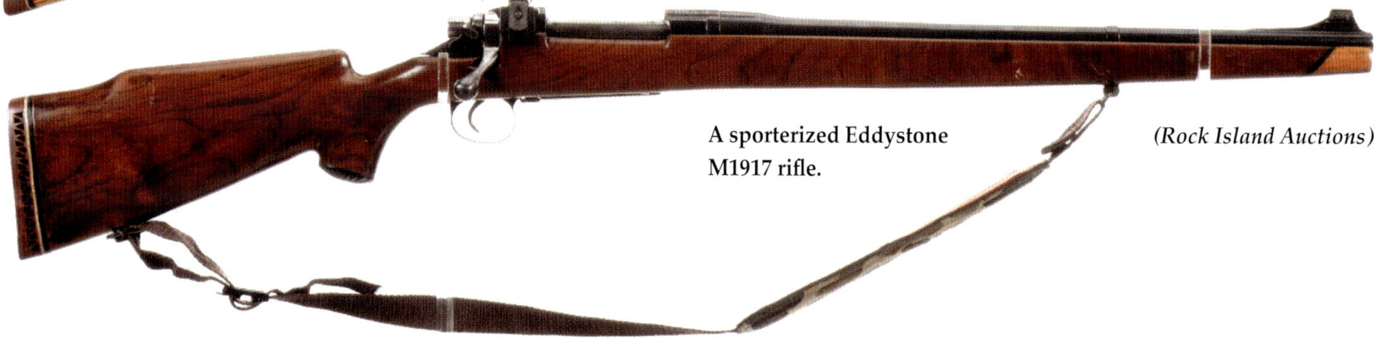

A sporterized Eddystone M1917 rifle. ***(Rock Island Auctions)***

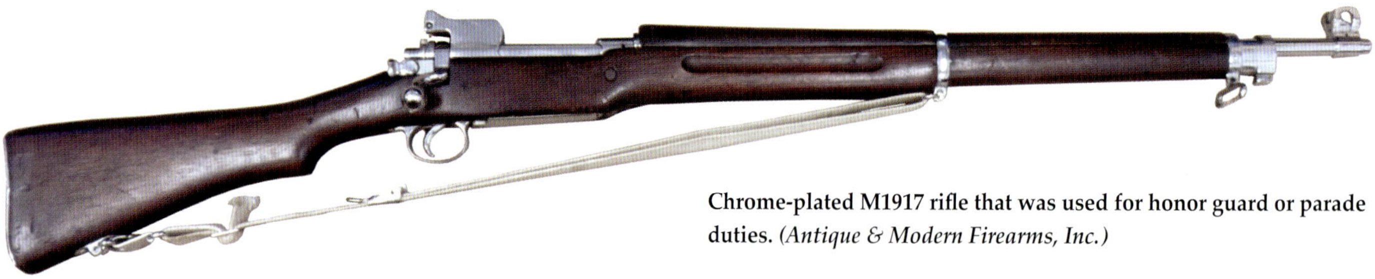

Chrome-plated M1917 rifle that was used for honor guard or parade duties. ***(Antique & Modern Firearms, Inc.)***

Model 1917 rifles for sale by the Civilian Marksmanship Program, March 2021. ***(CMP)***

Veterans' Groups — A number Veterans of Foreign Wars (VFW) posts and other veteran organizations' Honor Guards have used M1917 rifles with blanks to fire military salutes at funerals of deceased veterans and other ceremonial occasions, a practice that still occurs today. The rifles were "loaned" to the veterans' groups and still belonged to the government. After decades of use, many of the rifles were in poor condition with rusted bores from firing corrosive-primed blank ammunition without proper cleaning. In addition to the ceremonial salute use, some of the rifles were chrome or nickel plated for color guard use. While the latter are not representative of arms issued for combat use, they are legitimate collectible examples of quasi-official variants that may appeal to some collectors.

Civilian Marksmanship Program Sales — Although the days of the $5.00 DCM M1917 rifles are long gone, in recent years, limited numbers continue to be sold by the Civilian Marksmanship Program (CMP), successor of the Director of Civilian Marksmanship. As reflected in the current CMP sales brochure:

> "The CMP continues to receive a few dozen of these rifles each year. All of these rifles have been on loan from the U.S. Army to chartered veterans organizations for use in honor guards, funerals, and other ceremonial purposes. All are mix-masters and none are in original condition. We receive only a few of this model each year, and in overall poor condition."[20]

The CMP also offers repair and refinishing services as well as the sale of new commercial barrels for the rifles.[21]

Military firearms are now one of the most popular collecting themes around. For several years, such weapons as M1903 Springfields, M1 Garands, and M1 Carbines have become very sought-after acquisitions by collectors. However, for whatever reason(s), the M1917s rifles did not garner a great deal of collector interest. Within just the past few years, however, many collectors have taken a second look, and interest in the "American Enfields" is rapidly increasing.

Collectors naturally prefer examples that remain in their original military configuration. The large number of "sporterized" M1917 rifles has resulted in unmodified examples, by comparison, being surprisingly scarce, which has had the inevitable effect of steadily rising prices for collectible specimens. The vast majority of the rifles available today are those that were arsenal rebuilt from the 1920s through the 1940s. Rifles remaining in their WWI factory configuration, with original blued finishes and all matching components, are now bringing eye-raising prices on the collector market. The section of this book titled "Collecting the American Enfield" will have a rather lengthy discussion on the collecting aspects of the rifle. The story of the M1917 rifle continues into the 21st century as evidenced by its renewed popularity with shooters, collectors and arms enthusiasts. The old warhorse has a new lease on life.

Chapter 9

Variations

Most of the U.S. military rifles that saw service for a number of years went through several permutations, often designated by a different model number, intended to improve functionality, increase production rates or reduce manufacturing costs. For example, there were several variations of the U.S. .30-40 Krag, which was in active use for about a decade. These rifle and carbine variants included the Models of 1892, 1896, 1898, 1899, .22 caliber Gallery Practice rifle and M1899 Constabulary carbine.

The M1903 rifle had a much longer tenure of service, and there were many variants produced, including the .30-03 caliber "Rod Bayonet" Model 1903, .30-03 caliber Model 1903 with 1905 modifications for knife bayonet, Model 1903 altered for knife bayonet and M1906 (.30-06) cartridge, M1903 made by Remington in WWII, M1903 Mark I, M1903A1, M1903A2, M1903A3 (made by Remington and Smith-Corona) and a number of sniper rifles including the M1903A4, M1903 with M1908 and M1913 Warner & Swasey sights, M1903 with Winchester A5 (or Lyman 5A) scopes, M1903 with Unertl 8X USMC scopes as well as several types of .22 caliber and .30 caliber target and sporter rifles made by Springfield Armory.

The M1917 rifle, by comparison, was only in production for about two years, and soon after the end of the First World War, it was essentially withdrawn from service, and most of them were packed for long-term storage. Thus, there was no time, and there was no need for any standardized variations to be developed.

There were, however, a few experimental or prototype weapons based on the M1917 that were fabricated in limited numbers in WWI and then submitted for tests, but they were not put into production and saw no real use in the war. These will be discussed in this chapter along with an example of a M1917 rifle specially modified for use by an allied nation during the Second World War. In addition, there were some commercial sporting rifles based, to some extent, on the M1917 that were manufactured and marketed to civilian shooters, hunters, and sportsmen.

Model 1918 Sniper Rifle — Sniper rifles proved to be valuable for trench warfare. The standard U.S. Army sniper rifle at the time of America's entrance in the war was a Model 1903 rifle fitted with an unusual prismatic telescope sight made by the Warner & Swasey Company. A fair number of these rifles were fielded in the war, but they proved to be less than optimum sniping weapons and were not particularly well liked.

The Marine Corps utilized some Model 1903 rifles with Winchester A5 telescopic sights and special Mann-Niedner tapered mounting blocks for competitive marksmanship teams and as ad hoc sniping weapons during the war. While better than the Warner & Swasey sight, the A5 was a commercial scope that was not designed, nor intended, to encounter the rigors of combat, and it was prone to damage. However, some Model 1903 sniper rifles with Winchester A5 scopes (or the Lyman 5A) remained in Marine Corps use

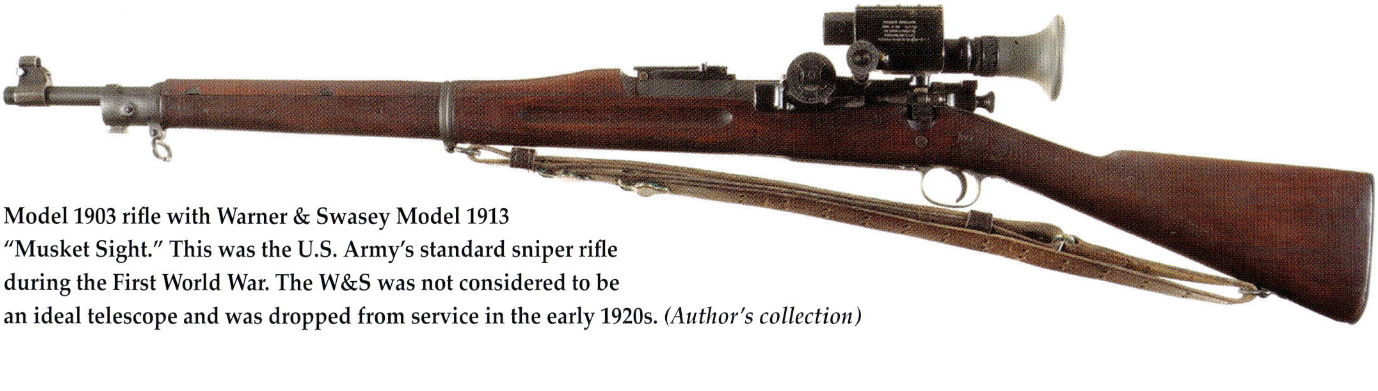

Model 1903 rifle with Warner & Swasey Model 1913 "Musket Sight." This was the U.S. Army's standard sniper rifle during the First World War. The W&S was not considered to be an ideal telescope and was dropped from service in the early 1920s. *(Author's collection)*

well into the early Second World War. In any event, given the deficiencies of both the Warner & Swasey and Winchester A5 sights, it was clear that a better sniper rifle was needed by our doughboys for use in the First World War.

In May of 1918, Major L.O. Wright of the U.S. Army's Engineering Bureau consulted with Winchester Vice President Henry Brewer to explore the feasibility of a sniping version of the Model 1917 rifle.[1] It is not known why Winchester was the only prime contractor to apparently have been considered to develop the proposed sniper rifle.

Of course, any new sniper rifle would only be as good as the telescopic sight used. To this end, the Ordnance Department tested a number of scopes. Eventually, a 2.6-power Winchester telescopic sight was selected and designated as "Telescopic Musket Sight, Model of 1918." Initially, there was some discussion about using the new scope on a Model 1903 rifle, but it was decided that the new sniper rifle should be based on the Model 1917. As discussed by author Clark S. Campbell:

> "By this time the M1917 rifle (the Enfield) being produced by Remington, Eddystone, and Winchester had become our principal service rifle. Since that rifle featured a bolt handle already concave and a safety lock (was) much more readily operated in that it was side mounted rather than close under the scope, it was apparent that the scopes would best be mounted on a rifle based on the M1917 action. This rifle, to be designated 'U.S. Telescopic Sight Rifle, Cal .30, Model of 1918' was to be produced by Winchester using receivers forged and machined to incorporate an integral rear scope block. The receivers were to be marked 'U.S.T.S.R MODEL OF 1918 WINCHESTER 000000' the serial number beginning with 50001."[2]

The 1919 publication *Handbook of Ordnance Data* contained the following information on the new sniper rifle:

> "*United States telescopic rifle, model 1918* — This rifle, which is derived from the United States rifle, model 1917, is adopted for the telescopic musket sight, model of 1918. It is intended primarily for sniping and most of its parts are interchangeable with the parts of the 1917 rifle. The exceptions are as follows:
>
> *Receiver* — There is no rear sight base and no rear sight. A dovetail base for the rear telescope mount takes the place of them.
>
> *Barrel* — There is no front sight, the barrel being adapted to receive screws for the front telescopic parts. The rifling is right-hand twist with one turn in 10 inches and the standard tolerances and type of land of the Springfield rifle are employed. The stock is of special construction with no hand guard, nor upper band, while the lower band is of special design.
>
> *Telescopic musket sight, model of 1918* — The telescopic musket sight, model of 1918, for use with the United States telescopic sight rifle, model of 1918, was formerly called the Winchester 2.6–2.7 telescopic sight, having been developed by the Winchester Repeating Arms Co. It was adopted as the United States Army standard in lieu of the (Warner & Swasey) model of 1913. It is mounted over the receiver and consists of a direct or nonreflecting telescope 2.7 power, with independent front and rear mounts. On the rear mount there is a windage and range adjustment. At 100 yards, the field is 44' 1". Aiming point is a post at the optical center of the telescope. A crosshair serves to level the instrument. This telescope possesses unusual optical and mechanical properties, and was adopted as standard after long competitive tests at Camp Meade. It is made by Winchester Repeating Arms Co."

Author and sniper historian Peter Senich, mentioned the following in his book *The Complete Book of U.S. Sniping*:

> "According to an article in the 13 April 1918 issue of *Arms and the Man*...a special (Model 1917) rifle mounting a M1918 Winchester sight appeared in civilian hands at an impromptu 1,000-yards shooting match held at the Congress Heights range near Washington, D.C., on April 4, 1918."[4]

On June 7, 1918, the Ordnance Department issued a requirement (No. 8514) to Winchester for the production of 52,635 "U.S. Telescopic Sight Rifles, Caliber .30, Model 1918."[5] Winchester archival records indicate that 59,472 Model 1918 sniper rifles were ordered (No. 16302, Supplement 2), but a subsequent Winchester document dated October 19, 1918, indicates that only 189 "Model 1918 Telescopic Sniper Rifles" were fabricated before the balance of the order was cancelled on January 10, 1919.[6]

Orders were placed with Winchester in the summer of 1918 for 32,000 Model 1918 telescopic sights to be manufactured under subcontract by the Eastman-Kodak Company of Rochester, New York. However, production difficulties resulted in only a few scopes being made before the Armistice. Most of the balance of the order was cancelled, although three of the scopes were sent to the Infantry School at Camp Benning, Georgia, for testing and evaluation. This testing resulted in the Infantry School recommending against adoption of the telescope.[7]

Despite the attempts to come up with the best telescope available, the Model 1918 proved to be disappointing. As discussed by E.C. Crossman in his authoritative *Book of the Springfield*, "The 1918 telescope was never a good glass and proved it promptly when subjected to use... I was one of the Department of Experiment (members)."[8]

The main deficiencies noted by Crossman were that the lenses became loose and useless after moderate firing, the rear mounting bracket was prone to breaking and the eye-relief was unsatisfactory.[9]

There are no known extant genuine Model 1918 sniper rifles complete with Model 1918 telescopic sights. However, there is what appears to be a Model 1918 sniper rifle, sans telescope, in the Buffalo Bill Center of the West Museum in Cody, Wyoming.[10] Although some testing of the Model 1918 sniper rifle continued into 1919, it was not a satisfactory

Model 1903 rifle with Winchester A5 telescopic sight.

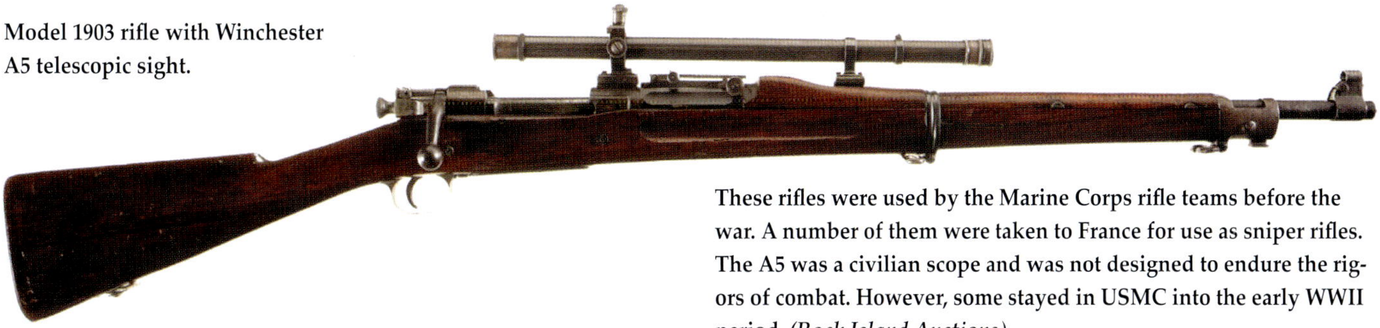

These rifles were used by the Marine Corps rifle teams before the war. A number of them were taken to France for use as sniper rifles. The A5 was a civilian scope and was not designed to endure the rigors of combat. However, some stayed in USMC into the early WWII period. *(Rock Island Auctions)*

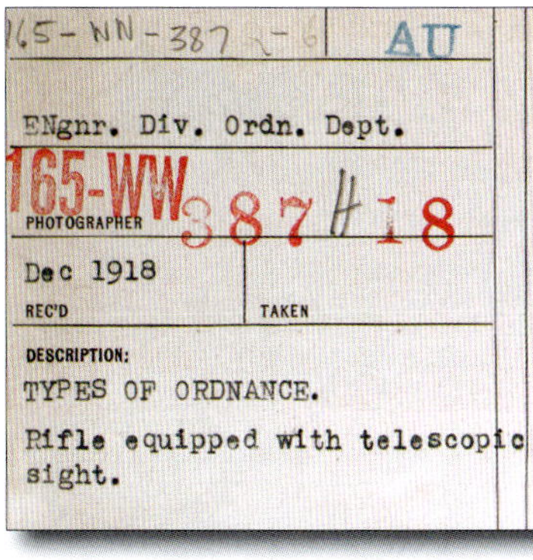

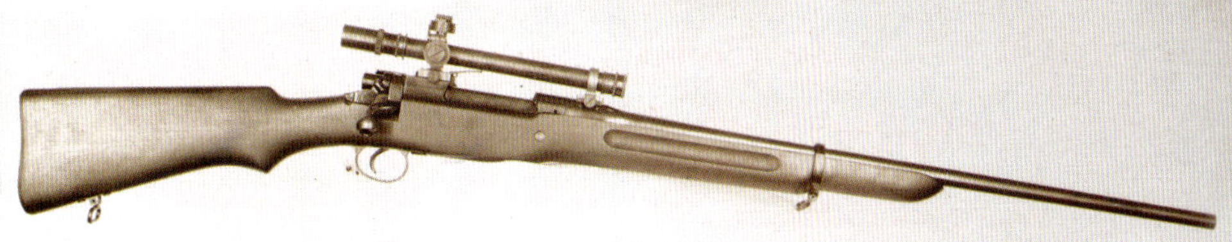

Ordnance photo dated December 1918 of a U.S. Rifle, Telescopic, Caliber .30, Model of 1918. *(U.S. Army Ordnance Department)*

Model 1918 Sniper rifle with Winchester Model 1918 telescopic sight.

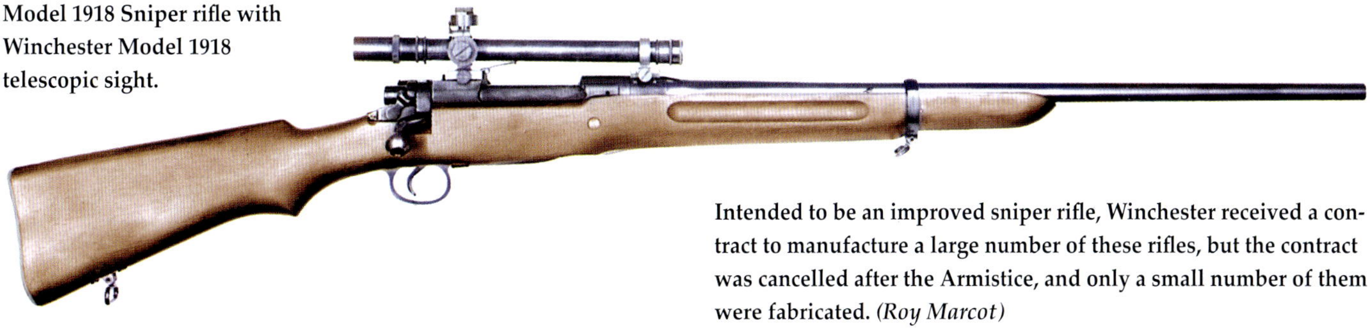

Intended to be an improved sniper rifle, Winchester received a contract to manufacture a large number of these rifles, but the contract was cancelled after the Armistice, and only a small number of them were fabricated. *(Roy Marcot)*

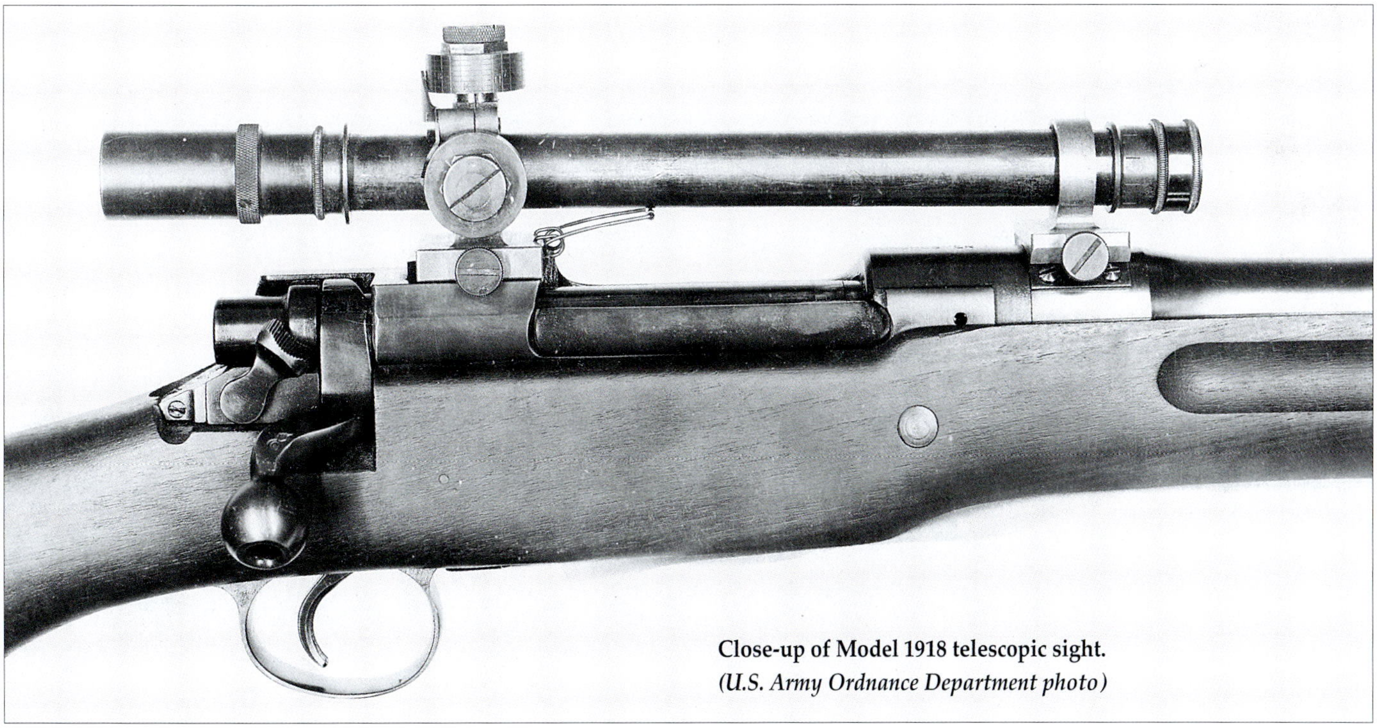

Close-up of Model 1918 telescopic sight. *(U.S. Army Ordnance Department photo)*

The rifle that is believed to be perhaps the only surviving specimen of the Model 1918 Sniper rifle, albeit with no telescope. *(Cody Firearms Museum. Courtesy of Danny Michael, Curator)*

Model 1917 rifle with special Mann-Niedner tapered mounts and Winchester A5 telescopic sight. *(Rock Island Auctions)*

Springfield Armory Museum photo of Model 1917 rifle with modified rear sight and A5 telescopic with tapered mounts. *(Springfield National Historic Site Museum. Courtesy of Peter Senich)*

weapon, and none were issued to American troops. The only sniper rifles to see action in the First World War were the Model 1903 rifles fitted with Warner & Swasey musket sights and a few with Winchester A5 telescopic sights.

Reportedly, there were a small number of Model 1917 rifles fitted with Winchester A5 telescopic and Mann-Niedner mounts during the war. These were presumably for the Marines to explore the feasibility of using some of the Model 1917 rifles as sniping weapons. The number of rifles fitted with A5 scopes under Marine Corps auspices is not known but was undoubtedly quite small.

There is an Eddystone Model 1917 rifle, ser. no. 69533, fitted with an A5 scope and an experimental "Rimkunas" rear sight in the Springfield Armory National Historic Site Museum, Catalog SPAR 5806. The accompanying documentation states: "Made by a member of the U.S. Marine Corps. Received from Maj. Besset, Ord. Dept., with instructions to be used as a study in developing rear sight with windage adjustment for U.S. Rifle Model 1917."[11]

The rifle is also pictured in Senich's excellent book *U.S. Marine Corps Scout-Sniper* with the caption, "A unique Model 1917 (Enfield) rifle with modified mount (tapered bases), Winchester A5 telescopic sight, and experimental Marine Corps rear sight."[12]

However, the concept of a scope-mounted Model 1917 did not find favor with the Marine Corps, likely due to preference for the Model 1903 and the fact that the Marines did not keep many of their Model 1917 rifles in inventory following the conclusion of the First World War. In the absence of compelling documentation, any example of a A5 scope-mounted Model 1917 rifle purported to be a genuine Marine Corps conversion should be approached with the utmost skepticism and caution.

"Trench Periscope" Rifles — The unique characteristics of trench warfare often required unconventional weapons. As discussed in the book *U.S. Infantry Weapons of the First World War*:

"The trenches of World War I were often a place where raising one's head above the parapet might well be one's last act on earth. The United States military was well aware of the horrific casualties inherent in trench warfare and looked for ways to reduce the threat. Many ideas were suggested, and several were eventually carried to the prototype stage and field tested. There were many periscopes issued that allowed for observation above the trenches without exposing the observer to enemy fire. While such items enabled observation to be done, it was not possible to return fire while viewing the enemy through a periscope. Other devices were considered such as metal shields, helmets with protective visors and various types of body armor. None of these devices proved to be practical. The most promising of these ideas was a device that allowed the rifle to be fired above a trench parapet while the rifleman was concealed below. Several nations experimented with this concept, but usage was quite limited and the concept never progressed much beyond the 'field trials' stage."[13]

Most of the belligerent nations, including Great Britain, France, Australia, Belgium, Russia, and Germany, experimented with such contrivances before the United States entered the war. Sighting was typically accomplished by mirrors or periscopes. A rifle, often the standard service rifle of the subject nation, was held in various types of frameworks ranging from rather crude wooden cradles to well-engineered metal supports.

When the United States came on the scene, there were a number of different ideas on how to build "trench periscope rifles," which were often referred to by the U.S. Army as "Sniperscopes." These included the Guiberson, Elder, Remington-Rioseco, and Cameron-Yaggi designs, any of which could be adapted to the Model 1917 as well as the Model 1903 rifle. The voluminous U.S. Army Ordnance Department reference publication *Handbook for Ordnance Data,* cited one particular model:

"*Periscopic attachment* — The Elder periscopic attachment, model of 1918, for attachment to United States rifles, model of 1903 and 1917... It is a device for supporting and permitting the operation of a rifle resting above the parapet, while the operator is below the trench. The device was tested and developed for the Ordnance Department, and as a result the 1918 model is lighter in weight, more substantial, and permits using sights upon the rifle. The experimental development of this device still in progress in the autumn of 1918. It consists of a rifle and shoulder brace with auxiliary mechanism, permitting the operation of the bolt by hand. It is not telescopic, and the rifle must be targeted. This device is not intended for use at long ranges."[14]

These unusual weapons were also the subject of an interesting article in the *American Rifleman* titled "Sniping from Below: Periscope Rifles in World War I":

"World War I saw the widespread introduction of many, what we would call 'modern' arms, including the aircraft, the machine gun, the armored vehicle, the submarine and even poison gas. When you drill down a little farther, there are several highly specialized arms used in the 'Great War' that were specific to the conflict.

One of these was the periscope rifle. Born in the battle of the trenches, the periscope rifle was a specialized tool for the deadly cat-and-mouse game of Great War snipers. The basic concept was to give the marksman the ability to aim and fire from the comparative safety of below ground—inside his trench position—without having to raise his head above the parapet and expose himself to enemy fire.

A standard rifle was cradled in one of many various style mounts, with a mirrored-periscope set up to provide the shooter with a reasonable approximation of a normal sight picture. A string or wire was attached to the trigger. The rest was the fine art of marksmanship.

Sniping was generally considered a dirty business, or a black art at best, on both sides and throughout most of WWI. Regardless of official sentiments, the snipers' deadly skills were readily cultivated and refined beginning in 1915. Allied sniper manuals from the Great War are quite detailed (and much of the information in those guides is still relevant today). The development of the periscope rifle was simply in keeping with the needs of the troops engaged in trenc h warfare, and periscope-sighted rifles could be found in service with the British, French, Belgian, Russian and German armies."[15]

As stated in an April 1917 Army War College publication:

"TRAINING IN THE USE OF THE RIFLE FOR TRENCH WARFARE"

Sniperscope: A device for aiming with the head below the line of sight. Reports indicate that with this device the tendency is to shoot high, and that they are only reasonably accurate up to ranges of 200 yards. They are usually used over parapets, between sandbags. Proper adjustment is essential and great care is necessary to prevent derangement."[16]

Later in the war, the following was published by the Equipment Section, Engineering Divisi on, A.E.F. 1918, which also referred to the weapon as "Sniperscopes:"

"The complexity of the present war, resulting in the various specialized vocations of the modern soldier, has created a need for many new items of equipment and necessitated a change in much of the old standard equipment. The Ordnance Department has conducted many tests of new inventions and devices, and investigations into the suitability of standard equipment. The following sections contain a brief synopsis of the more important items adopted and under consideration.

As the periscope is, comparatively speaking, a relatively new item of equipment, it was necessary to meet our requirements as completely as possible from French and British stocks, and owing to the scarcity of optical glass they have become more and more difficult of procurement. Everything possible is being done, however, to hasten production in the United States and increase the capacity of factories in Europe.

The sniperscope is a device consisting of a dummy gun or support in which the rifle rests and a periscope sighting arrangement. The type now being furnished is also equipped with a bolt-working mechanism and the operator can aim and fire an entire clip of cartridges without exposing himself to the view of the enemy. Effort is now being made to design a sniperscope that will function with either the 1903 or the 1917 rifle."[17]

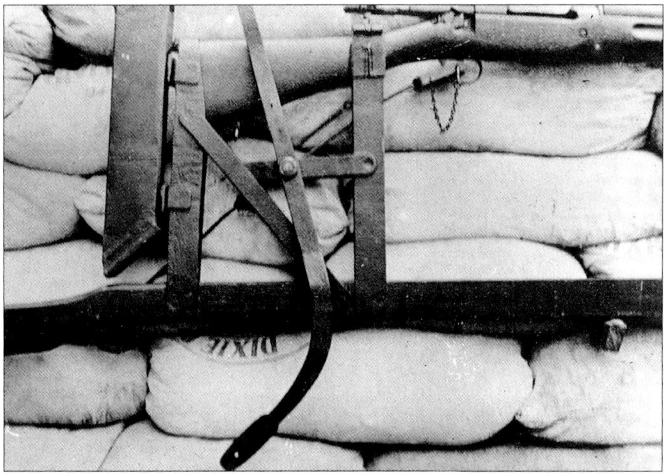

Unidentified trench periscope rifle contrivance for Model 1917 rifle. Note the box periscope at the rear, as well as the lever to manipulate the bolt and the chain to pull the trigger. *(Tom Laemlein)*

Photo of American solider sighting the rifle in a simulated trench position made with sandbags. *(Tom Laemlein)*

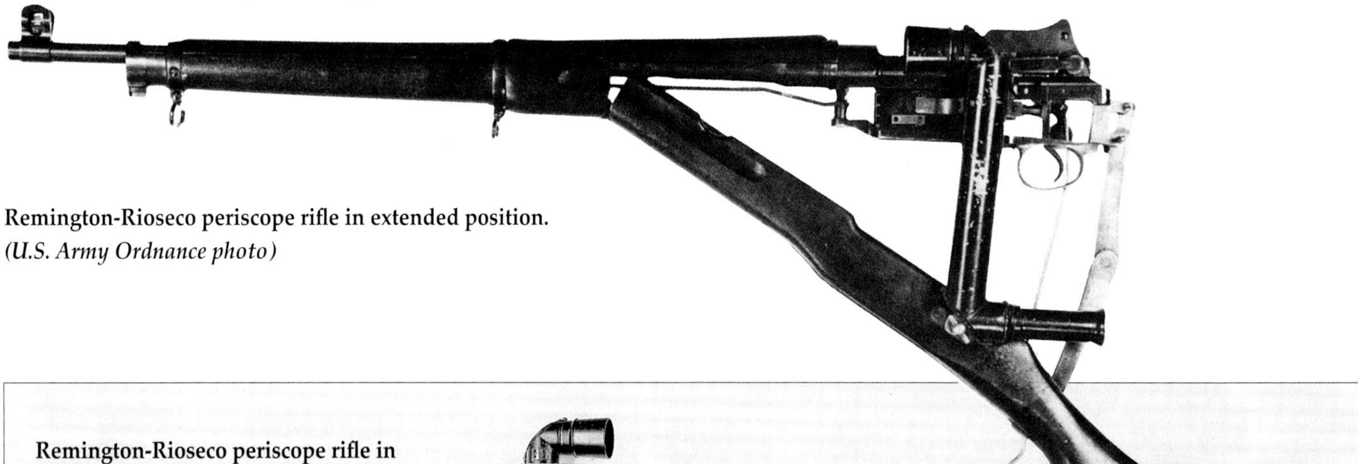

Remington-Rioseco periscope rifle in extended position.
(U.S. Army Ordnance photo)

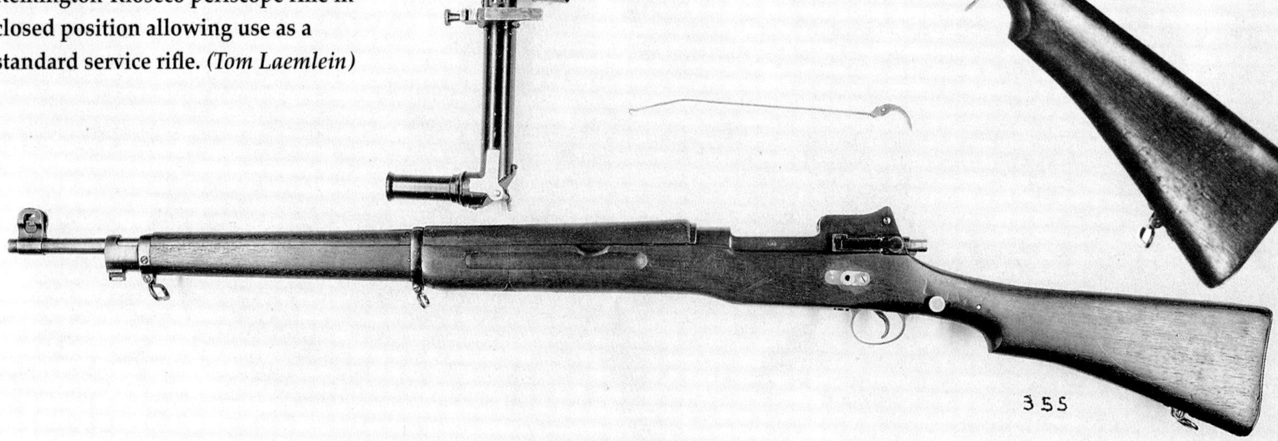

Remington-Rioseco periscope rifle in closed position allowing use as a standard service rifle. *(Tom Laemlein)*

Guiberson trench periscope rifle based on a Model 1903. Reportedly, there were some that were fabricated using Model 1917 rifles as well. *(Rock Island Auctions)*

Cameron-Yaggi trench periscope rifle incorporating Model 1903 rifle and fitted with the "Sightascope" optical periscope. The design was equally adaptable for both the M1903 and M1917 rifles, with virtually no modifications required. *(Author's collection)*

A February 9, 1918, memorandum from the Requirements Section, Control Bureau to the Engineering Bureau, Small Arms Division regarding "Periscope Rifles" made it clear that a weapon of this type should be available for both the Model 1903 and Model 1917 rifles. It also indicated that an improved design had been developed.

"1. Memorandum from the Small Arms Division of December 12, 1917 and January 5, 1918, asked for the proportion of Periscope Rifles, which should be provided for 1903 and 1917 rifles.

2. Since it is understood that the Periscope Rifle has been superseded by the Sitascope, and since the Sitascope can be applied to either the 1903 or 1917 rifle, it is believed that the information asked for is now unnecessary and that procurement can proceed.

3. Advice is requested if the above assumption is not correct."[18]

The "Sitascope" referred to was the prismatic optical sighting telescope developed in conjunction with the

Solider holding the assembly showing that the rifle was a standard Model 1917. *(Tom Laemlein)*

Left side of Cameron-Yaggi rifle showing the "Sightascope."

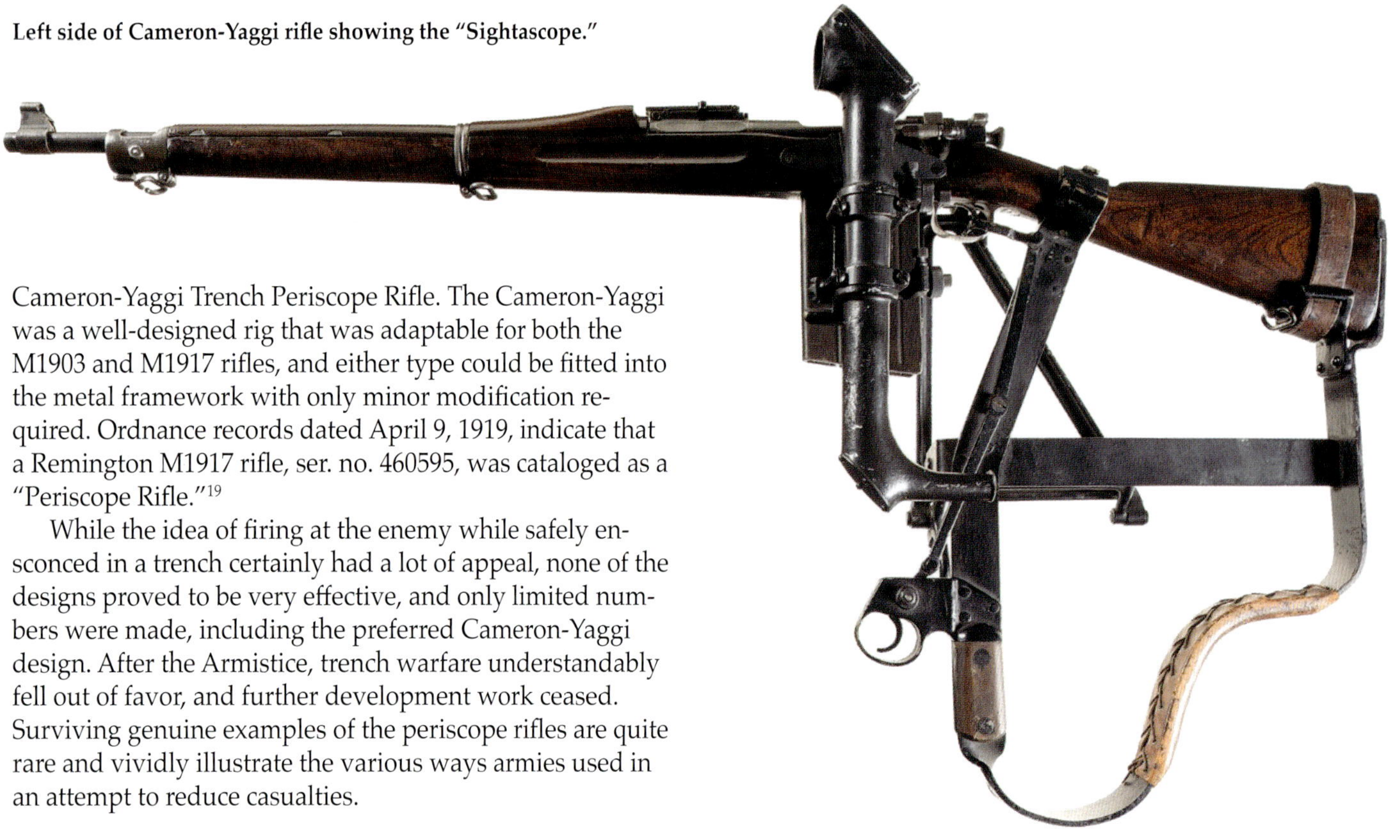

Cameron-Yaggi Trench Periscope Rifle. The Cameron-Yaggi was a well-designed rig that was adaptable for both the M1903 and M1917 rifles, and either type could be fitted into the metal framework with only minor modification required. Ordnance records dated April 9, 1919, indicate that a Remington M1917 rifle, ser. no. 460595, was cataloged as a "Periscope Rifle."[19]

While the idea of firing at the enemy while safely ensconced in a trench certainly had a lot of appeal, none of the designs proved to be very effective, and only limited numbers were made, including the preferred Cameron-Yaggi design. After the Armistice, trench warfare understandably fell out of favor, and further development work ceased. Surviving genuine examples of the periscope rifles are quite rare and vividly illustrate the various ways armies used in an attempt to reduce casualties.

Model 1917 Mark II Rifle (Prototype) for Pedersen Device — As briefly mentioned in Chapter 5, one of the most interesting items intended for use during World War I was developed by noted arms designer and engineer John Douglas Pedersen. Pedersen sought to come up with a mechanism that would address the two fundamental problems of all bolt-action service rifles of the period: the relatively slow rate of fire and the fact that the typical cartridge, such as the U.S. .30-06, could be overly powerful for some applications.

To this end, Mr. Pedersen devised an ingenious solution by developing an "automatic bolt" that permitted the rifle to fire a small .30 caliber cartridge, similar to a lengthened .32 ACP pistol cartridge or the later French 7.65mm MAS pistol round. The mechanism, widely known as the "Pedersen Device," operated in essentially the same manner as a "blow-back" pistol. The "barrel" of the device was rifled and was the same configuration as a .30-06 cartridge case, which enabled it to fit into the rifle's chamber. The device was easily installed by removing the rifle's bolt and inserting the device, which was locked in place by means of a slightly modified magazine cut-off lever. A small lever allowed the rifle's trigger to move forward with each pull to trip the device's sear and permitted the rifle to fire semiautomatically.

Prototype Model 1917 Mark II Rifle with Mark II Pedersen Device with a 40-round magazine and box of ammunition. ***(James Julia Auction Co.)***

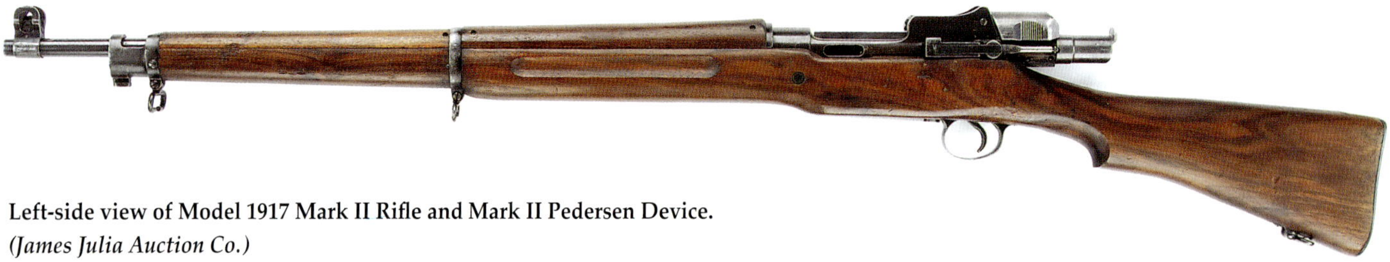

Left-side view of Model 1917 Mark II Rifle and Mark II Pedersen Device. *(James Julia Auction Co.)*

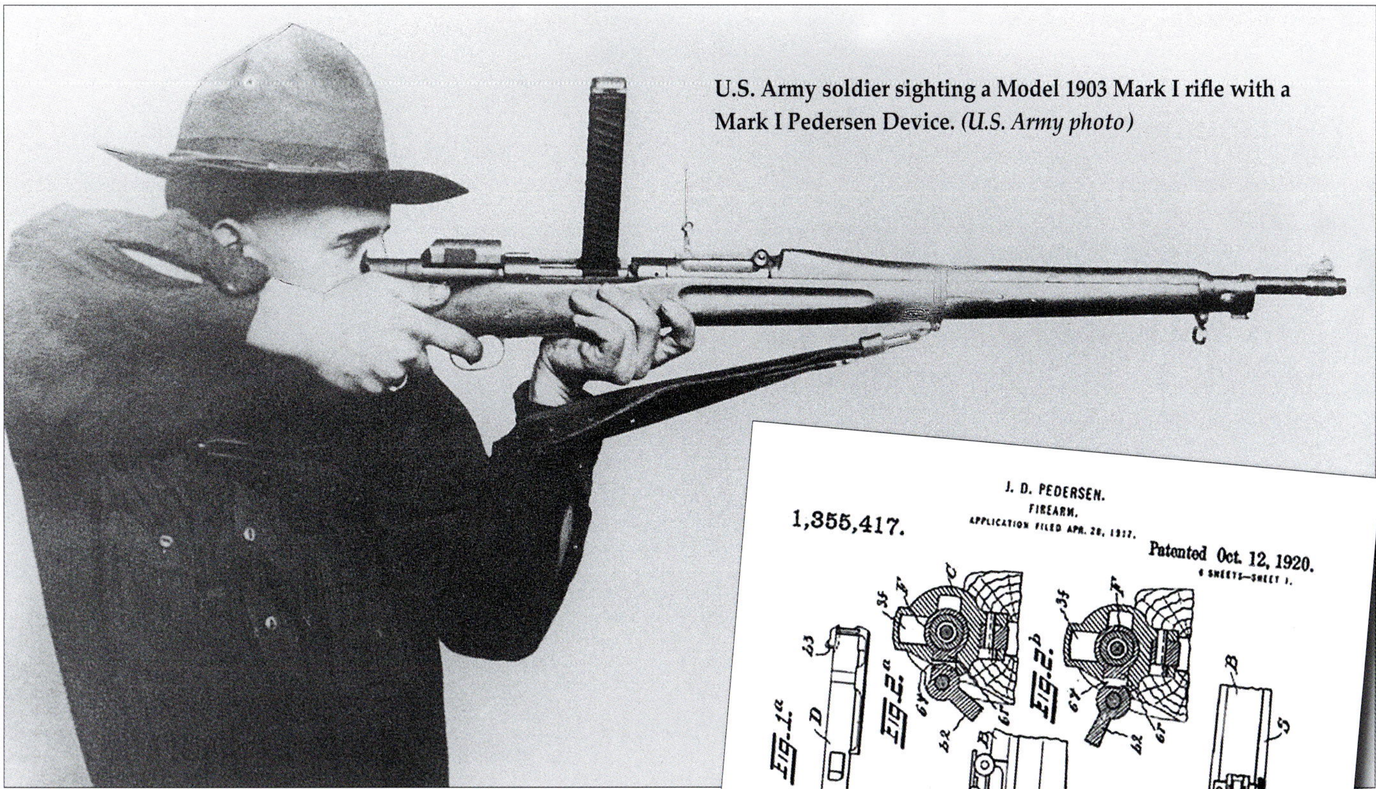

U.S. Army soldier sighting a Model 1903 Mark I rifle with a Mark I Pedersen Device. *(U.S. Army photo)*

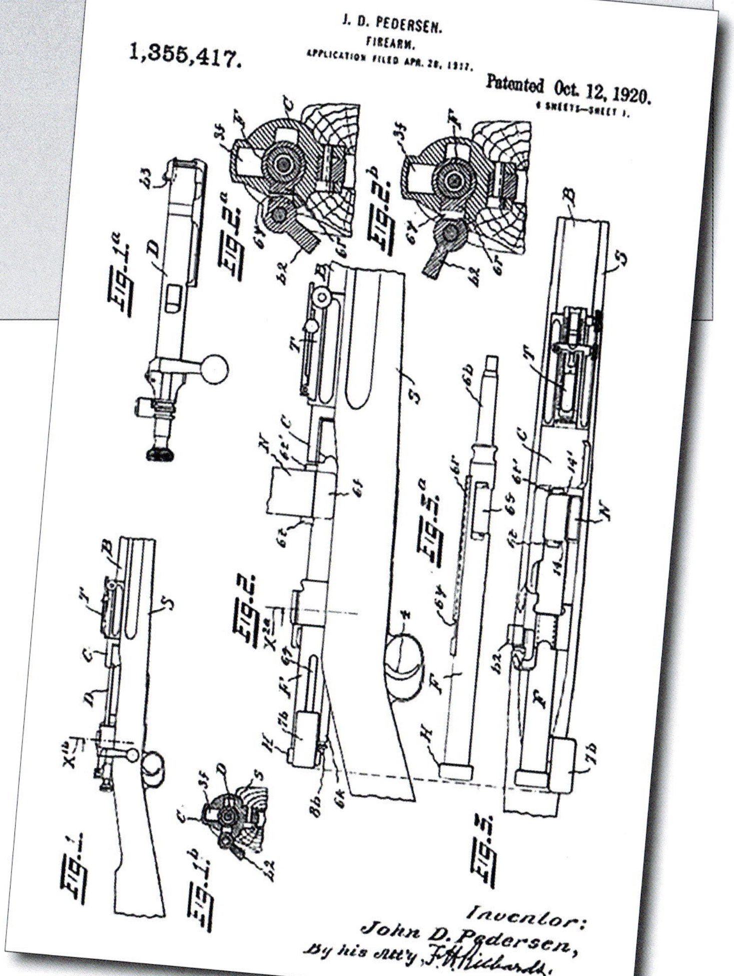

John D. Pedersen's patent application filed April 28, 1917, for his "device" and modifications to the Model 1903 rifle. The patent was not granted until October 12, 1920, likely due to the "Top Secret" classification during World War I.

A 40-round magazine was attached to the device at about a 45-degree angle and was secured by two spring-loaded "fingers." This allowed the rifle to be sighted in the normal manner with the magazine in place. Fired cartridge cases were ejected through an ejection port milled into the left side of the receiver. Besides the addition of the ejection port in the receiver wall, the only other modifications required to adapt a Model 1903 rifle for use with the Pedersen Device were a slightly modified magazine cut-off, trigger and a slight inletting of the stock underneath the ejection port.

U.S. M1906 (.30-06) cartridge (far right); M1918 Pedersen cartridge (near right). Although both are .30 caliber, the size disparity between the two cartridges is quite evident.

Mark I Pedersen Device. Note the "barrel" of the device, which is same configuration as a standard .30-06 cartridge. *(Author's collection)*

The device was carried in a sheet metal "scabbard," and a canvas case was provided to hold the rifle's bolt when it was removed to insert the device. A canvas pouch held five magazines. Cartridge belt hooks were fitted for attachment of the pouches and the metal scabbard to the standard infantry cartridge belt.

Mr. Pedersen demonstrated his invention to a group of select Army officers and Congressmen, all sworn to secrecy, at the Congress Heights Rifle Range near Washington, D.C., in the summer of 1917. The onlookers were amazed by the volume of fire and the ease with which the rifle could be switched back forth to fire the standard .30-06 round and the diminutive .30 caliber Pedersen cartridge. An ordnance officer who witnessed the demonstration extolled the virtues of the device for both offensive and defensive combat situations:

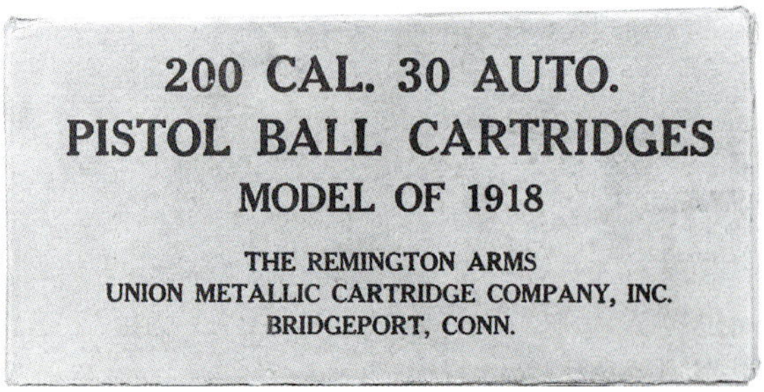

Pedersen Device ammunition. Loose cartridges, a 40-round box and a 200-round carton containing five 40-round boxes. One 40-round box would fill one magazine. This ammunition was to be used in both the Mark I and Mark II Pedersen Devices. ***(Author's collection)***

Collage of Pedersen Device showing a Model 1903 Mark I rifle with device attached, 200-round and 40-round boxes of ammunition, WWI infantry belt, bolt pouch with bolt and metal carrying case. The "Mark II" Pedersen Device for the Model 1917 rifle would utilize the same accessories and ammunition as the Mark I variant. ***(Rock Island Auction Co.)***

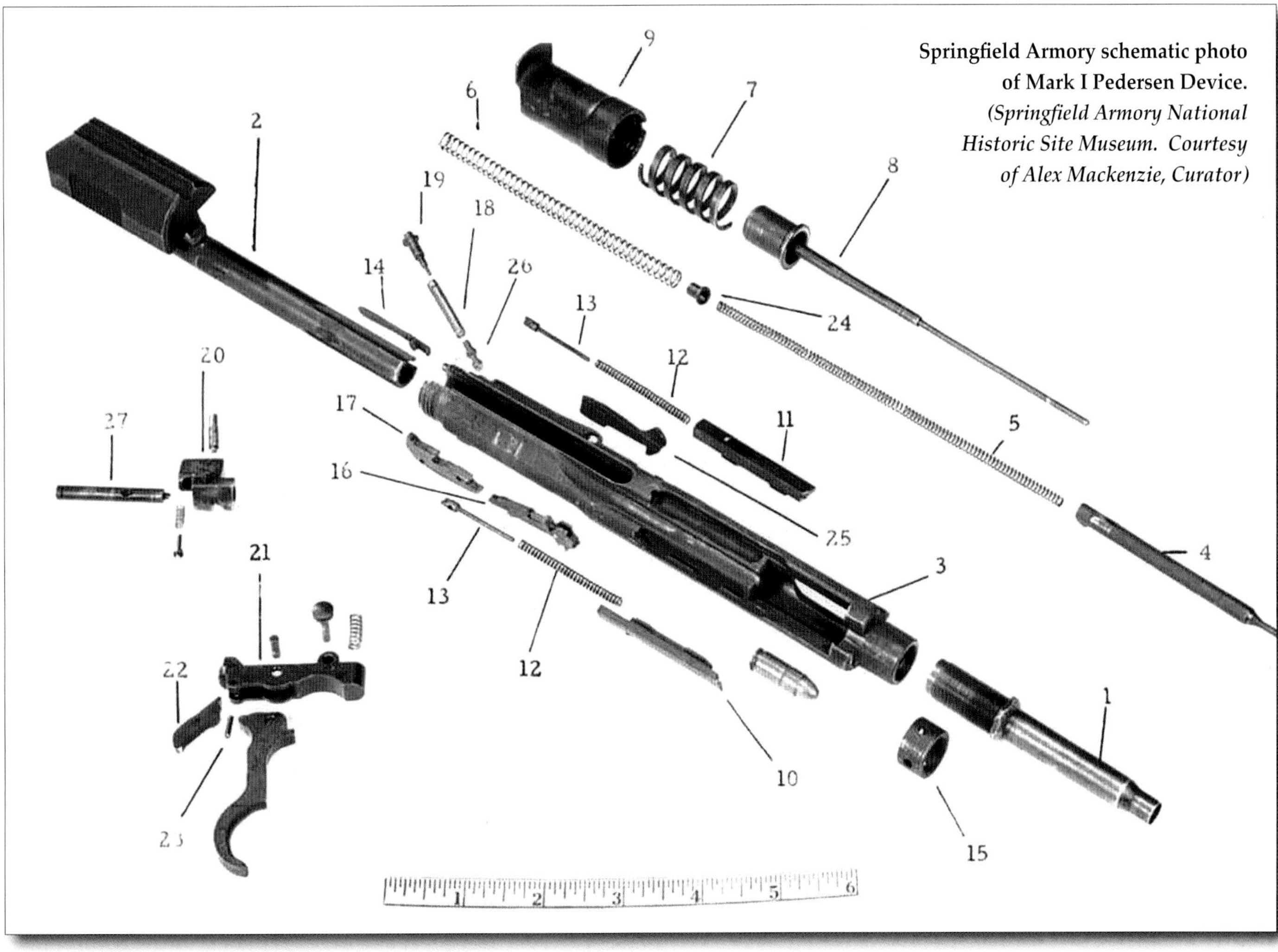

Springfield Armory schematic photo of Mark I Pedersen Device. *(Springfield Armory National Historic Site Museum. Courtesy of Alex Mackenzie, Curator)*

> "As the enemy came charging across No Man's Land each of our soldiers would start firing with this miniature machine gun and the entire zone in front of the trenches would be covered with such a whirlwind of fire that no attack could survive.
>
> ...a line of advancing across No Man's Land firing this device at the enemy trenches as they ran would make it extremely difficult for anyone in the trenches to show his head or any part of his body. Of course, fire while running or walking would not be so accurate, but the tremendous number of shots would more than make up for any inaccuracies and the whole of the enemy trench system would be smothered with a storm of bullets."[20]

The War Department was convinced that the Pedersen Device could be a "game changer" on the battlefields of France. In order to preserve secrecy, the device was given the purposely misleading name of "Automatic Pistol, Caliber .30, Model of 1918." An initial order for 100,000 Pedersen Devices for the Model 1903 rifle, designated as the "Mark I," was placed with the Remington Arms Company on March 26, 1918, followed by contracts for large numbers of magazines and cartridges. The original order was subsequently increased to 133,450 devices. Springfield Armory was ordered to go into production on a Model 1903 rifle with the necessary modifications for use with the Pedersen Device. The rifle was designated as the "U.S. Rifle, Caliber .30, Model of 1903, Mark I."

Despite repeated delays while Pedersen continued to revise his design, Remington and Springfield Armory were gearing up to manufacture the Pedersen Devices and Model 1903 Mark I rifles, and the War Department was already making plans to use the new weapon in the "Grand Offensive" that was expected to be launched in the spring of 1919. However, it was becoming apparent that Springfield Armory may not have been able to meet the required timeframe. To help alleviate any potential shortfall, John Pedersen was requested to develop a version of his device for use with the Model 1917 rifle. The new variant device would be designated "Mark II." On September 20, 1918, Remington was to be given a contract for the production of a large number of Mark II devices after the original Mark I contract was completed.

A July 31, 1918, cablegram from General John Pershing, General Headquarters, A.E.F. contained a passage to the Chief of Staff pertaining to the Pedersen Devices:

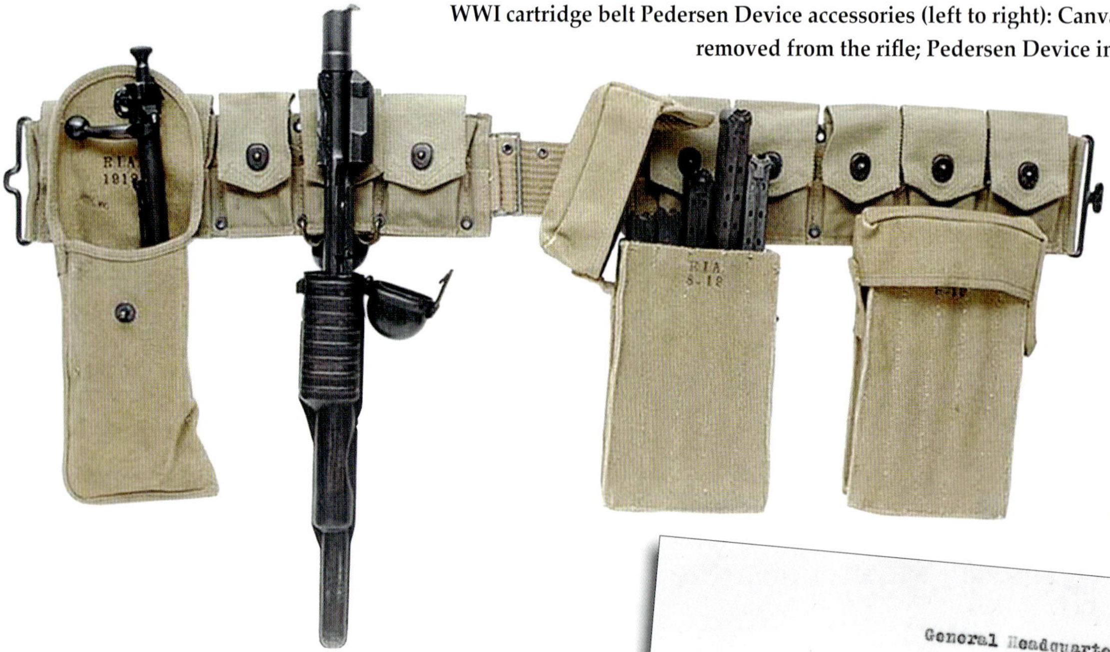

WWI cartridge belt Pedersen Device accessories (left to right): Canvas pouch holding bolt when removed from the rifle; Pedersen Device in metal carrying "scabbard;" Magazines and pouch. Each pouch holds five 40-round magazines. *(Author's collection)*

Cablegram dated July 31, 1918, from General John Pershing, General Headquarters, A.E.F., contained a passage to the Chief of Staff mentioning Pedersen Devices for the Model 1917 rifle. *(Courtesy of Frank Mallory)*

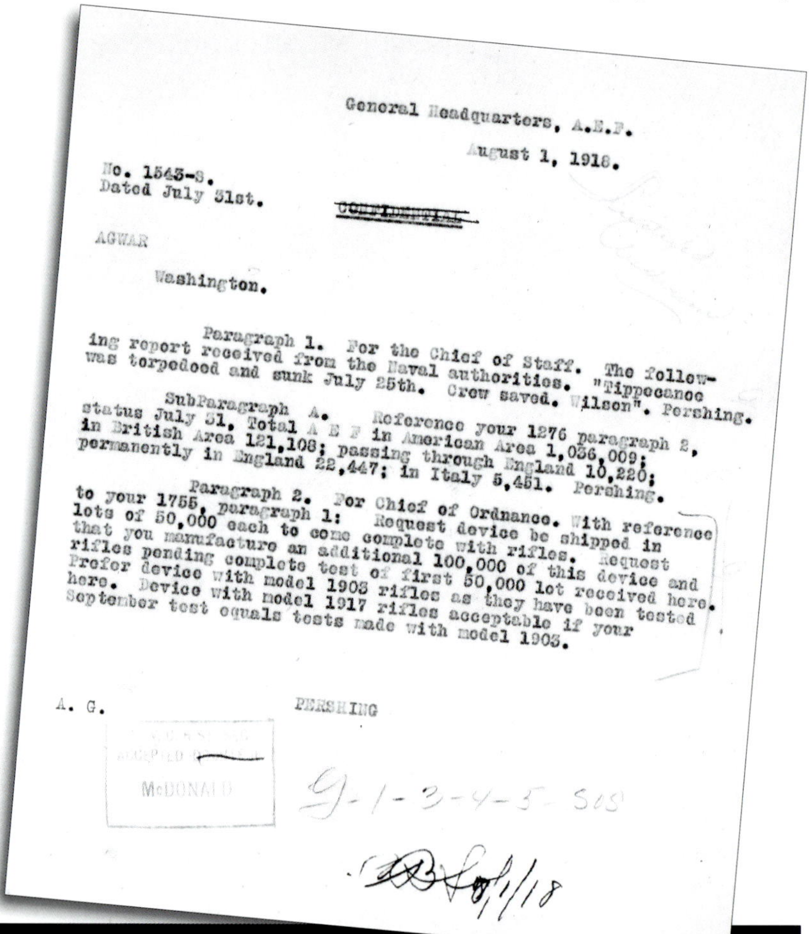

General Headquarters, A.E.F.

August 1, 1918.

No. 1543-S.
Dated July 31st.

AGWAR

Washington.

Paragraph 1. For the Chief of Staff. The following report received from the Naval authorities. "Tippecanoe was torpedoed and sunk July 25th. Crew saved. Wilson". Pershing.

SubParagraph A. Reference your 1276 paragraph 2, status July 31, Total A E F in American Area 1,036,009; in British Area 121,108; passing through England 10,220; permanently in England 22,447; in Italy 5,451. Pershing.

Paragraph 2. For Chief of Ordnance. With reference to your 1755, paragraph 1: Request device be shipped in lots of 50,000 each to come complete with rifles. Request that you manufacture an additional 100,000 of this device and rifles pending complete test of first 50,000 lot received here. Prefer device with model 1903 rifles as they have been tested here. Device with model 1917 rifles acceptable if your September test equals tests made with model 1903.

A. G. PERSHING

McDONALD

9-1-3-4-5-SOS

"Request device be shipped in lots of 50,000 each to come complete with rifles. Request that you manufacture an additional 100,000 of this device and rifles pending complete test of first 50,000 lot received here. Prefer device with model 1903 rifles as they have been tested here. Device with model 1917 rifles acceptable if your September test equals tests with model 1903."[21]

As can be seen, General Pershing had high hopes for the Pedersen Device. Results of the testing of the Model 1917 rifle with Pedersen Device have not been discovered. However, given that General Pershing referred to another test of the Model 1903 Mark I device scheduled for September, which was only two months before the Armistice, the testing of the Mark II device for the Model 1917 rifle may not have happened.

Close-up of prototype Model 1917 Mark II Rifle fitted with Mark II Pedersen Device. Note the lack of markings since Mark II devices were factory prototypes.

(James Julia Auction Co.)

Inset at left is a close-up of the cartridge ejection port that was milled into the receiver on the left side. This was also the case with the Model 1903 Mark I rifle used with the Mark I Pedersen Device. ***(Hayes Otoupalik Collection. Photograph by Bill O'Donnell)***

Another example of a prototype Model 1917 rifle with a Mark II Pedersen Device with magazine attached.

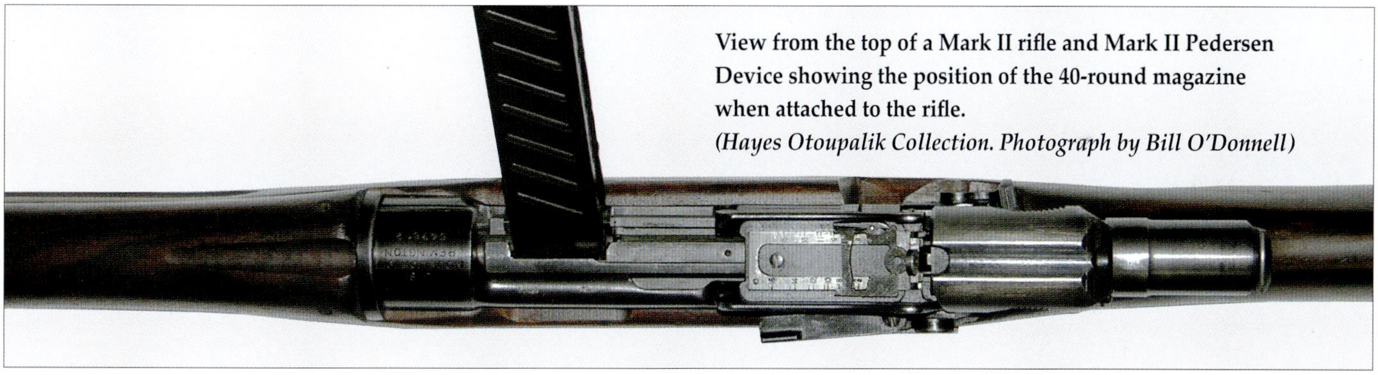

View from the top of a Mark II rifle and Mark II Pedersen Device showing the position of the 40-round magazine when attached to the rifle.
(Hayes Otoupalik Collection. Photograph by Bill O'Donnell)

Likely getting ready for what was assumed to be a large contract for the Mark II Pedersen Devices, Remington fabricated a few, probably no more than half a dozen, along with a like number of Model 1917 rifles altered for use with the device. Plans were formulated to go into production of the purpose-built Model 1917 rifle, modified for use with the Pedersen Device, which would be designated as the "U.S. Rifle, Caliber .30, Model 1917, Mark II" when the contract was granted. Interestingly, Remington had previously fabricated a very small number, possibly only one or two, Pedersen Devices for the Russian Mosin-Nagant rifle. However, the fall of the Czar quickly put an end to that program.

Before production of the Model 1917 Mark II rifles and Mark II Pedersen Devices began, the Armistice was declared, and all outstanding contracts were cancelled soon after. By this time, production came to an end on March 1, 1919, and Remington had made some 65,000 Mark I Pedersen Devices. Springfield Armory produced 97,520 Model 1903 Mark I rifles before the production order was also cancelled. The Mark I Pedersen Devices were never issued and remained in storage until April 1931 when they were ordered to be destroyed. Surviving examples are rare and very desirable martial collectibles. Most of the Model 1903 Mark I rifles subsequently had the special Pedersen Device-related components replaced by standard parts and were issued as regular service rifles.

The Pedersen Device was a remarkable feat of engineering, but it was never issued and is now consigned to the scrapheap of U.S. military history. Surviving Mark II Pedersen Devices and modified Model 1917 Mark II rifles are incredibly rare. The number of genuine examples extant can be counted on one hand, likely with a couple of fingers left over. Despite the fact that it was never fired in anger, the Pedersen Device vividly illustrates the imaginative thinking that went into attempts to give our doughboys an edge in the savage trench combat of the Western Front in 1917–1918.

Model 1917 "Carbine" — As briefly cited in Chapter 7, there was a little-known but very interesting modification of the M1917 rifle created during WWII, which was intended to be used by our Chinese allies. For a long time, the origins of this seldom-seen weapon have been the subject of much conjecture and unfounded theories. Fortunately, thanks to information compiled by David Albert, the story is now complete. In 1987, Mr. Albert acquired a shortened M1917 rifle. Others who saw the rifle were skeptical of it having any sort of official military connection and believed it was simply the creation of a talented gunsmith. Some speculated that it was perhaps modified for use in the Philippines, but this could not be verified, and no reference sources mentioned such rifles. In 2013, Mr. Albert was made aware of another rifle that appeared identical to his specimen. The owner of the other example also gave him a lead as to where to find some authoritative, reliable information on the subject. While never any sort of official nomenclature, he dubbed the weapon the "M1917 Carbine."

The following is excerpted from Mr. Albert's 2014 monograph titled "The U.S. M1917 Carbine: Mystery Solved":

> "It turned out the answer existed under the author's nose for almost two years, and the individual who made contact via e-mail provided information that shed light down the correct trail. The 'M1917 Carbine' (author designated name) was indeed a variation that saw service, and was modified at the direction of the U.S. Army. It saw WWII service with American-trained Chinese troops who were fighting the Japanese in the China-Burma-India (CBI) theatre. The Chinese had an urgent need for weapons, and the allies provided and trained the Chinese with a variety of them through the efforts of the U.S. Army Services of Supply, China Theatre. (SOS CT).
>
> The M1917 rifle became the main weapon issued to Chinese troops. The standard M1917 is a long rifle, and its ergonomics did not match well with the stature of most Chinese soldiers. As a result, 18,000 M1917s were experimentally shortened by 4 inches in barrel length and had their stocks also slightly shortened (shorter length of pull). The initial experimental carbines were likely modified by U.S. Army armorers at a Ramgarh, India training camp.
>
> Ramgarh was a U.S. Army facility for a large Chinese troop training effort, located in eastern India, close to the Nepalese crossing route from China.

Photo dated 1945 of surplus weapons in India, consisting of a number of M1928A1 Thompson submachine guns and several shortened M1917 rifles. *(David Albert)*

Standard M1917 Rifle.

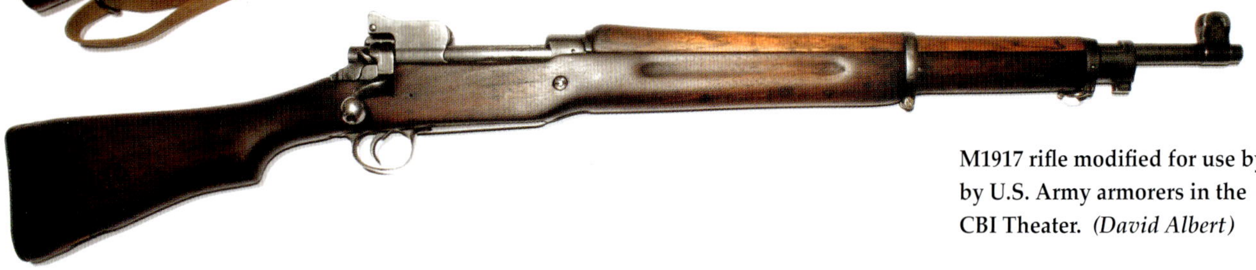

M1917 rifle modified for use by Chinese by U.S. Army armorers in the CBI Theater. *(David Albert)*

Period photographs, training documents, and the 1945 book, *Ramgarh: Now It Can Be Told,* confirm the M1917 carbines were modified under contract. The 2011 book *Soldiers of the White Sun*, by Philip Jowett contained two photos of the M1917 carbine, but they had gone unnoticed by the author, as the book was purchased to research Chinese Thompsons, and the M1917s just did not register during the author's perusal.

The U.S. Army base at Ramgarh was converted from a British POW camp and initially served to rehabilitate Chinese soldiers following their May 1942 withdrawal from Burma. Its purpose soon expanded as additional Chinese troops were flown in for training, until it became the prime training site for the CBI Theatre in late 1942. Weapons and other training courses were taught by U.S. Army personnel in a challenging physical environment. In addition to the M1917 rifle, many at Ramgarh were trained on the M1928A1 Thompson Submachine Gun, Bren Light Machine Gun, and the M1917 Browning Machine Gun, and also some larger support weapons, including mortars, pack howitzers, as well as 105mm and 155mm howitzers. Courses of instruction were published for each weapon.

The modification of the standard M1917 rifle to the M1917 carbine involved shortening the barrel by 4 inches, shortening the front of the stock and handguard, and removing about 1 inch of the buttstock. The short rifle measures 41⅛ inches overall, while the standard rifle is 46¼ inches overall. Both rifles documented were manufactured by Eddystone. However, Remington examples were almost certainly also modified, based on the circumstances of their in-theatre modification. Both example rifles have markings on the stock, indicating they were reworked at Ishapore Arsenal in India. This mark went unnoticed by the author until the rifle was displayed at a 2014 meeting of the American Society of Arms Collectors, where a fellow member inspected the stock very closely, and noticed the mark. Upon discovery, the other M1917 carbine example was checked and was confirmed as having the same arsenal mark. The Ishapore Arsenal was only about 27 miles from Ramgarh, making it a convenient resource for the modifications.

Various photographs exist of Chinese troops using the M1917 carbine in WWII, one of which is proof by itself of the rifle's use in WWII. The photo shows a U.S. Army captain instructing a Chinese major general how to shoot an M1917 carbine. Other photographs show the rifle in training and field use.

The M1917 carbine is indeed a scarce U.S. variation that deserves further historical documentation and collector awareness. Over 70 years after its official modification, it is quite possibly the only remaining undocumented U.S. small arms variation of WWII. Today, the M1917 has less of a collector following than other U.S. rifles and can still be purchased at reasonable prices. Be on the lookout for its short version, as it surely saw combat use and comes as close as any firearm to fulfilling the commonly heard wish of "If only this rifle could talk..." It was used quite effectively against the Japanese by American-trained Chinese troops."[22]

Genuine M1917 rifles shortened for use by our Chinese allies under U.S. Army auspices are quite rare today. It would not be particularly difficult for someone with reasonably competent gunsmithing skills to alter a standard M1917 rifle to such a similar configuration today, thus any rifle of this type offered for sale should be approached very warily. In any event, these shortened M1917 rifles are an almost unknown part of the story of the M1917 rifle in World War II and further illustrate how these WWI-vintage rifles helped serve the forces of freedom over a quarter century after they were manufactured.

Rare photo (albiet, poor quality) of a U.S. Army captain coaching a Chinese Army Major General with marksmanship techniques using a shortened M1917 rifle. *(David Albert)*

Interesting collage of four different views of a shortened M1917 rifle. *(C&RSENAL)*

Winchester Bolt-Action rifles, Models A thru E, designed by Thomas C. Johnson.

(Cody Firearms Museum.
Courtesy of Danny Michael and Jathan Clark)

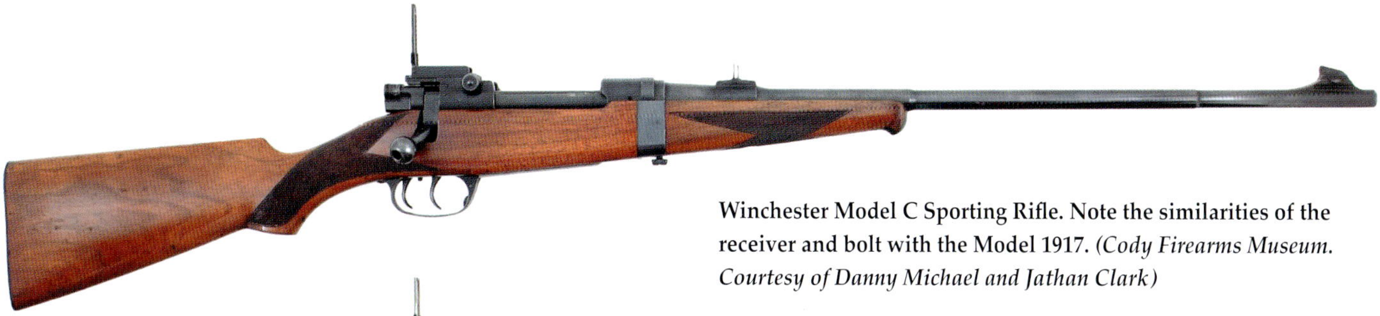

Winchester Model C Sporting Rifle. Note the similarities of the receiver and bolt with the Model 1917. *(Cody Firearms Museum. Courtesy of Danny Michael and Jathan Clark)*

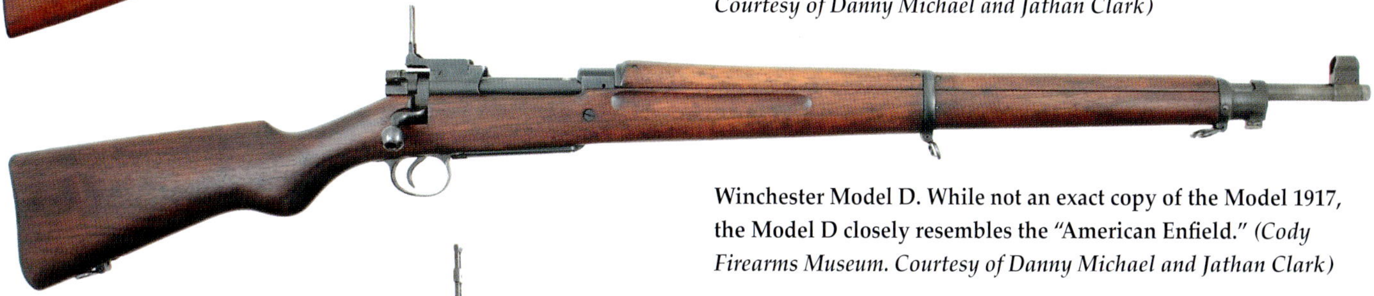

Winchester Model D. While not an exact copy of the Model 1917, the Model D closely resembles the "American Enfield." *(Cody Firearms Museum. Courtesy of Danny Michael and Jathan Clark)*

Winchester Model 51 rifle.
(Cody Firearms Museum. Courtesy of Danny Michael)

Commercial Sporting Rifles — As alluded to in Chapter 6 (Between the Wars), shortly after the Armistice, production of the Model 1917 r ifle would soon be discontinued. Winchester and Remington sought to utilize the experienced personnel and manufacturing tooling and equipment on hand by developing rifles for the commercial market based, in some measure, on the Model 1917. Eddystone did not embark on such a project because the entity was no longer manufacturing firearms.

Winchester — Thomas C. Johnson, a noted firearms designer at Winchester, designed several bolt-action rifles intended for both civilian and military markets. The model designations ranged from "A to E." The "Model A" and "Model B" rifles were similar in appearance to the Model 1903 Springfield. The "Model C" more closely resembled the Model 1917 but also had some features comparable to the Model 1903 and was made in military and civilian configurations.

The "Model D" more closely resembled the Model 1917 rifle and was intended primarily for the military market. As related in Chapter 4, in November of 1916, Russia ordered 500 "Model D" rifles, which were sent in March of 1917 for evaluation and testing, but the Bolshevik Revolution spelled an end to any more orders.

The final Johnson-designed bolt-action rifle, was the "Model E." Only two examples were made, but the design evolved into the "Model 51 Imperial" rifle, which was the first bolt-action rifle introduced by the company following the First World War. However, only 25 or so Model 51s were made before production was cancelled.[23]

While the Model 51 never reached mass production, Winchester still harbored hopes that the Model 1917 production equipment still on hand, but owned by the government, could be utilized for the manufacture of civilian rifles. These hopes were bolstered in the summer of 1919. As referenced by author and researcher Herbert Houze:

> "At the time, the United States Government began making overtures to the Winchester Company concerning a possible exchange of the government's

Winchester Model 54 Rifle. *(Rock Island Auction Co.)*

Model 1917 Rifle machinery in the New Haven factory for the firm's Browning Automatic Rifle tooling. By late autumn, these negotiations were well advanced, and it was clear that the firm would acquire full ownership of the Model 1917 machinery. Realizing that this would put Winchester in an almost immediate position to begin production of a sporting rifle based upon the Enfield action, Remington decided to strike first."[24]

Winchester's discontinued Model 51 was then followed by the Model 54, which was manufactured using some modified Model 1917 production machinery, but it bore little resemblance to the Model 1917 rifle. It stayed in the Winchester product line until c.1936 when it was discontinued in favor of the company's superb Model 70 rifle.

Remington — Of course, like Winchester, Remington was well aware of the advantages that could accrue from the utilization of the Model 1917 machinery being used to manufacture rifles for the commercial market. Remington decided to develop a bolt-action rifle heavily based on the Model 1917.

Model 30 and 30 Express Sporting Rifles — As discussed by Remington researcher and author Roy Marcot:

> "As early as November 1917, Rem-UMC officials envisioned a sporting rifle based on the Model 1917 military action. In April 1921, Remington Arms Co., Inc., announced its first high-power bolt-action hunting rifle since the Remington-Lee rifle had been discontinued in 1909 – the Remington Model 30 Rifle. Remington designers Crawford C. Loomis and C.H. Barnes utilized leftover receiver forgings and other parts from Model 1917 Enfield rifles that Remington had built during the war. Like the military configuration, the Model 30 rifle was chambered for the versatile .30-06 Springfield centerfire cartridge. Higher quality versions with fine wood, checkering, and engraving were available upon request."[25]

The Model 30 was superseded by the improved "Model 30 Express" rifle, and it remained in the company's product line until around 1940, available in several variations, but using a conventional cock-on-opening bolt. The basic level of the Model 30 Express was the Model 30A, with a 22-inch barrel and barrel-mounted sights. These were available in seven different calibers, but .30-06 was most popular. About 22,800 Model 30 Express rifles were made in all variations.

Right and left-side views of the Remington Model 30 Express rifle. *(Rock Island Auction Co.)*

Remington Model 30 Express receiver markings.

Close-up of Model 30 Express action. The configuration of the bolt and safety are virtually identical to the Model 1917 rifle.

Remington M720 rifle. *(Rock Island Auction Co.)*

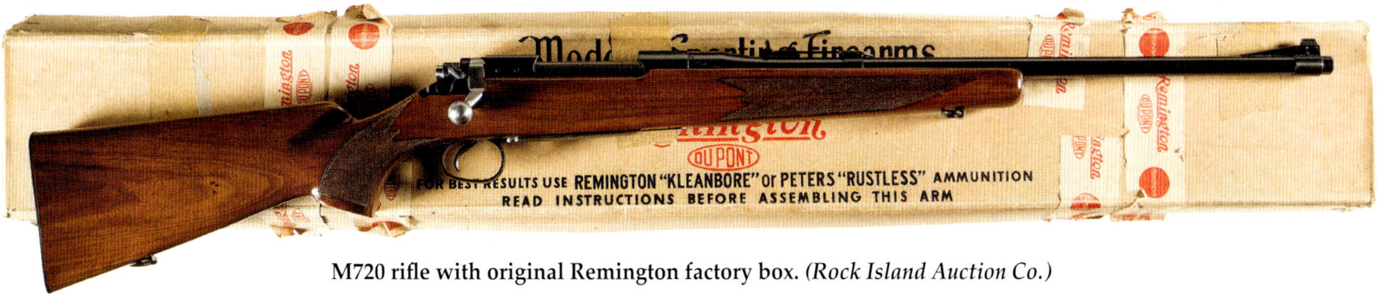

M720 rifle with original Remington factory box. *(Rock Island Auction Co.)*

Markings on M720 rifle showing model designation, serial number and caliber. *(American Rifleman)*

Model 720 — In 1940, the Model 30 was replaced by the more attractive Model 720, which stopped during WWII after only 2,427 were made. About 1,000 of the Model 720 rifles were procured from Remington inventory at the start of WWII for anticipated U.S. Navy needs, but they remained in storage at NAD Crane. In 1963, the rifles were slated to be destroyed. Fortunately, this policy was reversed, and it was decided that the rifles would be awarded as trophies for Navy and Marine Corps marksmanship competition. They were awarded c.1971–1998 to the top shooter at major fleet or National Matches. Some but not all were inscribed on the floorplate with the match information. M1 Garand rifles were also awarded as trophies for other matches. The Model 720 was the last descendant of the Model 1917 rifle program.[26]

Model of 1934 and Model of 1935 Military Bolt-Action Rifles — As is the case with any successful business, Remington was looking for additional markets for its products, including rifles made using the Model 1917 manufacturing machinery and leftover parts. As discussed by Roy Marcot:

> "The total sale of military firearms and ammunition to the U.S. and foreign governments between 1920 and 1934 had been less than two percent of the total sales volume of Remington. In fact, the sole contract for military weaponry during this period was for 9,444 Model 1905–15 Bolt-action rifles for France between 1922 and 1931, to augment those made in the years prior to America's entry into the war.
>
> In 1933, Loomis (C.C. Loomis) and other Remington designers fabricated the first prototype Remington Model of 1933 Military Rifle in 7x57 Mauser caliber utilizing an Enfield Model 1917 action and other parts. The following year Remington received a contract to make 3,000 of these military bolt-action rifles for Honduras, each chambered 7x57 Mauser centerfire ammunition. These were called Model of 1934 Military Rifles. About 500 were fabricated in 1934, and 2,500 were made in 1935, utilizing the same receivers as those in the Remington Model 30S Express Rifles. Each were fitted with the standard Springfield-pattern bayonet. Soon after, Nicaragua placed a large order for these rifles in .30-06 Springfield caliber, but the contract was never consummated.
>
> Remington Model of 1934 Military Rifles were referred to in the factory as Model 40 rifles. In mid-

Left and right-side views of a Remington Model 1934 Honduran contract Military Rifle – 7x57 Mauser. *(Rock Island Auction Co.)*

Remington Model 1935 Military Rifle. .30-06 Caliber. Note the differences between the Model 1934 and Model 1935 rifles in the configuration of the stock and the front sight. *(Cody Firearms Museum)*

1934 Remington made up ten prototype Model 40 military-style rifles in different calibers and barrel lengths, hoping to interest buyers in Central and South America. Some were stamped Model of 1934 Military Rifle and others marked Model of 1935 Military Rifle. Remington salesmen attempted, in vain, to interest military entities in Ecuador, Bolivia, Cuba, and the Republic of Haiti in these military rifles. Calibers fabricated included .30-06, 7x57 Mauser, 7.65 Bolivian, 7.65 Brazilian, 7.92mm Mauser, 8mm (7.99 Mauser Special), 8mm (7.9 Mauser), and 8mm Mannlicher-Schoenauer."[27]

Surviving examples of the 3,000 Model 1934 Honduran contract rifles are seldom seen today, and most are in rather poor condition. Little is known about the small number of "Model of 1935" rifles intended to be marketed to the various Central and South American countries. One example, chambered for the .30-06 cartridge, is in the Cody Firearms Museum Collection.

Other than caliber, the Model 1934 and Model 1935 rifles are quite similar, including the same pattern barrel-mounted, folding-leaf rear sight. The Model 1934 has the Model 1917-style pistol grip stock, but it has a front sight blade without protective ears. The Model 1935 has a straight grip stock similar to the Model 1903 Springfield, but the front sight appears to be identical to the Model 1917's front sight.

Both the Model 1934 and 1935 had the trigger guard and stock straightened to eliminate the "hump," which was originally needed for the rimmed .303 cartridge. Other than the relatively small order of 3,000 Model 1934 rifles for Honduras, Remington's attempt to market a military bolt-action rifle based on the Model 1917 action proved to be a disappointment.

U.S. Coast Guard .30 Caliber Shoulder Line-Throwing Guns (M1903 and M1917 Types) — Line throwing guns are used mainly for nautical emergencies to pass a lightweight "shot line" from shore to a vessel in distress, or from one vessel to another. That line is used to haul over a larger line for towing, or to rescue personnel. They are also used in non-emergency situations to pass mooring lines when approaching a pier, or between ships to rig hoses or lines to transfer fuel, cargo or personnel while underway. There are also firefighting and various commercial uses for line throwing guns, but those are not related to M1917 rifles.[28]

In 1935, the U.S. Coast Guard adopted modified M1903 Springfield rifles for line throwing use. These ".30 Caliber Shoulder Line Throwing Guns" are still used by the Coast Guard today… Although primarily a Coast Guard item, some were used by the U.S. Navy. One set was issued to most USCG vessels and shore stations, along with a rewinding machine to reuse shot lines. The Model 1903 version is familiar to many collectors from their mention in Brophy's *The M1903 Springfield Rifles*:[29]

> "The new type of shoulder line-throwing equipment now being issued to the service is a Coast Guard development, and is the result of several years of experimental tests. Several officers and men have contributed ideas that have been incorporated in the equipment as issued."[30]

The 1935 and 1942 Coast Guard manuals specifically (and only) refer to modified Model 1903 Springfield rifles. By 1980, the Coast Guard revised their description to state that the "shoulder line-throwing rifle may be either a modified Springfield M1903 or a Lee [sic] Enfield model 1917." The M1917 rifles had their overall length shortened to match the M1903 rifles, allowing use of the same wooden chests.

While their 1980 description below mentions removal of the rear sight, this was not done on the Model 1917s, only the Model 1903s.

LINE-THROWING GUN EQUIPMENT

The .30 caliber (CAL.) shoulder line-throwing rifle, which was developed by the Coast Guard, is issued as a kit. When issued as original allowance, the kit contains:

a. One carrying case
b. Seven light projectiles (13 ounce)
c. Three heavy projectiles (15 ounce)
d. Two illuminated buoyant projectiles
e. One .30 CAL. Rifle modified with canister attached
f. One .30 CAL cleaning rod
g. Cleaning brushes and patches
h. Five shot lines on spindles
i. One wooden mallet
j. Forty .30 CAL. Grenade cartridges
k. One rewinding machine per unit.

DESCRIPTION OF EQUIPMENT
"THE SHOULDER LINE THROWING RIFLE

Your shoulder line-throwing rifle may be either a modified Springfield M1903 or a Lee [sic] Enfield model 1917. The barrel has been rebored and the lands removed, making the rifle a smooth bore gun. The front sight movable stud assembly and fixed stud, as well as the rear sight assembly, have been removed. The stock has been cut down and a canister has been attached. Two pounds of lead have been placed in the butt stock, and a recoil pad has been placed in [sic] the butt in lieu of the buttplate."[31]

The 13-ounce projectile can be fired about 350 feet with the rifle against the shoulder, but the 15-ounce (red) projectiles were to be used only in an emergency when the 13-ounce failed to reach the target.[32] The cartridge used with the M1903 and M1917 line throwing guns is the standard .30 caliber Grenade Launching cartridge M3.[33]

One example is located in the U.S. Coast Guard Museum in New London, Conn. Another example shown below, Eddystone ser. no. 597755. was sold at auction in 2013.

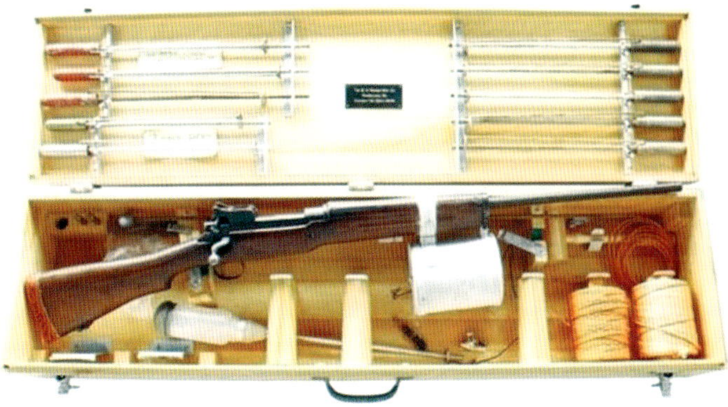

U.S.C.G. .30 cal. Shoulder Line-Throwing Gun made from M1917 Eddystone, ser. no. 597755. *(Courtesy Poulin Antiques & Auctions)*

Identification plaque for .30 cal. Shoulder Line-Throwing Gun converted for the U.S.C.G. by Price Brothers, Frederick, Md. *(Courtesy of Amoskeag Auction Co.)*

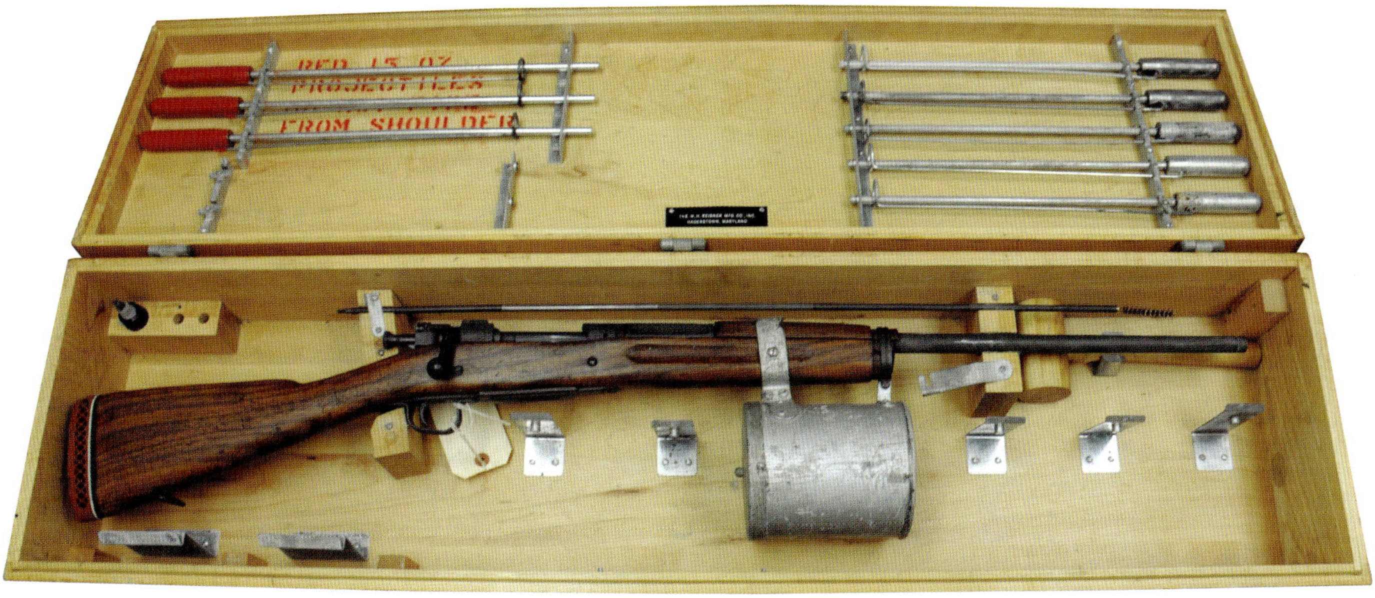

U.S. Coast Guard .30 Caliber Shoulder Line-Throwing Gun made from M1903 Springfield, ser. no. 1316819. Converted for line throwing c.1935, tested at USCG Station, Point Adams, at the mouth of the Columbia River, in January of 1941. Excessed by the Coast Guard in the early 1990s to the Tennessee Valley Authority, then donated to the USCG Historian's Office in 2014. *(Courtesy U.S. Coast Guard Historians Office)*

Most of the conversions were done by W.H. Reisner Manufacturing of Hagerstown, Md., with a few by Price Brothers in Frederick, Md., and Van Karner Chemical Arms Corporation of New York.

The number of rifles converted to ".30 Caliber Shoulder Line Throwing Guns" is unknown, but it is estimated at around 500 to 750 units, with perhaps 10 to 20 percent being M1917s and 80 to 90 percent based on M1903 rifles. Many of these .30 Caliber Shoulder Line-Throwing Guns remain in use, and some of those disposed of in the past have likely been converted into hunting rifles, making either model hard for collectors to find. But, amazingly, some M1917 rifles are still being used by the U.S. military more than a century after they were made.[34]

Identification plaque for .30 cal. Shoulder Line-Throwing Gun converted for the U.S. Navy by the Van Karner Chemical Arms Corp. of New York. *(Courtesy of National Museum of American History)*

Chapter 10

Accessories, Appendages and Accouterments

Although the rifle was the basic tool of the infantryman, there were other items that were required that would enable the weapon to be used with maximum efficiency. Some of these items, such as bayonets and grenade launchers, increased the lethality of the weapon. Others such as slings and magazine depressors made it easier to carry or manipulate. Still others, such as various manuals, oiler and thong cases, canvas breech covers and spare parts chests assisted in keeping the weapon properly maintained. All are important, interesting and worthy of consideration but are sometimes overlooked or minimized when studying the Model 1917 rifle. This chapter will discuss the most prominent of these items that are an integral part of the story of the "American Enfield." Many of these items are also becoming valued martial collectibles in their own right.

Bayonets and Scabbards

Like the rifle, the Model 1917 bayonet is rapidly growing in collector interest, and collectible specimens are getting more difficult to find. It is interesting to note that the Model 1917 bayonet remained in service for decades after the Model 1917 rifles were purged from the government inventory, and bayonet production had to be restarted during the Vietnam War era.

Pattern 1913 Bayonet — The bayonet is unquestionably the most important accessory for any rifle and, arguably, the most interesting. As has been discussed in previous chapters, the forerunner of the Model of 1917 rifle was the British Pattern 1913 .276 caliber rifle, which was made in limited numbers by the Royal Small Arms Factory.[1] The Pattern 1913 rifle was soon superseded by the .303 caliber Pattern 1914 rifle, and the Pattern 1913 bayonet was retained. Large numbers of Pattern 1914 rifles were subsequently manufactured in the United States by Winchester, Remington, and Eddystone.

Production contracts for the Pattern 1914 rifle in the United States also included the manufacture of the Pattern 1913 bayonets. The only two American firms to manufacture the Pattern 1913 bayonets were Winchester and Remington. Interestingly, Eddystone, the firm that made the largest number of Patten 1914 rifles, did not make any bayonets. As mentioned in Chapter 4, Remington made the corporate decision to have the Eddystone and Ilion factories concentrate on rifle production, and the Bridgeport plant was to manufacture bayonets.

The Pattern 1913 bayonets made by Remington and Winchester had an overall length of 21.75 inches with a 17-inch blade and two-piece walnut grips. Winchester bayonets were stamped with a small "W" on the left side of the blade between the fuller and crossguard along with the "1913" (the pattern number) and the month and year of manufacture. The Remington bayonets had "REMINGTON" enclosed in a circle on the left side along with "1913" and the month and year of production. The right side was

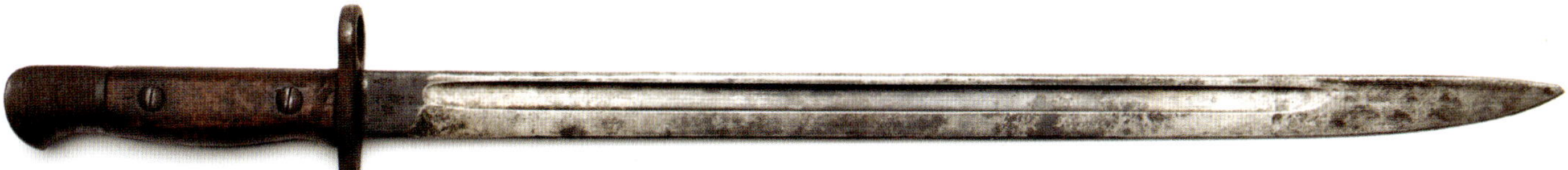

Note the similarity of the bayonet as with the Pattern 1913 bayonet seen on page 180.

Pattern 1907 bayonet and scabbard as issued with the British SMLE rifle.

stamped with British proof/acceptance markings including the" broad arrow" British government ownership marking.[2] The Pattern 1907 scabbards made for the British Pattern 1913 bayonets were manufactured under subcontract by the Jewell Belting Company (Hartford, Conn.) and Graton & Knight Company (Worchester, Mass.)[3]

Model 1917 Bayonet — As stated by Asst. Sec. of War Benedict Crowell in his book *America's Munitions 1917–1918*:

> "(Another) large ordnance operation was the production of bayonets for the service rifles. The British bayonet had proved to be highly satisfactory in the war; and since it was already designed to fit the Enfield rifle, which we had adopted for our own, we took the British bayonet as it was and, with only one slight alteration, set to produce it in this country.
>
> The Government found both the Remington Arms-Union Metallic Cartridge Co. at Bridgeport, Conn. Works, and the Winchester Repeating Arms Co. building these bayonets for the English Government. The latter's bayonet needs by 1917 were being well supplied by home manufacturers, and this permitted us to buy approximately 545,000 bayonets which had already been manufactured for the British.
>
> The Ordnance Department at once started out these two concerns on contracts for bayonets for the American Government, Remington with orders for 2,820,803 bayonets and Winchester with orders for 672,000. Remington delivered in all 1,565,644 bayonets and Winchester 395,894. This was a total

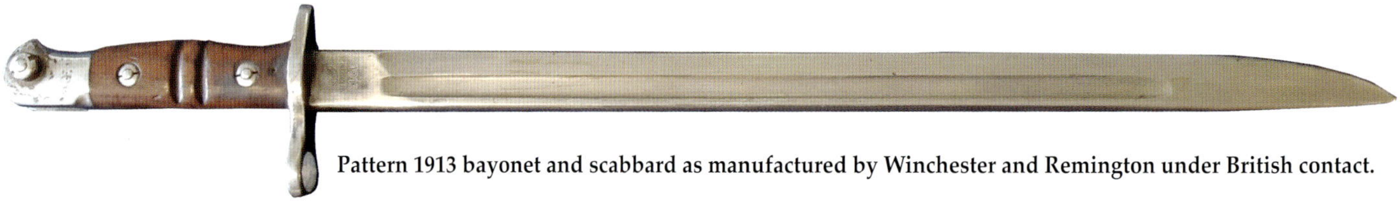
Pattern 1913 bayonet and scabbard as manufactured by Winchester and Remington under British contact.

Pattern 1913 bayonet and scabbard.

The scabbards were manufactured in the United States by Jewell Belting and Graton & Knight. Note the design of the hanger on the scabbard.

Close-up of the wooden grip on Pattern 1907 bayonet. The grips are smooth as compared to the Pattern 1913 bayonet grips, which have two vertical grooves.

1 British property markings on a Pattern 1913 bayonet.
2 Pattern 1913 bayonet manufactured by Winchester. The markings denote the Pattern "1913", the maker "W" and the date of manufacture "November 1916."
3 Pattern 1913 bayonet manufactured by Remington, April 1917.

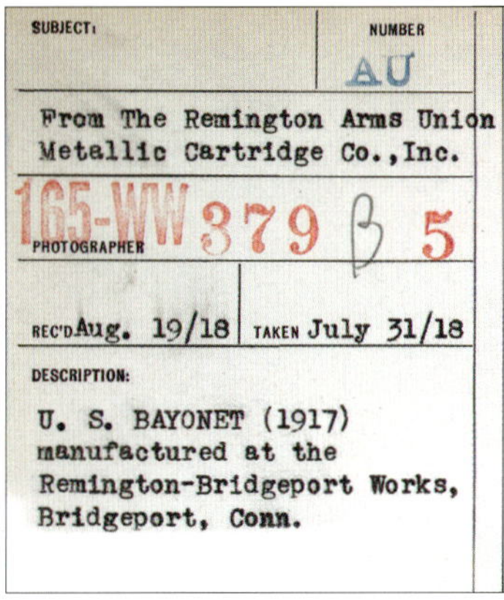
SUBJECT: NUMBER AU

From The Remington Arms Union Metallic Cartridge Co., Inc.

165-WW 379 B 5

PHOTOGRAPHER

REC'D Aug. 19/18 TAKEN July 31/18

DESCRIPTION:

U. S. BAYONET (1917) manufactured at the Remington-Bridgeport Works, Bridgeport, Conn.

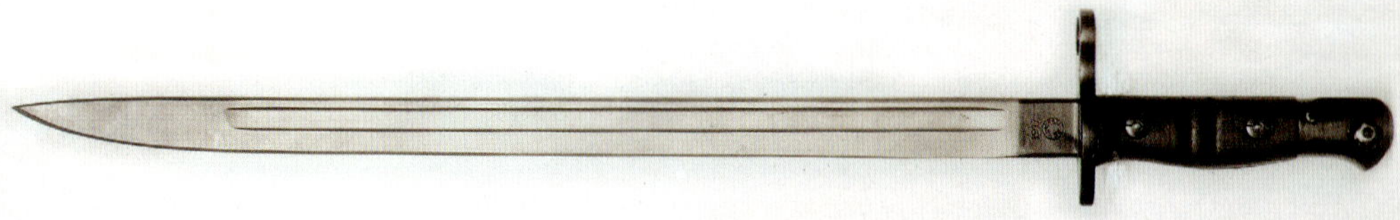
Ordnance photo of Remington-made M1917 bayonet. *(National Archives)*

of 1,961,500 bayonets. These figures indicate that we were short over 500,000 bayonets at the time hostilities ceased; and as a matter of fact, this shortage had already become acute, especially in the training camps.

The (Model 1917) bayonets had not come as rapidly as we had expected, because to produce them at the rate originally planned, we would have interfered with the more essential production of rifles by the same companies.

(In) 1918 additional (bayonet) contracts were made. Landers, Frary & Clark, of New Britain, Conn., engaged to manufacture 500,000 bayonets, and the National Motor Vehicle Co., 225,000. These latter contracts, however, were suspended after the armistice was signed. The additional orders had made it certain there would be no bayonet shortages by the spring of 1919."[4]

Model 1917 bayonet and scabbard.

It was also planned that the American Fork and Hoe Company would go into production with the Model 1917 bayonets, but no examples were produced, and the proposed contract was cancelled. In any event, adding the number of Pattern 1913 bayonets purchased from the British to the totals that were cited by Crowell would more than make up for the stated shortfall. As related by noted author and bayonet scholar Gary Cunningham:

"Other records do not show that there was any shortage, and neither Remington nor Winchester noted any problems in producing the number of bayonets being ordered. It is believed that the error (in citing shortages) was that the purchase of the 545,500 Pattern 1913 bayonets from the British was not factored in the figures of rifles and bayonets being delivered, which created on paper a shortage of 'over 500,000' bayonets."[6]

Photo dated 1918 of U.S. soldier with M1917 rifle. Note the M1917 bayonet and scabbard attached to his pack. *(National Archives)*

Photo dated 1918, captioned "Soldiers of the 3rd Training Camp in Porto Rico at bayonet practice." (National Archives)

Also, the production totals cited by Secretary Crowell did not include those made after the Armistice ("...at the time hostilities ceased"), so the total number manufactured by the time production stopped in early 1919 would have been higher. This is confirmed by data in the Ordnance publication *Project Supporting Paper No. 44 – Bayonets, Knives and Scabbards – Development, Standardization and Production*, which contained a table detailing a 1939 inventory of bayonets and scabbards in order to estimate the need for additional items should the United States become actively involved in the war that was breaking out in Europe. The category for the "Bayonet, M1917 (limited standard)" gave a figure of "2,048,016 (condition questionable)."[7]

If, after the inevitable loss of bayonets during training and overseas combat during WWI and the transfer of thousands to Mexico and the Philippines in the 1920s and '30s, over two million bayonets were still in inventory 21 years after WWI was over, then Secretary Crowell's quoted figures definitely left out the sizable number of bayonets made after the Armistice. Also, as mentioned in Chapter 4, these factors may partially account for the substantial difference between numbers cited by Secretary Crowell and figures contained in Winchester factory archives regarding Model 1917 bayonet production totals for that company.

Description — An excellent summary of the characteristics of the M1917 bayonet is contained in the book *The American Bayonet 1776–1964*:

> "The straight blade is 17⅛ inches long, 1$\frac{1}{32}$ inches wide, and ⅛ inch thick. The back is perfectly straight until it begins to taper to the semi-spear point about 2 inches from the end. The edge is parallel with the back for about 12 inches when it begins its gradual upward sweep to the point. A deep, narrow, square-edge, unstopped fuller runs for 12½ inches beginning 3⅜-inch inches from the point. There is no false edge as such, the entire back being slightly rounded. The true edge is sharply 'V' ground for 14½ inches. The M1917 bayonets are not serially numbered like their M1905 cousins Bayonets made for the U.S. have the obverse ricasso vertically stamped with the ordnance escutcheon over 'U.S.', and the reverse ricasso is stamped with '1917' over the manufacturer's mark or name. This model blade is variously finished bright, blue, brown or Parkerized.
>
> The 4$\frac{13}{16}$-inch-long hilt fits the hand well. The guard has a configuration reminiscent of the Lee Straight Pull; a high, narrow upper guard with only a rudimentary lower guard. The bore is .609 inch in diameter. The black walnut scales are secured to the tang by two machine screws passing through the obverse scale and held by round, slotted, recessed nuts in the reverse scale. Two heavy vertical grooves centrally located on each scale provide a more positive grip, although their prime purpose had been as a means of identification which was applicable only to their British service. The pommel is a; small variation of the bird's-head pattern. On some models, notably U.S., a ⅜-inch diameter hole is bored transversely through the upper half of the pommel just above the scales. The function of this hole is to afford a means of cleaning dirt from the undercut groove. The self-contained, transversely operated catch device is of the Mauser pattern. There are no hilt markings. The hilt is finished dark blue and the smoothly sanded grips are linseed-oiled. The arm weighs 1.17 pounds."[8]

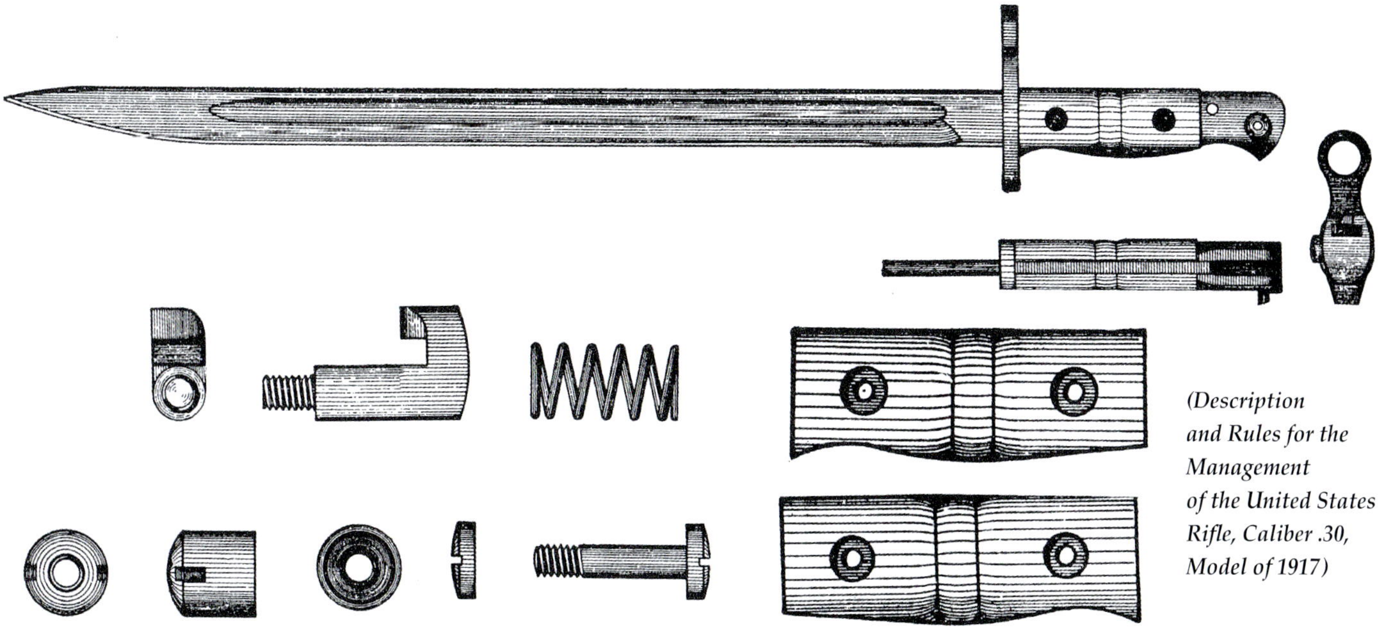

(Description and Rules for the Management of the United States Rifle, Caliber .30, Model of 1917)

Typically, the hilt and about ¾ inches (+ or -) of the blade (below the guard) were blued. As stated above, the balance of the blade could bright, blued or Parkerized.

Since the Pattern 1913 bayonets made for the British would fit the Model 1917 rifle with no alterations, those subsequently acquired by the Americans were issued "as is." The only modifications performed were to deface the British property markings and stamp "U.S." in close proximity. Two methods were used to "cancel" the British markings; they were either overstamped with a series of lines or obscured by stippling. There was, however, an exception. As related by Cunningham:

U.S. Army troops at marksmanship training with M1917 rifles and bayonets. *(Tom Laemlein)*

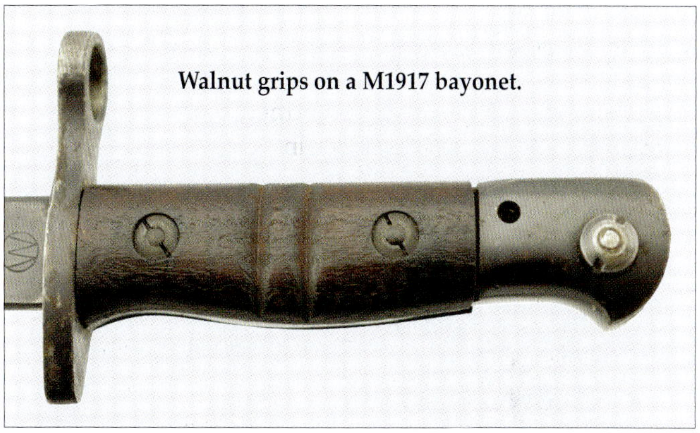

Walnut grips on a M1917 bayonet.

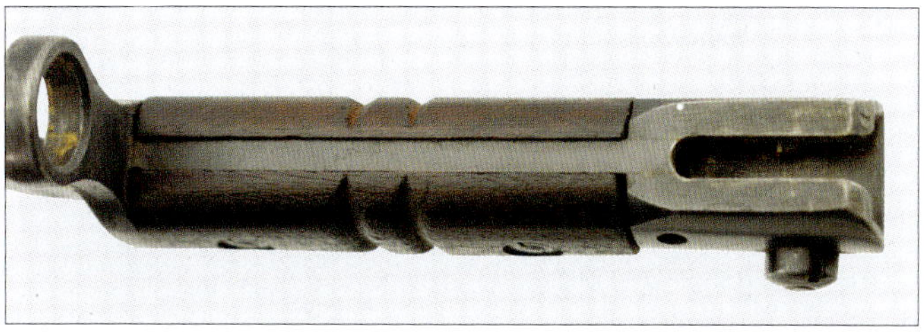

Close-up view of the M1917 bayonet Mauser-influenced locking mechanism.

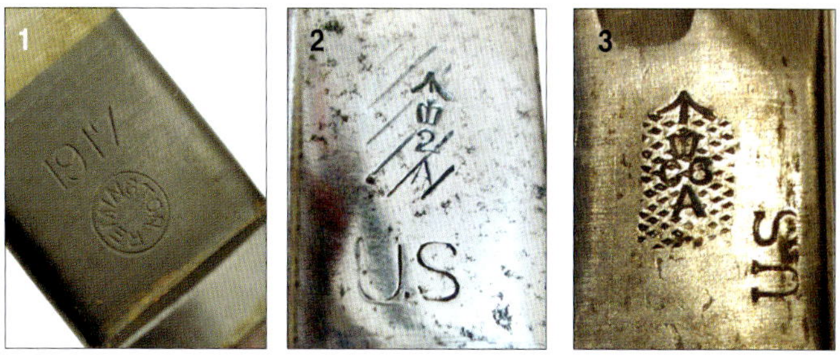

1 M1917 bayonet blade showing the blued portion below the guard, which is a hallmark of a bayonet that has not been refinished.
2 "Lined-out" British property markings and "US" marking added.
3 British property markings obscured by stippling with "US" marking added.

"The first bayonets produced by Remington on the US contract in September 1917 used the standard British marks as shown above. These then had the British proofs marked out and the US marking added. Why the British proof was used is not known as the British inspectors were no longer at the plant. A very few dated 10-17 have also been noted. Some of the 9-17 and 10-17 marked bayonets do not have the British proofs on the ricasso, but instead have the standard US marks of the Ordnance Shell and Flame, US, and a stylized eagle head with number underneath. The Ordnance Department adopted the eagle head over number mark in 1917 so that the individual inspectors could be made responsible for their work."[9]

Production of the Model 1917 bayonet was begun around September of 1917 by Winchester and Remington. The only significant change from the Pattern 1913 bayonet was the addition of the 3/8-inch diameter hole that was bored through the upper half of the pommel.

This change occurred in the fall of 1917, so a few very early-production bayonets lacked this feature, although some had the hole retrofitted later.[10] This modification was added to permit the removal of dirt or other debris from the undercut groove. It was officially designated as a "clearing hole" but is sometimes, incorrectly, referred to as an "oil hole."

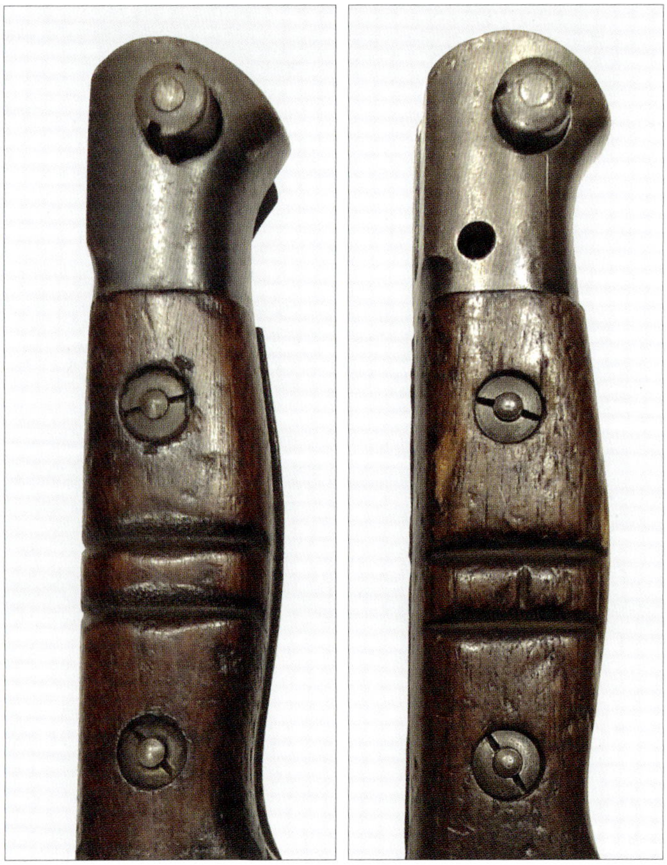

Pattern 1913 bayonet (left) and M1917 bayonet (right). Note the "clearing hole" added to the later-production M1917 bayonet.

Markings — The Remington bayonets were marked on the left ricasso with the model date of "1917" above "REMINGTON" enclosed in a circle. The right ricasso was marked with the ordnance flaming bomb insignia over "US". To the left was an eagle-head inspection stamp as well as an inspector identification number. Some also had an "X" stamped to the right. Occasionally, a Remington-made bayonet can be found marked "1918". It has been surmised that Remington changed their marking die in early 1918 to reflect the actual year of manufacture, but they were

1 Markings on left ricasso of Remington M1917 showing the model year "1917" and the circled "REMINGTON" logo.

2 Right ricasso markings consisting of ordnance flaming bomb insignia over "US" and an "eagle-head" inspection stamp and inspector identification number.

3 Slightly different marking format showing "US", flaming bomb insignia, faint eagle head and "x" stamp.

4 "1918" marking on Remington M1917 bayonet. Remington was instructed by Ordnance to use the "1917" model designation instead, so relatively few bayonets had the "1918" marking.

5 Left ricasso markings on Winchester M1917 bayonet.

6 Right ricasso markings on Winchester M1917 bayonet consisting of a flaming bomb above "US", an eagle-head inspection stamp and inspector code number on the left.

7 Scarce Winchester-made M1917 bayonet marked with just a circled "W" logo. These are believed to have been made shortly after WWI for sale to various domestic law enforcement agencies that received surplus M1897 trench guns from the U.S. Army but lacked bayonets.

Early 1920s photo of the Denver Police Department examining Model 1897 trench guns received from the U.S. Army. Many of these lacked bayonets, which resulted in Winchester tooling up to make limited numbers. ***(Denver Public Library Collection)***

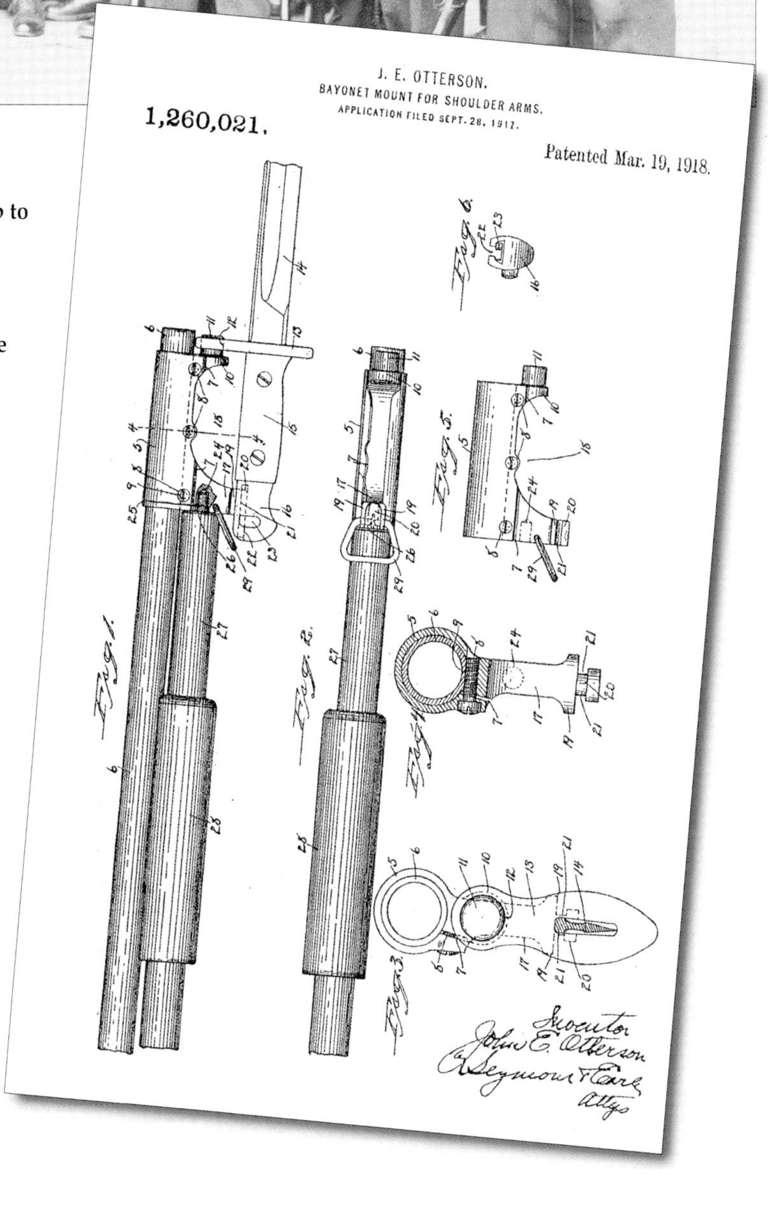

Patent drawing for handguard/bayonet adapter assembly designed by Winchester to permit short-barrel shotguns to be able to mount M1917 bayonets.

instructed by Ordnance to revert to using the "1917" model designation instead. Remington bayonets marked "1918" are relatively uncommon.[11]

Winchester bayonets were marked on the left ricasso "1917" above the circled "W" logo. The right ricasso was marked in much the same manner as the Remington-made bayonets — a flaming bomb above "US" and an eagle-head inspection stamp and inspector code number on the left. Occasionally, a M1917 bayonet may be encountered that has only the circled "W" on the left ricasso but with no other markings on either side of the blade. These seldom-seen bayonets are the subject of some conjecture. The most likely explanation is that these were made by Winchester in the early 1920s for domestic law enforcement agencies who received a rather substantial number of surplus Winchester Model 1897 short-barrel shotguns fitted with bayonet adapters from the U.S. Army, ostensibly for riot control purposes. Since these bayonets were not the property of the government, no ordnance, "US" or inspection stamps were applied.

As with the Model 1917 rifle, large numbers of Model 1917 bayonets were cleaned and refurbished after World War I and early in World War II. If required, the blades were generally refinished by Parkerizing. Bayonets refinished in the 1920s typically exhibited Parkerizing that was relatively smooth in texture with a fairly dark tint. Later examples often had Parkerizing that was a bit rougher in texture and lighter in tint. Some of the bayonets refurbished by other nations were stamped with government ownership markings specific to that country.

Model 1917 Bayonets for Military Shotguns — As previously alluded to, in addition to the Model 1917 rifle, the Model 1917 bayonet was approved for use with bayonet-equipped Winchester Model 1897 and Remington Model 10 short-barrel shotguns (trench guns) during World War I. The probable reasons for the selection of the M1917 bayonet, rather than the M1905 bayonet as used with the M1903 Springfield rifle, were twofold. First, the M1917 bayonet was in mass production at a rate that greatly exceeded the manufacture of the M1905 bayonet. Also, the fact that Winchester and Remington were the firms that made the trench guns and also produced the M1917 bayonet likely had some bearing as well. Relatively few of the Model 1897 and Model 10 trench guns made it overseas before the Armistice, but the M1917 remained the authorized bayonet for U.S. military shotguns after the war.

When the United States became actively involved in the Second World War, contracts were given for the production

WWII photo of U.S. Marine attaching a M1917 bayonet to his Winchester Model 97 trench gun. *(U.S. Marine Corps photo)*

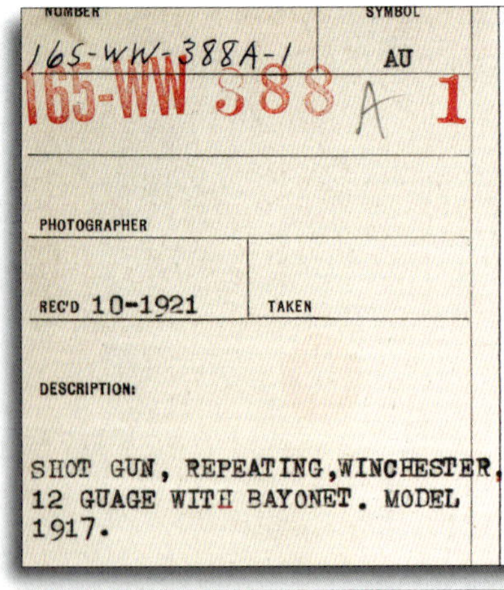

Ordnance photo of WWI-vintage Winchester Model 1897 trench gun with M1917 bayonet.

(U.S. Army Ordnance Department)

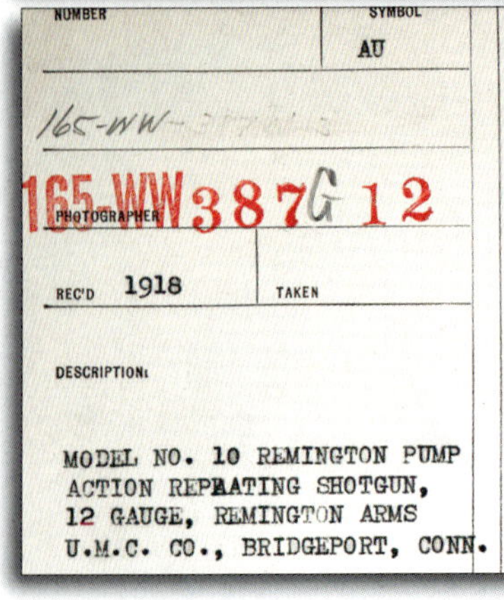

Ordnance photo of WWI-vintage Remington Model 10 trench gun with M1917 bayonet.

(U.S. Army Ordnance Department)

World War II production Winchester Model 12 trench gun with M1917 bayonet.

(Guns from author's collection)

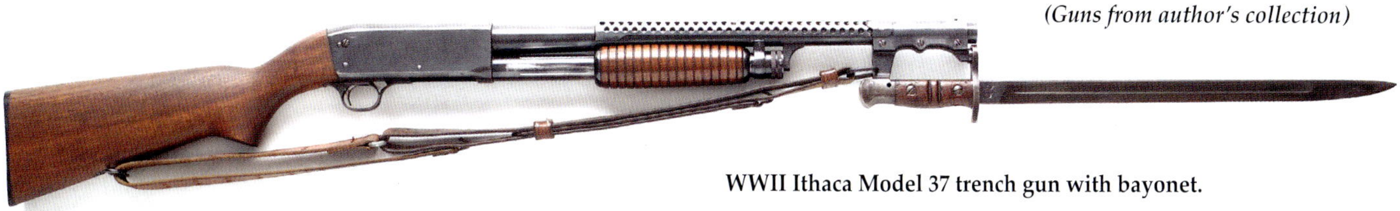

WWII Ithaca Model 37 trench gun with bayonet.

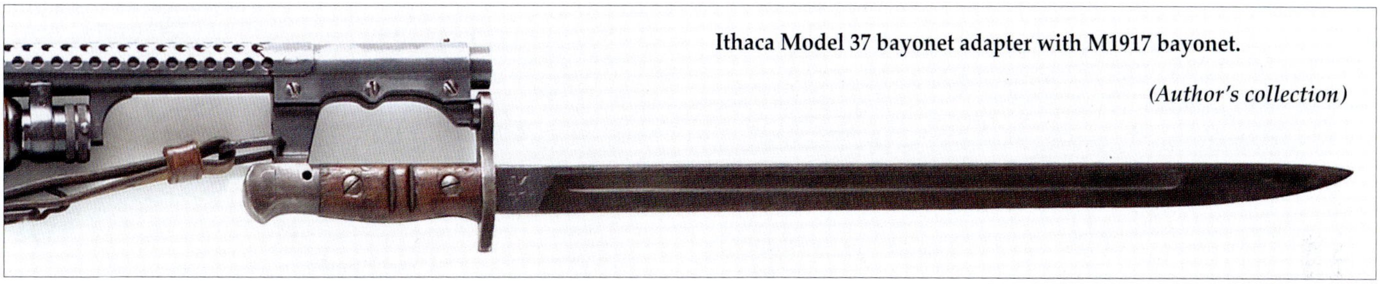

Ithaca Model 37 bayonet adapter with M1917 bayonet.

(Author's collection)

of several additional types of trench guns fitted with adapters for the M1917 bayonet. The M1917 bayonet was not put back into production during WWII, which bolsters the assumption that there was not a shortage for use with the M1917 rifles and the bayonet-equipped shotguns. In addition to the First World War-era Model 1897 and Model 10 shotguns still on hand, the following trench guns adapted for the M1917 bayonet were manufactured in World War II: Winchester Model 97, Winchester Model 12, Ithaca Model 37, Stevens Model 520-30, and Stevens Model 620A.

Vietnam War-Era M1917 Bayonets — While having nothing to do with the M1917 rifle, no discussion of the M1917 bayonet would be complete with mentioning the bayonets made in the late 1960s. As discussed in *Complete Guide to United States Military Combat Shotguns*:

> "By the time of America's increased military involvement in Vietnam in the early 1960s, the M1917 bayonet and scabbard remained the standard issue for the U.S. 'trench guns.' The new production Ithaca Model 37 and Winchester Model 1200 'trench guns' made in the late 1960s were still designed for use with the 50+ year-old M1917 bayonet.
>
> As the war in Vietnam continued to drag on, the supply of First World War production M1917 bayonets became exhausted. The U.S. Government gave contracts to two firms for renewed production of M1917 bayonets. These firms were General Cutlery of Fremont, Ohio and Canadian Arsenals Limited, Long Branch, Canada. The identity of the maker and 'US M1917' was stamped on top of the guard. The Canadian Arsenal bayonets were marked with a small capital 'A' enclosed by a larger 'C'. The General Cutlery bayonets were marked 'GEN CUT' and also were generally stamped with the 'stars over eagle' Defense Acceptance Stamp' (DAS) the guard.

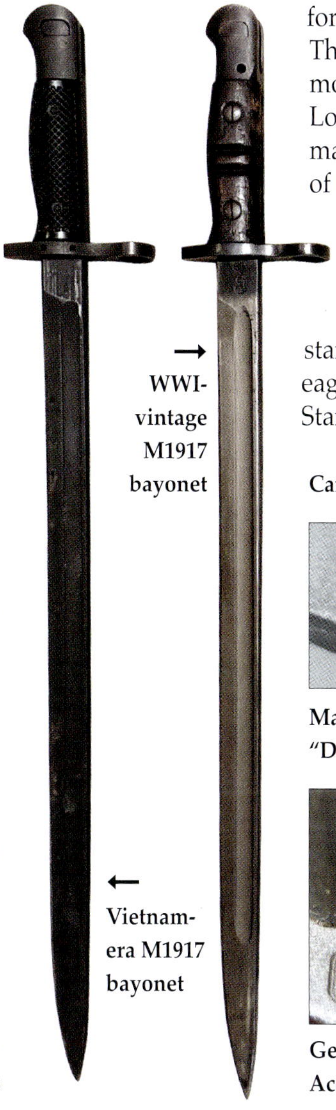

→ WWI-vintage M1917 bayonet

← Vietnam-era M1917 bayonet

Canadian Arsenals, Ltd. marking.

Markings on Vietnam-era M1917 bayonet. Note the "Defense Acceptance" stamp on the left.

General Cutlery manufacture marking and Defense Acceptance stamp.

1960s-production Winchester M1200 trench gun with Vietnam-era M1917 bayonet.

Some of the Canadian Arsenal, Ltd. bayonets had a smaller DAS applied to the guard.

As compared to the WWI-vintage M1917 bayonets, the Vietnam-era M1917 bayonets were noticeably less well made and finished. The blades were slightly shorter than the earlier blades (16½ inches vs. 17 inches) with a correspondingly shorter overall length. The blades were not as well machined as the WWI-vintage bayonets and were finished in roughly textured medium gray Parkerizing. Coarsely checkered plastic grips were substituted for the WWI-era walnut grips. The guard was attached by two metal roll pins."[12]

Shotguns adapted for the M1917 bayonet, especially the Winchester M1200, remained in active use by the U.S. Army through the Iraq and Afghanistan campaigns, more than a century after first being carried into combat by American troops.

Late 1960s-era ordnance manual drawing showing bayonet adapter for M1200 trench gun, illustrating how to attach the M1917 bayonet.

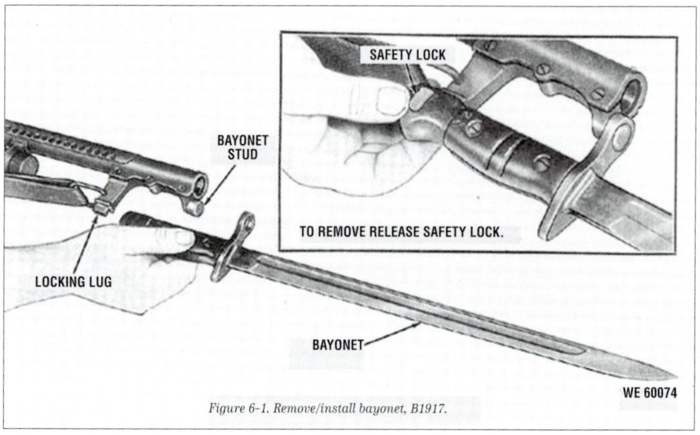

Vietnam-era M1917 bayonet grips (below). Note the checkered plastic grip as compared to the walnut grip on the WWI example.

WWI-era bayonet grips.

Note also the pinned guard on the Vietnam-era bayonet.

Model 1917 Bayonet Scabbards

Obviously, bayonets required scabbards. As stated by Secretary Crowell:

> "We not only had to provide bayonets but also the scabbards to hold them in. The scabbard of the 1917 bayonet was of simple manufacture and there were no difficulties in securing sufficient quantities. The Jewell Belt Co. delivered 1,810,675 of them; Graton & Knight delivered 1,669,581; while Rock Island Arsenal produced 3,000. This gave us a total of 3,480,000 scabbards. A quantity greatly in excess of the production to either bayonets or rifles."[13]

There were two basic types of M1917 bayonet scabbards that differed primarily in the design of the hanger assembly, both adapted to the U.S. M1910 cartridge belt "double hook" assembly. The first pattern, patented and manufactured by firearms designer Hiram P. Maxim and patented to the Maxim Silencer Co. of Hartford, Conn., had a leather hanger attached to the body of the scabbard by two spot-welded metal studs.[14] The leather body was essentially the same as the British Pattern 1907 scabbard and was painted green and was fitted with a metal tip at the bottom.

It is likely that this design was an expedient measure to use British-style throat pieces for an easy modification.

Markings on back of M1917 bayonet scabbard showing the manufacturer "JEWELL–1918" (year of production) and inspector's initials "H.E." It is believed that the "M.S." marking on back of the metal mouthpiece indicates that this component was made by the Maxim Silencer Co.

However, it was quickly seen that altering the frog stud, adding the leather chape, belt hook and riveting was a lot more work and expense than simply eliminating the leather hanger and forming an extension of the metal mouthpiece (throat) to which the hook assembly could be fastened. The design for the belt attachment on the M1917 scabbard was later used on the M3 and M7 scabbards for the M1905 and M1 bayonets. It has been postulated that some of the early-production examples used Pattern 1907 leather scabbard bodies made for the British contract as was the case with the British-marked Pattern 1913 bayonets, but this is unconfirmed.

Noted researcher Frank Mallory indicated that the change to the second pattern scabbard occurred in October of 1917.[15] This change resulted in a more durable scabbard

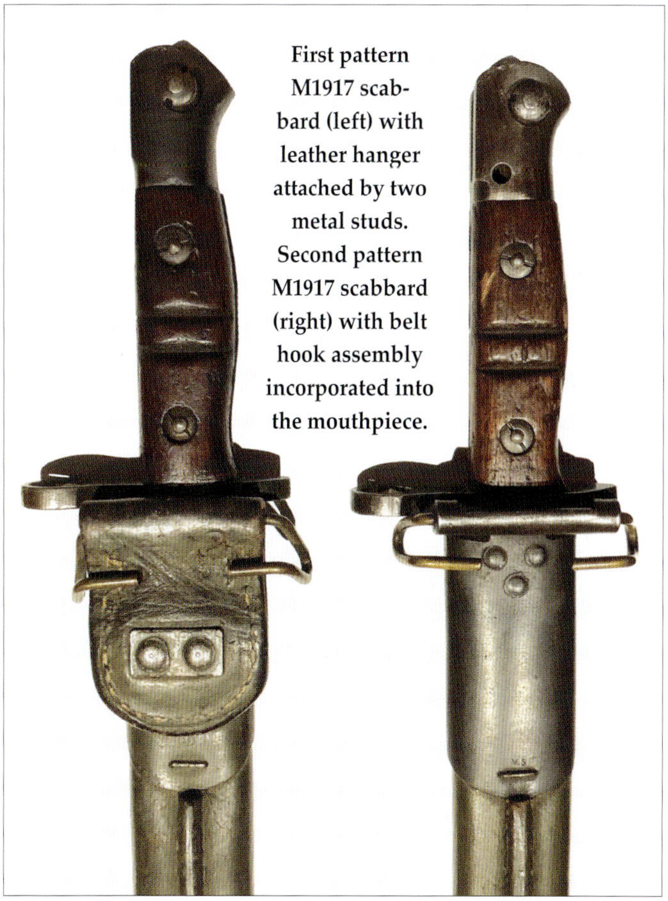
First pattern M1917 scabbard (left) with leather hanger attached by two metal studs. Second pattern M1917 scabbard (right) with belt hook assembly incorporated into the mouthpiece.

Front view of second pattern M1917 scabbard mouthpiece assembly along with the metal tip on the bottom of the M1917 bayonet scabbard.

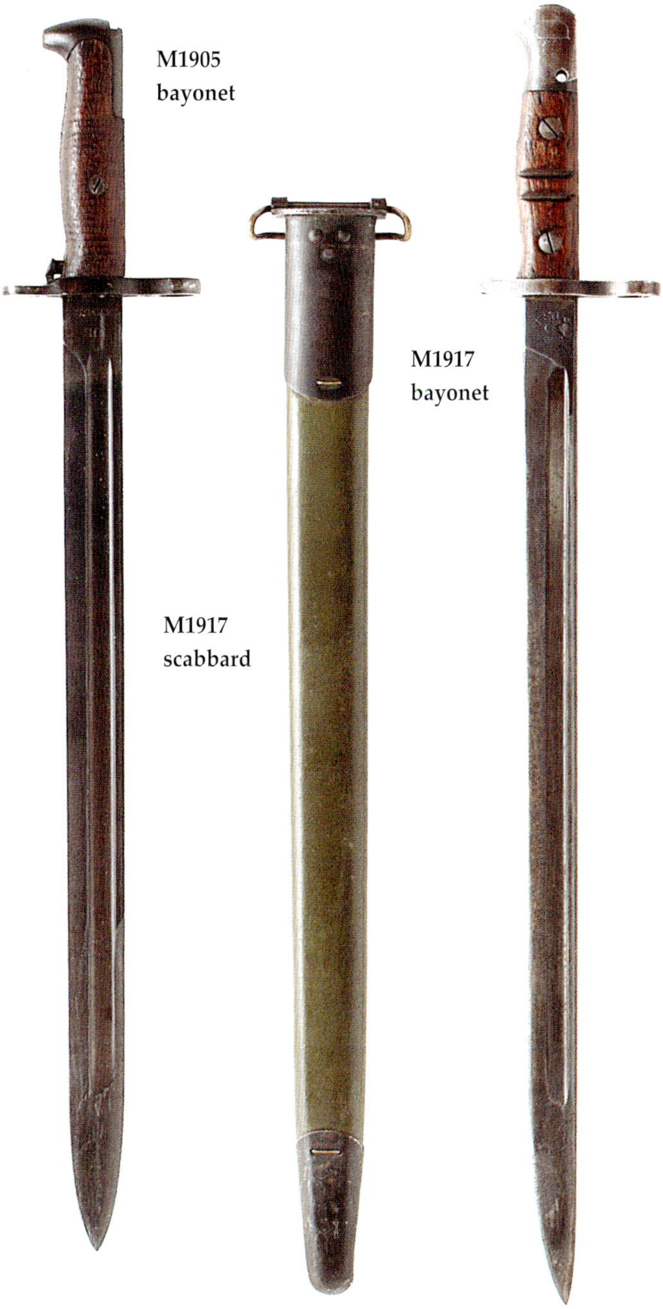

Photo at left — In 1920, the M1917 scabbard was authorized for use with the M1905 and M1917 bayonets, and many were used well into the Second World War.

that could be made cheaper and faster. Both types of scabbards were made by Jewell Belting Co. (Hartford, Conn.) and Graton & Knight Co. (Worcester, Mass.). It is unclear which type of scabbard Rock Island Arsenal manufactured, but given the extremely low production total of 3,000, extant examples are very rare. Due to the fewer numbers made, surviving examples of the first pattern scabbard are relatively uncommon as compared to the second pattern scabbard.

There were more than sufficient numbers of M1917 scabbards to supply the demand after the First World War, but this was not the case with the M1910 bayonet scabbard used with the M1905 bayonet. As discussed in *Project Supporting Paper Relating to Bayonets, Knives and Scabbards 1917–August 1945*:

> "In 1920, authorization was given for the issue of M1917 Scabbard to troops armed with the M1903 Rifle when necessary to complete their equipment, provided M1910 Scabbards were not available; attention being called to the fact that the M1917 Scabbard was interchangeable between the M1905 and M1917 Bayonets. In 1924, the M1917 Scabbard was made substitute standard for the M1910 Scabbard for use by the Army in time of war, because it was less difficult to manufacture and considerably cheaper."[16]

It would seem that the 3,480,000 M1917 scabbards on hand in late 1918, as reported by Crowell, would be ample to meet any anticipated future demand. Unfortunately, this was not the case. A 1939 ordnance report revealed that there were 1,691,048 M1917 bayonet scabbards in inventory, which represents a reduction of 1,788,952 scabbards in 21 years! One reason for this decline was the fact that the thousands of M1917 bayonets and scabbards supplied to Mexico in the 1920s and the Philippines in the 1930s obviously reduced the supply. Another significant reason was that the leather body of the M1917 scabbard proved to be prone to deterioration, particularly when stored in less-than-ideal conditions, and large numbers had to be subsequently discarded.

WWII photo of U.S. Army soldier with a M1903 rifle with a fixed M1905 bayonet. Note the close-up view of the M1917 bayonet scabbard attached to his pack, which was issued as "Substitute Standard." *(U.S. Army Signal Corps photo)*

Throat markings on the WWII M1917 plastic scabbard.

Markings on M1917 late 1960s scabbard. These were virtually identical to the WWII production plastic scabbards except for the markings.

The WWII composition (plastic) M1917 bayonet scabbard.

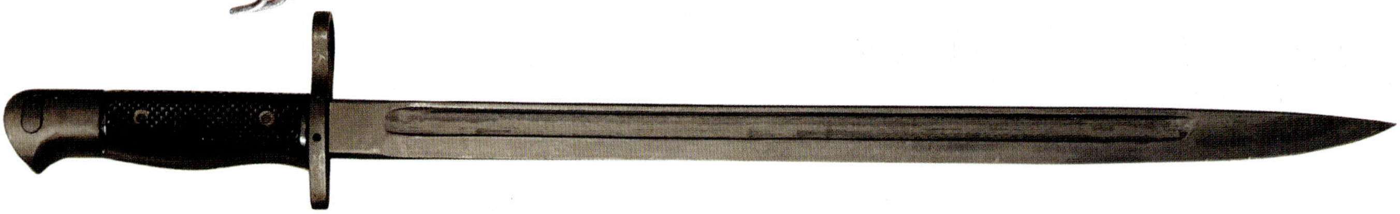

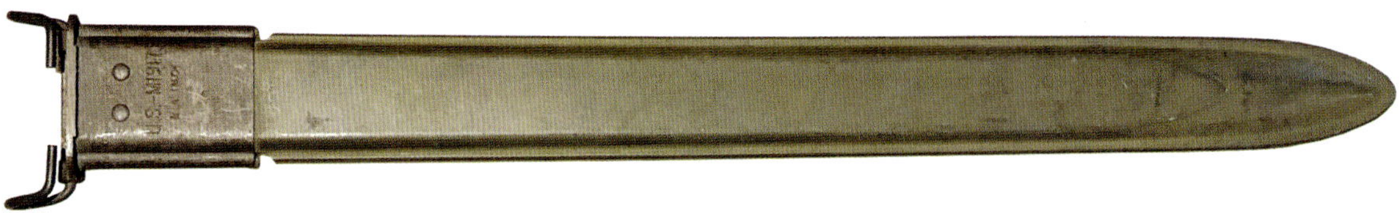

Vietnam-era M1917 bayonet and scabbard.

M1917 bayonet and scabbard with British leather frog as issued for Home Guard use in WWII. These frogs could be used with P1907, P1913 and M1917 bayonet scabbards.

(Brittonatwar.com)

The same 1939 survey cited above revealed that there were 2,048,016 M1917 bayonets on hand, which meant there were 356,968 fewer scabbards than bayonets available. Regardless, there are a number of WWII-era photographs that depict M1905 bayonets being carried in M1917 scabbards.

The shortage in M1917 scabbards due to their use for both the M1917 and M1905 bayonets was exacerbated when significant numbers of bayonet-equipped shotguns went into production beginning in 1942. This situation was discussed in the above-cited ordnance publication:

> "In 1943, there were still a number of the M1917 Bayonets in use or in stores, left from World War I for which, in many instances, the scabbards had deteriorated to such an extent as to necessitate replacement before the M1917 Bayonets could be issued. To meet this need, a plastic scabbard was developed which radically changed in design but, since either the old or the new plastic scabbards were approved for issue, no new model (number) was necessary."[17]

The new scabbard was very similar to the composition (plastic) M3 bayonet scabbard made for the M1905 bayonet. The only substantive difference was the model designation on the metal mouthpiece. Ordnance records indicate that 179,000 plastic M1917 bayonet scabbards were made between November of 1944 and of February 1945.[18] Even though a rather sizable number were produced, surviving examples are relatively uncommon today.

The M1917 bayonets made during the Vietnam War era mentioned above required additional scabbards, and contracts were given to several firms for manufacture of a new plastic scabbard that was essentially identical to the late WWII variety, except for the markings on the mouthpiece. The WWII-made scabbards were marked with 'U.S.' inside a flaming bomb and 'M 1917', whereas the later Vietnam-era scabbards were marked 'US-1917' above the initials of the maker."[19]

A number of M1917 scabbards were supplied to several foreign countries that received M1917 rifles during and after World War II. As related by Cunningham:

> "Like the Model 1917 rifles, the bayonets and scabbards were widely distributed during WW2 to other countries…the bayonets can be found with Canadian and Danish markings. Rifles (and bayonets) were also supplied to England, China, and France during the WW2, and other nations post war Rebuilds and refinishes by these countries were common, and today a lot of the scabbards are found repainted in all shades of olive drab as well as black and other colors.
>
> When the rifles and bayonets went to England, they used their canvas frogs to support the scabbard on their web gear. Sometimes they were used with the double hook remaining on the scabbard, while at other times the extension holding the double hook was ground off. Following WW2, Denmark was given a number of the Model 1917 Rifles by England, and the Danes used a double hook adapter for attaching equipment to their belts. They made a new top plate with double hooks, and brazed this to the top of the throat of the scabbards that were missing them."[20]

Flaming Bayonet — An interesting and quirky item to conclude the story of the M1917 bayonet harkens back to the perilous time of the First World War when the United States was scrambling to develop weapons that might aid our doughboys in their fight against the Central Powers. Such examples included the Pedersen Device, V-B rifle grenade launcher and trench periscope rifles. Another much lesser-known but extremely interesting item was not really a bayonet but was given the imaginative moniker "Flaming Bayonet." The device was actually a miniature flamethrower that attached to the service rifle. A monograph written by the Historian of the U.S. Army Chemical School, discussed the concept of the flaming bayonet:

> "With rare exceptions, surprise has been a key element in achieving tactical success in the use of chemical warfare – including the related area of flame. Historical examples show that this surprise isn't easy to achieve. Efforts to develop a flaming bayonet during World War I provide an example.
>
> …The main value of flame seemed to be the psychological shock. The U.S. Army tried to find an effective way to secure this shock. Out of this effort, beginning in late 1917, came a project to develop a flaming bayonet. The basic idea was simple. Armed with a small, inconspicuous device mounted on

Extremely rare photo showing a M1917 rifle with the "flaming bayonet" being test fired.

War Department, Chemical Warfare Service Research Division, schematic drawing of a "flaming bayonet." Presumably, this is the six-shot "Cartridge type, Mark III," which was the final prototype of the "flaming bayonet" before the project was abandoned in early 1919.

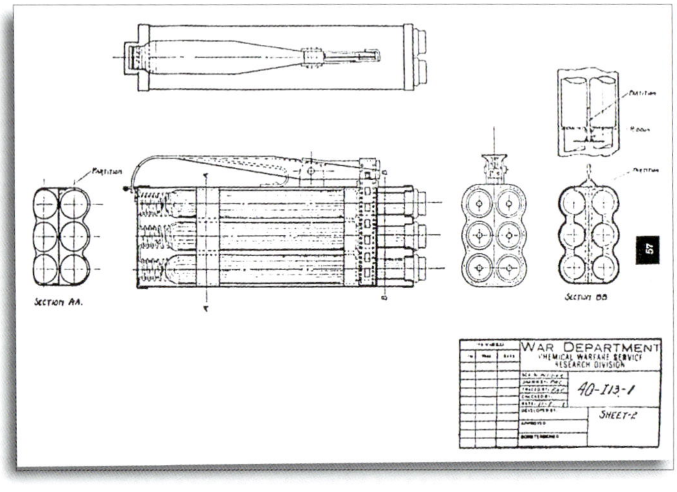

Chemical Warfare Service photo of a U.S. soldier holding a M1917 rifle with later pattern "Mark III" flaming bayonet attached. *(Photos from Chemical Warfare Service, National Archives collection)*

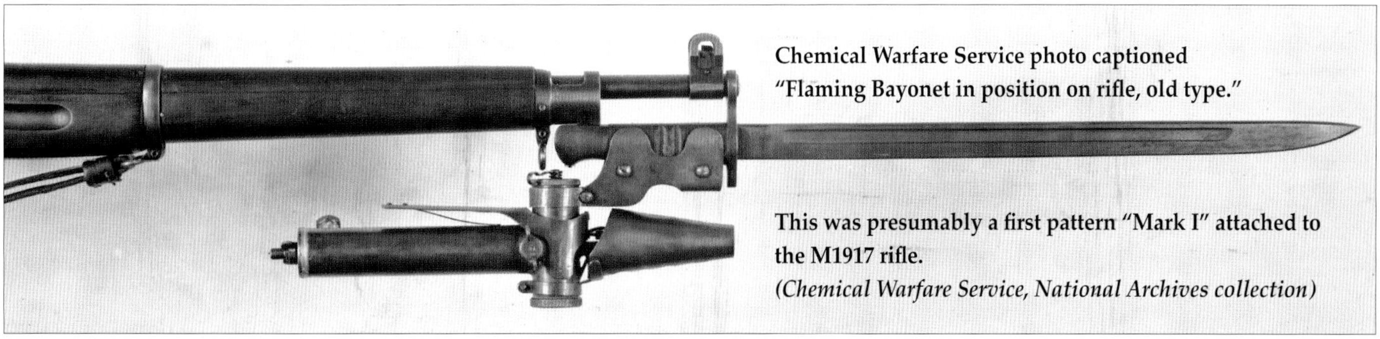

Chemical Warfare Service photo captioned "Flaming Bayonet in position on rifle, old type."

This was presumably a first pattern "Mark I" attached to the M1917 rifle.
(Chemical Warfare Service, National Archives collection)

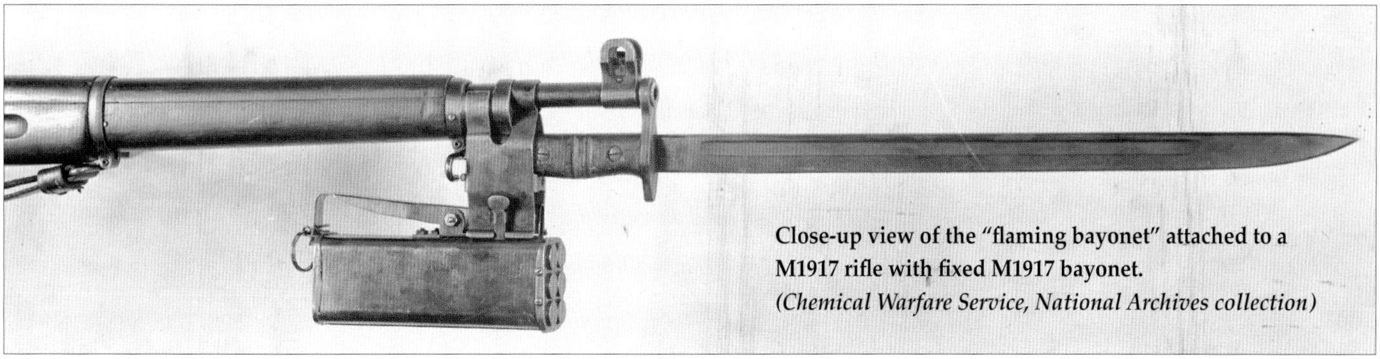

Close-up view of the "flaming bayonet" attached to a M1917 rifle with fixed M1917 bayonet.
(Chemical Warfare Service, National Archives collection)

his bayonet, an attacking infantryman would close to within eight or ten feet of an opponent. At that instant, as his enemy would be preparing to engage, the attacker would shift his hand slightly and trigger 'a flood of fire in the face of his opponent.' The 'disconcerting effect' would be enough to throw the defender off guard, making him easy prey for the naked steel."[21]

The development of the flaming bayonet was described in *History of Trench Warfare Material, Army Ordnance 1917–1919*:

"Flaming bayonet, liquid type, Mark I - The total weight charged 7/8 pound. To be used as an accessory to the military rifle, by the projection of a spurt of flame 5 to 15 feet in length. The shape and character of the flame obtained from the liquid type of flaming bayonet was not satisfactory and it was abandoned in favor of a cartridge type.

Flaming bayonet, cartridge type, Mark I – Weight loaded, (5 shots), 5/8 pound). To be attached to the muzzle of a rifle in convenient position for operation when rifle is held in position for bayonet fighting. Intended for the projection of a burst of flame from 5 feet to 15 feet long. This was deemed superior to the liquid type, but was abandoned in favor of flaming bayonet, cartridge type, Mark III, which new model is entirely enclosed and has a trigger which is operated by compressing the forefinger of the left hand."[22]

[Author's note: There was no "Mark II" flaming bayonet mentioned in ordnance documentation]

The Mark III variant was further described in *Chemical Warfare Monographs, Volume 43*:

"When loaded with six cartridges known as 'F.B. Revolver Cartridges,' .44 caliber, the device weights 5/8 pounds.

...The device is attached between the muzzle of the rifle and the back of the bayonet handle and is held in place by the bayonet. The trigger is located so that it can be readily grasped and operated when the rifle is held in the proper position for bayonet fighting. The cartridge which is the same as used in the 'Flaming Bayonet, Cartridge Type, Mark I' ejects a solid flame 6 to 10 feet long and 3 feet in diameter, and ejects white hot burning particles 20 to 30 feet. The weapon is intended for use at close range to temporarily disconcert the enemy immediately before the bayonet attack."[23]

The extensive tests regarding the composition of the incendiary material and the various dispersal mechanisms were apparently all conducted using the Model 1917 rifle as revealed in the ordnance document *Flaming Bayonet – Partial Report Number 3 on Investigation #40*: "The present (attachment) clips are made to fit the 1917 model modified Enfield, and are not adaptable for the Springfield type. A structural modification would be necessary for such rifle type."[24]

Limited testing and evaluation continued after the war, but it was apparent that the device looked more impressive on the drawing board than in the field. In early 1919, the concept was dropped by the Chemical Warfare Service, never to be resurrected.

It is nothing short of amazing that use of the M1917 bayonet spanned the period from wooden fabric-covered biplanes to the space shuttle. Like the rifle, the Model 1917 bayonet is rapidly growing in collector interest, and collectible specimens can sometimes be surprisingly difficult to find.

Grenade Launchers

V-B Launchers — As mentioned in Chapter 5, when the growing number of American doughboys were deployed overseas and began to experience the unfamiliar aspects of trench warfare, it was soon realized that some types of weaponry not in Uncle Sam's arsenal would be needed. One of these was the requirement for rifle grenades. The standard American rifle grenade at the time was the "Babbitt," adopted in 1911. It had long rod (stem) that was inserted into the rifle barrel and launched by means of a special blank-type cartridge. It was cumbersome to carry, could cause bore damage to the rifle and was unsuited for modern warfare. It was obvious that something better was needed. As was the case with several other types of weapons, we turned to our French allies.

The standard French rifle grenade was the "V-B," named after its developers, Jean Viven and Gustave Bessieres. The launcher, called a "tromblon," was a heavy metal cup that was designed to fit over the muzzle of the rifle. The cylindrical grenade had a hole through the middle and, when inserted into the launcher, was propelled by means of a standard ball cartridge. The bullet passed through the hole in the grenade and armed the fuse. The escaping gas of the cartridge permitted the grenade to be fired at distances up to 200 yards. Two V-B grenades could be launched at the same time, although range would be limited and the recoil increased.

The V-B saw wide use in the war by the French and performed satisfactorily. It was envisioned that the French pattern rifle grenades could be used with the American V-B launchers.

The V-B grenade and launcher were discussed in *U.S. Infantry Weapons of the First World War*:

> "Early in 1917, it was decided to adopt the French V-B launcher and grenade for use with our standard M1903 and M1917 rifles. The V-B was standardized by the Chief of Ordnance on July 16, 1917. While the adoption and production of American-made V-B launchers were in process, a number of Lebel or Berthier rifles fitted with French V-B launchers were issued to American infantry units. Despite the fact that this introduced non-standard weapons into American infantry units, the consensus was that this practice was better than having no rifle grenade launchers at all."[25]

It didn't take long for some of the French V-B launchers to be "unofficially" modified to fit the American service rifles. While good in theory, this practice caused problems, as revealed in a report from the U.S. Army 42nd Infantry "Rainbow" Division:

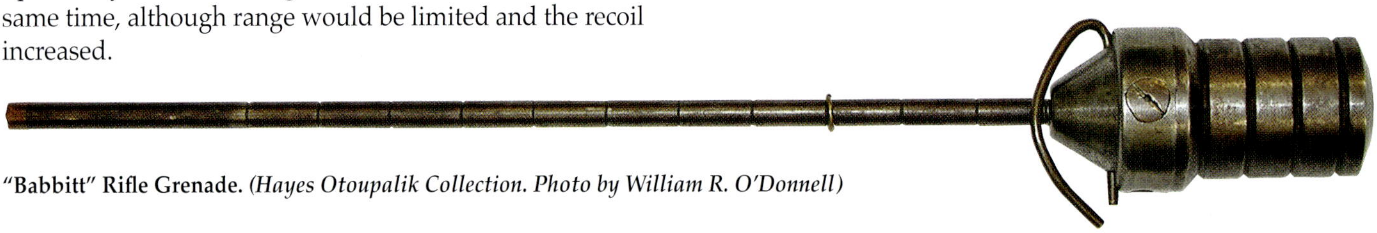

"Babbitt" Rifle Grenade. *(Hayes Otoupalik Collection. Photo by William R. O'Donnell)*

Very rare photo of U.S. soldiers training with M1903 rifles and Babbitt grenades. *(Paul Hastings. Courtesy of Hayes Otoupalik)*

French Army officer demonstrating a V-B grenade launcher on a Lebel rifle. *(National Archives)*

WWI photo of a French V-B launcher attached to a Lebel rifle. Note the VB grenades on the ground and the leather carrying case on the belt of the Poilu at the left. This was likely a training exercise, as neither solider has their canteen, and the cartridge belts appear to be empty. *(National WWI Museum and Memorial)*

U.S. Troops of the 32nd U.S. Infantry Division with French weapons. The soldier on the right is firing a Chauchat Automatic Rifle (note the bolo on his belt) and the Doughboy at left is preparing to fire a rifle grenade from a French V-B launcher attached to a Berthier rifle. *(U.S. Army photo. Courtesy of Konrad F. Schreier Jr.)*

American troops receiving training from French instructors on the V-B grenade launchers. *(National WWI Museum and Memorial)*

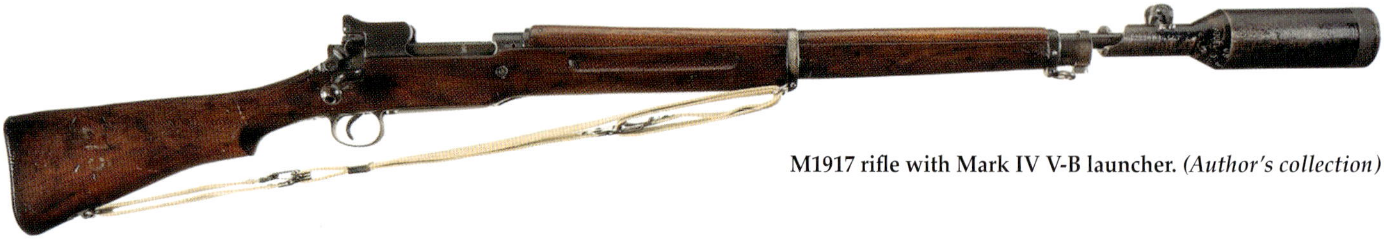

M1917 rifle with Mark IV V-B launcher. *(Author's collection)*

"…Someone at the ordnance base re-designed the base of the French tromblon so that it would fit the muzzle of the Springfield rifle, but they failed to take into consideration the great difference in pressure developed by the propelling charge of the American cartridge. It seems that the Rainbow was the first to receive this new brainchild and they were promptly issued to all the infantry squads in the division. The next day many of the men were in the hospital and their rifles damaged beyond repair."[26]

The development of a V-B launcher specifically made for the American Army was discussed in the Ordnance publication *History of Trench Warfare Material 1917–1919*:

"A discharger similar to the French with a special shim to fit behind the front sight band and the upper band, having two vent holes drilled in the base, was tested by the American Expeditionary Force and proved satisfactory. Fifty thousand of these were manufactured in France and issued to American troops. Drawings were forwarded to the United States with the request that this discharger be immediately put into production."[27]

However, adapting the French V-B rifle grenade and launcher for use with the American M1903 and M1917 rifles proved to be more challenging than anticipated. As related in the above-cited Ordnance document:

"The V.B. grenade fired from the discharger has apparently been very satisfactory for use in the trenches. Considerable difficulty, however, was experienced in adapting this article for use with the United States rifle. The American rifle ammunition is more powerful than the French, the result being that the pressure exerted in the discharger was excessive, thereby causing the rifle stocks to split as a result of continuous firing of the grenades.. In view of the fact that the supply of rifle grenades of the French and Americans are pooled for issue, it was found necessary to drill two ventholes in the American discharger to permit venting the excess pressure…The net result of this practice is that 30 yards less range is obtained with the American grenade than the French. It was directed that all dischargers manufactured in the United States have two ventholes referred to above drilled in them."[28]

The V-B launcher weighed 3 lbs., 3 ozs. The earliest launchers were made in France for the U.S. Army and had blued finishes and hand-stamped markings. Most of the launchers manufactured in the United States were finished in dark phosphate or Parkerizing and had machine-stamped markings. Most of the American V-B launchers were made by the Westinghouse Corp. and were marked with "W" on the body.[30] Westinghouse, along with eight other manufacturers, made large numbers of V-B rifle grenades as well.[29] There were several variants of V-B launchers used by the U.S. in World War I:

French V-B launcher as used with the Lebel and Berthier rifles. *(IMA)*

Vent hole in V-B launcher for M1917 rifle. *(Author's collection)*

Mark I V-B launcher (left) made in France under U.S. contract. Note the gas vent hole drilled in the side; Standard French V-B launcher (right) for Lebel and Berthier rifles. Other than the length of the section that slides on the barrel, both launchers are essentially identical. *(Rick Larson. Courtesy of Hayes Otoupalik)*

***The Original French V-B Launcher* —** These were initially used with Lebel or Berthier rifles that were supplied by the French, which worked satisfactory with those rifles. However, as mentioned, the "unofficial" modification of such launchers for use with the American service rifles resulted in serious problems.

***Mark I* •** "Mark I" was apparently the designation given to the 50,000 V-B launchers made in France under U.S. contract for the AEF. The Mark I launchers were withdrawn from service in favor of the U.S.-made launchers when they became available.

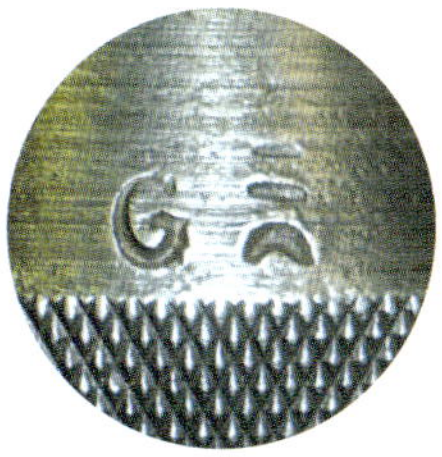

Only markings on French-made Mark I launcher consist of a single hand-stamped letter "G" and an unidentified symbol.

***Mark II* •** Although assigned a "mark" designation by Ordnance, this variant was not put into production.[27]

***Mark III* •** The Mark III V-B launchers were made in the United States for both the M1917 and M1903 rifles. The model of the rifle for which the launcher was intended was stamped on the outside of the body "For Use Only on Rifle Model of 1917 (or Model of 1903"). The Mark III launchers are characterized by a straight slot into which the barrel was inserted. A sheet metal shim was used in order to maintain a secure fit and proper alignment of the bore with the hole through the V-B grenade while attached to the rifle. The shims for the M1903 rifles were of different length from the M1917 variety and were marked with the identity of the respective rifles. The Mark III launcher made for the M1903 rifle had a smooth body while the type made for the M1917 rifle had a knurled band near the top.

***Mark IV* •** These launchers were also made for both the M1903 and M1917 with the type of rifle stamped on the outside of the body. The Mark IV launcher featured a spiral slot, which permitted it to be securely attached to the barrel without the necessity for a separate shim. As with the Mark III, the Mark IV launcher for the M1903 rifle had a smooth body, and the launcher for the M1917 had the knurled band.

Mark III V-B launcher for M1917 rifle

A Mark IV V-B launcher for M1917 rifle. This example still retains part of the original factory paper wrapping.

M1917 rifles with Mark IV V-B launchers at a U.S. stateside demonstration in Washington, D.C., 1918. *(Tom Laemlein)*

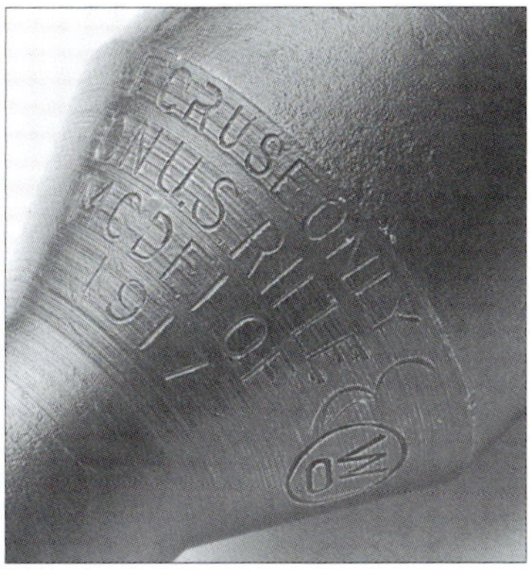

Markings on a Mark IV V-B launcher for the M1917 rifle. *(Author's collection)*

Mark III V-B launcher mounted on 1917 Model rifle. Note the metal shim marked for the M1917 rifle. In use, the shim was placed on the barrel behind the front sight before sliding the launcher in place. Mark IV V-B launcher (shown below) installed on M1917 rifle. *(Hayes Otoupalik Collection. Photos by William R. O'Donnell)*

Although the Mark IV V-B launcher was intended to supersede the Mark III, both types saw action during World War I.

The American V-B launchers were carried in a leather case with M1910-pattern belt hooks for attachment to the standard infantry cartridge belt. The leather case was not particularly sturdy and prone to deterioration and, in the immediate postwar period, was superseded by a simpler and less expensive canvas carrying case. Rock Island Arsenal was the primary, if not sole, maker of the canvas cases and most examples are dated 1919 or 1920.

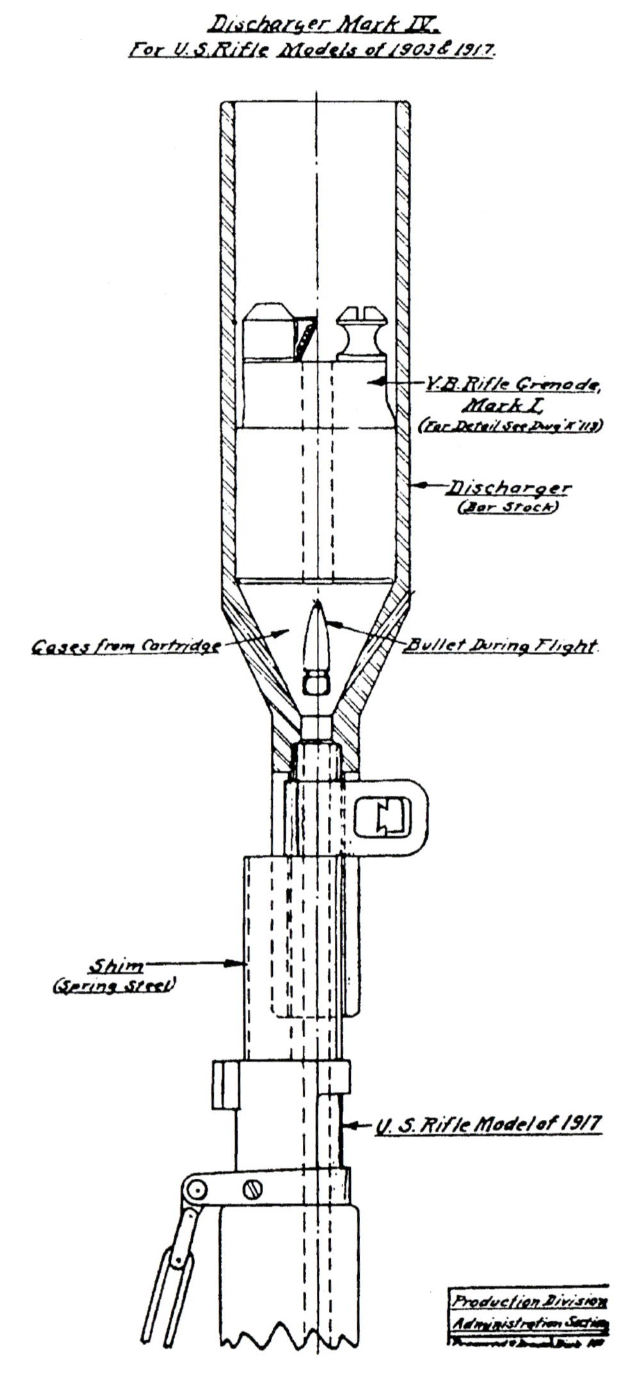

FIG. 94.—Rifle grenade discharger, Mark IV

Front and back views of the leather carrying case as used with the America V-B launchers. Note the cartridge belt hooks. Also shown is a Mark IV launcher. *(Author's collection)*

1918 photo of U.S. troops undergoing training with hand grenades and M1917 rifles with V-B launchers. *(National Archives)*

Ordnance drawing of a sectionalized V-B launcher attached to a M1917 rifle. Note the orientation of the V-B grenade and the bullet in the launcher. The launcher is incorrectly identified in the drawing as a Mark IV, but the shim indicates it is actually a Mark III. *(Handbook of Ordnance Data)*

Two doughboys undergoing training in France during WWI with M1917 rifles and Mark IV V-B launchers, under the watchful eye of an instructor. *(U.S. Army Signal Corps photo)*

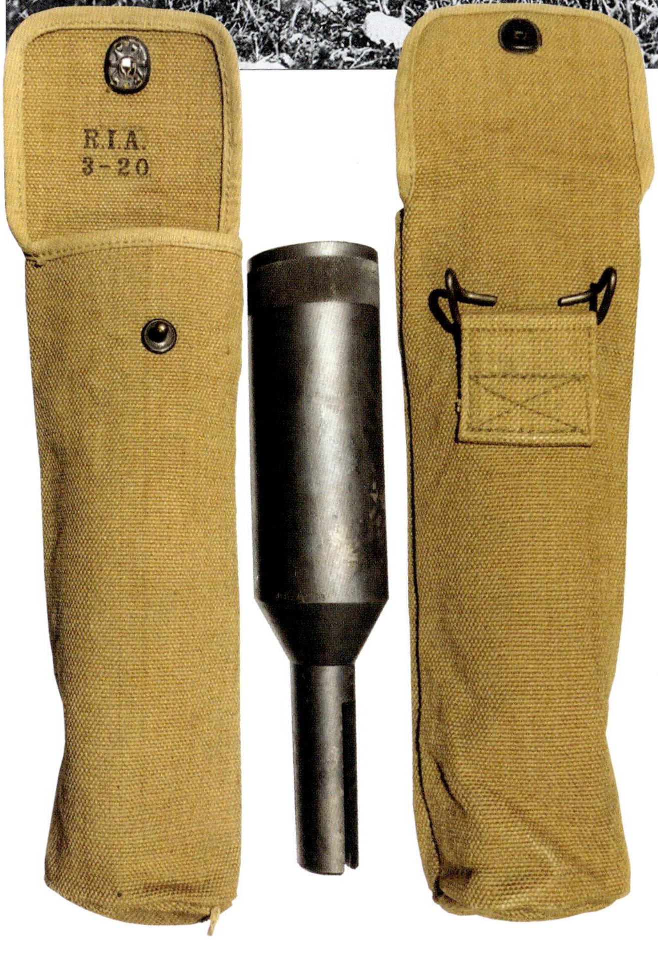

Front and back views of canvas carrying case for V-B launcher made by Rock Island Arsenal in March of 1920. *(Author's collection)*

V-B Grenades — The primary grenade used with the American V-B launchers was the "Mark I" fragmentation and was very similar to the French-made variant. Other than slight changes in the dimensions and configuration, the main difference was that the American Mark I grenades had a malleable body instead of the cast iron type of the French V-B grenade. The Mark I was about 2½ inches long and 2 inches in diameter and weighed approximately 12⅓ ounces. The maximum range was about 200 yards. The "danger zone" was stated to be approximately 75 yards. The Mark I was designated as both an "offensive" and "defensive" grenade since the user was generally out of range of the exploding fragments. The Mark I grenades were serrated internally for maximum fragmentation effect while leaving a smooth surface on the outside of the body.

There were initially a number of problems encountered with the American-made launchers due to several factors, including not making the necessary allowances to account for the differences between the French and American rifle cartridges. The problems with the American V-B launchers were discussed in the book *America's Munitions*:

> "...in the anxiety of producing this weapon (the V-B launcher), in the shortest time possible the models were not sufficiently tested...The result was that the French grenade did not function well with our ammunition... In May, 1918, several months after manufacture of this grenade had been in progress, the entire undertaking was canceled pending the development of new designs, and the 3,500,000 completed grenades had to be salvaged."[31]

The American-made V-B grenade was redesigned to be compatible with the U.S. M1906 .30 caliber (.30-06) cartridge, and orders were placed for over 30 million grenades. The grenade was designated as the "V-B Rifle Grenade, Mark I." By August of 1918, daily production of the Mark I V-B grenades reached 130,000, and they had increased to about 250,000 by the time of the Armistice in November.[32]

An eleven-pocket cloth bandoleer, sometimes referred to as a "grenade vest," was manufactured to carry V-B grenades. Sizeable numbers were made, and specimens, often in excellent condition, can still be found today. Interestingly, a similar vest was put back into production in WWII for the Marine Corps V-B grenades. These were made of a lighter canvas material than the WWI-era vests. When the USMC V-B launchers were withdrawn from service around 1943, these vests were sometimes used by the Marines to carry shotgun shells.

The recoil of the V-B grenade was quite pronounced,

Mark I V-B Grenade (left); Mark I V-B grenade, sectionalized (right). *(Hayes Otoupalik Collection. Photo by William R. O'Donnell)*

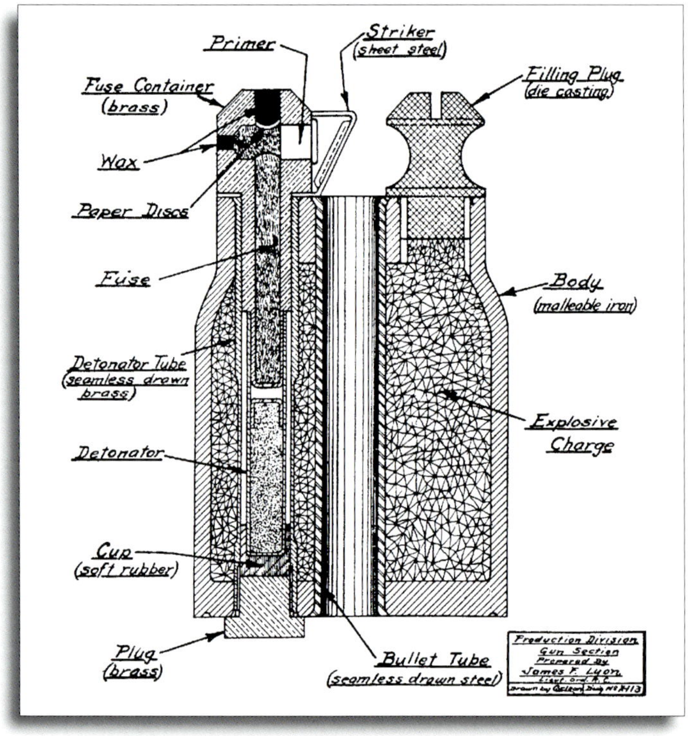

Ordnance drawing of sectionalized Mark I V-B rifle grenade. *(Handbook of Ordnance Data)*

Ordnance photo dated May 18, 1918, showing a shipping crate for V-B Mark I rifle grenades. *(National Archives)*

Mark I V-B rifle grenades being manufactured at the Westinghouse Electric & Manufacturing Co. Works at East Pittsburgh, Penn., August 22, 1918. *(National Archives)*

WWI photo showing a U.S. solider wearing a V-B grenade vest. Note the M1917 trench knife. *(U.S. Army Military History Institute)*

Markings on back of the WWI V-B grenade vest. This example was made in May of 1918 by the firm of Earl & Wilson and was inspected by John E. Wright.

WWI eleven-pocket V-B grenade "vest" with Mark I V-B grenades.

(Far left) V-B practice grenade.

(Near left) V-B "Message" grenade.

(Hayes Otoupalik Collection. Photo by William R. O'Donnell)

and broken stocks were not an uncommon occurrence when the M1903 and M1917 rifles were used as grenade launching platforms. The butt of the rifle was placed on the ground and the aiming was essentially by "Kentucky windage," i.e., visually estimated.

In addition to the standard V-B Mark I fragmentation grenade, there were other variants:

- Inert dummy practice grenade was characterized by an "hourglass-shaped" body.
- V-B rifle "message grenade" intended to convey written communications across the battlefield. According to *History of Trench Warfare Material 1917–1919,* several hundred thousand of these were made for the French shipped to the front lines. One hundred were obtained by the AEF and tested at the First Corps School in France. However, it was determined that the message grenade was not practical for field use and was officially disapproved by the Trench Warfare Section of the AEF.[33]
- "V-B Star cartridge, Mark I" and "V-B Parachute cartridge, Mark I" for signaling and illumination.[34]

Despite the inevitable "teething pains" that resulted from adapting the French V-B launchers and grenades to be compatible with the American rifles, the V-B rifle grenade proved to be a useful addition to the doughboys' arsenal. The official "TO&E" requirements for a First World War U.S. Army Infantry Division in 1918 was 1,560 V-B rifle grenade launchers (Grenade Dischargers – Rifle).[35]

While an integral part of the AEF infantry units, utilization of the V-B grenades could be dangerous to the user without proper training and constant vigilance. A good example of what could go wrong was related in the book *Fighting Soldier – The AEF in 1918*:

> "A man named Gilbert, a few yards from me, had a rifle grenade that he clumsily was trying to fire. I do not know where he got it; there was not one in the platoon when we started down the road after the supposed machine gun nest; finally, he got it into the cup on the muzzle of his rifle and to my horror and before I could stop him, he fired it straight up; I drew in a knot waiting for it to fall on us; and it fell a few yards behind us with a loud crash; fortunately, without damage."[36]

In addition to carelessness and/or improper use, the previously cited manufacturing defects of the early grenades also resulted in some tragic accidents, particularly when training. One example was revealed in the book *Over There*:

> "…a rifle grenade hung fire and exploded in the horn, killing instantly the first-lieutenant instructor and wounding a student. Only yesterday afternoon, that same instructor gave me my first lesson in the use of the rifle-grenade… It has cast quite a gloom over the whole school, but I suppose we ought to expect some accidents occasionally — for there is much work going on the range all the time."[37]

Despite these unfortunate incidents, there were numerous cases in which the V-B grenades proved to be valuable assets to our troops. A dispatch from the U.S. Army's 89th Infantry Division stated:

> "Every infantry soldier was required to carry two rifle grenades in addition to his other equipment. The 177th Brigade, leading the attack against machine gun nests in the Bois de Barricourt and just beyond…
>
> The rifle grenade proved to be one of the most effective weapons for the work of reducing machine gun nests, when operated by determined and courageous grenadiers. For example, a rifle grenadier, Private First-Class Cecil E. Reed, of Company 'E' 353rd Infantry, advanced out of the forest into the open ground to obtain a range for the grenades, though the place was under withering machine gun fire and his comrades were seeking what cover they could find. Kneeling in the open he fired his entire supply of ammunition, then dashed back and obtained more from the other men. He returned to his firing position and when all his grenades were again fired, returned the second time to get more. This time he moved forward to a still more advanced position, from which he succeeded in placing shots so close to the enemy machine gunners that he drove them from their shelters and they were killed by other men of the company."[38]

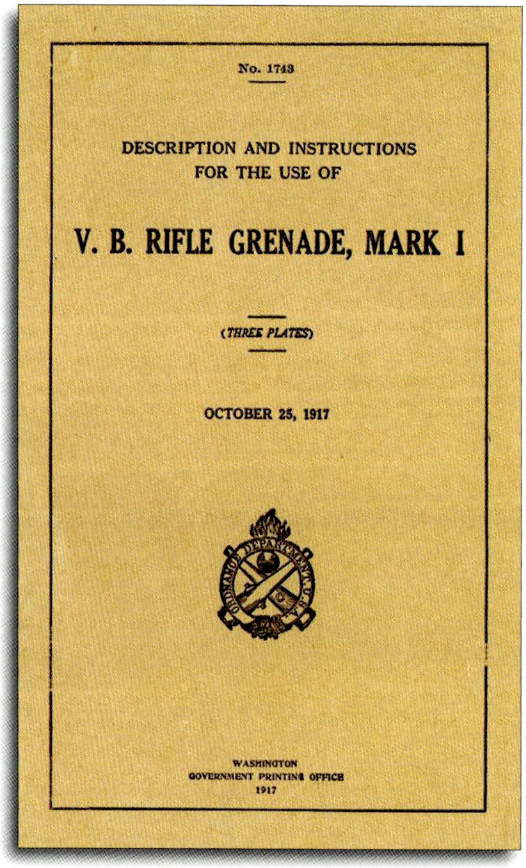
No. 1743

DESCRIPTION AND INSTRUCTIONS FOR THE USE OF

V. B. RIFLE GRENADE, MARK I

(THREE PLATES)

OCTOBER 25, 1917

WASHINGTON
GOVERNMENT PRINTING OFFICE
1917

WWI Manual for V-B Rifle Grenade, Mark I Launcher.

There were many similar instances in which the V-B grenades were used to good effect by the doughboys. It is reported that some stateside Army recruiters would display a V-B launcher and tell impressionable potential recruits that the weapon would permit them to fight the war in the safety of their trench positions while lobbing grenades at the distant German positions. It is not known how many young men contemplating joining the service were gullible enough to buy that story!

After the Armistice, despite the proven effectiveness of rifle grenades in the war, there was still some bias against continued issuance of the weapons. Criticisms included limited range, inaccuracy, exposure of the firer to the enemy, small danger radius and weight. There was also concern that the standard service rifle was not rugged enough to withstand the rigors of grenade launching for an extended period. There seemed to be a consensus in post-WWI Ordnance circles that a rifle grenade may not be necessary, and if a range greater than that of hand grenade was required, then trench mortars could be employed with greater effectiveness.[39]

In 1923, the Chief of Infantry recommended that all rifle grenades be dropped from the Table of Basic Allowance and declared obsolete.[40] This obviously signaled an end to the issuance of the V-B grenade launchers to the U.S. Army. The V-B launchers made for the M1917 rifle, and most of those made for the M1903 rifle, were ordered destroyed and sold for scrap metal. The relatively few remaining V-B launchers still on hand in the Army's inventory were reclassified as "Discharger, Pyrotechnic, Mark IV, for Rifle, model 1903" and were to be used solely to launch flares and other pyrotechnic devices but saw little, if any, actual use.

Unlike the Army, the U.S. Marine Corps held on to their Mark IV V-B launchers for use with their '03 rifles and saw use in the early days of World War II, primarily during the Guadalcanal campaign. Surviving examples of any type of V-B launcher are not common and are sought-after collectibles today. A V-B launcher would be a great addition to a collection. Some reproductions have been made but, fortunately, are generally easy to distinguish from the genuine article.

M2 Launcher — After the abandonment of the V-B launcher and grenade, there was sporadic developmental work done on developing a more effective rifle grenade launcher in the late 1920s and well into the 1930s. However, the lack of urgency and the ambivalence of the American military establishment during this period regarding rifle grenades resulted in little real progress being made. With the adoption of the semiautomatic M1 rifle in 1936, most of the developmental work shifted to the new rifle rather than the '03. Of course, with most of the M1917 rifles in storage, there was no substantive discussion about a launcher for the "American Enfield."

When war broke out in Europe in 1939, and with the success of Germany's Blitzkrieg tactics using tanks and other armored vehicles, the U.S. War Department became acutely aware of the need for some type of anti-tank rifle grenade. The development of the "shaped-charge" technology by Henry Mohaupt led to the subsequent adoption of the M9 and M10 anti-tank rifle grenades. However, there still was not an effective launcher in inventory.

Developing a launcher compatible with the M1 rifle's semiautomatic mechanism proved to be quite vexing. The manually operated bolt-action M1903 and M1917 rifles made it much easier to develop a grenade launcher and, as an interim measure until a workable launcher for the M1 rifle could be found, a simple tube that clamped to the end of the barrel was developed by Ordnance for the bolt-action rifles.

Launchers of this type were fabricated at Springfield Armory for the M1903 and M1917 rifles. Subsequent testing at Aberdeen Proving Ground determined that the design worked satisfactorily with both rifles. The launcher for the M1903 rifle was originally designated as the "T3," later changed to "T9." On September 9, 1941, the T9 was stan-

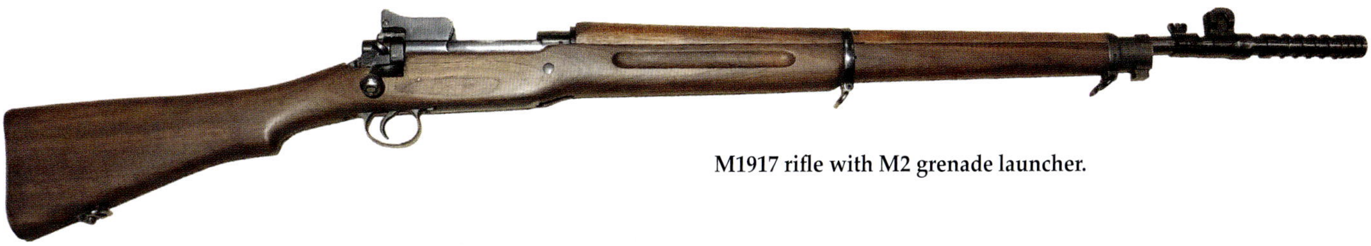

M1917 rifle with M2 grenade launcher.

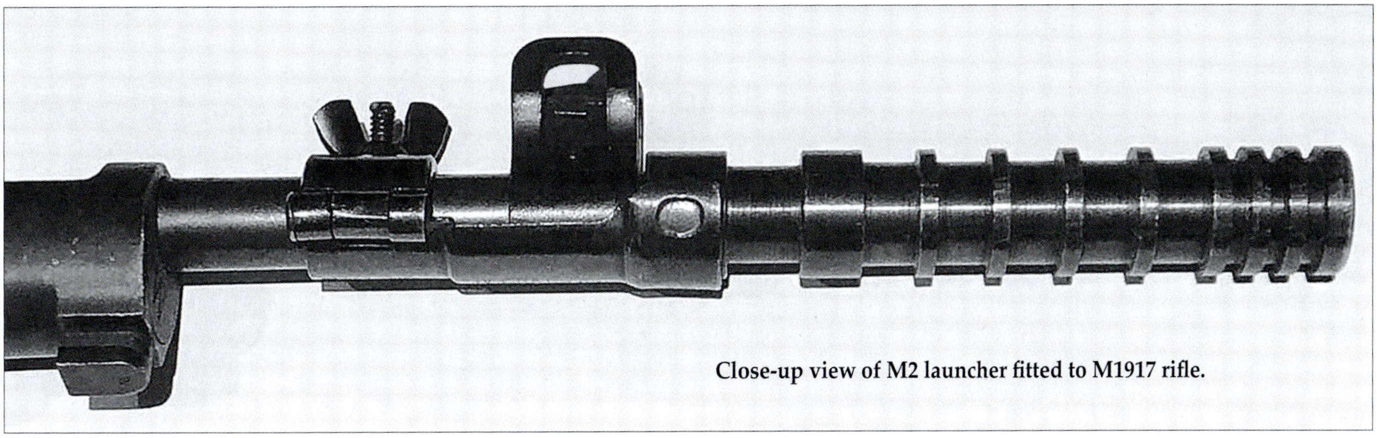

Close-up view of M2 launcher fitted to M1917 rifle.

dardized as the "Launcher, Grenade, M1." The launcher for the M1917 rifle was initially designated as the "T7," later changed to the "T10." The T10 was adopted on September 9, 1941, as the "Launcher, Grenade, M2."

Information regarding manufacture and subsequent disposition of these launchers was related in the Ordnance publication *Project Supporting Paper Relating to Grenade Launchers 1917–August 1945:*

> "Even before the new launchers had been perfected sufficiently to justify standardization, immediate requirements for large quantities of launchers were established, the indication being that 150,000 of the T3 (M1903 Rifle) Launchers and 50,000 of the launchers for the T7 (M1917 Rifle) would be required in the very near future. Accordingly, steps were taken to set up a contract at Knapp Monarch Company for the manufacture (of the launchers),
>
> ...Orders at Knapp Monarch Company had been revised to call for 100,000 M1 Grenade Launchers (for M1903 Rifle) of which 50,000 were for the U.S. Army and 50,000 for Defense Aid; all 100,000 M2 Launchers (for M1917 Rifle) which were for Defense Aid, mostly for use by the British Lee Enfield which was approximately the same as the U.S. Rifle M1917."[42]

Perhaps, surprisingly, priority was initially given to production of the M2 launcher for the M1917 rifle rather than the M1 launcher for the '03 , but after Pearl Harbor, this policy changed as explained in the above-referenced Ordnance publication:

> "In the beginning, it had been believed Defense Aid should be given preference in production, with emphasis on the M2 Launcher, since that was the type desired by the British, to whom most of the Launchers were to go. In December of 1941, 31,980 of the M2 launchers were produced, as compared to 3,520 M1 launchers for the U.S. Army. The active participation of the United States in the war, after December 7, 1941, created an immediate need for M1 launchers for the U.S. Army, causing a directive to be sent for a change-over to production of the M1 launchers. In January of 1942, 63,360 of that model were produced and 35,040 in February. Immediate requirements having been met, heavy production swung back to the M2 launcher in March, the production in that month practically winding up the established requirement of 100,000.
>
> Peak production on the M1 grenade launcher was in January of 1942 and on the M2 grenade launcher in March of 1942. Final production of the M2 launchers was in July of 1942, with a total of 112,327 having been produced. **Of that number, 39,793 were converted into M1s as there was a reduction in requirements for the M2 Launchers.** Production on the M1 Launcher continued until May 1943, a total of 322,892 having been produced."[43]
>
> *[Bold print added by author]*

Efforts continued to develop a satisfactory launcher for the M1 rifle resulted in the "M7" launcher being standardized on February 11, 1943, followed by the adoption of the "M8" launcher for the M1 carbine on the same date.[44] The availability of these launchers for the two primary U.S. infantry weapons resulted in the M1 and M2 launchers becoming less important.

In November of 1944, the M1903 rifle was reclassified as "Limited Standard," which resulted in the M1 launcher being similarly reclassified, although many remained in use until the end of the war.[45] During this same period, the M1917 rifle and its M2 launcher became expendable and were reclassified as "obsolete," as discussed in the above-referenced ordnance publication:

> "Because M1917 Rifles were on hand, at the beginning of World War II they were issued to American troops, and the M2 Grenade Launcher was developed for use with that rifle. Later on, when full production of the M1 Rifle made issuance of the M1917 Rifle no longer necessary, naturally there was no further requirement for the M2 Grenade Launcher so it was reclassified as Obsolete...on September 7, 1944."[46]

As events transpired, few, if any, M2 launchers were actually used in combat since most were earmarked for the British Home Guard for potential use with their Lend-Lease M1917 rifles. There are no known accounts of M2 launchers being utilized by American troops in WWII except, presumably, for training exercises. As stated above, almost 40,000 M2 launchers were converted to M1 launcher specifications given the greater demand for the latter. All M1 and M2 launchers were originally blued, but most, if not all, of the M2s converted to M1 specs were Parkerized as part of the

Marking on an M2 launcher for the M1917 rifle. ***(John Spangler)***

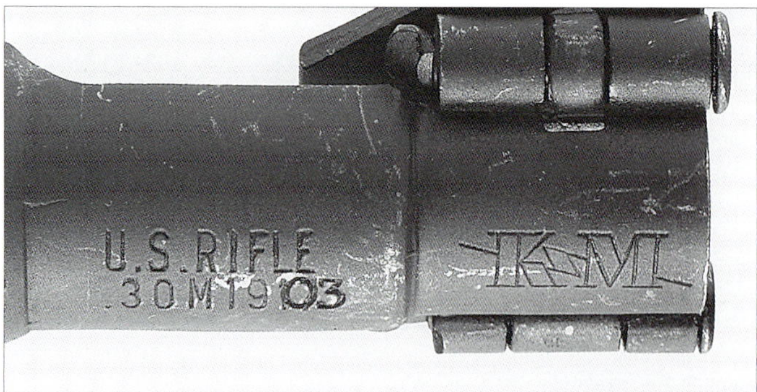

M2 launcher converted to "Launcher, Grenade, M1" specifications. The original "17" in the designation was overstamped "03." Note the "KM" marking indicating production by the Knapp-Monarch Co.

conversion procedure. Also, the "17" on the M2 launchers converted to M1 was overstamped "03" to denote the change in nomenclature. Despite rather sizable numbers made, genuine M1 grenade launchers are scarce. Unaltered M2s are even more difficult to find given the smaller numbers made and the fact that over one-third were modified to M1 configuration. Any genuine example would be a valued collectible today.

Slings

Model 1907 — When the Model 1903 rifle was standardized, a new pattern leather sling was also adopted at the same time and designated as the "Model 1903." While marginally adequate as a carrying sling, it was not long enough to properly serve as a useful shooting aid, and a better sling design was needed. A 1906 ordnance report discussed the Model 1903 sling, and it was stated, "…the sling in question had been subject to "considerable criticism from expert shots." The report further remarked that an experimental sling was in development that was intended to "…support the gun in firing." This new sling was designed by a Colonel Wright of the Ohio National Guard and was

M1917 rifle with M1907 sling.

Close up of the markings on the 1918-production M1907 sling.

M1907 sling. Note the brass "frogs" and the series of holes for adjustment.

Photo depicting the use of a M1907 sling as a marksmanship aid with a M1903 rifle. The same arrangement would be used with the M1917 rifle as well. *(Western News Union)*

subsequenly adopted as the Model 1907."[47] It proved to be a serviceable and popular design.

The new sling consisted of two leather straps 1¼ inches wide with a series of holes for adjusting the length. One strap is 24 inches long and the other 45 inches long. Both sections had brass adjustment "frogs" secured by rivets, and one of the sections had a metal "D" ring to join the sections together and permit easy adjustment. The sling attached to the rifle by means of the upper and lower sling swivels. Most, if not all, of the initial production slings were fabricated by Rock Island Arsenal.

When the Model 1917 rifle was adopted, the Model 1907 sling was approved for use with the new rifle. Of course, the enormous number of rifles being made for the First World War would require many more slings than could be produced by Rock Island, so production contracts were given to a number of commercial firms. These slings were often marked with the identity of the maker, the year of production and inspector initials.

In addition to being used with both the Model 1903 and Model 1917 rifles in World War I, the Model 1907 sling was also authorized for use with the 12-gauge "trench guns." As well as being used to carrying the rifle, the M1907 sling had application as a marksmanship aid to support the rifle and increase stability while shooting.

Doughboy with M1917 rifle without a sling. Rifles were not shipped from the factory with slings attached, as they were fitted by the receiving/issuing unit. *(Charlie Williams)*

Marking on M1907 sling manufactured in 1942 by the Boyt Harness Co.

Rifles were not shipped from the factory with slings attached, and the type subsequently utilized was dependent on what type of slings the supply system was able to provide. The supply of WWI-manufactured Model 1907 slings was sufficient to meet the demand until the outbreak of the Second World War.

The M1907 sling was put back into production in 1942. As with the WWI-era slings, the WWII slings were typically marked with the name of the maker, year of production and, sometimes, inspector initials. The only substantive difference from the 1917–1918 vintage slings was the substitution of steel hardware for the brass hardware. There were a great many WWI-era M1907 slings used during WWII concurrent with the newly made variety. It was not uncommon to find M1907 slings used in WWII with one strap having WWI brass hardware and dates and the other with WWII steel hardware. In mid-1943, the leather M1907 sling was superseded by a simpler and less expensive web sling (to be discussed) although the M1907 sling remained in service throughout WWII and well into the Korean War period. Genuine "GI" M1907 slings are becoming rather scarce and are sought-after by collectors. There are a number of copies of the slings being made today, some well-made and some shoddy, but these can generally be distinguished from the originals without a great deal of difficulty.

M1917 "Kerr No Buckle" — While the Model 1907 was the predominate service rifle sling used in World War I with the Model 1917 rifle (and Model 1903 rifle), there was another sling that was adopted as a supplemental sling to meet the huge demand. An excellent summation of the story of this sling was related by author Lt. Col. (Ret.) William S. Brophy in his excellent book *The Springfield 1903 Rifles*:

> "The 1913 report of the Chief of Ordnance suggested that a small number of web woven gun slings from a private manufacturer be made and that they be issued to three companies of infantry at the School of Musketry for a test of one year in order to determine the relative merit of a web sling in comparison with one made of leather.
>
> The result of the test was the adoption of the Kerr sling, which was a product of the Kerr Adjustable Strap Company, Inc., 40 Cedar Street, New York City, and was patented July 21, 1914. Upon adoption (in 1917), it was designated 'Gun Sling, M1917 (webbing). The new sling was authorized for use on the '03, **but most of them were used on the Model 1917 rifle** during and after World War I. Captain Roy S. Tinnney, musketry instructor, described the Kerr sling as follows: *[Bold print added by author]*
>
> 'Until recently leather was the universal material for rifle slings. Every now and then someone would suggest the use of webbing, but none of us could figure out the problem of adjustment on a cotton strap, so we clung to our strip of bull hide and did the best we could with it. Then just when we had given up all

hope of a radical improvement, along came the Kerr extension sling, and solved the problem. As usual, the first to adopt it were those premier riflemen, the Marine Corps; the Navy followed, and last summer the Army fell in line, so today the Kerr has become the standard sling of the service, a position it holds purely as a matter of merit. Just as webbing has replaced leather straps in the soldier's field harness, the Kerr sling with its extension principle replaced the cumbersome bit of bull hide that formerly cluttered up the army rifle.

Markings Kerr sling metal adjustment tab.

M1917 Kerr sling.

M1917 rifle with Kerr sling attached.

The advantages of webbing over leather for this particular purpose are many; it is equally strong and very much lighter; it wears better; it costs much less to manufacture and conserves leather; it is softer and more pliable; and most of all because of the patented principle involved, it is more readily adjustable to suit the individual using it. In a word, it fully meets every requirement demanded of a perfect sling. These are briefly the main reasons for the adoption of the Kerr sling. Like most improvements in firearms, it is an American invention and the War Department is to be congratulated upon the promptness with which is appreciated these points of superiority."[48]

The cost of a Kerr sling in 1918 was $1.05 each. Some people have assumed that the Kerr sling was developed for use with the M1917 rifle due to its M1917 designation and the fact the "American Enfield" rifle was pictured in company factory brochures, but the 1917 designation referred to the year of adoption. In addition to being authorized for the M1903 and M1917 rifles, the sling was also adopted for use

Two U.S. soldiers with M1917 rifles in a trench dugout position. Note the M1917 Kerr sling on the right. *(National Archives)*

with the Krag rifle. The sling had the identity of the manufacturer and patent markings on the metal adjustment tabs.

It has been theorized that a different variant of the M1917 sling was made specifically for the M1903 rifle, as some examples have been noted that were shorter than the norm or lack the metal link for the short strap to snap on to, and instead have to snap over the metal tab. While company's brochure that accompanied the sling illustrated how to attach the sling to the M1917 rifle, it also included instructions on how to install it on M1903 and Krag rifles in a slightly different manner from that used with the M1917 rifle. Thus, it would seem there would be no need for a different sling for use with the '03. However, researcher Frank Mallory discovered a document revealing that the Kerr sling was adopted by the U.S. Marine Corps in April of 1917. It also stated that the Marines "...made some of their own slings at the Marine Corps Depot of Supplies."[50] While the document did not specifically state that these slings were of the M1917 pattern, it is not unreasonable to infer that this was the case. The Marines had adopted the Kerr sling in 1917, and the '03 was the primary rifle used during that period, so perhaps it would have been logical to fabricate a sling specially for that rifle.

Despite Captain Tinney's effusive praise of the Kerr sling, it never came close to supplanting the M1907 sling. It is interesting to note that the above-cited document also stated that the leather M1907 was "re-adopted" by the Marines in November of 1920, so use of the Kerr sling by the USMC was rather short-lived despite the fact the Marines were the first to adopt it in 1917.[51] The Kerr sling actually saw relatively little use after WWI, although it did remain listed as a substitute for the M1907 sling as late as World War II.[52]

Although the Kerr sling did have some advantages, the M1907 sling was clearly preferred by the American armed forces as evidenced by the fact it was put back into production in 1942 and saw widespread service into the 1950s.

A sling similar to the Kerr M1917, designated as the "M3," was adopted in WWII for the Thompson Submachine Gun. The M3 sling functioned the same way as the M1917 sling but was shorter and had a different color to the webbing, and the metal adjustment tabs were not marked. Some of the standard Kerr M1917 slings were also used on WWII weapons, including some M1917 and M1903 rifles, a few M1 rifles and some Thompson submachine guns. However, the vast majority of these rifles used either M1907 leather slings or M1 web slings (discussed below). At one time, original Kerr M1917 slings were relatively common but are getting more difficult to find with correspondingly higher prices.

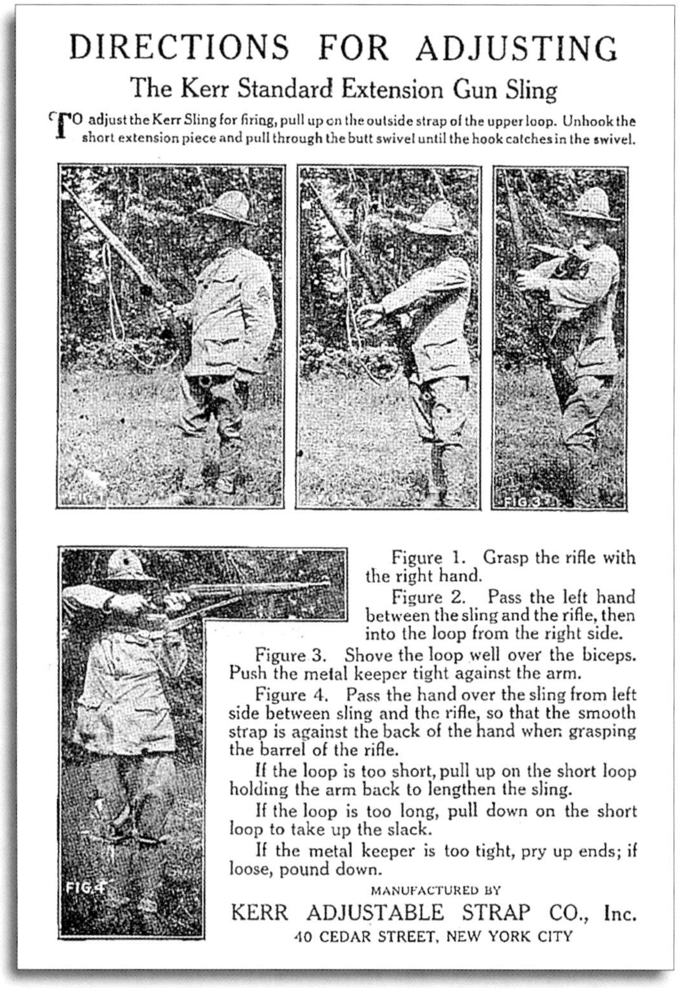

DIRECTIONS FOR ADJUSTING

The Kerr Standard Extension Gun Sling

To adjust the Kerr Sling for firing, pull up on the outside strap of the upper loop. Unhook the short extension piece and pull through the butt swivel until the hook catches in the swivel.

Figure 1. Grasp the rifle with the right hand.

Figure 2. Pass the left hand between the sling and the rifle, then into the loop from the right side.

Figure 3. Shove the loop well over the biceps. Push the metal keeper tight against the arm.

Figure 4. Pass the hand over the sling from left side between sling and the rifle, so that the smooth strap is against the back of the hand when grasping the barrel of the rifle.

If the loop is too short, pull up on the short loop holding the arm back to lengthen the sling.

If the loop is too long, pull down on the short loop to take up the slack.

If the metal keeper is too tight, pry up ends; if loose, pound down.

MANUFACTURED BY

KERR ADJUSTABLE STRAP CO., Inc.

40 CEDAR STREET, NEW YORK CITY

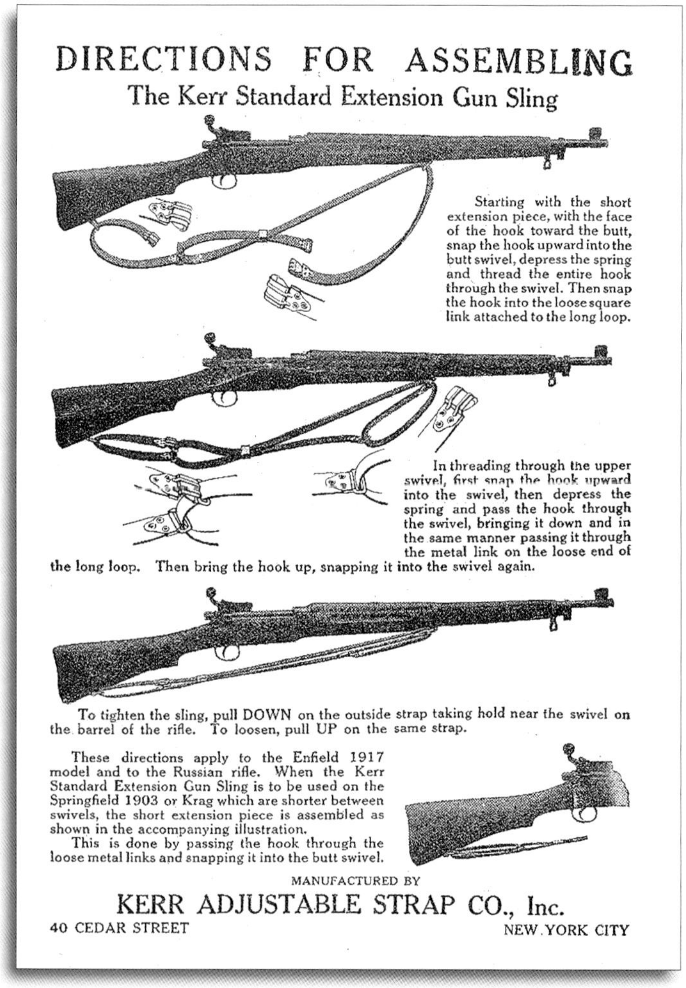

DIRECTIONS FOR ASSEMBLING

The Kerr Standard Extension Gun Sling

Starting with the short extension piece, with the face of the hook toward the butt, snap the hook upward into the butt swivel, depress the spring and thread the entire hook through the swivel. Then snap the hook into the loose square link attached to the long loop.

In threading through the upper swivel, first snap the hook upward into the swivel, then depress the spring and pass the hook through the swivel, bringing it down and in the same manner passing it through the metal link on the loose end of the long loop. Then bring the hook up, snapping it into the swivel again.

To tighten the sling, pull DOWN on the outside strap taking hold near the swivel on the barrel of the rifle. To loosen, pull UP on the same strap.

These directions apply to the Enfield 1917 model and to the Russian rifle. When the Kerr Standard Extension Gun Sling is to be used on the Springfield 1903 or Krag which are shorter between swivels, the short extension piece is assembled as shown in the accompanying illustration.

This is done by passing the hook through the loose metal links and snapping it into the butt swivel.

MANUFACTURED BY

KERR ADJUSTABLE STRAP CO., Inc.

40 CEDAR STREET NEW YORK CITY

Kerr sling factory brochure, pages 1 and 2. One brochure was included in each package of ten slings.

Photo dated June 3, 1918, of Marines at the Mare Island Naval Training Station with M1903 rifles fitted with M1917 Kerr slings. The Marines were the first branch of service to adopt the Kerr sling for their M1903, rifles but it was phased out of USMC service c.1920 in favor of the M1907 sling. ***(National Archives)***

M1923 — Like the Kerr sling, the M1923 was adopted as an alternative to the leather M1907 sling. It was adopted in 1923 as the Sling, Gun, M1923 (webbing) but was not put into quantity production until c.1941–42. Initially, the sling was intended to be issued with M1903 rifles modified to .303 caliber for the British, but that project never materialized. The M1923 slings were reportedly utilized on some of the M1917 and M1903 rifles and Thompson Submachine Guns supplied to the British under Lend-Lease. Limited numbers were also issued to American troops for use with the same weapons.

In theory, the M1923 was judged to be better than either the M1907 or M1917 Kerr slings. The web material was lighter and less prone to deterioration and, supposedly, easier to adjust than the leather M1907. In reality, any advantages were greatly offset by the fact that the sling was absurdly complicated to install and adjust. Even today, some collectors scratch their heads trying to figure out how to attach the supposedly "superior" sling to their rifle. It was equally disliked by the troops during WWII for the same reasons. As stated by Brophy:

> "The Model 1923 sling was made of webbing having two different slotted buckles and a clamp-type keeper. It was not popular, because it was difficult to adjust and because, if taken apart for cleaning,

Buckle section of a M1923 sling.

Close-up view of a rare photo of a M1923 sling in use on a M1917 rifle carried by a U.S. Army Combat Engineer in WWII. Original color photo. ***(U.S. Army Military History Institute)***

few soldiers could remember how the combination of buckles and straps went together. The Standard Nomenclature List, SNLB-3 (Organizational Spare Parts and Equipment) dated 30 January, 1944, listed the 1907, 1917, and 1923 slings as authorized, in that order.

The metal parts were of blackened brass. It could be adjusted in the same way as the 1907 sling; however, placing it in the 'loop' position required a different method of keeping the tail end from slipping through the lower swivel.

ORD8 SNL B-3 HESP & E (Addendum), dated 3 November 1944, lists under 'Maintenance Equipment Authorized for Unit Replacement' the Model 1923 sling (webbing) for issue until the supply is exhausted. This means that when all were issued, for worn out and scrapped, no more would be procured for the supply system!"[53]

Early examples of the M1923 sling were made of khaki-colored material, and later examples were olive green. Most were marked with the initials of the maker and the year of production. The most common was "S.M. Co.", which represents the Schlegel Manufacturing Co. of Rochester, N.Y. It is unlikely many remained in service after WWII.

Despite the M1923 sling's unpopularity with most of its users, it is and very collectible item today and are getting difficult to find, especially with all the buckles and clamps intact. It remains a little-known accessory of American service rifles, including the M1917.

Drawing showing how to install and adjust a M1923 sling. A M1903 rifle is shown, but the procedure for a M1917 rifle would be the same.

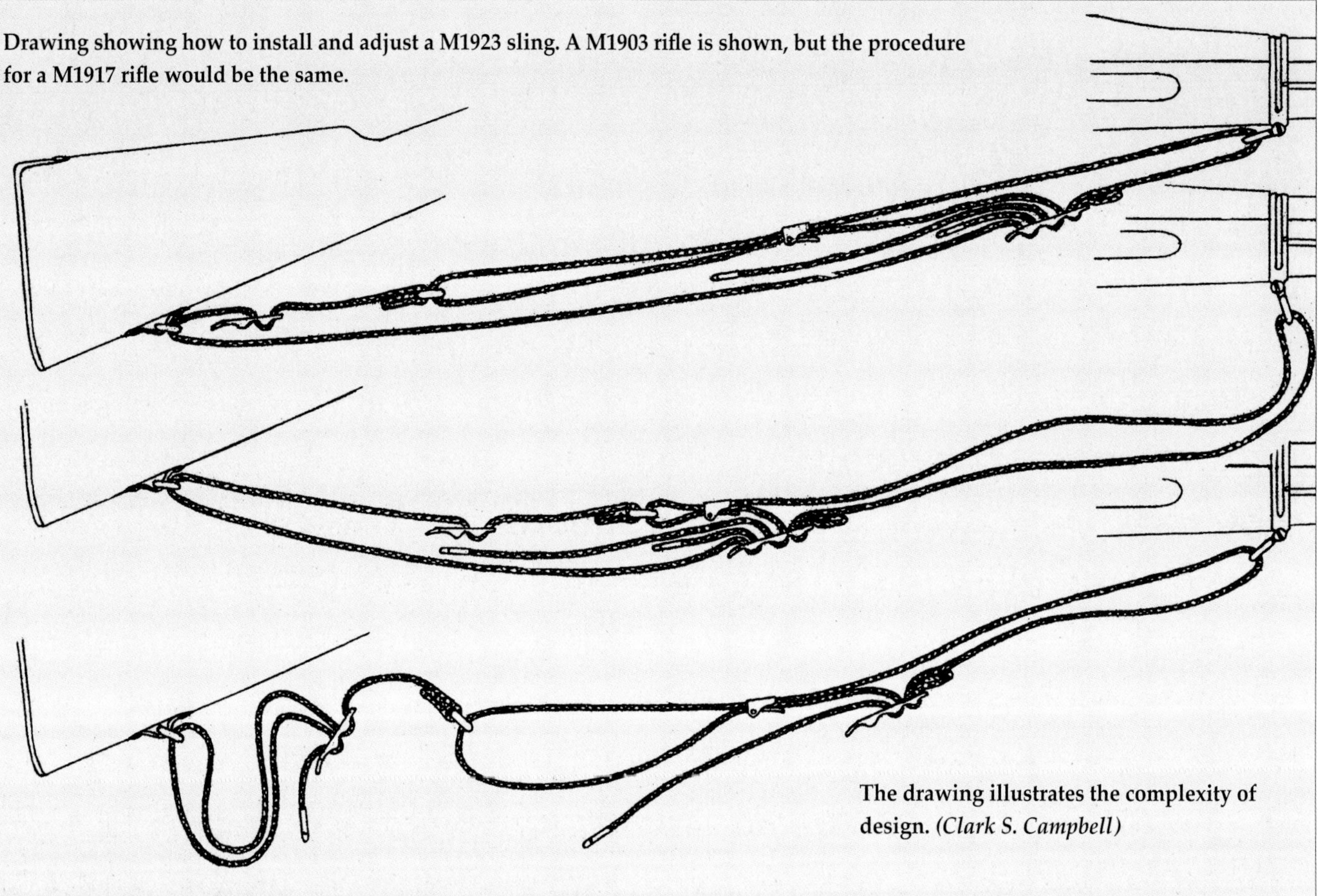

The drawing illustrates the complexity of design. ***(Clark S. Campbell)***

M1 — The final sling to be discussed is the "Sling, Gun, M1 (webbing)". By early 1942, the War Department sought a sling that was less expensive, simpler and lighter than either the M1907, M1917 Kerr, or M1923 slings. After evaluating several designs, the M1 sling was adopted and put into production by a number of commercial firms beginning circa early 1943. The sling was made of web material with sheet-metal adjustment buckles. It worked satisfactorily for carrying the rifle, but it was not as well suited for use as a marksmanship aid as the M1907 sling. It could be used with any U.S. rifle, including the M1917.

M1 slings began to be issued in 1943, but they never fully replaced the M1907 in service during WWII. Early versions were made of khaki-colored web material, and later examples were olive green. Many were ink-stamped with the initials of the maker and the year of production. Some are found stamped "MRT" with a date, indicating that "Mildew Resistant Treatment" had been applied on that date, even though the sling may have been made much earlier.

After the war, the M1 was retained as the standardized U.S. military rifle sling, and the basic design continues in use today. The WWII production M1 sling had a flat sheet metal adjustable buckle, a smaller adjustment loop and a finer weave to the material than the postwar slings.

Genuine WWII-production M1 slings have become rather scarce, but the postwar variety are still quite common. There were undoubtedly some M1 slings used with M1917 rifles during WWII, but by the postwar period, the "American Enfield" was no longer an item of issue to the U.S. military, thus a post-1945 M1 sling would not be technically correct for display with a M1917 rifle.

WWII-production M1 rifle sling.

Markings on M1 sling (left) manufactured in 1943 by the Schlegel Manufacturing Co. MRT "Mildew Resistant Treatment" markings (right) applied to sling in the postwar period (Nov. 1952).

Post-WWII M1 sling hardware (at left, top); WWII production M1 sling (at left, bottom). Note the different configuration of the buckles and the weave of the material.

Oiler and Thong Cases

The "Case, Oiler and Thong" was developed concurrent with the adoption of the M1903 rifle and was later issued with the M1917 as well. Designed to be carried in a butt trap recess accessed by a hinged metal cover, the case consisted of a nickel-plated cylindrical brass tube, about .75 inches in diameter and 6 inches long. It had two separate compartments, one for a supply of oil for lubrication and the other for a thong that was used as a "pull-through" for cleaning the rifle bore.

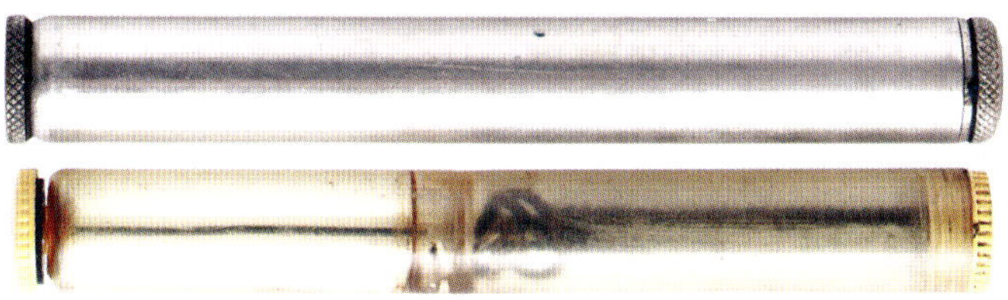

(Top) Nickel-plated brass oiler and thong case; (Bottom) WWII-vintage plastic oiler and thong case.

The section containing the oil reservoir was closed with a threaded screw cap fitted with a leather washer to prevent leakage and had an "oil dropper." The other section was closed with a threaded screw cap with a small leather pad on the end, which was intended to reduce rattling while in the butt trap recess.

Close-up view of the oiler and thong case as it is partially inserted into a M1917 rifle butt trap recess.

The thong consisted of a cord with a brass weight on one end to assist in passing it through the bore and a "thong tip" fastened to the other end, which had a slot to hold a cleaning patch. It was also threaded for a brass bristle brush that was stored in the same compartment as the thong.

The oiler and thong case saw widespread use during both World Wars with the M1903 and M1917 rifles. In mid-1942, the M1 Garand rifle stock was modified to include a butt trap recess to accommodate the case.

Circa mid-1943, the nickel-plated brass container was superseded by a plastic oiler and thong case, which began to be issued but never replaced the earlier metal case. The plastic variety remained in use by the U.S. military well into the 1960s.

"Exploded view" of the oiler and thong cases showing the contents.

Magazine Follower Depressors

As briefly mentioned in Chapters 3 and 5, unlike the Model 1903, the Model 1917 rifle did not have a magazine cutoff. When the magazine cutoff on the '03 was engaged, the magazine follower was held in a lowered position, which enabled the bolt to be retracted and closed without inference from the follower. The lack of a cutoff in the Model 1917 required that the follower be manually depressed in order to close the bolt on an empty magazine, which proved to be somewhat cumbersome in close order drill and inspection.

In order to rectify this problem, a stamped sheet metal device was fabricated, which was placed on top of the magazine follower, and the concave surface enabled the bolt to slide back and forth over an empty magazine. The device was first developed for the Pattern 1914 rifle and designated as the "Magazine Platform Depressor." A similar device, slightly longer than the Pattern 1914 type, was developed for the Model 1917 rifle and termed "Magazine Depressor or Magazine Follower Depressor." For some reason, these devices were not initially available when the Model 1917 rifles were first issued which sometimes resulted in the field-expedient practice of inserting a coin or empty stripper clip to keep the follower depressed.

The introduction of the Model 1917 rifles to new recruits at the beginning of WWI, listed some of the likes and dislikes of the weapon, as discussed in an *American Rifleman* article:

> "Another gripe encountered was the lack of a magazine cutoff like the one found on the M1903. While of dubious combat effectiveness, cutoffs have the pleasant effect of making the rifle much easier to use during the manual of arms, inspections and other training, as the bolt can be run freely back and forth without being stopped by the follower. Capt. Don Preussner (noted Army marksman and Ordnance 'Demonstrator' of the Model 1917 rifle assigned to training instructors posted to various training camps across the country) instructed his students that they were absent for a reason, 'in that a man cannot become confused and forget he has a magazine full of cartridges.'

To ease the manual of arms, soldiers were to be issued with follower depressors, although these often seemed to be in short supply. The same effect can be achieved with a strategically placed coin, a trick that many a Doughboys surely made use of. At least one soldier got lucky, however, as Capt. Don Preussner

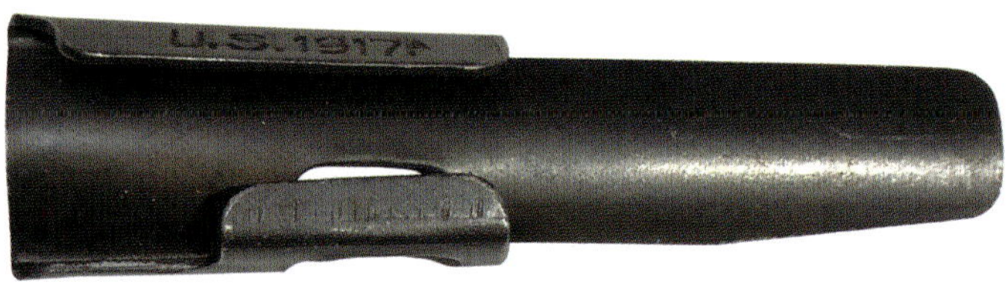

Top and bottom views of the magazine follower depressor.

"A" marking on depressor believed to have been the type used by Winchester.

'found a follower of the British model assembled in Eddystone rifle No. 111397 and it was a source of gratification to the owner for it did not lock his bolt open when it was opened for 'inspection arms.' "[55]

Production figures for the number of magazine depressors for the Model 1917 rifle are sketchy, as two separate contracts for the devices were placed with Winchester: August 12, 1918, contract for 400,000 and a contract on October 19, 1918, for 200,000. Despite the fact that 600,000 devices were contracted for, Winchester archival records indicate that the company made a total of only 215,512 magazine platform depressors.[56] Remington factory records reflect a July 12, 1917, contract for 250,000 magazine platform depressors with an additional 50,000 to follow after completion of the first contract.[57] It is not known if any additional contracts were placed for the devices. Eddystone contracts and production totals for magazine platform depressors are not known.

The Remington and Eddystone examples are marked "R" and "E". There are no known "W"-marked examples, although many have been observed marked "A", which are assumed to have been the type used by Winchester. It has been theorized that these examples were made by a subcontractor.[58] While unconfirmed, this was likely the case, since Winchester delivered over 200,000 depressors and, as stated, no known "W"-marked examples have been noted.

Breech Covers

In order to help shield the receiver/bolt mechanism from mud and dirt or in inclement weather, as evidenced by period photographs, a canvas cover was issued for use with the Model 1917 rifle and the Model 1903. The cover was secured by four-snap fasteners and a leather lace. Most WWI-era photos depict the breech cover being used on more Model 1917 rifles than Model 1903s, but this disparity was likely due to the larger numbers of "American Enfields" in use as compared to the '03s. Genuine examples are rather uncommon today, and some reproductions have been made.

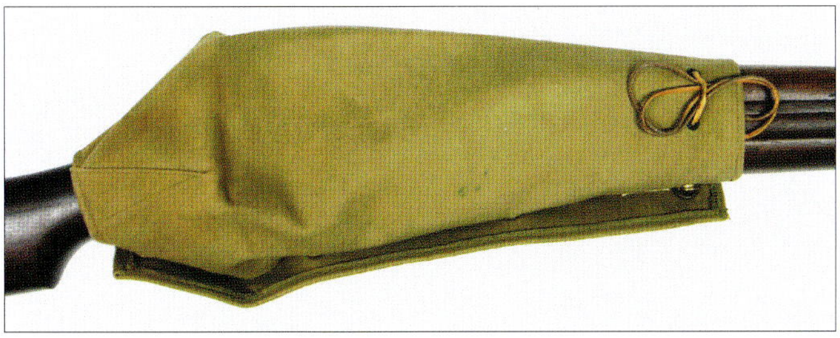

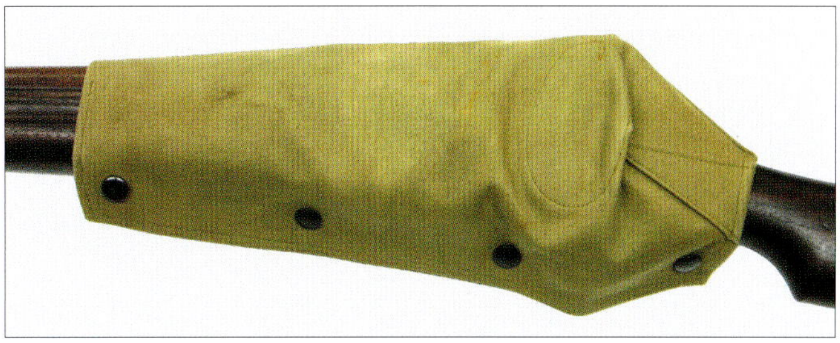

Left and right-side views of the canvas breech cover on M1917 rifle. Note the leather lace. *(Hayes Otoupalik Collection. Photo by William R. O'Donnell)*

Weary American doughboys taking a break, France 1918. Note the canvas breech covers on their M1917 rifles. Photo on the right shows a U.S. soldier chatting with a YMCA representative at Cuisy, France, October of 1918. His Model 1917 rifle is fitted with a breech cover. *(U.S. Army photos. Courtesy of Tom Laemlein)*

Rifle Carrying Case

Prior to the start of WWI, the Army had adopted the Model 1916 "Rifle Cover" (drawing number 20-17-67), a sturdy canvas case with a leather flap covering the butt end. This was in the system when the M1917 rifle was adopted, so naturally it was used for both 1903 and 1917 rifles. In 1918, a Rifle Cover Model 1918 (drawing number 38-1-18) was adopted, basically identical to the 1916 cover except the end flap was a continuation of the canvas body instead of leather. When the M1918 Browning Automatic Rifle was adopted, it also fit these cases, and they were issued as needed for any of these arms.

Even though not exclusively made for the M1917 rifle, these rifle covers (or carrying cases) are a nice addition to several different collecting themes. While not rare today, genuine examples are becoming harder to find, and most examples are in less-than-pristine condition. Reproductions have been made, so caution should be exercised when considering a purchase.

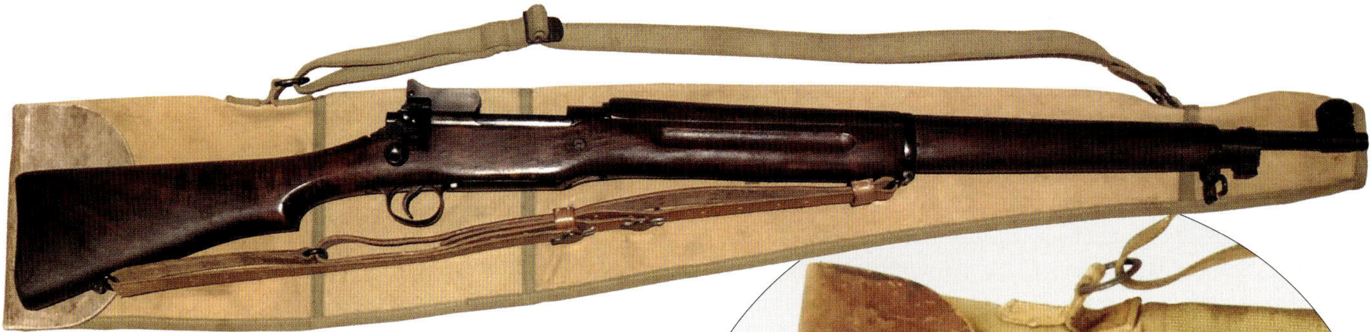

M1916 Rifle carrying case.
At right is a close-up view of the M1916 carrying case showing the leather cover flap with the metal closure buckle.

Below is shown an M1918 carrying case. This case differed from the M1916 as it had a canvas cover flap rather than leather. Both cases were approved for use with the M1903 and M1917 rifles as well as the M1918 Browning Automatic Rifle.

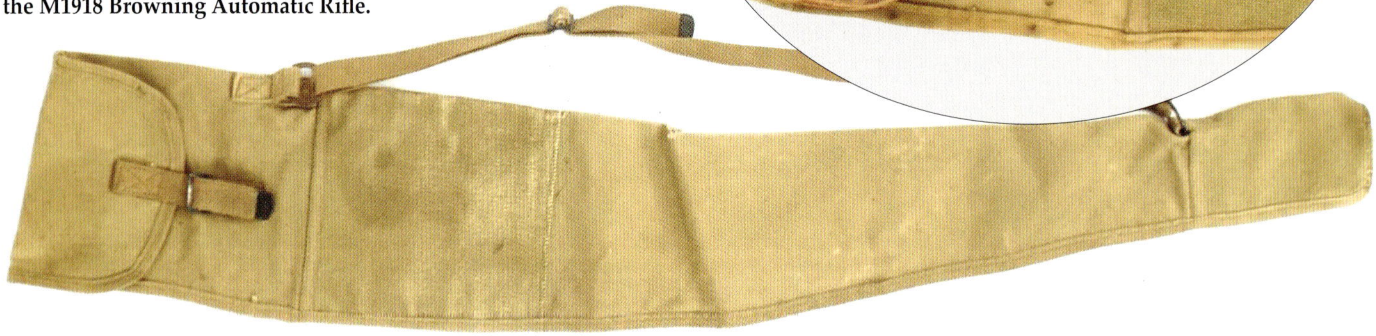

Arms Repair Chest

Like any mechanical object, firearms require routine cleaning, maintenance, and the occasional replacement of broken or worn parts. This is particularly true with military weapons, which are typically subjected to much more wear and tear far beyond most sporting guns. The Model 1917 rifle was no exception. Soldiers were issued items for basic cleaning and lubrication such as the oiler and thong cases discussed above. However, there was a need for more extensive cleaning supplies and spare parts.

Top view showing an arms repair chest devoid of contents. This is typically the condition of most M1917 arms repair chests found today. ***(IMA)***

View from the top showing the contents. Extant examples complete with the original contents are quite uncommon today.

(Hayes Otoupalik Collection. Photos by William R. O'Donnell)

To this end, the "Arms Repair Chest, Model 1917," was adopted and produced in relatively large numbers. A similar chest, designated as the "Model 1910," had been previously adopted for the Model 1903 rifle. As described in *Handbook for Ordnance Data*: "Arms-repair chests, model 1917 – This chest contains tools and spare parts for the United States Rifle, model 1917. It is intended for use by the company's mechanics."[59]

The *Description and Rules for the Management of the United States Rifle, Caliber .30, Model of 1917*, stated the following: "Arms Repair Chest – Model of 1917. An arms repair chest will be issued to each organization armed with the rifle, and to each regimental or post ordnance officer, for cleaning, repairing, and preserving arms and equipment."[60]

The sturdy wooden chest had a hinged cover with two locking clasps and separate compartments to hold the various implements. It was painted green with the designation stenciled on the right side. The bottom of the cover had a listing of the items contained in the chest. Included were various tools, cleaning materials consisting of cotton patches and six pint cans of "sperm oil," and two pint cans of "cosmic," when heavier lubrication was needed. There were also numerous spare parts for the rifle judged most likely to need repair or maintenance or replacement. There were also extra catches, springs, and thumbpieces for the M1917 bayonet.[61]

These chests can still be found today from time to time but are usually missing most, if not all, of the components. A specimen would be an interesting addition for display with a Model 1917 rifle.

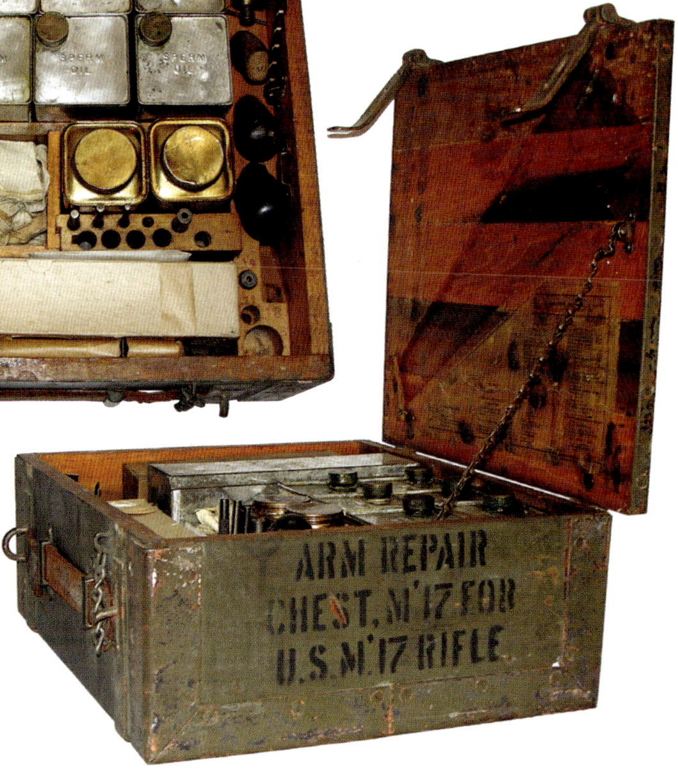

Left side of the M1917 arms repair chest.

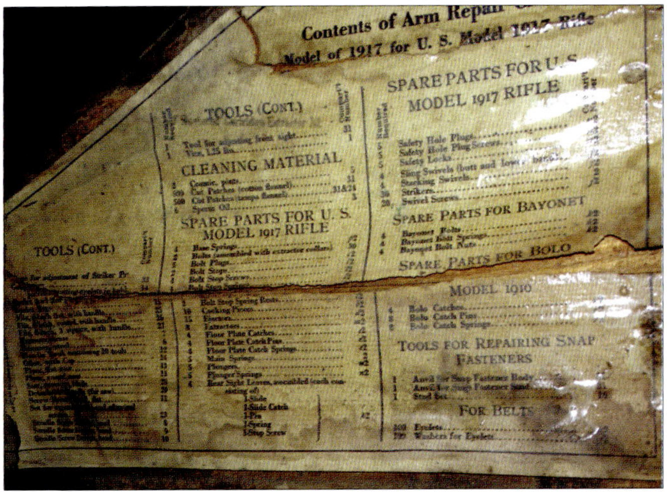

Shown at right is a label detailing the contents of the M1917 arms repair chest. *(John Spangler)*

"Hollifield Dotter"

As has been previously been discussed, gaining marksmanship proficiency was a vital part of the training for all soldiers. In 1875, the New York State National Guard was using a rifle practice training device developed by General George W. Wingate, the general inspector of rifle practice for the guard. The purpose of the device was to enable a shooter's aim to be improved without firing the rifle.[62]

When the Model 1903 rifle was adopted, General Wingate's basic idea was dusted off, and a number of devices intended to accomplish the same purpose were developed. One design resulting in a greatly improved device, which proved very effective, was the "Hollifield Target Practice Rod Indicator" or "Dotter," manufactured by the Hollifield Target Practice Rod Co. of Middletown, New York. It was adopted by the U.S. Army, Navy and Navy Militia as well as the Canadian military. The Chief of Ordnance authorized the purchase of fifty units in 1908.[63]

The Hollifield "Dotter" kits were packed in a wooden box with a built-in sliding target holder and could be mounted on the wall. These were made in several variations, including the Krag, the Model 1903, and the Model 1917 Enfield rifles. The main difference was the length of the rods to match the rifle barrel length, and a corresponding difference in length of the wooden box, and the inclusion of .30 Krag or .30-06 cartridges, as appropriate. The label on the outside of the box indicated the rifle for which it was intended. The instruction label on the inside of the box was the same for all three rifles and used a Model 1903 in the illustration.

In addition to the box, the set also included the following:

- Long rod, used for slow fire training when dummy cartridges are not used.
- Short rod, used with the dummy cartridges for rapid fire practice.
- A book of paper "A" targets
- A book of paper "200 yard and 300 yard" targets
- An envelope containing Aim Fixing Target Slip and Marksmanship Disks on a staff.
- Dummy cartridges in a clip. The clip was the same

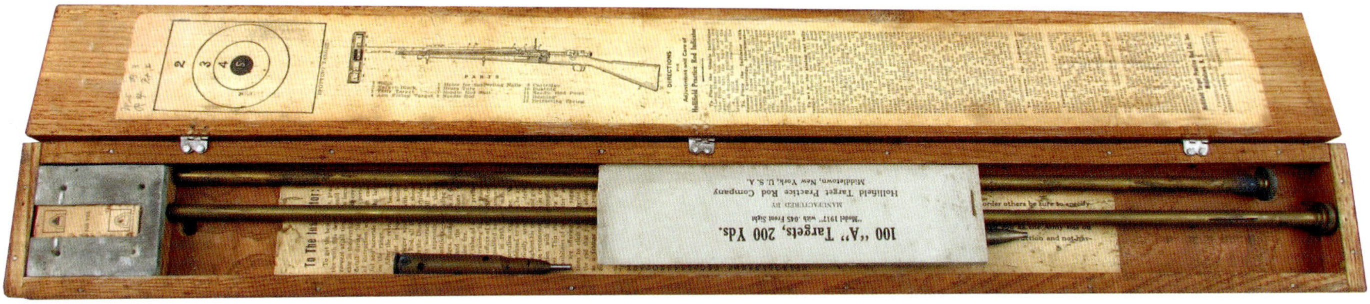

Hollifield "Dotter" kit open revealing interior label, rods, cartridges, and paper targets. *(JoeSalter.com)*
Contents of the Hollifield "Dotter" kit. *(John Spangler)*

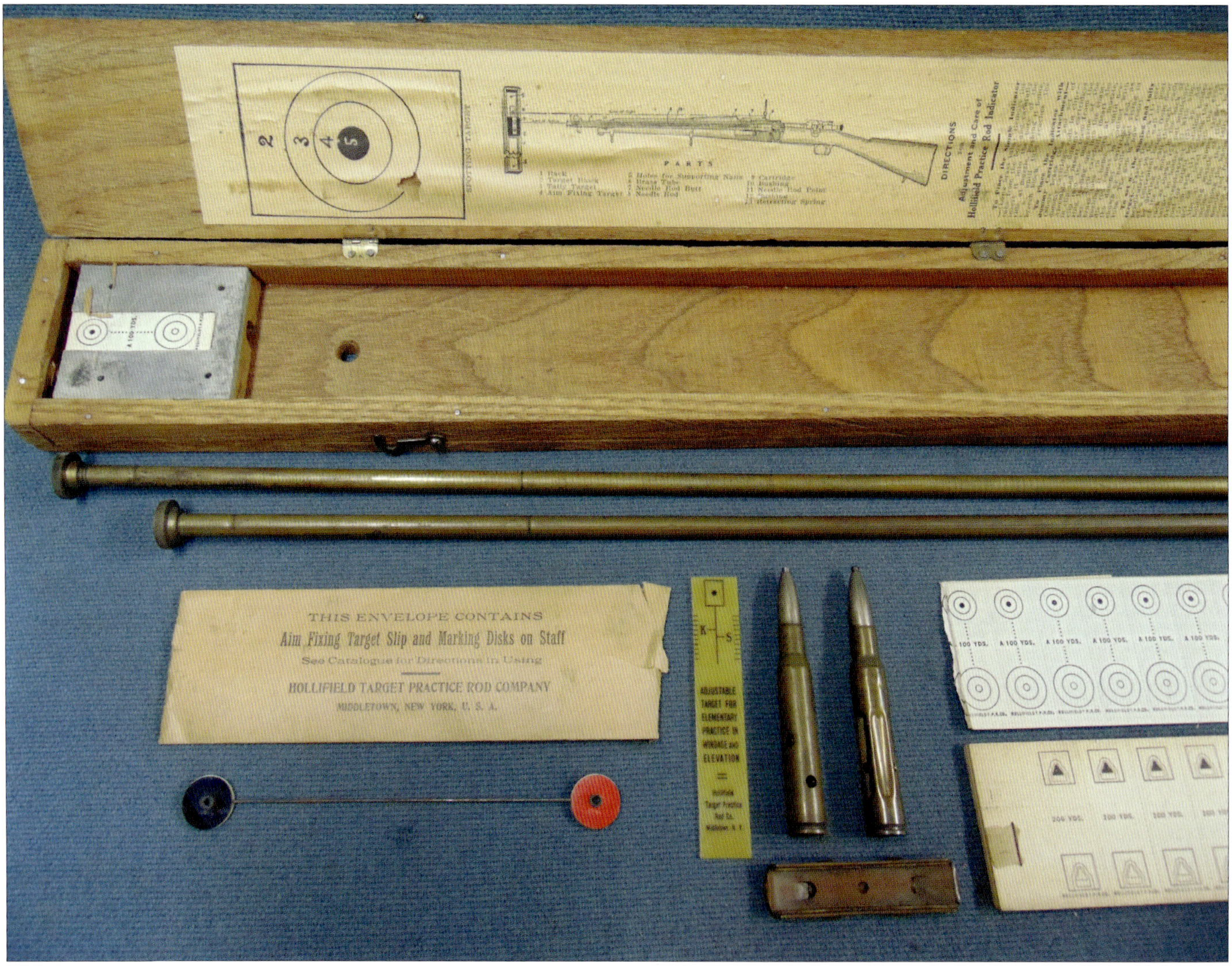

five-round charger ("stripper clip") used with the regular service cartridges.
- Instructions pasted on the cover of the box.
- A target holder to be used with a bayonet."[64]

When the United States entered the First World War, such a device was viewed as being an asset in training the large numbers of troops entering service to improve their marksmanship capability. An article in the May 26, 1917, issue of *Arms and the Man* (forerunner of the *American Rifleman)* published an article that described the device and its intended use:

"The improved dotter method provides a complete course of instruction in all mechanical details of shotting with service small arms. The shooter loads the gun, cocks it, and aims at a target reduced to a proper size corresponding with the distance, and when his aim is satisfactory, pulls the trigger. When this is done a needle rod licks out of the muzzle and

World War I U.S. Army troops at Camp Dix, N.J., with Model 1917 rifles being instructed in marksmanship techniques with what appears to be Hollifield "Dotters." *(National Archives)*

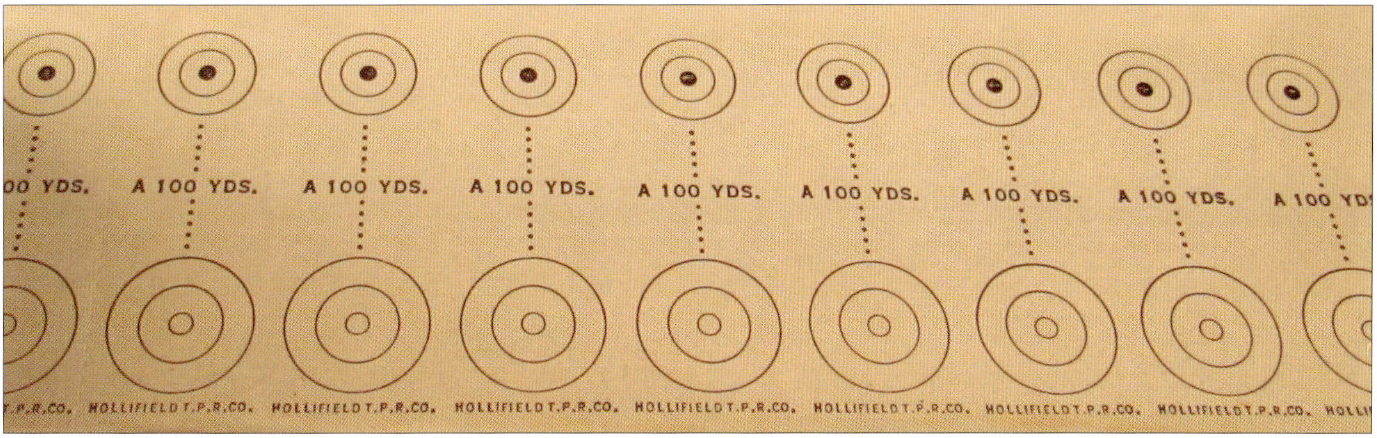

Close-up view of the paper targets. *(JoeSalter.com)*

punctures a hole in a miniature target also reduced to the proper size…It consequently gives the same experience in holding a pull-off as is had in actual shooting."[65]

The article also contained the following quote by the noted marksman, Captain E.C. Crossman:

> "For the instruction of the recruit the device is most valuable, as not only does he learn to aim and pull the trigger correctly, but he learns the proper handling of cartridges at the same time. Rapid fire works perfectly with the dummies, which is not true when the cut-off is turned down, and at the same time the shots are spotted on the tell-tale target. No use trying to deceive yourself as to your shooting and to tell yourself that such and such a shot would have been a bull. You've got to show this little Missouri device.
>
> I can see no escape from the conclusion that if a man practices at all faithfully with the rod, he will make a good shot of himself."[66]

During the First World War, the government purchased a significant number of the Hollifield "Dotter" kits, which could be used for either the Model 1903 or Model 1917 rifles. There was also a Hollifield "Dotter" kit made for the Model 1911 .45 pistol that functioned in essentially the same manner as the device made for the rifles except that it was slow fire only, with no dummy cartridges involved. The "Dotter" was nicely summed up by author Lt. Col. (Ret.) William S. Brophy:

> "Used as a training aid by shooting teams and units instructing recruits, the Dotter enjoyed a good reputation and was acclaimed by many to have increased the ability of the soldier. Over 500,000 outfits were in use before WWI. It was priced at $5.87 in 1914. The Hollifield Company discontinued operations in 1927."[67]

Hollifield "Dotter" kits, especially complete with all components, are uncommon today. An example can be an interesting addition to a World War I collection featuring Model 1917 or Model 1903 rifles.

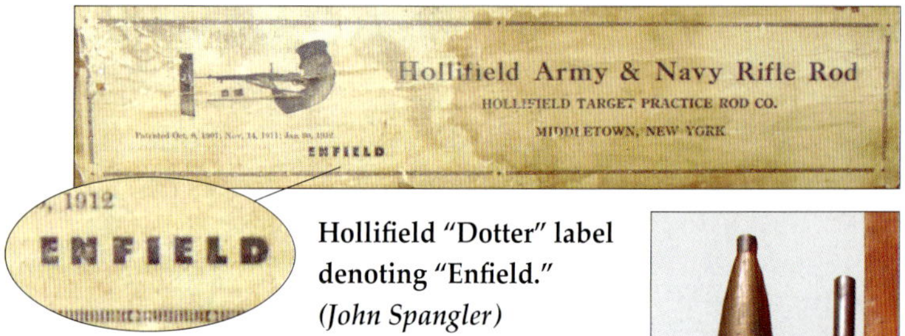

Hollifield "Dotter" label denoting "Enfield." *(John Spangler)*

Close-up view of the "Dotter" cartridges. *(JoeSalter.com)*

Musketry Rule

Also known as a "stadia," the musketry rule was an instrument used to estimate the distance to a target. The stadia had been first introduced in 1858. During the "Trapdoor Springfield" era, elaborate stadia were awarded for marksmanship excellence.[68] There were three different variations used by the United States military in the First World War era:

- Model 1916. This variant was made of aluminum and did not have the brass "slide rule" or peep hole as did the subsequent models.
- Model 1917. Constructed of brass or aluminum, this model had a slide rule "peep hole" feature.
- Model 1918. This final variant was made of brass. In addition to the slide rule feature of the Model 1917, it also had a simplified sliding "peep hole."[69]

The Models 1917 and 1918 were the types primarily used during the First World War and were equally applicable to the Model 1917 and Model 1903 rifles. The instrument was mentioned in the *Handbook for Ordnance Data*:

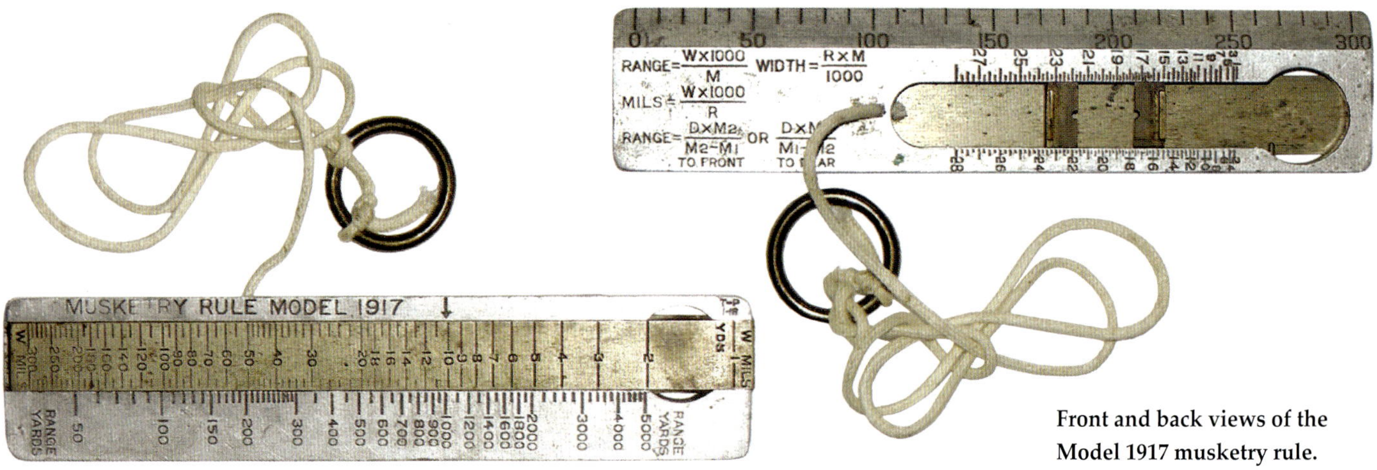

Front and back views of the Model 1917 musketry rule.

The musketry rule was carried in a common first aid canvas pouch.

"Musketry Rule, Model 1918 — This consists of a mil scale with built-in slide rule for estimating and computing ranges, etc. It also includes an inverted sight for determining sight sets for the direct fire. The 1918 model is an improvement on that of the 1917, but is identical is principle. It was manufactured by the Gorham Manufacturing Co. and Graff, Washburne & Dunn. It is used in fire control when type E.E. field glasses are not available."[70]

The procedure for using the musketry rule was described by Brophy:

"While in use the rule was held fifteen inches from the eye and perpendicular or horizontal to the line of sight. To be sure the rule is held at the correct distance from the eye, a string is attached to the rule. A knot (Model 1916) or brass ring (Models 1917 and 1918) is held just under the eye or by the teeth. With the cord stretched lightly, the scale to be used is superimposed on the area being observed.

Musketry and the musketry rules became passé in the 30s, when infantry tactics and warfare were modernized by introduction of new technology and methodology. The new methods of fire-direction commands and target identification speeded up the placement of accurate fire on a target to the point that the old WWI Musketry Rules were too slow to be used."[71]

The musketry rule was often carried in a common first aid canvas pouch. The Model 1917 was the most widely used variant in World War I, but any example would be an interesting item for display in a World War I collection along with a Model 1917 rifle.

Model 1918 Periscope

Not every accessory used with the Model 1917 rifle was intended to make the weapon deadlier, such as the bayonet and V-B rifle grenade launcher. And not every accessory was as technologically advanced as the Pedersen Device or required more than a modicum of training for proper use, such as the Hollifield "Dotter" and the musketry rule. One decidedly "low-tech" item that didn't fit into any of these categories was the "M1918 Periscope."

Front view of the Model 1918 periscope.

Although designated as a "periscope," it was actually nothing more than a mirror in a metal frame that could be attached to the bayonet to permit viewing over a trench parapet without exposure to enemy fire. It could be easily attached to either the Model 1917 bayonet or Model 1905 bayonet. The item is sometimes unofficially referred to by the more descriptive designation of "Bayonet, Trench Mirror" or some derivation thereof. Most of the belligerent nations also used similar contrivances during the war.

Rear view of the Model 1918 periscope. *(Hills militaria)*

The M1918 periscope was described in the ordnance publication *Fire Control and Sighting Equipment, Small Arms and Special Infantry Weapons*:

> "It is fastened to the bayonet with the mirror facing the object to be observed and inclined at an angle of about 45 (degrees) from the horizontal. The mirror is then elevated above the edge of the dugout, etc., with the back of the observer toward the object to be observed and the mirror turned in the direction necessary to view the object."[72]

These items were apparently made in rather large numbers, and they are not difficult to find today. However, they most likely saw little or no use, as they are frequently encountered in virtually new condition. They are sometimes mistakenly assumed to be Army shaving mirrors, although some may have been used as such.

There is no known official Army documentation detailing the efficacy of this item. Brophy made the logical comment: "It is wondered if it was practical and ever used successfully."[73]

British troops near Ypres, France, September 1917. The British "trench mirror" is attached to the bayonet of a SMLE rifle. The M1918 periscope was used in precisely the same manner with the Model 1917 and Model 1903 rifles by U.S. troops. *(Tom Laemlein)*

Manuals

There were various types of manuals produced for all U.S. military firearms that assisted in becoming familiar with the basics of the weapon, provided instructions on marksmanship techniques as well as details on how to clean and maintain the arm. Such manuals for the Model 1917 rifle were produced from c.1917 until 1942. There are newly made reprints for most of the manuals as well, but of course, they do not have the value or collectability of the original examples. The following are not the only government printed materials on the Model 1917 rifle but are among the most common and were the most widely used in service:

Description and Rules for the Management of the United States RifleCaliber .30, Model of 1917 October 8, 1917 – Revised January 16, 1918 – Revised May 7, 1918

This was the basic manual for the Model 1917 rifle during World War I and included the following:

- Three-page line drawing of the rifle.
- Complete list and accompanying drawings of all components for the rifle, bayonet, scabbard, appendages and accessories including oiler and thong case and magazine depressor.
- General description
- Instructions on maintenance, cleaning and simple repairs.
- Marksmanship Instruction
- Numerous tables on ballistics

There was also a ***Supplement No. 1917*** that had instructions for fitting strikers to cocking pieces to the rifle. Both the Basic Manual and Supplemental Manual had blue paper covers and were printed at the Government Printing Office, Washington, D.C., in 1917 and 1918.

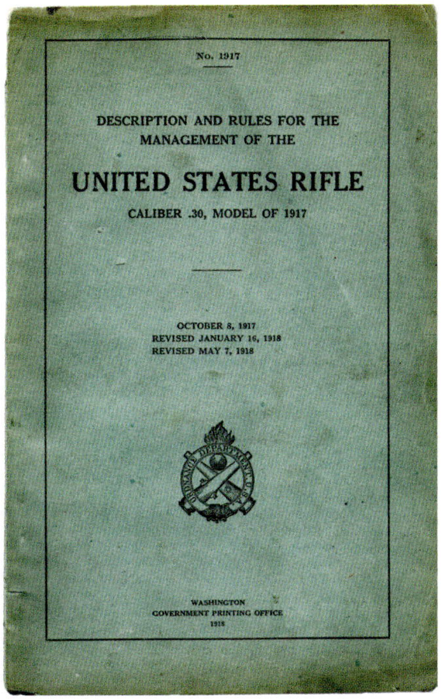

1

2

3

1 Description and Rules of Management of the United States Rifle, Caliber .30, Model of 1917.

2 Supplement No. 1917 to above manual.

3 Soldier's Handbook of the Rifle, and Score Book for Special Course "C". Arranged for the United States Rifle Model of 1917.

Soldier's Handbook of the Rifle and Score Book for Special Course "C" Arranged for the United States Rifle Model of 1917 – Edited at the Army War College – November 1917

This handbook had cloth covers and contained detailed information and illustrations regarding marksmanship techniques along with a number of diagrams that enabled the shooter to record his results. It was printed at the Government Printing Office, Washington, D.C., in 1919. A handbook almost identical in format and material was printed for the Model 1903 rifle and had blue cloth covers.

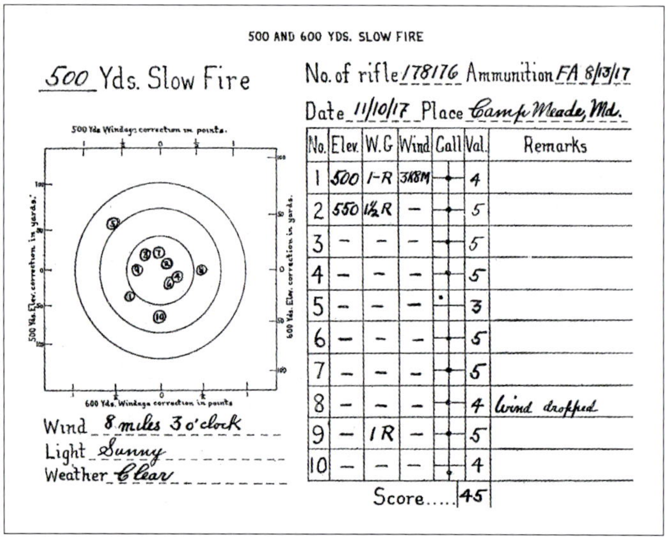

SOLDIER'S HANDBOOK OF THE RIFLE

500 AND 600 YDS. SLOW FIRE

500 Yds. Slow Fire

No. of rifle 178176 Ammunition FA 8/13/17

Date 11/10/17 Place Camp Meade, Md.

No.	Elev.	W.G	Wind	Call	Val.	Remarks
1	500	1-R	3K8M		4	
2	550	1½R	–		5	
3	–	–	–		5	
4	–	–	–		5	
5	–	–	–		3	
6	–	–	–		5	
7	–	–	–		5	
8	–	–	–		4	Wind dropped
9	–	1R	–		5	
10	–	–	–		4	
				Score.....	45	

Wind 8 miles 3 o'clock

Light Sunny

Weather Clear

Sample page from the Soldier's Handbook showing the correct way to fill out the data for the 500-yard slow fire target.

SOLDIER'S HANDBOOK OF THE RIFLE. 35

Firing from standing trench.

Target practice from an improvised "trench position."

100, 200 AND 300 YDS. SLOW FIRE

____Yds. Slow Fire

Target 4 feet square.
Bullseye 8 inches in diameter

No. of rifle____ Ammunition________

Date_____ Place____________

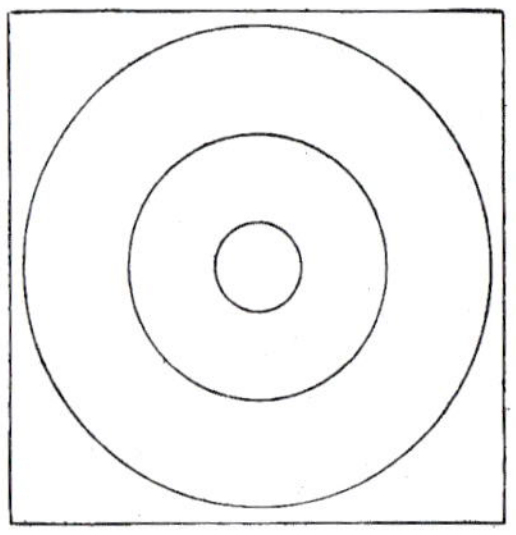

No.	Elev.	W.G.	Wind	Call	Val.	Remarks.
1						
2						
3						
4						
5						
6						
7						
8						
9						
10						
				Score....		

Wind________________

Light________________

Weather______________

Page for the shooters to fill out after completing the target session.

Ordnance Field Service
Base Shop Data U.S. Rifle, Cal. .30, M1917

Prepared at Rock Island Arsenal and printed in September of 1942, with a later edition in February 1943, this publication contains detailed information and instruction regarding the maintenance and repair for the M1917 rifle.

Such materials as the handbooks, field manuals and supply catalogs are sometimes overlooked as collectibles but are interesting items and can enhance any collection.

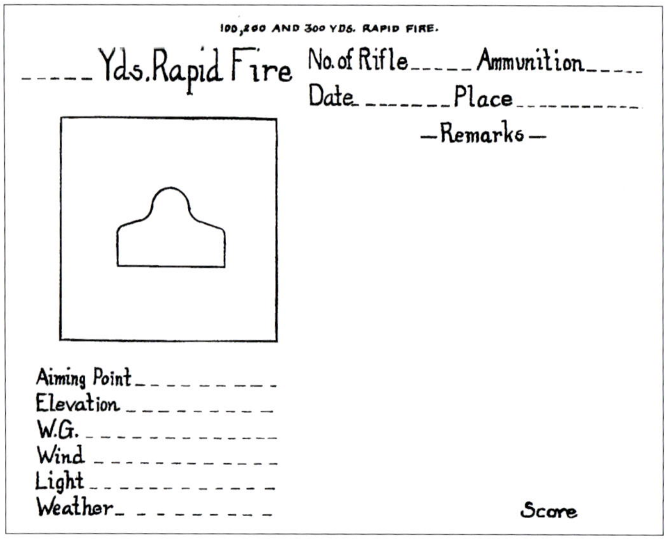

100, 200 AND 300 YDS. RAPID FIRE.

_____Yds. Rapid Fire

No. of Rifle_____ Ammunition_____

Date_______ Place__________

—Remarks—

Aiming Point__________

Elevation__________

W.G.____________

Wind____________

Light____________

Weather__________

Score

Page for rapid fire target.

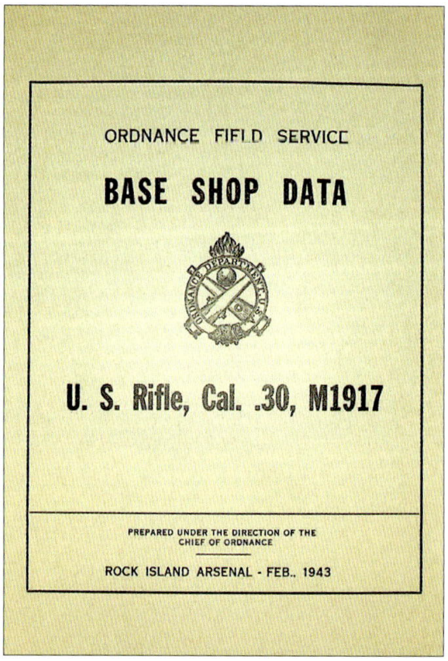

Ordnance Field Service – Base Shop Data, February 1943.

FM 23-6 • Basic Field Manual U.S. Rifle, Caliber .30 M1917 (Enfield)

This was the basic World War II manual for the M1917 rifle. The first edition was 230 pages and was dated August 3, 1942. This manual, with tan paper covers, included material on mechanical training (disassembly, maintenance, etc.), marksmanship and advice to instructors. It is interesting to note that the designation "M1917" was used rather than "Model of 1917" as was the case in the various World War I manuals. Also, the term "(Enfield") is included in the nomenclature, which is confirmation that, although often thought to be an informal, if not incorrect, designation, it was nevertheless a term used by the War Department. The second edition, dated October 28, 1943, was expanded to 316 pages with the addition of a lot more material on marksmanship training. This was originally classified as "restricted," but that classification was cancelled in 1944. The "RESTRICTED" marking likely caused most of these to be needlessly destroyed making them scarcer on the collector market.

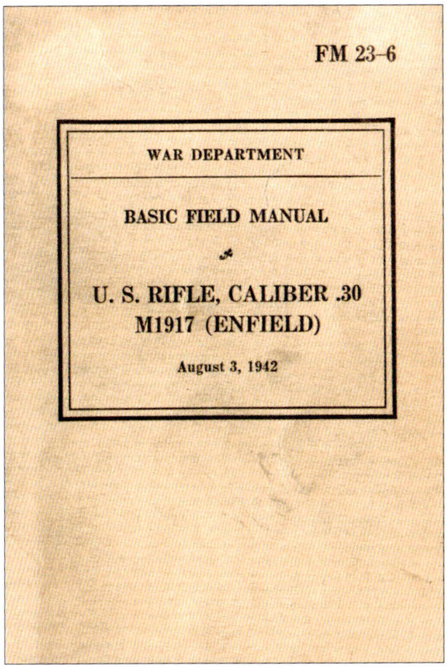

Basic Field Manual – U.S. Rifle, Caliber .30 M1917 (Enfield), 1942.

Army Service Force Catalog – ORD 9 SNL B-4 Ordnance Supply Catalog

Although not actually a manual, as the title implies, this publication, dated January 24, 1944, lists all parts of the M1917 rifle. It is interesting to note that such material was still being printed for the weapon at this relatively late date in the war.

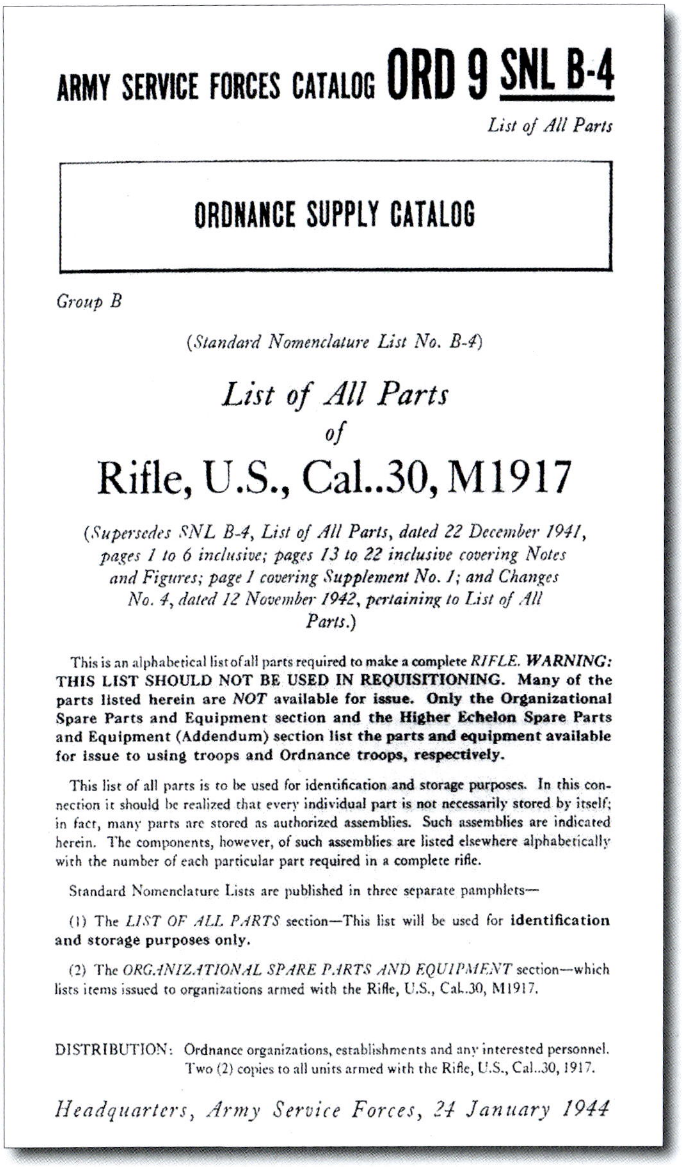

ARMY SERVICE FORCES CATALOG ORD 9 SNL B-4

List of All Parts

ORDNANCE SUPPLY CATALOG

Group B

(*Standard Nomenclature List No. B-4*)

List of All Parts
of

Rifle, U.S., Cal..30, M1917

(*Supersedes SNL B-4, List of All Parts, dated 22 December 1941, pages 1 to 6 inclusive; pages 13 to 22 inclusive covering Notes and Figures; page 1 covering Supplement No. 1; and Changes No. 4, dated 12 November 1942, pertaining to List of All Parts.*)

This is an alphabetical list of all parts required to make a complete *RIFLE*. ***WARNING:* THIS LIST SHOULD NOT BE USED IN REQUISITIONING. Many of the parts listed herein are *NOT* available for issue. Only the Organizational Spare Parts and Equipment section and the Higher Echelon Spare Parts and Equipment (Addendum) section list the parts and equipment available for issue to using troops and Ordnance troops, respectively.**

This list of all parts is to be used for identification and storage purposes. In this connection it should be realized that every individual part is not necessarily stored by itself; in fact, many parts are stored as authorized assemblies. Such assemblies are indicated herein. The components, however, of such assemblies are listed elsewhere alphabetically with the number of each particular part required in a complete rifle.

Standard Nomenclature Lists are published in three separate pamphlets—

(1) The *LIST OF ALL PARTS* section—This list will be used for **identification and storage purposes only.**

(2) The *ORGANIZATIONAL SPARE PARTS AND EQUIPMENT* section—which lists items issued to organizations armed with the Rifle, U.S., Cal..30, M1917.

DISTRIBUTION: Ordnance organizations, establishments and any interested personnel. Two (2) copies to all units armed with the Rifle, U.S., Cal..30, 1917.

Headquarters, Army Service Forces, 24 January 1944

WWII Ordnance Supply Catalog - 1944.

Collecting the "American Enfield"

Gun collecting as we know it today is a relatively recent phenomenon. Like many hobbies, it is continuously evolving. For example, in the 1950s and early 1960s, there was a lot of interest in vintage Colt revolvers and Winchester lever-action rifles. The interest in these guns was due, in part, to the popularity of television programs of the era such as *Gunsmoke, The Rifleman, Have Gun Will Travel, Rawhide,* and *Bonanza* (among others), as well as numerous western-themed movies.

As collectible specimens of these guns began to become harder to find and more expensive, would-be collectors gravitated to other firearms such as "Pre-64" Winchesters, German Lugers and P38 pistols, Japanese Nambu pistols, certain Smith & Wesson revolvers and Civil War-era weapons, just to name a few. There were a lot of American 20th-century military weapons around, and gun fanciers might pick up a U.S. Krag rifle, an M1903 rifle, an M1 Garand rifle, an M1911 / M1911A1 pistol and maybe a Model 1917 rifle. However, to most collectors, these weapons were not deemed particularly collectible. Also, as discussed previously, many of the rifles were modified into low-cost "sporters," which reduced the availability of unmodified specimens.

This was generally the case until the mid-1960s, when about 250,000 M1 carbines were sold by the Director of Civilian Marksmanship (DCM) to members of the NRA for the very attractive price of $20. The allotment of carbines sold out in short order, but there was generally a sentiment of "you've seen one, you've seen them all." The carbines were not particularly suitable candidates for sporterizing and most remained in their more or less "military" configuration. During the next decade or so, M1 Garand rifles began to become more available through the DCM and its successor, the CMP, and interest in those weapons started to increase.

Many people also began to reassess the old surplus M1903 rifles they had on hand as possible collecting material. U.S. military firearms eventually evolved into perhaps the most popular collecting genre around today. As was the case with the Colt and Winchester guns mentioned above, there is no doubt that a number of television programs in the 1960s such as *Combat, The Gallant Men* and *Rat Patrol,* as well as many movies, helped fuel the tremendous surge in collector interest for American military weapons. This trend continued for years with popular movies and television series such as *The Longest Day, Saving Private Ryan, The Thin Red Line, Band of Brothers* and *The Pacific*. Anyone doubting the impact of movies and television on firearms collecting for the past 60+ years hasn't been paying attention. If anyone is still skeptical, they need only to consider the popularity and value of the Smith & Wesson Model 29 .44 Magnum revolver before and after the release of the 1971 *Dirty Harry* movie.

However, despite large numbers being sold after WWII via the DCM, the M1917 "U.S. Enfield," for whatever reason(s), never caught on as did most of the other American military firearms. Some collectors might pick one up at a gun show as a "representative example," but it did not have the popularity or value of the M1 rifle, the M1 carbine and M1903 rifle.

This brings up the obvious question, why didn't the M1917 rifle gain the same level of collector attention as the other weapons mentioned? As referred to several times in this book, the M1917 was widely considered to be rather ugly as compared to the trim and svelte M1903. To use a bit of literary personification, the '03 was the pretty high school cheerleader, and the M1917 was the nerdy bookworm who was solid, dependable and always willing to lend a helping hand but found herself at home on prom night.

Aesthetics aside, there were a number of factors in the bias against the M1917 rifle as compared to other U.S. military firearms. For example, one attribute of popular military weapons collectability is the number of variations. In comparing the M1903 to the M1917, let's look at the disparity between the two rifles regarding variations. There were four manufacturers of Model 1903 and variants: Springfield Armory, Rock Island Arsenal, Remington Arms Co., and Smith-Corona Co., while there were three makers of the Model 1917, so the difference in the number of manufacturers is insignificant.

However, there were only a few true variants of the Model 1917 such as the Model 1918 Sniper Rifle, the various Trench Periscope rifles, the Model 1917 Mark II (with or without the Pedersen Device) and the WWII-modified M1917 carbine. Since these were made in very limited numbers and are all but unobtainable (some literally unobtainable), virtually no collector will ever have a chance to own an example. The only other recognizable variants (albeit rather minor) are the early Winchester Model 1917s with the "W"-marked receiver (rather than the full nomenclature) and the late WWI factory Parkerized Eddystones, but even these are rather uncommon today.

On the other hand, there is a vast array of interesting and desirable variants of the '03 beyond just the four manufacturers. These include the original Model 1903 .30-03 caliber rod bayonet rifle, Springfield Armory and Rock Island Arsenal Model 1903 rifles with modification of 1905 still in .30-03 caliber, SA and RIA Model 1903s in .30-06 caliber, Model 1903 Mark I, Model 1903A1, Model 1903 made by Remington, the Model 1903A3 made by Remington and

Smith-Corona and numerous sniper variants including the Model 1903A4, Model 1903 fitted with M1908 and M1913 Warner & Swasey musket sights, Model 1903 with Winchester A5 (and 5A) telescopes and Model 1903A1 with Unertl 8X USMC telescope, and several Springfield Armory target and sporting rifles in .22 caliber and .30-06 caliber. While becoming harder to find, the M1903A4 sniper rifle is still relatively obtainable today (some 28,000 were made) without having to take out a second mortgage, but most of the other sniper rifles are, by comparison, more difficult to find and more expensive, but still avidly sought after by collectors. Clearly, the '03 trumps the M1917 in the number of variations available.

Nevertheless, despite these disadvantages and prejudices, the Model 1917 is now coming into its own as a desirable and sought-after collectible. There are numerous things for a would-be collector to evaluate when determining the suitability of a particular example as a collectible. Naturally, there are some who are only interested in shooting their rifle and don't care about the collecting aspects, so the factors to be discussed are not germane. The most important attributes for a shooter are the proper headspace, condition of the bore and any signs of damage such as cracks in the receiver. Of course, collecting the rifle and shooting the rifle are not mutually exclusive, so a rifle can be both a shooter and a collectible.

There is no one correct way to collect guns, and categories can be as narrow or as broad as desired. An example of a very broad category is the collector who wishes specialize in U.S. military weapons. Such a field can cover guns, accouterments, and accessories from Revolutionary War flintlock muskets to today's military weapons and everything in between. A narrow category could be Colt 1873 single-action army revolvers dating from the Custer era. Most collecting themes are somewhere between these two extremes. American military weapons of World War II would be a good example of a rather broad theme but with discernible parameters. Others may wish to narrow the focus and collect specific types of weapons such as M1 rifles, M1 carbines, M1903 rifles, M1911/M1911A1 pistols, et al. The Model 1917 rifle could fit into its own category or be part of a collection of WWI or WWII military small arms, depending on the desires of the collector.

For the sake of this discussion, it is assumed that an individual wishes to acquire a M1917 rifle that fits his chosen field of collecting. After that decision is made, there are numerous other things that must be taken into consideration. Is the goal to acquire one example from each manufacturer or just a single specimen? What type of condition is desired? Is an arsenal rebuilt rifle okay or should it be as close as possible to the configuration and condition as when it left the factory?

Many collectors say they want a rifle that is in its "original" or "as issued" condition, but those terms are rather nebulous. For the sake of clarification, let's examine the two terms separately.

Original — This term generally refers to a rifle that is in the same configuration, but not necessarily the same condition, as when it left the factory. Such rifles do exist but are quite uncommon and are eagerly sought after by collectors. Examples can be found ranging from absolutely pristine to well-worn but otherwise unmolested condition. Obviously, there will be a significant price differential between these two extremes. In order to qualify as "original," a M1917 rifle will exhibit several distinctive features.

The rifle should retain the factory rust-bluing on the external metal parts. The only exceptions are the very late-production, factory-Parkerized Eddystone rifles. Since the overwhelming majority of M1917 rifles were Parkerized when ordnance rebuilt in the 1920s to the 1940s, any example retaining the original rust-blued factory finish is rather uncommon. With sufficient talent, knowledge, and money, it might be theoretically possible to have a Parkerized finish removed and refinished with rust bluing, but this would be a difficult task and would be generally easy to detect. Such a gun would have a much lower value and garner less interest among collectors than one retaining its original finish.

Identifying a genuine factory-Parkerized Eddystone rifle is rather straightforward. The first thing to confirm is a proper correlation of barrel date and serial number. As mentioned, the barrel date is the date that the component was manufactured and not the date of assembly. However, given the rapid pace of manufacture during WWI, with few exceptions, the difference between the barrel date and the date of rifle assembly was generally only a month or two. Eddystone began Parkerizing rifles in September of 1918. By this time, Eddystone serial numbers were in the approximate 1,100,000 range. While there could be some slight deviation, a barrel date between August 1918 and March 1919 (when Eddystone production ceased) would be expected for a factory-Parkerized example. Most of the factory-Parkerized Eddystone rifles were subsequently refinished during overhaul, so it is probable that an example encountered today has actually been refinished. While there is no one correct tint of Parkerizing, during rebuild, the external metal parts were often sand-blasted before Parkerizing, which resulted in a rather rough texture. When originally manufactured in late 1918 and early 1919, the Eddystone factory did not sandblast the metal before applying Parkerizing, which resulted in a noticeably smoother finish as compared to most of the arsenal-Parkerized guns.

Components should be made by the same manufacturer. In other words, Winchester rifles should have "W"-marked parts, Remington would have "R"-marked parts, and Eddystone would have "E"-marked parts. During overhaul, the typical procedure was for the rifles to be completely disassembled and all usable parts sorted, cleaned, repaired and then stored in bins until needed. When a rebuilt rifle was to be reassembled, parts were taken from the bins at random. There would have been no attempt, and no reason, to try to use parts consistent with the manufacturer of the receiver, and any parts that might match would be strictly happenstance. Thus, a rebuilt rifle with a Remington receiver, for example, would likely have a mixture of "R" "W" and "E"-marked parts. It has been theorized that perhaps the manufacturers shared parts during production, but there is no evidence that this happened. Mismatched parts are de facto proof that the rifle in question is not all original. See the discussion below regarding mismatched rifles that have been "restored."

As Issued — The term "as issued" is all but meaningless. A new, pristine rifle, fresh from the factory in 1918 and handed to a doughboy departing for France was clearly an as-issued rifle. By the same token, a M1917 rifle that had gone through two arsenal overhauls and fitted with a replacement barrel and supplied to a soldier in 1942 was an as-issued weapon as well. While the former example would obviously be more valuable and more desirable than the latter for collecting purposes, both are authentic U.S. military weapons that can be great additions to a collection. In fact, to a collector interested in WWII weapons, a 1942 rebuild could be considered as more of an as-issued M1917 rifle than a pristine example that had been untouched since 1918.

The Dilemma of Restorations — Given the disparity in value and desirability of original M1917 rifles as compared to the typical rebuilt examples, it is common practice for collectors today to attempt to make a rifle "correct" by replacing any parts that don't correspond with the manufacturer of the receiver. Of course, a rifle can only be "original" once, so a rifle fitted with "matching" replacement parts is generally referred to as a "restoration." While a widespread practice today, which may sound innocuous, there can be some potential ethical issues as well as financial considerations. In the past, M1917 rifles have not be plagued with the often-contentious issue of restorations nearly as much as was the case with some other U.S. military weapons, especially the M1 Garand rifle, but this is changing.

The first thing that must be evaluated when considering whether or not to attempt a restoration is the condition and configuration of the "host" gun. The ideal candidate for a restoration project would be:

- Rifle with the correct barrel manufacturer and date appropriate for the vintage of the receiver
- Retains the original blued finish on the receiver and barrel (except for very late Eddystones)
- Has a number of correctly coded blued parts (the more, the better)
- Genuine WWI-vintage "GI" stock with proper initial on the tip ("E", "R", or "W"), correct eagle-head inspection stamps, matching condition hand guards, and no rebuild markings

The next step is to obtain replacements for whatever parts are on the rifle that are not "correct" for the make and vintage of the rest of the gun. Finding the proper parts with which to complete the rifle would take varying degrees of difficulty but would likely be an obtainable goal. However, the effort involved in obtaining the desired parts would be compounded by the fact that the parts would need to be blued to match the rest of the gun since a rifle with mixed blued and Parkerized parts is clearly not "correct" regardless of the code markings. Also, the condition of the replaced parts would need to be consistent with the rest of the gun. In other words, a rifle that is in pristine condition would not be expected to have obviously worn parts such as a bolt or barrel bands, etc. The reverse is also true, since a well-worn rifle shouldn't have parts in noticeably better condition than the rest of the gun.

If correct parts are found and installed on the rifle, few would object to calling it a legitimate and reasonable restoration. It might make sense from a financial standpoint, assuming the replacement parts could be found at prices that weren't exorbitantly expensive.

The issue now becomes a bit less clear from an ethical viewpoint. A person is absolutely free to do whatever he or she wishes with property they own. However, when a rifle subsequently changes hands, which it eventually will someday, the fact that is has been restored and is not actually original may be lost. For example, when the owner passes away, the wife or executor of the estate may have no idea that the rifle had been restored years before. There's no ethical dilemma in such a case because there was no attempt to mislead or defraud the subsequent buyer, but the fact that the gun is not all original is lost in perpetuity. However, if the rifle is restored by someone who intends to sell the piece, then he would be ethically obligated to disclose that fact. Unfortunately, this is not always the case.

Non-disclosure of a restoration is not the only troubling factor when it comes to collecting. Collectors, understandably, wish to have the proper markings on the weapons they collect. A prime example is the final inspection stamp (cartouche) on M1 Garand rifle stocks. This is a hallmark of the rifle, and stocks with the proper stamp are significantly more valuable and sought after than those without. Not surprisingly, this has resulted in some individuals fabricating bogus stamping dies to apply the correct markings to a stock. This can be done on genuine replacement "GI" stocks that are unmarked or, in some cases, applied to newly made "replica" stocks.

Sometimes, the origin of the newly marked stock is disclosed, but more often, it is not. If a collector is informed that the stock has fake markings but still wants it, that is certainly his prerogative. However, will a subsequent owner of the rifle be informed of the stock's true background? Fortunately, in the case of the M1917 rifle, this has not been a widespread problem, as the only inspection markings applied at the factory were the small eagle-head stamps, which, thus far, have not attracted the attention of the con artists to any great extent. However, with the increase in collector interest in the American Enfield, this is likely to change. At one time, genuine "GI" unmarked M1917 replacement stocks were not particularly difficult to find, but that is not the case today. There are newly made M1917 stocks on the market, some of excellent quality and some rather shoddy. If fake eagle-head stamps become more common, it is a certainty that some will be used to apply "correct," but fake, markings to replica stocks. This makes it especially important to compare the inspector eagle-head code numbers on the stocks originally made by each manufacturer as presented in this and other books to confirm that the code is correct for the rifle in question.

Another practice that is sometimes engaged in is the removal of legitimate arsenal rebuild stamps on stocks that went through overhaul. Obviously, a rifle with a stock with rebuild markings is not all original, and removal of the markings could help disguise its previous history. The removal of markings would require that the stock be sanded and refinished but, if skillfully done, it might take close scrutiny to spot.

There are some instances when a M1917 rifle has been modified in such a manner that would make any sort of a realistic restoration extremely difficult, if not impossible. A good example is a rifle that has been sporterized. As mentioned previously, the extent of sporterizing can vary widely. The only modification made on some rifles was cutting off part of the fore-end, but everything else was left more or less intact. This would be an easy "fix" if a genuine stock could be found, but this can prove to be harder than might be expected. Some cut-down stocks have been restored by splicing the undamaged fore-end portion of a "donor" stock that was otherwise broken or in poor condition to the cut-off portion of the sporterized stock. If a rifle with a suspect stock is encountered, it would be a good practice to slide the lower barrel band forward to determine if the fore-end has been spliced together.

Other the other hand, if the rifle has been drastically modified as part of the sporterizing procedure, the ability to restore it to any semblance of genuineness can be insurmountable. For example, if the protective "ears" of the rear sight have been ground down or milled off, trying to put the rifle back into its original configuration would be essentially out of the question. Other modifications such as cutting down or replacing the barrel with a non-standard barrel or removing the rear sight and drilling the receiver to install a telescopic-sight mount could possibly be undone to some extent but would be far from an easy or inexpensive task.

Even with the potential pitfalls, determining the originality of a M1917 rifle is a much easier task than is the case with many other types of U.S. military firearms. The main reason is the fact that all were originally blued (with the one exception of the late Eddystones), so a Parkerized example is a sure sign that the rifle has been refinished, either as part of a post-WWI ordnance rebuild, or perhaps by a subsequent civilian owner. Since rifles still in their unmodified factory configuration are highly desired by collectors, rebuilt examples are generally deemed less desirable, hence the popularity of restorations.

Collecting Rebuilds — Collectors, or potential collectors, who don't wish to wade through the morass of restoring or "correcting" M1917 rifles, should perhaps consider arsenal-overhauled rifles in a different light. It is usually obvious if a rifle has gone through an overhaul no matter how many matching parts may have been subsequently added by a collector. An arsenal-rebuilt rifle is every bit as much of a legitimate and historic example of an American military small arm as an example that appears to have just come off the factory assembly line in 1918. It can be reasonably argued that a restoration actually does more harm than good, since it alters the provenance and historical lineage of a classic U.S. military weapon.

Collectors of other U.S. martial arms, especially M1 Garand rifles, have been gravitating toward rebuilds as an interesting and worthwhile collecting theme for several valid reasons. A legitimate arsenal-overhauled M1 rifle is almost always much easier to find and less expensive than a purported original example. Since a con artist is unlikely to apply a bogus rebuild stamp, such a rifle can be purchased today with little trepidation that the gun has been enhanced by a fake inspection stamp or other bogus parts added to turn the rifle into something it never was. Garand collectors who have embraced rebuilds as collectibles have several options. One of the most popular approaches is to obtain an example from each of the arsenals or depots that rebuilt M1 rifles from WWII into the late 1960s. This can prove to be more challenging than might be imagined, since examples from some of the facilities are much less common than others. Ten years ago, such a rifle would not elicit a great deal of interest and would most likely have been widely viewed as just another ho-hum, nondescript rebuild rather than possibly a recognizable and desirable collectible example.

This same approach can be followed when collecting M1917 rifle rebuilds, as there is a surprisingly fertile field of rebuilt examples from which to choose. As discussed in Chapters 6 and 7 of this book, there were a number of arsenals or armories that rebuilt American Enfields. Collectors could focus on acquiring examples of rebuilds from one or more of the arsenals ranging from the 1920s to the WWII or immediate postwar era.

Also, as mentioned in Chapter 6, M1917 rifles were rebuilt in the pre-WWII period by Springfield Armory (1919–1922), Rock Island Arsenal (1920–1932), and Columbus General Depot (1933–1938). Unfortunately, the type of rebuild markings applied to these rifles, if any, have not been confirmed. In WWII and in the immediate postwar period, a number of M1917 rifles were rebuilt by several arsenals and by one civilian corporation. The rebuild markings used by these entities have been identified, with the exception of the rifles overhauled by the Wright Engineering Co. in 1945. The type of rebuild stamp, if any, used by Wright has not been positively established. Also, Springfield Armory rebuilt a rather small number of M1917 rifles in FY 1942, but the type of rebuild stamp is not confirmed. To add a bit of mystery and uncertainty to the topic, as discussed in Table #4 of this book, there is a baffling rebuild stamp that remains unidentified. There were some variations in the format of the rebuild inspection stamps used by some of the ordnance facilities. In many cases, there may be different rebuild stamps from more than one arsenal, indicating multiple overhauls. See Table #4 also for photos and information on rebuild / inspection stamps of the WWII and immediate postwar period.

While remaining on the subject of collecting rebuilds, another approach might be to collect examples overhauled in WWII that had replacement barrels made by Johnson Automatics, Rock Island Arsenal, and / or High Standard. Adding a specimen with one of the non-Marine Corps "USMC"-marked replacement bolts made by United Shoe Machinery Corp. would be an interesting addition to a collection as well.

There is no doubt that restorations will continue to be popular but, frankly, it would be difficult to argue that an ill-advised or poorly rendered restoration is preferable to an honest and unmolested rebuild.

Collecting the M1917 rifle can be an interesting and fulfilling endeavor and one that is sure to continue gaining in popularity. Whether it's a genuine and original example or a rebuilt specimen, the M1917 rifle is a historic American military weapon that served the interests of our nation and allies for over three decades. The American Enfield does not have to take a backseat to any other military weapon as a worthwhile and valuable collectible. If you find a nice example today, be prepared to open your wallet a lot wider than you would have had to do just a few years ago!

Table 1

WWII and Postwar Rebuild/Inspection Stamps for M1917 Rifles

Augusta Arsenal

Inspection rebuild stamps used by Augusta Arsenal typically consisted of "AA" followed by a single letter representing the inspector who performed or supervised the overhaul. A number of different inspector initials may be found.

1 M1917 stock with two Augusta Arsenal stamps ("AAX" and "AAP") indicating the rifle was rebuilt on two separate occasions at the arsenal. **2** Augusta Arsenal rebuild stamp "AAM". **3** Ogden Arsenal rebuild/inspection stamp, boxed. **4** Ogden Arsenal rebuild/inspection stamp, unboxed. The stock had been on a M1917 rifle that was rebuilt at Rock Island Arsenal in the WWII era and was later cycled through Ogden Arsenal.

Ogden Arsenal

There are two variations of rebuild / inspection stamps observed on M1917 rifles that had been cycled through Ogden Arsenal during, or shortly after, World War II: "O.G.E.K." enclosed in a rectangular box and "OGEK" without the box.

It is confirmed that the boxed "OG EK" stamp was used by Elmer Keith, a well-known firearms author previously mentioned in Chapter 7. Purportedly the unboxed "OG EK" stamp was used by another inspector at Ogden with the same initials, Ed Klouser. This has been presented ad infinitum for many years as undisputed fact and has been cited in countless publications, but there appears to be reasonable doubt that this information is correct. Apparently, the identification of Klouser came from the prolific and respected author Col. William S. Brophy. However, there has been no known documented confirmation regarding Klouser's purported tenure at Ogden Arsenal. As cited in the well-researched book *U.S. Military Arms Inspector Marks* by Charles Pate and Anthony Daum, "Brophy attributes the unboxed version of the "OG EK" mark to an individual named Ed Klouser who presumably worked at the Ogden Arsenal alongside Keith during WWII. The author has been unable to verify this or to uncover any information about Ed Klouser."

Other than this brief citation by Pate and Daum (two excellent researchers), there has been little, if any, questioning whether the unboxed "OG EK" stamp actually denoted Ed Klouser. John Spangler, a very thorough and meticulous researcher, has delved into the question and related the following:

> "I dug into the unboxed 'EK' Ed Klouser, and I am 100% certain that Brophy was wrong about this. The only Ed Klousers in the public ancestry family trees are one born in 1864 and died in 1908 and the other was born in 1905 and only lived one year. Not a single Klouser grave or voting registration was found in Utah. There have been maybe a dozen or two members of a Klouser family in Pennsylvania, but no indication of any named Edward or anything similar. The Social Security death index for 1935–2014 has only seven E. Klousers; Earl, Elvin, Emma, two Eugenes, and two Evas, and all those are from Pennsylvania. The *only* Klousers in the 1940 census are the Pennsylvania folks, along with only two WWII draft registration cards, with no Ed. There is not a single mention of any Klouser found in any Utah newspaper prior to 1974. My conclusion is that the lack of a box has no meaning, and I theorize that boxes were commonly used in WWII and unboxed seem to be more common post-WWII."

Thus, it appears that both variants of the "OG EK" stamp can perhaps be attributed solely to Elmer Keith. This is not intended to be an exercise in "revisionist history" but rather to point out that even the most respected and highly regarded writers sometimes make mistakes. Unfortunately, when erroneous information is repeated long enough it acquires a patina of unquestioned fact by subsequent writers who don't bother to verify information before repeating it. Even the most accomplished and trusted authors can occasionally fall prey to it. Thanks to Mr. Spangler for his efforts in providing this information. Readers can make of it what they will.

Raritan Arsenal

The primary rebuild stamp during this period denoting Raritan Arsenal was "RA-P". "RA", of course, stands for Raritan Arsenal, and the "P" represented Harry Petersen, Chief Inspector. There were at least two variances of the "RA-P" stamp, which differed in the size of the letters. This marking is distinctly different than the Remington Arms factory "RA" stamp on WWII M1903, M1903A3, and M1903A4 rifles made by the company.

5 **M1917 stock with two inspection stamps indicating it was overhauled twice at Raritan Arsenal. The "RA" represents Raritan Arsenal and "-P" is Harry Petersen, Chief Inspector. Note the difference of the size of the lettering of the two stamps. 6 Rock Island Arsenal inspection stamp of Frank Krack (1941–1946). 7 Rock Island Arsenal inspection stamp of Elmer Bjerke (1947–1958).**

Rock Island Arsenal

Although Rock Island Arsenal rebuilt large numbers of Model 1917 rifles in the 1920s and 1930s, the type of rebuild inspection stamp applied, if any, has not been confirmed. However, as discussed in Chapter 7, the rifles that were rebuilt at Rock Island during WWII were stamped "RIA / FK", to denote that the work was done under the supervision of Frank Krack, Assistant Foreman of the Small Arms Inspection Division at the arsenal from September 27, 1941 to July 19, 1946. Weapons rebuilt from 1947 to 1958 were stamped "RIA / EB", the initials of Krack's successor, Elmer J. Bjerke. It is doubtful if any of Bjerke's inspection stamps will be found on M1917 rifles, given that his tenure began two years after Rock Island ceased rebuilding M1917 rifles.

San Antonio Arsenal

San Antonio Arsenal used rebuild stamps with several different formats. One had the initials of the Arsenal "SAA" and a single initial of the inspector who supervised the overhaul applied on the right, both markings enclosed in separate boxes. Others had a single initial as a suffix or prefix to the "SAA" marking.

8 Inspection stamp on M1917 rifle rebuilt at San Antonio Arsenal. The boxed "S" is believed to be the initial of the inspector that supervised the overhaul. 9 M1917 rifle stock denoting three rebuilds, two at San Antonio Arsenal. One of the San Antonio stamps has the inspector initial "S" preceding "SAA" and the other has a boxed "E" after "SAA". The "AAO" stamp indicates the rifle was also overhauled at Augusta Arsenal.

Springfield Armory

As related previously in this book, Springfield Armory overhauled a sizeable number of Model 1917 rifles in fiscal years 1919–1921, but as was the case with Rock Island Arsenal, the type of rebuild inspection stamp used, if any, is not known. Springfield Armory rebuilt just 859 Model 1917 rifles in FY 1942 and a relatively small number (5,559) in FY 1947. Again, there is no known information regarding the type of rebuild inspection stamps applied. When other weapons were rebuilt by Springfield in WWII and the postwar era, there was typically an "SA" (boxed or unboxed) stamp applied, which may or may not have been the case with the Model 1917 rifles.

The Enigmatic "3 GM–K"

There is another stamp that is rather commonly encountered, but the identity of the entity has not yet been discovered despite the efforts of a number of dedicated and talented researchers over the years. The identification as a rebuild marking is bolstered by the fact that all observed weapons with the marking were arsenal overhauled.

There have been a lot of theories bandied about, but none have been confirmed, and most can be effectively debunked. It may seem a bit unusual that the following rather lengthy discussion over an obscure inspection stamp is presented in this book. However, the frequency with which the stamp is encountered, and the imaginative theories, some vociferously advanced, warrant further study.

10 "3GM–K" inspection stamp. 11 "3GM–K" stamp on a 1942 vintage Remington 1942 rifle. 12 "3GM–D" stamp on a M1917 rifle.

It is often assumed that this stamp was only found on M1917 rifles. While it is true that the vast majority of the stamps observed have been found on "American Enfields," there have been a few examples found on M1903 rifles. This stamp has been observed on some 1942-vintage Remington M1903 rifles, which confirms the widely held belief that the stamp is of the WWII or the immediate postwar era.

Conjecture as to the origin of the "3GM-K" stamp is widespread. The most widely proffered theory is that it was associated with General Motors given the "GM" component of the marking. Some imaginative individuals have gone further and stated that the marking actually meant

"General Motors, Plant No. 3, Kelvinator." This can be easily debunked since the Kelvinator Corporation was never an affiliate of General Motors. It is sometimes assumed that the "K" in the marking refers to an inspector, which is most likely a correct conclusion. While the majority of these stamps do contain the "K" marking, a few have been observed having a different initial such as "3GM-D", which would suggest that more than one inspector was assigned to the rebuild project.

In any event, the entire General Motors connection is almost certainly not correct. While General Motors manufactured a wide array of weaponry for the United States during WWII, including firearms, tanks, and numerous others items, all of the production contracts were issued to GM subsidiaries. Small arms made by GM affiliates included Inland (carbines), Saginaw (carbines and machine guns), Frigidaire (machine guns), and Guide Lamp (M3/M3A1 submachine guns and FP-45 "Liberator" pistols). No contracts were given to the generic General Motors Corporation parent company. All of these General Motors subsidiaries' weapons production contracts are well documented, but there are no known contracts to any GM affiliate for the rebuilding of firearms during this period. Clearly, the preponderance of evidence argues against the "GM = General Motors" conjecture.

One theory that is occasionally tendered is that "GM" actually indicated Fort George Meade, an army base in Maryland. During WWII, Fort Meade was a training base and a prisoner of war camp. It would not be expected that such a facility would have the capacity to rebuild firearms when that had never been its mission. Since there is no known documentation to suggest such an unlikely role, the "GM = George Meade" connection can effectively be discarded as well.

The types of inspection stamps that were used by the various government arsenals and armories during WWII and the immediate postwar period are, for the most part, well documented. This fact would point to the "3GM-K" stamp as being associated with a commercial enterprise. The only such entity confirmed to have rebuilt M1917 rifles was Wright Engineering in 1944–1945. There is no indication that Wright's contract included rebuilding other weapons such as M1903 rifles, some of which bear the "3GM-K" stamp. Even though it is the only civilian entity known to have rebuilt M1917 rifles during this period, it is difficult to decipher how "3GM-K" could be connected to Wright, but perhaps it is.

The "3GM-K" stamp is an integral part of the story of the M1917 rifle, and the fact that its actual origin is still elusive today continues to frustrate researchers and collectors who are interested in the "American Enfield." Perhaps someday authoritative documentation will be uncovered, but until or unless that happens, it will remain a mystery. Undoubtedly, the conjecture will continue.

Table 2

Winchester Factory Polishing Shop Ledgers for "G17R" Rifles (Model 1917)

Courtesy of Herbert G. Houze, below are the dates that serial numbers were stamped on Model 1917 Rifle receivers at the Winchester factory. They are not the dates of rifle assembly. In addition to the information related below that was provided by Mr. Houze, a later review of the Polishing Shop Ledger by John Spangler included the following additional information: May 11, 1917 "New Enfield" with Model designation G.17.R book stated:

"Eleven receivers marked A to K inclusive on May 11, 1917 and continued through Z on May 12, 1917.

2A to 2Z marked May 12th to 15th

3A to 3Z marked 15-16 May

Last use of Model 1917 "old fashioned" markings (U.S./W/serial number) was on ser. no. 4663.

(No daily entries made July 13 to Aug. 22)

Serial number "1" was stamped on May 21, 1917.

The daily records show number used at the start of each day working 6 days a week."

1917

May	Last Serial #
21	45
22	105
23	180
24	255
25	345
28	509
29	660
31	705

June	Last Serial #
1	795
2	990
4	1125
5	1275
6	1335
20	1673
21	1838
23	1988
29	2333

July	Last Serial #
2	2558
3	2753
7	3053
9	3173
10	3423
12	4025

Aug.	Last Serial #
23	4491
24	4843
25	5203
27	5719
28	6079
29	6529
30	7054
31	7459

Sept.	Last Serial #
4	7789
5	8599
6	9578
7	10332
8	10930
10	11845
11	12750
12	13856
13	14892
14	15865
15	16270
17	17148
18	18034
19	19158
20	20038
21	20864
22	21044
24	21179
25	22265
26	23407
27	24578
28	25896
29	25971

Oct.	Last Serial #
1	27111
2	27966
3	28898
4	30203
5	31178
6	31911
8	32916
9	34442
10	34952
11	36122
12	37397
13	37772
15	38537
16	39587
17	40997
18	41687
19	42722
20	42932
22	43772
23	45167
24	46406
25	47351
26	49256
29	49931
30	50901
31	52386

Nov.	Last Serial #
1	53646
2	53831
3	55131
5	56846
6	56846
7	58346
8	59396
9	60836
10	61571
12	62587
13	63742
14	64882
15	66112
16	67722
17	68547
19	70017
20	71515
21	72880
22	74290
23	75775
24	76780
25	78490
27	80125
28	81715
30	83410

Dec.	Last Serial #
1	84430
3	86290
4	87850
5	89365
6	90670
7	92155
8	92890
10	94525
11	96070
12	97855
13	99580
14	100735
15	101680
17	103270
18	105070
19	106540
20	108145
21	109570
22	110515
24	112030
26	113250
27	114660
28	116055
29	117135
31	118725

1918

Jan.	Last Serial #	*Feb.*	Last Serial #	*March*	Last Serial #	*April*	Last Serial #	*May*	Last Serial #	*June*	Last Serial #
2	120575	1	158215	1	199085	1	242520	1	284970	1	321900
3	122055	2	159505	2	199945	2	243990	2	286335	3	322845
4	123975	4	160690	4	201415	3	245990	3	288255	4	324160
5	124965	5	162310	5	203275	4	248090	4	289275	5	325105
7	126675	6	163915	6	205090	5	249980	6	291270	6	325945
8	128025	7	166075	7	207220	6	251460	7	292950	7	327415
9	129885	8	167935	8	209225	8	252615	8	294150	8	328750
10	131370	9	169465	9	210715	9	254295	9	295620	10	330010
11	132915	11	171115	11	212610	10	255765	10	297090	11	331585
12	133800	12	172630	12	214525	11	257190	11	298245	12	333317
14	135120	13	174550	13	216545	12	259050	13	299295	13	335365
15	136860	14	176080	14	218540	13	260415	14	300450	14	336835
16	138480	15	178070	15	220430	15	262095	15	302340	15	338725
17	140222	16	179140	16	222385	16	263880	16	304440	17	339640
18	141930	18	180850	18	224260	17	265350	17	305700	18	340240
19	143085	19	182830	19	225975	18	266760	18	307245	19	341860
21	144105	20	184750	20	227670	19	268380	20	308610	20	343120
22	145545	21	186460	21	229590	20	270165	21	310605	21	344530
23	146955	22	188440	22	231510	22	271740	22	312390	22	345955
24	148345	23	189535	23	233025	23	273315	23	313965	24	346795
25	150025	25	191425	25	234975	24	274995	24	314910	25	348160
26	150895	26	193360	26	236550	25	276675	25	315645	26	349525
28	152365	27	195205	27	236635	26	278355	27	316905	27	350890
29	153220	28	197110	28	240420	27	279930	28	318585	28	352045
30	155170					29	281715	29	319740	29	353200
31	156670					30	283290	31	321315		

July	Last Serial #	*Aug.*	Last Serial #	*Sept.*	Last Serial #	*Oct.*	Last Serial #	*Nov.*	Last Serial #	*Dec.*	Last Serial #
1	354250	1	386705	3	416785	1	449155	1	481860	2	512625
2	355825	2	388070	4	418045	2	450520	2	482910	3	513990
3	357190	3	389120	5	419410	3	451570	4	484380	4	515145
5	358345	5	390590	6	420565	4	452725	5	485745	5	516405
6	359500	6	391925	7	421615	5	453850	6	487185	6	517875
8	361075	7	392915	9	423085	7	455053	7	488970	7	518610
9	362545	8	393995	10	424630	8	456388	8	490230	9	520095
10	364435	9	394910	11	426175	9	458070	9	491250	10	521250
11	365590	10	395855	12	427465	10	459225	11	491460	11	522195
12	366955	12	396695	13	428410	11	460485	12	493620	12	523395
13	367900	13	397640	14	429355	12	461280	13	495510	13	524370
15	369265	14	398430	16	430780	14	462315	14	496770	14	525315
16	370780	15	399105	17	432670	15	463110	15	497715	16	526365
17	371965	16	399855	18	433930	16	464055	16	498450	17	527250
18	373225	17	400515	19	435505	17	465105	18	500130	18	528300
19	374225	19	401715	20	437395	18	465735	19	501495	19	529005
20	374750	20	403020	21	438445	19	467055	20	502860	20	530160
22	376475	21	404265	23	439390	21	468210	21	504120	21	530685
23	377105	22	405000	24	440965	22	469575	22	505275	23	531630
24	378470	23	406765	25	442435	23	471150	23	506010	24	532575
25	380135	24	407530	26	443905	24	472410	25	507165	26	533460
26	381620	26	408475	27	445495	25	473670	26	508425	27	534017
27	382415	27	410050	28	446425	26	474300	27	509790	30	534631
29	383750	28	411730	30	447896	28	475350	29	510885	31	535576
30	384590	29	413425			29	477030	30	511620		
31	385472	30	414640			30	478395				
		31	415735			31	480075				

1919

Jan.	Last Serial #	Feb.	Last Serial #	March	Last Serial #
2	535996	1	547235	25	548255
3	536731	3	547311		
4	537106	4	547506		
6	537631	5	547611		
7	538576	6	547716		
8	538786	7	547776		
9	539521	8	547963		
10	540361	11	548060		
11	540991	12	548150		
13	541306	13	548230		
14	541726				
15	542146				
16	542566				
17	543301				
18	543871				
20	544396				
21	545026				
22	545926				
23	546169				
28	546426				
29	546635				
30	546845				
31	547040				

Table 3

Component Markings for Model 1917 Rifle

Part	Eddystone	Remington	Winchester
Barrel	E	R	W
Barrel Band Upper	E	R	W
Barrel Band Lower	E	R	W
Bolt	E	R	W
Bolt Stop	E	R	W
Bolt Stop Spring	E	R	W
Bolt Stop Rest	E	R	W or blank
Bolt Sleeve	E	R	W
Buttplate	Not marked	R	W
Buttplate Trap	E	Not marked	W
Cocking Piece	E or "EN"	R or "EN"	W
Ejector	E	R	W
Extractor	E	R	W
Firing Pin	E or "EN"	R or "EN"	W
Floor Plate	E	R	W
Floor Plate Release	E	R	W
Follower	E	R	W
Sight Base Front	E	R	W
Sight Insert Front	E or blank	R	W
Handguard Lower (early only)	E	R	W
Handguard Ring	E	R or blank	W
Magazine Box	E	R	W
Magazine Spring	E	R	W
Sight Elevator	E	R	W
Sight Leaf Spring Rear	E	R	W
Sight Slide Bar	E	R	W
Sight Slide Tab Rear	E	Not marked	W
Safety	E	R	W
Safety Cover	E	R	W
Sear	E	R	W
Stock (at tip of the forearm)	E	R	W
Swivel Rear Base	E	R	W
Swivel Sling Front	E	R or blank	W or blank
Swivel Sling Rear	E	R or blank	W or blank
Swivel Stacking	E	R or blank	W or blank
Trigger	E	R	W
Trigger Guard	E	R	W

Compiled by Mark Wade. Used with permission.

TABLE 4

SPRINGFIELD RESEARCH SERVICE MODEL 1917 RIFLE SERIAL NUMBERS

SERIAL NUMBERS OF M1917 RIFLES CITED IN GOVERNMENT DOCUMENTS

This data was complied by the noted researcher the late Frank Mallory who generously gave permission for use of this data. Collectors and researchers owe a great debt to Mr. Mallory for all of his exhaustive work over the years.

SER. NO.	MODEL	MM/DD/YY	USAGE
1	W	01/23/18	PRESENTED TO PRES WILSON
4	R	03/26/19	INSP OF ORDNANCE WRA CO
26		07/02/17	SIGHT CALIBRATION
27		07/02/17	SIGHT CALIBRATION
30		07/02/17	SIGHT CALIBRATION
31		07/02/17	SIGHT CALIBRATION
32		07/02/17	SIGHT CALIBRATION
68	R	03/26/19	INSP OF ORDNANCE WRA CO
83	R	03/26/19	INSP OF ORDNANCE WRA CO
117	R	03/26/19	INSP OF ORDNANCE WRA CO
118	R	03/26/19	INSP OF ORDNANCE WRA CO
126	R	03/26/19	INSP OF ORDNANCE WRA CO
150		04/25/18	HQ 27TH ENGR CAMP MEADE (INQ)
237	E	06/24/18	LOWEST NUMBER IN 322ND INF
244		04/25/18	HQ 27TH ENGR CAMP MEADE (INQ)
267		04/25/18	HQ 27TH ENGR CAMP MEADE (INQ)
268	R	03/26/19	INSP OF ORDNANCE WRA CO
291	W	10/04/18	CAMP HUMPHRIES
304	W	10/04/18	CAMP HUMPHRIES
315	W	07/01/18	324TH INF CAMP SEVIER
372	W	09/04/17	WRA CO (PACKED - CASE 162)
372	W	12/04/17	CO B 302ND ENGR AEF (DRAWN)
373	R	03/26/19	INSP OF ORDNANCE WRA CO
397	E	05/20/18	165TH DEPOT BRIG CAMP TRAVIS
416	R	09/24/35	US NW PENITENTIARY LEWISBURG
431		04/25/18	HQ 27TH ENGR CAMP MEADE (INQ)
464		04/25/18	HQ 27TH ENGR CAMP MEADE (INQ)
618	R	04/09/18	PETERS CTGE CO AMMO TESTING
660	R	09/24/35	US NW PENITENTIARY LEWISBURG
690	W	07/05/18	CO B 311TH ENGR CAMP GRANT
742	R	09/24/35	US NW PENITENTIARY LEWISBURG
757	E	02/25/18	349TH INF
783	R	09/24/35	US NW PENITENTIARY LEWISBURG
795		09/22/18	CO C 27TH ENGR AEF
831		04/25/18	HQ 27TH ENGR CAMP MEADE (INQ)
878		01/23/18	CO B 303RD ENGR AEF (ON HAND)
913		04/25/18	HQ 27TH ENGR CAMP MEADE (INQ)
1000	W	08/12/18	CO I 22ND ENGR FT BENJ HARRISON
1012	R	09/24/35	US NW PENITENTIARY LEWISBURG
1014	R	09/24/35	US NW PENITENTIARY LEWISBURG
1017	W	10/04/18	CAMP HUMPHRIES
1067	R	09/24/35	US NW PENITENTIARY LEWISBURG
1095	W	08/12/18	CO I 22ND ENGR FT BENJ HARRISON
1171	R	09/24/35	US NW PENITENTIARY LEWISBURG
1189	W	07/01/18	324TH INF CAMP SEVIER
1196	W	07/05/18	CO B 311TH ENGR CAMP GRANT
1212		07/14/47	6TH INF
1232		04/25/18	HQ 27TH ENGR CAMP MEADE (INQ)
1302		02/21/18	CO A 303RD ENGR CAMP DIX (ON HAND)
1333	W	09/04/17	WRA CO (PACKED - CASE 162)
1333	W	12/04/17	CO B 302ND ENGR AEF (DRAWN)
1337	E	02/19/18	349TH INF
1337	E	02/25/18	349TH INF
1364	W	08/12/18	CO I 22ND ENGR FT BENJ HARRISON
1435	W	07/05/18	CO B 311TH ENGR CAMP GRANT
1440	R	09/24/35	US NW PENITENTIARY LEWISBURG
1453	W	08/12/18	CO I 22ND ENGR FT BENJ HARRISON
1481	R	09/24/35	US NW PENITENTIARY LEWISBURG
1492	R	09/24/35	US NW PENITENTIARY LEWISBURG
1651	R	03/26/19	INSP OF ORDNANCE WRA CO
1654		08/21/18	CO D 123RD INF
1698	E	02/25/18	349TH INF
1725		04/25/18	HQ 27TH ENGR CAMP MEADE (INQ)
1772	R	03/26/19	INSP OF ORDNANCE WRA CO
1773	W	09/04/17	WRA CO (PACKED - CASE 162)
1773	W	12/04/17	CO B 302ND ENGR AEF (DRAWN)
1801	R	09/24/35	US NW PENITENTIARY LEWISBURG
1805		04/25/18	HQ 27TH ENGR CAMP MEADE (INQ)
1837	W	10/04/18	CAMP HUMPHRIES
1840	W	09/04/17	WRACO (PACKED - CASE 162)
1840	W	12/04/17	CO B 302ND ENGR AEF (DRAWN)
1851	W	02/20/18	309TH FA (BROKEN XTRACTOR)
1877		02/21/18	CO A 303RD ENGR CAMP DIX (ON HAND)
1891	W	08/12/18	CO I 22ND ENGR FT BENJ HARRISON
1898	R	09/24/35	US NW PENITENTIARY LEWISBURG
1915		04/25/18	HQ 27TH ENGR CAMP MEADE (INQ)
1924		04/25/18	HQ 27TH ENGR CAMP MEADE (INQ)
1942	W	09/04/17	WRA CO (PACKED - CASE 162)
1942	W	12/04/17	CO B 302ND ENGR AEF (DRAWN)
1961		04/25/18	HQ 27TH ENGR CAMP MEADE (INQ)
2015	W	10/04/18	CAMP HUMPHRIES
2120	W	08/12/18	CO I 22ND ENGR FT BENJ HARRISON
2190		04/25/18	HQ 27TH ENGR CAMP MEADE (INQ)
2196	W	10/04/18	CAMP HUMPHRIES
2227	W	08/12/18	CO I 22ND ENGR FT BENJ HARRISON
2231	R	09/24/35	US NW PENITENTIARY LEWISBURG
2242	R	09/24/35	US NW PENITENTIARY LEWISBURG
2313		04/25/18	HQ 27TH ENGR CAMP MEADE (INQ)
2338		01/25/18	CO A 343RD INF
2357		04/25/18	HQ 27TH ENGR CAMP MEADE (INQ)
2375	W	08/12/18	CO I 22ND ENGR FT BENJ HARRISON
2391		05/11/31	APG
2400	R	09/24/35	US NW PENITENTIARY LEWISBURG
2491	R	09/24/35	US NW PENITENTIARY LEWISBURG
2543		04/25/18	HQ 27TH ENGR CAMP MEADE (INQ)
2588	E	06/19/18	LOWEST NUMBER IN 321ST INF
2638	W	10/04/18	CAMP HUMPHRIES
2643	R	04/12/18	CO E 353RD INF CAMP FUNSTON
2753	W	09/04/17	WRACO (PACKED - CASE 162)
2753	W	12/04/17	CO B 302ND ENGR AEF (DRAWN)
2762		05/11/31	APG
2770		04/25/18	HQ 27TH ENGR CAMP MEADE (INQ)
2813	W	08/12/18	CO I 22ND ENGR FT BENJ HARRISON
2841	W	10/04/18	CAMP HUMPHRIES
2844		04/25/18	HQ 27TH ENGR CAMP MEADE (INQ)
2924		01/23/18	COB 303RD ENGR AEF (ON HAND)
2975		01/23/18	CO B 303RD ENGR AEF (ON HAND)
3007	W	08/12/18	CO I 22ND ENGR FT BENJ HARRISON
3048	W	12/04/17	CO B 302ND ENGR AEF (DRAWN)
3048	W	09/04/17	WRACO (PACKED - CASE 162)
3057		02/21/18	CO A 303RD ENGR CAMP DIX (ON HAND)
3064	W	08/12/18	CO I 22ND ENGR FT BENJ HARRISON
3098		04/25/18	HQ 27TH ENGR CAMP MEADE (INQ)
3123	W	08/12/18	CO I 22ND ENGR FT BENJ HARRISON
3178	W	08/12/18	CO I 22ND ENGR FT BENJ HARRISON
3179		01/23/18	CO B 303RD ENGR AEF (ON HAND)

SER. NO.	MODEL	MM/DD/YY	USAGE
3281	W	08/12/18	CO I 22ND ENGR FT BENJ HARRISON
3296	W	07/05/18	CO B 311TH ENGR CAMP GRANT
3337	W	09/13/17	WRA CO VELOCITY TEST
3346	W	10/04/18	CAMP HUMPHRIES
3380	R	04/25/18	INTERCHANGEABILITY TEST
3381		05/11/31	APG
3553	W	08/12/18	CO I 22ND ENGR FT BENJ HARRISON
3622	W	07/05/18	CO B 311TH ENGR CAMP GRANT
3800	W	08/12/18	CO I 22ND ENGR FT BENJ HARRISON
3913	W	09/04/17	WRA CO (PACKED - CASE 162)
3913	W	12/04/17	CO B 302ND ENGR AEF (DRAWN)
3956	W	08/12/18	CO I 22ND ENGR FT BENJ HARRISON
3985	E	06/27/18	LOWEST NUMBER IN 323RD INF
4050	W	05/15/18	SA (TARGETED)
4052	W	05/15/18	SA (TARGETED)
4053	W	05/15/18	SA (TARGETED)
4056	W	05/15/18	SA (TARGETED)
4059	W	05/15/18	SA (TARGETED)
4060	W	05/15/18	SA (TARGETED)
4061	W	05/15/18	SA (TARGETED)
4062	W	05/15/18	SA (TARGETED)
4063	W	05/15/18	SA (TARGETED)
4064	W	05/15/18	SA (TARGETED)
4121	W	10/04/18	CAMP HUMPHRIES
4156	W	08/12/18	CO I 22ND ENGR FT BENJ HARRISON
4183	W	08/12/18	CO I 22ND ENGR FT BENJ HARRISON
4228		07/23/18	CAMP DEVENS (STOCK TEST)
4305	W	09/04/17	WRA CO (PACKED - CASE 162)
4305	W	12/04/17	CO B 302ND ENGR AEF (DRAWN)
4311	E	02/11/18	CO E 342ND INF CAMP GRANT
4408	E	04/17/18	INTERCHANGEABILITY TEST
4480	W	08/12/18	CO I 22ND ENGR FT BENJ HARRISON
4480	W	09/04/17	WRA CO (PACKED - CASE 162)
4480	W	12/04/17	CO B 302ND ENGR AEF (DRAWN)
4620	W	08/12/18	CO I 22ND ENGR FT BENJ HARRISON
4622	R	09/24/35	US NW PENITENTIARY LEWISBURG
4732	W	08/12/18	CO I 22ND ENGR FT BENJ HARRISON
4781	W	07/12/18	CO K 34TH ENGR FT BENJ HARRISON
4788	W	08/12/18	CO I 22ND ENGR FT BENJ HARRISON
4813	E	04/17/18	INTERCHANGEABILITY TEST
4846	W	08/12/18	CO I 22ND ENGR FT BENJ HARRISON
4915	W	08/12/18	CO I 22ND ENGR FT BENJ HARRISON
4966	W	08/12/18	CO I 22ND ENGR FT BENJ HARRISON
4969	R	04/25/18	INTERCHANGEABILITY TEST
4985	W	03/26/19	INSP OF ORDNANCE WRA CO
5005	W	08/12/18	CO I 22ND ENGR FT BENJ HARRISON
5037	W	08/12/18	CO I 22ND ENGR FT BENJ HARRISON
5068	W	08/12/18	CO I 22ND ENGR FT BENJ HARRISON
5195	R	04/25/18	INTERCHANGEABILITY TEST
5217	W	10/04/18	CAMP HUMPHRIES
5237	W	08/12/18	CO I 22ND ENGR FT BENJ HARRISON
5258	W	02/09/46	MIL GOVT HAWAII (RELEASED)
5322	R	09/24/35	US NW PENITENTIARY LEWISBURG
5330	W	07/12/18	CO K 34TH ENGR FT BENJ HARRISON
5381	W	09/03/18	METALS CORP
5381	W	09/03/18	GORHAM MFG CO
5442	W	10/04/18	CAMP HUMPHRIES
5476	W	08/12/18	CO I 22ND ENGR FT BENJ HARRISON
5497	W	08/12/18	CO I 22ND ENGR FT BENJ HARRISON
5537	W	08/12/18	CO I 22ND ENGR FT BENJ HARRISON
5546	W	08/12/18	CO I 22ND ENGR FT BENJ HARRISON
5790	R	04/12/18	CO F 353RD INF CAMP FUNSTON
5867	W	08/12/18	CO I 22ND ENGR FT BENJ HARRISON
5891	W	08/12/18	CO I 22ND ENGR FT BENJ HARRISON
5976	W	08/12/18	CO I 22ND ENGR FT BENJ HARRISON
5985	W	08/12/18	CO I 22ND ENGR FT BENJ HARRISON
6109	W	10/04/18	CAMP HUMPHRIES
6124	W	08/12/18	CO I 22ND ENGR FT BENJ HARRISON
6155	W	08/12/18	CO I 22ND ENGR FT BENJ HARRISON
6164		06/25/18	CO A 309TH ENGR CAMP SHERMAN
6218	R	04/25/18	INTERCHANGEABILITY TEST
6236	W	08/12/18	CO I 22ND ENGR FT BENJ HARRISON
6278	W	08/12/18	CO I 22ND ENGR FT BENJ HARRISON
6355	R	04/24/18	CO G 353RD INF CAMP FUNSTON
6363	W	08/12/18	CO I 22ND ENGR FT BENJ HARRISON
6372	W	08/12/18	CO I 22ND ENGR FT BENJ HARRISON
6399	R	04/11/18	CO A 353RD INF CAMP FUNSTON
6415	W	08/12/18	CO I 22ND ENGR FT BENJ HARRISON
6600	W	08/12/18	CO I 22ND ENGR FT BENJ HARRISON
6604	W	10/04/18	CAMP HUMPHRIES
6612	W	08/12/18	CO I 22ND ENGR FT BENJ HARRISON
6636	W	08/12/18	CO I 22ND ENGR FT BENJ HARRISON
6693	W	08/12/18	CO I 22ND ENGR FT BENJ HARRISON
6746	W	02/12/18	309TH FA CAMP DIX
6787		02/21/18	CO A 303RD ENGR CAMP DIX (ON HAND)
6865	W	10/04/18	CAMP HUMPHRIES
6937	W	08/12/18	CO I 22ND ENGR FT BENJ HARRISON
6965	W	08/12/18	CO I 22ND ENGR FT BENJ HARRISON
7019	R	04/12/18	CO E 353RD INF CAMP FUNSTON
7191		03/04/18	REG DET 309TH ENGR CAMP Z TAYLOR
7235	W	08/12/18	CO I 22ND ENGR FT BENJ HARRISON
7325	W	08/12/18	CO I 22ND ENGR FT BENJ HARRISON
7377	R	04/12/18	HQ CO 353RD INF CAMP FUNSTON
7423	W	08/12/18	CO I 22ND ENGR FT BENJ HARRISON
7432	W	08/12/18	CO I 22ND ENGR FT BENJ HARRISON
7504	R	04/12/18	CO F 353RD INF CAMP FUNSTON
7549		01/12/18	CO D 309TH ENGR CAMP Z TAYLOR
7566	W	08/12/18	CO I 22ND ENGR FT BENJ HARRISON
7567	R	04/25/18	INTERCHANGEABILITY TEST
7617	W	08/12/18	CO I 22ND ENGR FT BENJ HARRISON
7625	W	07/04/43	USS SC728
7718		01/12/18	CO D 309TH ENGR CAMP Z TAYLOR
7797		12/03/17	CO B 117TH ENGR AEF
7800		12/03/17	CO B 117TH ENGR AEF
7826	R	04/20/18	CO I 353RD INF CAMP FUNSTON
7868	W	08/12/18	CO I 22ND ENGR FT BENJ HARRISON
8111	W	08/12/18	CO I 22ND ENGR FT BENJ HARRISON
8120	E	03/26/19	INSP OF ORDNANCE WRA CO
8169	E	04/17/18	INTERCHANGEABILITY TEST
8215	W	10/04/18	CAMP HUMPHRIES
8219		01/12/18	CO D 309TH ENGR CAMP Z TAYLOR
8320	W	10/04/18	CAMP HUMPHRIES
8428	W	10/04/18	CAMP HUMPHRIES
8431		01/23/18	CO B 303RD ENGR AEF (ON HAND)
8482	W	02/04/18	317TH INF CAMP LEE
8482	W	02/09/18	317TH INF CAMP LEE
8501	W	08/12/18	CO I 22ND ENGR FT BENJ HARRISON
8924		06/25/18	CO A 309TH ENGR CAMP SHERMAN
8960	R	04/11/18	CO A 353RD INF CAMP FUNSTON
8973	W	10/04/18	CAMP HUMPHRIES
9018	R	05/18/18	CO C 353RD INF CAMP FUNSTON
9032		05/15/18	115TH INF (STRIKER TEST)
9049	R	04/12/18	HQ CO 353RD INF CAMP FUNSTON
9115	W	10/04/18	CAMP HUMPHRIES
9188	W	08/12/18	CO I 22ND ENGR FT BENJ HARRISON
9190		01/23/18	CO B 303RD ENGR AEF (ON HAND)
9294	E	04/11/18	328TH INF
9294	E	05/01/18	WATERTOWN ARS (TESTED)
9303	W	08/12/18	CO I 22ND ENGR FT BENJ HARRISON
9478	W	04/16/18	PETERS CTGE CO
9478	W	04/09/18	PETERS CTGE CO AMMO TESTING
9499	W	02/04/18	317TH INF CAMP LEE
9499	W	02/09/18	317TH INF CAMP LEE
9508		02/02/18	REG DET 309TH ENGR CAMP Z TAYLO
9522	W	08/12/18	CO I 22ND ENGR FT BENJ HARRISON
9566	R	04/11/18	CO H 353RD INF CAMP FUNSTON
9605		04/13/18	CO A 309TH ENGR
9615	W	08/12/18	CO I 22ND ENGR FT BENJ HARRISON
9640	W	08/12/18	CO I 22ND ENGR FT BENJ HARRISON
9687		01/23/18	CO B 303RD ENGR AEF (ON HAND)
9695		01/23/18	CO B 303RD ENGR AEF (ON HAND)
9771	W	08/12/18	CO I 22ND ENGR FT BENJ HARRISON
9849	R	06/11/18	337TH INF CAMP CUSTER
9860	R	04/11/18	CO H 353RD INF CAMP FUNSTON
9893	W	10/04/18	CAMP HUMPHRIES
9908		01/23/18	CO B 303RD ENGR AEF (ON HAND)
9954	R	11/10/22	AUGUSTA ARS (OVERHAULED)
9995		01/23/18	CO B 303RD ENGR AEF (ON HAND)
10076		01/23/18	CO B 303RD ENGR AEF (ON HAND)
10092		02/21/18	CO A 303RD ENGR CAMP DIX (ON HAND)
10108		01/23/18	CO B 303RD ENGR AEF (ON HAND)
10159		01/12/18	CO D 309TH ENGR CAMP Z TAYLOR
10203	R	04/12/18	HQ CO 353RD INF CAMP FUNSTON
10209		01/23/18	CO B 303RD ENGR AEF (ON HAND)

SER. NO.	MODEL	MM/DD/YY	USAGE
10235	R	04/12/18	CO E 353RD INF CAMP FUNSTON
10321		06/25/18	CO A 309TH ENGR CAMP SHERMAN
10335		01/23/18	CO B 303RD ENGR AEF (ON HAND)
10367		01/23/18	CO B 303RD ENGR AEF (ON HAND)
10431		01/12/18	CO D 309TH ENGR CAMP Z TAYLOR
10514		01/23/18	CO B 303RD ENGR AEF (ON HAND)
10515		09/22/18	CO C 27TH ENGR AEF
10600	E	07/01/18	324TH INF CAMP SEVIER
10616		02/21/18	CO A 303RD ENGR CAMP DIX (ON HAND)
10657	E	05/16/40	VETERANS ADMIN
10672	W	08/31/18	BRIDGEPORT BRASS CO
10683	R	05/18/18	CO C 353RD INF CAMP FUNSTON
10708		06/25/18	CO A 309TH ENGR CAMP SHERMAN
10892		02/21/18	CO A 303RD ENGR CAMP DIX (ON HAND)
10931		02/21/18	CO A 303RD ENGR CAMP DIX (ON HAND)
10940	W	09/03/18	PENNA STAMPING CO
10941		06/25/18	CO A 309TH ENGR CAMP SHERMAN
10989	E	06/19/18	321ST INF CAMP SEVIER
11028	R	04/24/18	CO G 353RD INF CAMP FUNSTON
11088		02/21/18	CO A 303RD ENGR CAMP DIX (ON HAND)
11113		02/21/18	CO A 303RD ENGR CAMP DIX (ON HAND)
11176		01/23/18	CO B 303RD ENGR AEF (ON HAND)
11226	E	04/17/18	INTERCHANGEABILITY TEST
11253		01/23/18	CO B 303RD ENGR AEF (ON HAND)
11307		02/21/18	CO A 303RD ENGR CAMP DIX (ON HAND)
11358		02/21/18	CO A 303RD ENGR CAMP DIX (ON HAND)
11398		02/02/18	REG DET 309TH ENGR CAMP Z TAYLO
11400	E	07/01/18	324TH INF CAMP SEVIER
11431	W	10/04/18	CAMP HUMPHRIES
11454	R	04/12/18	HQ CO 353RD INF CAMP FUNSTON
11459	R	04/11/18	CO A 353RD INF CAMP FUNSTON
11509		02/21/18	CO A 303RD ENGR CAMP DIX (ON HAND)
11524		01/23/18	CO B 303RD ENGR AEF (ON HAND)
11554		02/21/18	CO A 303RD ENGR CAMP DIX (ON HAND)
11559	W	10/04/18	CAMP HUMPHRIES
11740	W	02/09/18	317TH INF CAMP LEE
11740	W	02/04/18	317TH INF CAMP LEE
11798		02/21/18	CO A 303RD ENGR CAMP DIX (ON HAND)
11835		01/23/18	CO B 303RD ENGR AEF (ON HAND)
11887	W	10/04/18	CAMP HUMPHRIES
11949		01/23/18	CO B 303RD ENGR AEF (ON HAND)
11997		01/23/18	CO B 303RD ENGR AEF (ON HAND)
12048	E	05/22/18	CAMP FUNSTON (RUSTY)
12095		01/23/18	CO B 303RD ENGR AEF (ON HAND)
12100		01/23/18	CO B 303RD ENGR AEF (ON HAND)
12117	W	08/21/35	US SW REFORMATRY EL RENO TO RIA
12260		01/23/18	CO B 303RD ENGR AEF (ON HAND)
12286		02/21/18	CO A 303RD ENGR CAMP DIX (ON HAND)
12465		02/21/18	CO A 303RD ENGR CAMP DIX (ON HAND)
12500	E	07/01/18	324TH INF CAMP SEVIER
12559		02/21/18	CO A 303RD ENGR CAMP DIX (ON HAND)
12560		01/23/18	CO B 303RD ENGR AEF (ON HAND)
12563		02/21/18	CO A 303RD ENGR CAMP DIX (ON HAND)
12609	W	10/04/18	CAMP HUMPHRIES
12639		02/21/18	CO A 303RD ENGR CAMP DIX (ON HAND)
12654		02/21/18	CO A 303RD ENGR CAMP DIX (ON HAND)
12694	W	04/00/18	ACCURACY TEST
12741	W	10/04/18	CAMP HUMPHRIES
12770	W	10/04/18	CAMP HUMPHRIES
12801		01/12/18	CO D 309TH ENGR CAMP Z TAYLOR
12857		02/21/18	CO A 303RD ENGR CAMP DIX (ON HAND)
12897		01/23/18	CO B 303RD ENGR AEF (ON HAND)
12971		06/25/18	CO A 309TH ENGR CAMP SHERMAN
12982		02/21/18	CO A 303RD ENGR CAMP DIX (ON HAND)
13028	W	03/26/19	INSP OF ORDNANCE WRA CO
13031	W	04/20/18	TO SMALL ARMS SECT
13078		02/21/18	CO A 303RD ENGR CAMP DIX (ON HAND)
13090		01/23/18	CO B 303RD ENGR AEF (ON HAND)
13115	W	02/20/18	309TH FA (BROKEN XTRACTOR)
13183		03/04/18	REG DET 309TH ENGR CAMP Z TAYLO
13213		01/23/18	CO B 303RD ENGR AEF (ON HAND)
13356	W	10/04/18	CAMP HUMPHRIES
13373	W	10/04/18	CAMP HUMPHRIES
13380		02/21/18	CO A 303RD ENGR CAMP DIX (ON HAND)
13393		01/23/18	CO B 303RD ENGR AEF (ON HAND)
13408		02/21/18	CO A 303RD ENGR CAMP DIX (ON HAND)
13415		01/23/18	CO B 303RD ENGR AEF (ON HAND)
13476	W	10/04/18	CAMP HUMPHRIES
13524		02/21/18	CO A 303RD ENGR CAMP DIX (ON HAND)
13525	W	10/04/18	CAMP HUMPHRIES
13577		02/21/18	CO A 303RD ENGR CAMP DIX (ON HAND)
13579	W	10/04/18	CAMP HUMPHRIES
13607	R	04/12/18	HQ CO 353RD INF CAMP FUNSTON
13660	R	05/02/18	SPECIAL FIRING PIN TEST
13678	W	10/04/18	CAMP HUMPHRIES
13696	W	03/26/19	INSP OF ORDNANCE WRA CO
13777		02/21/18	CO A 303RD ENGR CAMP DIX (ON HAND)
13811	W	02/12/18	309TH FA CAMP DIX
13842	W	11/20/18	EXTRACTION TESTS
13842	W	03/26/19	INSP OF ORDNANCE WRA CO
13847	W	02/12/18	309TH FA CAMP DIX
13905		02/21/18	CO A 303RD ENGR CAMP DIX (ON HAND)
13927		02/21/18	CO A 303RD ENGR CAMP DIX (ON HAND)
13933	W	10/04/18	CAMP HUMPHRIES
13961		01/23/18	CO B 303RD ENGR AEF (ON HAND)
13969		02/21/18	CO A 303RD ENGR CAMP DIX (ON HAND)
14150		01/23/18	CO B 303RD ENGR AEF (ON HAND)
14163		02/21/18	CO A 303RD ENGR CAMP DIX (ON HAND)
14178		01/23/18	CO B 303RD ENGR AEF (ON HAND)
14202		01/23/18	CO B 303RD ENGR AEF (ON HAND)
14302		01/23/18	CO B 303RD ENGR AEF (ON HAND)
14303		01/23/18	CO B 303RD ENGR AEF (ON HAND)
14311	W	10/04/18	CAMP HUMPHRIES
14419		02/21/18	CO A 303RD ENGR CAMP DIX (ON HAND)
14437		02/21/18	CO A 303RD ENGR CAMP DIX (ON HAND)
14475		01/23/18	CO B 303RD ENGR AEF (ON HAND)
14505		01/23/18	CO B 303RD ENGR AEF (ON HAND)
14540		02/21/18	CO A 303RD ENGR CAMP DIX (ON HAND)
14558		01/23/18	CO B 303RD ENGR AEF (ON HAND)
14565	W	10/04/18	CAMP HUMPHRIES
14600		01/23/18	CO B 303RD ENGR AEF (ON HAND)
14608		01/23/18	CO B 303RD ENGR AEF (ON HAND)
14621	E	04/11/18	328TH INF
14639		01/23/18	CO B 303RD ENGR AEF (ON HAND)
14646	W	10/04/18	CAMP HUMPHRIES
14727	W	10/04/18	CAMP HUMPHRIES
14734		02/21/18	CO A 303RD ENGR CAMP DIX (ON HAND)
14781		02/21/18	CO A 303RD ENGR CAMP DIX (ON HAND)
14802		02/21/18	CO A 303RD ENGR CAMP DIX (ON HAND)
14828	W	10/04/18	CAMP HUMPHRIES
14892	W	10/04/18	CAMP HUMPHRIES
14929	E	01/17/18	CO H 350TH INF (DEFECTIVE)
14929	E	03/07/18	CO C 313TH AMMO TRAIN
14980		01/23/18	CO B 303RD ENGR AEF (ON HAND)
14999		01/23/18	CO B 303RD ENGR AEF (ON HAND)
15114		01/23/18	CO B 303RD ENGR AEF (ON HAND)
15124		01/23/18	CO B 303RD ENGR AEF (ON HAND)
15132		06/25/18	CO A 309TH ENGR CAMP SHERMAN
15136		02/21/18	CO A 303RD ENGR CAMP DIX (ON HAND)
15258		02/21/18	CO A 303RD ENGR CAMP DIX (ON HAND)
15282		02/21/18	CO A 303RD ENGR CAMP DIX (ON HAND)
15286		02/21/18	CO A 303RD ENGR CAMP DIX (ON HAND)
15292	E	04/11/18	328TH INF
15319		01/23/18	CO B 303RD ENGR AEF (ON HAND)
15344	W	10//04/18	CAMP HUMPHRIES
15357	W	02/21/38	BA (TO BE SCRAPPED)
15389		02/21/18	CO A 303RD ENGR CAMP DIX (ON HAND)
15501	R	05/11/18	CAMP GRANT TO MIDVALE
15636	W	10/04/18	CAMP HUMPHRIES
15676		02/21/18	CO A 303RD ENGR CAMP DIX (ON HAND)
15755		01/23/18	CO B 303RD ENGR AEF (ON HAND)
15783		02/21/18	CO A 303RD ENGR CAMP DIX (ON HAND)
15841		01/23/18	CO B 303RD ENGR AEF (ON HAND)
15852	W	10/04/18	CAMP HUMPHRIES
15881		02/21/18	CO A 303RD ENGR CAMP DIX (ON HAND)
15894		01/23/18	CO B 303RD ENGR AEF (ON HAND)
15895		02/21/18	CO A 303RD ENGR CAMP DIX (ON HAND)
15917		01/23/18	CO B 303RD ENGR AEF (ON HAND)
15936		01/23/18	CO B 303RD ENGR AEF (ON HAND)
15947		02/21/18	CO A 303RD ENGR CAMP DIX (ON HAND)
15974	W	03/26/19	INSP OF ORDNANCE WRA CO
15992		02/21/18	CO A 303RD ENGR CAMP DIX (ON HAND)
16061		01/23/18	CO B 303RD ENGR AEF (ON HAND)
16151	E	04/17/18	CAMP JACKSON TO MIDVALE

SER. NO.	MODEL	MM/DD/YY	USAGE
16205		01/23/18	CO B 303RD ENGR AEF (ON HAND)
16268		02/21/18	CO A 303RD ENGR CAMP DIX (ON HAND)
16289		02/21/18	CO A 303RD ENGR CAMP DIX (ON HAND)
16459		02/21/18	CO A 303RD ENGR CAMP DIX (ON HAND)
16472		02/21/18	CO A 303RD ENGR CAMP DIX (ON HAND)
16483	W	10/04/18	CAMP HUMPHRIES
16489	E	03/26/19	INSP OF ORDNANCE WRA CO
16498		01/23/18	CO B 303RD ENGR AEF (ON HAND)
16566		02/21/18	CO A 303RD ENGR CAMP DIX (ON HAND)
16617	W	12/04/17	CO B 302ND ENGR AEF (DRAWN)
16687	R	06/11/18	340TH INF CAMP CUSTER
16694		02/21/18	CO A 303RD ENGR CAMP DIX (ON HAND)
16701	E	04/17/18	INTERCHANGEABILITY TEST
16762		01/23/18	CO B 303RD ENGR AEF (ON HAND)
16781	E	04/23/18	REVIEWED BY ENGR BUREAU
16781	E	05/01/18	WATERTOWN ARS (TESTED)
16781		09/17/18	FAILED IN SERVICE
16795	R	05/11/18	CAMP GRANT TO MIDVALE
16797	W	10/04/18	CAMP HUMPHRIES
16860		02/21/18	CO A 303RD ENGR CAMP DIX (ON HAND)
16946		02/06/18	CO E 306TH ENGR CAMP JACKSON
17105	E	01/27/34	GRIFFIN & HOWE (SALE)
17145	W	08/12/18	CO I 22ND ENGR FT BENJ HARRISON
17196		01/23/18	CO B 303RD ENGR AEF (ON HAND)
17221	W	02/09/18	317TH INF CAMP LEE
17357	W	08/12/18	CO I 22ND ENGR FT BENJ HARRISON
17466	W	08/12/18	CO I 22ND ENGR FT BENJ HARRISON
17505	E	03/26/19	INSP OF ORDNANCE WRA CO
17521	E	03/26/19	INSP OF ORDNANCE WRA CO
17626	R	05/11/18	CAMP GRANT TO MIDVALE
17669	E	02/25/18	349TH INF
17965	W	02/09/18	317TH INF CAMP LEE
18193	W	10/04/18	CAMP HUMPHRIES
18237	W	10/04/18	CAMP HUMPHRIES
18243	W	10/04/18	CAMP HUMPHRIES
18255		01/23/18	CO B 303RD ENGR AEF (ON HAND)
18311	W	10/04/18	CAMP HUMPHRIES
18521	R	05/11/18	CAMP GRANT TO MIDVALE
18652	W	02/09/18	317TH INF CAMP LEE
18666	W	11/20/18	EXTRACTION TESTS
18666	W	03/26/19	INSP OF ORDNANCE WRA CO
18666	W	04/00/18	ACCURACY TEST
18701	W	10/04/18	CAMP HUMPHRIES
18829	W	03/26/19	INSP OF ORDNANCE WRA CO
18830	W	10/04/18	CAMP HUMPHRIES
18893	W	02/09/18	317TH INF CAMP LEE
19062	W	02/09/18	317TH INF CAMP LEE
19107	R	06/11/18	340TH INF CAMP CUSTER
19205	W	02/09/18	317TH INF CAMP LEE
19243	E	01/23/18	CO D 313TH AMMO TRAIN CAMP DODGE
19243	E	03/20/18	CO M 350TH INF
19307	E	03/26/19	INSP OF ORDNANCE WRA CO
19308	R	05/11/18	CAMP GRANT TO MIDVALE
19312	W	10/04/18	CAMP HUMPHRIES
19331	W	03/26/19	INSP OF ORDNANCE WRA CO
19597	W	02/09/18	317TH INF CAMP LEE
19795	W	10/04/18	CAMP HUMPHRIES
20004	W	08/12/18	CO I 22ND ENGR FT BENJ HARRISON
20021	W	10/04/18	CAMP HUMPHRIES
20067	W	10/04/18	CAMP HUMPHRIES
20068	W	10/04/18	CAMP HUMPHRIES
20087	W	10/04/18	CAMP HUMPHRIES
20117	W	10/04/18	CAMP HUMPHRIES
20208	W	10/04/18	CAMP HUMPHRIES
20311	W	10/04/18	CAMP HUMPHRIES
20341		02/21/18	CO A 303RD ENGR CAMP DIX (ON HAND)
20435		02/21/18	CO A 303RD ENGR CAMP DIX (ON HAND)
20446		12/03/17	CO B 117TH ENGR AEF
20477	R	05/11/18	CAMP GRANT TO MIDVALE
20529		04/25/18	HQ 27TH ENGR CAMP MEADE (INQ)
20636	W	10/04/18	CAMP HUMPHRIES
20887		01/23/18	CO B 303RD ENGR AEF (ON HAND)
20986	W	12/04/17	CO B 302ND ENGR AEF (DRAWN)
21031	W	02/09/18	317TH INF CAMP LEE
21087	E	02/19/18	349TH INF
21160		12/17/17	326TH INF
21187	E	03/26/19	INSP OF ORDNANCE WRA CO
21340	W	03/26/19	INSP OF ORDNANCE WRA CO
21341	W	02/09/18	317TH INF CAMP LEE
21418	W	10/04/18	CAMP HUMPHRIES
21820		12/03/17	CO B 117TH ENGR AEF
21918	W	07/12/18	CO K 34TH ENGR FT BENJ HARRISON
21962	R	05/11/18	CAMP GRANT TO MIDVALE
21996	W	08/12/18	CO I 22ND ENGR FT BENJ HARRISON
22212	W	10/04/18	CAMP HUMPHRIES
22226	W	08/12/18	CO I 22ND ENGR FT BENJ HARRISON
22269	E	04/18/18	307TH INF
22333	R	05/11/18	CAMP GRANT TO MIDVALE
22450	W	10/04/18	CAMP HUMPHRIES
22514	E	02/25/18	349TH INF
22660	R	05/11/18	CAMP GRANT TO MIDVALE
22717	W	08/12/18	CO I 22ND ENGR FT BENJ HARRISON
23104	W	08/12/18	CO I 22ND ENGR FT BENJ HARRISON
23303	W	10/04//18	CAMP HUMPHRIES
23626		09/22//18	CO C 27TH ENGR AEF
23638	E	05/01//18	WATERTOWN ARS (TESTED)
23638	E	04/23//18	REVIEWED BY ENGR BUREAU
23638		12/17//17	327TH INF
23638	E	04/22//18	CAMP GORDON
23640	W	12/04//17	CO B 302ND ENGR AEF (DRAWN)
23715		12/17//17	327TH INF (LOOSE BBL)
23715	E	04/23//18	REVIEWED BY ENGR BUREAU
23715	E	05/20//18	CAMP GORDON (DEFECTIVE)
23715	E	05/23//18	ORD DEPT INSP DIV
23715	E	11/21//18	CAMP GORDON
23838	W	06/30//34	SHANGHAI VOLUNTEER CORPS
24140	E	04/23/18	REVIEWED BY ENGR BUREAU
24149	E	10/29/18	WATERTOWN ARS (EXAMINED)
24149		03/04/18	81ST DIV
24149	E	04/23/18	CAMP JACKSON
24149	E	10/19/18	FAILED IN SERVICE
24273	E	02/25/18	349TH INF
24273	E	02/19/18	349TH INF
24504	E	03/26/19	INSP OF ORDNANCE WRA CO
24586	W	08/12/18	CO I 22ND ENGR FT BENJ HARRISON
24719	E	02/25/18	349TH INF
24807	E	02/19/18	349TH INF
24870	E	03/20/18	CO M 350TH INF
24870	E	02/08/18	CO B 313TH AMMO TRAIN
24892	W	12/04/17	CO B 302ND ENGR AEF (DRAWN)
25050	E	03/20/18	CO M 350TH INF
25050	E	01/17/18	CO H 350TH INF
25102	W	02/04/18	317TH INF CAMP LEE
25161	W	10/04/18	CAMP HUMPHRIES
25204	W	12/04/17	CO B 302ND ENGR AEF (DRAWN)
25295	W	08/12/18	CO I 22ND ENGR FT BENJ HARRISON
25361	E	04/16/18	PETERS CTGE CO
25361	E	04/09/18	PETERS CTGE CO AMMO TESTING
25375	W	10/04/18	CAMP HUMPHRIES
25431	E	02/19/18	349TH INF
25467	W	12/04/17	CO B 302ND ENGR AEF (DRAWN)
25485	W	10/04/18	CAMP HUMPHRIES
25501		09/22/18	CO C 27TH ENGR AEF
25533	W	12/04/17	CO B 302ND ENGR AEF (DRAWN)
25549	W	12/04/17	CO B 302ND ENGR AEF (DRAWN)
25647	R	05/11/18	CAMP GRANT TO MIDVALE
25690	W	12/04/17	CO B 302ND ENGR AEF (DRAWN)
25736	E	03/26/19	INSP OF ORDNANCE WRA CO
25982		09/22/18	CO C 27TH ENGR AEF
26039	W	10/04/18	CAMP HUMPHRIES
26045	R	05/11/18	CAMP GRANT TO MIDVALE
26107		02/06/18	CO E 306TH ENGR CAMP JACKSON
26394	W	02/09/18	317TH INF CAMP LEE
26394	W	02/04/18	317TH INF CAMP LEE
26470	R	05/11/18	CAMP GRANT TO MIDVALE
26520	R	05/11/18	CAMP GRANT TO MIDVALE
26522	W	10/04/18	CAMP HUMPHRIES
26666	W	10/01/18	CAMP HUMPHRIES
26670	R	05/11/18	CAMP GRANT TO MIDVALE
26685	W	08/12/18	CO I 22ND ENGR FT BENJ HARRISON
26743	W	10/04/18	CAMP HUMPHRIES
26835	W	10/04/18	CAMP HUMPHRIES
26852	E	03/26/19	INSP OF ORDNANCE WRA CO
26965		09/22/18	CO C 27TH ENGR AEF

SER. NO.	MODEL	MM/DD/YY	USAGE
26985	W	08/12/18	CO I 22ND ENGR FT BENJ HARRISON
27115	W	10/04/18	CAMP HUMPHRIES
27161	W	10/04/18	CAMP HUMPHRIES
27179	R	05/11/18	CAMP GRANT TO MIDVALE
27233	W	08/12/18	CO I 22ND ENGR FT BENJ HARRISON
27268	E	03/26/19	INSP OF ORDNANCE WRA CO
27311	R	05/11/18	CAMP GRANT TO MIDVALE
27400	E	02/19/18	349TH INF
27408	E	02/19/18	349TH INF
27408	E	02/25/18	349TH INF
27440	W	10/04/18	CAMP HUMPHRIES
27443	W	10/04/18	CAMP HUMPHRIES
27688	W	10/04/18	CAMP HUMPHRIES
27713	W	08/12/18	CO I 22ND ENGR FT BENJ HARRISON
27759	R	05/11/18	CAMP GRANT TO MIDVALE
27772	E	02/25/18	349TH INF
27936	W	08/12/18	CO I 22ND ENGR FT BENJ HARRISON
28005		12/03/17	CO B 117TH ENGR AEF
28111	R	06/07/18	330TH FA CAMP CUSTER
28146	W	10/04/18	CAMP HUMPHRIES
28186	W	12/04/17	CO B 302ND ENGR AEF (DRAWN)
28193	W	08/12/18	CO I 22ND ENGR FT BENJ HARRISON
28256	R	05/11/18	CAMP GRANT TO MIDVALE
28364	W	10/04/18	CAMP HUMPHRIES
28368	W	08/12/18	CO I 22ND ENGR FT BENJ HARRISON
28370		12/03/17	CO B 117TH ENGR AEF
28405	E	05/16/40	VETERANS ADMIN
28424	W	08/12/18	CO I 22ND ENGR FT BENJ HARRISON
28528	W	07/12/18	CO K 34TH ENGR FT BENJ HARRISON
28555		12/03/17	CO B 117TH ENGR AEF
28690	W	08/12/18	CO I 22ND ENGR FT BENJ HARRISON
28753	W	08/12/18	CO I 22ND ENGR FT BENJ HARRISON
28764		12/03/17	CO B 117TH ENGR AEF
28787	E	02/19/18	349TH INF
28804	W	07/12/18	CO K 34TH ENGR FT BENJ HARRISON
28807	R	04/12/18	CO F 353RD INF CAMP FUNSTON
28818	E	04/18/18	CO B 312TH INF CAMP DIX
28818	E	04/18/18	309TH INF
28843	R	05/11/18	CAMP GRANT TO MIDVALE
28992	W	10/04/18	CAMP HUMPHRIES
29022	W	10/04/18	CAMP HUMPHRIES
29071	W	08/12/18	CO I 22ND ENGR FT BENJ HARRISON
29078	W	10/04/18	CAMP HUMPHRIES
29106	W	10/04/18	CAMP HUMPHRIES
29110		12/17/17	360TH INF
29110	E	05/01/18	WATERTOWN ARS (TESTED)
29120	E	02/19/18	349TH INF
29125	E	02/19/18	349TH INF
29230	W	08/12/18	CO I 22ND ENGR FT BENJ HARRISON
29299	W	10/04/18	CAMP HUMPHRIES
29320	W	08/12/18	CO I 22ND ENGR FT BENJ HARRISON
29384	W	10/04/18	CAMP HUMPHRIES
29546		12/03/17	CO B 117TH ENGR AEF
29620	W	10/04/18	CAMP HUMPHRIES
29656	R	05/11/18	CAMP GRANT TO MIDVALE
29662	W	08/12/18	CO I 22ND ENGR FT BENJ HARRISON
29815	R	05/11/18	CAMP GRANT TO MIDVALE
29905	W	10/04/18	CAMP HUMPHRIES
29933	W	08/12/18	CO I 22ND ENGR FT BENJ HARRISON
29947	W	08/12/18	CO I 22ND ENGR FT BENJ HARRISON
30028	W	12/04/17	CO B 302ND ENGR AEF (DRAWN)
30072	W	08/12/18	CO I 22ND ENGR FT BENJ HARRISON
30187	R	05/11/18	CAMP GRANT TO MIDVALE
30305	W	12/04/17	CO B 302ND ENGR AEF (DRAWN)
30325	W	10/04/18	CAMP HUMPHRIES
30328	W	07/12/18	CO K 34TH ENGR FT BENJ HARRISON
30341	E	02/25/18	349TH INF
30342	R	05/11/18	CAMP GRANT TO MIDVALE
30409	W	10/04/18	CAMP HUMPHRIES
30430	W	08/12/18	CO I 22ND ENGR FT BENJ HARRISON
30455	R	05/11/18	CAMP GRANT TO MIDVALE
30470	W	10/04/18	CAMP HUMPHRIES
30482		12/03/17	CO B 117TH ENGR AEF
30506		12/03/17	CO B 117TH ENGR AEF
30556	W	10/04/18	CAMP HUMPHRIES
30570	W	10/04/18	CAMP HUMPHRIES
30691	W	10/04/18	CAMP HUMPHRIES
30731	W	08/12/18	CO I 22ND ENGR FT BENJ HARRISON
30750	W	10/04/18	CAMP HUMPHRIES
30770	E	02/19/18	349TH INF
30770	E	02/25/18	349TH INF
31032	E	02/19/18	349TH INF
31032	E	02/25/18	349TH INF
31073		09/22/18	CO C 27TH ENGR AEF
31121		09/22/18	CO C 27TH ENGR AEF
31162	W	08/12/18	CO I 22ND ENGR FT BENJ HARRISON
31280	W	08/12/18	CO I 22ND ENGR FT BENJ HARRISON
31289	W	10/04/18	CAMP HUMPHRIES
31317	W	08/12/18	CO I 22ND ENGR FT BENJ HARRISON
31387	W	10/04/18	CAMP HUMPHRIES
31589	W	12/04/17	CO B 302ND ENGR AEF (DRAWN)
31599	W	02/09/18	317TH INF CAMP LEE
31786		08/12/18	CO I 22ND ENGR FT BENJ HARRISON
31795	W	08/12/18	CO I 22ND ENGR FT BENJ HARRISON
31801	W	12/04/17	CO B 302ND ENGR AEF (DRAWN)
31802	W	10/04/18	CAMP HUMPHRIES
31874		09/22/18	CO C 27TH ENGR AEF
31890	E	03/05/18	349TH INF
31952	W	08/12/18	CO I 22ND ENGR FT BENJ HARRISON
31968	W	08/12/18	CO I 22ND ENGR FT BENJ HARRISON
32061		09/22/18	CO C 27TH ENGR AEF
32065	W	08/12/18	CO I 22ND ENGR FT BENJ HARRISON
32136	W	12/04/17	CO B 302ND ENGR AEF (DRAWN)
32140	W	10/04/18	CAMP HUMPHRIES
32144	W	08/12/18	CO I 22ND ENGR FT BENJ HARRISON
32173	E	03/05/18	349TH INF
32192	E	02/25/18	349TH INF
32192	E	02/19/18	349TH INF
32214	W	07/12/18	CO K 34TH ENGR FT BENJ HARRISON
32230	E	03/20/18	CO M 350TH INF
32230	E	02/08/18	CO B 313TH AMMO TRAIN
32306	W	10/04/18	CAMP HUMPHRIES
32357	W	10/04/18	CAMP HUMPHRIES
32376	W	08/12/18	CO I 22ND ENGR FT BENJ HARRISON
32518	E	02/19/18	349TH INF
32525	E	02/19/18	349TH INF
32641	W	02/09/18	317TH INF CAMP LEE
32641	W	02/04/18	317TH INF CAMP LEE
32647	E	03/05/18	349TH INF
32697	W	08/12/18	CO I 22ND ENGR FT BENJ HARRISON
32740	R	08/12/18	CO I 22ND ENGR FT BENJ HARRISON
32753	W	08/12/18	CO I 22ND ENGR FT BENJ HARRISON
32865	E	02/25/18	349TH INF
32942	W	07/12/18	CO K 34TH ENGR FT BENJ HARRISON
33065		12/03/17	CO B 117TH ENGR AEF
33172	E	02/19/18	349TH INF
33189	W	07/12/18	CO K 34TH ENGR FT BENJ HARRISON
33215		12/03/17	CO B 117TH ENGR AEF
33241	W	02/04/18	317TH INF CAMP LEE
33241	W	02/09/18	317TH INF CAMP LEE
33260		12/03/17	CO B 117TH ENGR AEF
33289	W	08/12/18	CO I 22ND ENGR FT BENJ HARRISON
33339	W	10/04/18	CAMP HUMPHRIES
33347	W	08/12/18	CO I 22ND ENGR FT BENJ HARRISON
33358	W	10/04/18	CAMP HUMPHRIES
33387		09/22/18	CO C 27TH ENGR AEF
33503	E	02/25/18	349TH INF
33522		12/03/17	CO B 117TH ENGR AEF
33541	W	07/12/18	CO K 34TH ENGR FT BENJ HARRISON
33544	W	08/12/18	CO I 22ND ENGR FT BENJ HARRISON
33602	W	08/12/18	CO I 22ND ENGR FT BENJ HARRISON
33635	W	10/04/18	CAMP HUMPHRIES
33704	W	08/12/18	CO I 22ND ENGR FT BENJ HARRISON
33723	E	03/20/18	CO M 350TH INF
33879	E	05/20/18	165TH DEPOT BRIG CAMP TRAVIS
33889	E	05/20/18	165TH DEPOT BRIG CAMP TRAVIS
33893	E	05/20/18	165TH DEPOT BRIG CAMP TRAVIS
33898	E	05/20/18	165TH DEPOT BRIG CAMP TRAVIS
33900	E	05/20/18	165TH DEPOT BRIG CAMP TRAVIS
33906	E	05/20/18	165TH DEPOT BRIG CAMP TRAVIS
33907	E	05/20/18	165TH DEPOT BRIG CAMP TRAVIS
33908	E	05/20/18	165TH DEPOT BRIG CAMP TRAVIS
33909	E	05/20/18	165TH DEPOT BRIG CAMP TRAVIS
33916	E	05/20/18	165TH DEPOT BRIG CAMP TRAVIS

SER. NO.	MODEL	MM/DD/YY	USAGE
33921	E	05/20/18	165TH DEPOT BRIG CAMP TRAVIS
33925	E	05/20/18	165TH DEPOT BRIG CAMP TRAVIS
33929	E	05/20/18	165TH DEPOT BRIG CAMP TRAVIS
33932	E	05/20/18	165TH DEPOT BRIG CAMP TRAVIS
33937	E	05/20/18	165TH DEPOT BRIG CAMP TRAVIS
33940	E	05/20/18	165TH DEPOT BRIG CAMP TRAVIS
33945	E	05/20/18	165TH DEPOT BRIG CAMP TRAVIS
33954	E	05/20/18	165TH DEPOT BRIG CAMP TRAVIS
33955	E	05/20/18	165TH DEPOT BRIG CAMP TRAVIS
33956	E	05/20/18	165TH DEPOT BRIG CAMP TRAVIS
33957	E	05/20/18	165TH DEPOT BRIG CAMP TRAVIS
33961	E	05/20/18	165TH DEPOT BRIG CAMP TRAVIS
33962	E	05/20/18	165TH DEPOT BRIG CAMP TRAVIS
33963	E	05/20/18	165TH DEPOT BRIG CAMP TRAVIS
33964	E	05/20/18	165TH DEPOT BRIG CAMP TRAVIS
33969	E	05/20/18	165TH DEPOT BRIG CAMP TRAVIS
33977	E	05/20/18	165TH DEPOT BRIG CAMP TRAVIS
33977	W	10/04/18	CAMP HUMPHRIES
33978	E	05/20/18	165TH DEPOT BRIG CAMP TRAVIS
33979	E	05/20/18	165TH DEPOT BRIG CAMP TRAVIS
33985	E	05/20/18	165TH DEPOT BRIG CAMP TRAVIS
33989	E	05/20/18	165TH DEPOT BRIG CAMP TRAVIS
33994	E	05/20/18	165TH DEPOT BRIG CAMP TRAVIS
33995	E	05/20/18	165TH DEPOT BRIG CAMP TRAVIS
33997	E	05/20/18	165TH DEPOT BRIG CAMP TRAVIS
34003	E	05/20/18	165TH DEPOT BRIG CAMP TRAVIS
34009	E	05/20/18	165TH DEPOT BRIG CAMP TRAVIS
34010	E	05/20/18	165TH DEPOT BRIG CAMP TRAVIS
34011	E	05/20/18	165TH DEPOT BRIG CAMP TRAVIS
34013	E	05/20/18	165TH DEPOT BRIG CAMP TRAVIS
34014	E	05/20/18	165TH DEPOT BRIG CAMP TRAVIS
34019	E	05/20/18	165TH DEPOT BRIG CAMP TRAVIS
34025	E	05/20/18	165TH DEPOT BRIG CAMP TRAVIS
34029	E	05/20/18	165TH DEPOT BRIG CAMP TRAVIS
34037	E	05/20/18	165TH DEPOT BRIG CAMP TRAVIS
34051	W	10/04/18	CAMP HUMPHRIES
34057	E	05/20/18	165TH DEPOT BRIG CAMP TRAVIS
34064	E	05/20/18	165TH DEPOT BRIG CAMP TRAVIS
34065	E	05/20/18	165TH DEPOT BRIG CAMP TRAVIS
34081	E	05/20/18	165TH DEPOT BRIG CAMP TRAVIS
34085	E	05/20/18	165TH DEPOT BRIG CAMP TRAVIS
34086	E	05/20/18	165TH DEPOT BRIG CAMP TRAVIS
34087	E	05/20/18	165TH DEPOT BRIG CAMP TRAVIS
34089	E	05/20/18	165TH DEPOT BRIG CAMP TRAVIS
34094	E	05/20/18	165TH DEPOT BRIG CAMP TRAVIS
34097	E	05/20/18	165TH DEPOT BRIG CAMP TRAVIS
34101		12/03/17	CO B 117TH ENGR AEF
34101	E	05/20/18	165TH DEPOT BRIG CAMP TRAVIS
34103	E	05/20/18	165TH DEPOT BRIG CAMP TRAVIS
34112	E	05/20/18	165TH DEPOT BRIG CAMP TRAVIS
34113	E	05/20/18	165TH DEPOT BRIG CAMP TRAVIS
34119	E	05/20/18	165TH DEPOT BRIG CAMP TRAVIS
34121	E	05/20/18	165TH DEPOT BRIG CAMP TRAVIS
34126	E	05/20/18	165TH DEPOT BRIG CAMP TRAVIS
34133	E	05/20/18	165TH DEPOT BRIG CAMP TRAVIS
34152	E	05/20/18	165TH DEPOT BRIG CAMP TRAVIS
34153	E	05/20/18	165TH DEPOT BRIG CAMP TRAVIS
34163	E	05/20/18	165TH DEPOT BRIG CAMP TRAVIS
34166	E	05/20/18	165TH DEPOT BRIG CAMP TRAVIS
34174	E	05/20/18	165TH DEPOT BRIG CAMP TRAVIS
34176	E	05/20/18	165TH DEPOT BRIG CAMP TRAVIS
34185	E	05/20//18	165TH DEPOT BRIG CAMP TRAVIS
34197	W	07/12/18	CO K 34TH ENGR FT BENJ HARRISON
34230	E	05/20/18	165TH DEPOT BRIG CAMP TRAVIS
34233	E	05/20/18	165TH DEPOT BRIG CAMP TRAVIS
34245		12/03/17	CO B 117TH ENGR AEF
34281	E	05/20/18	165TH DEPOT BRIG CAMP TRAVIS
34315	E	05/20/18	165TH DEPOT BRIG CAMP TRAVIS
34318	E	05/20/18	165TH DEPOT BRIG CAMP TRAVIS
34352	E	04/17/18	CAMP JACKSON TO MIDVALE
34413		09/22/18	CO C 27TH ENGR AEF
34416	W	10/04/18	CAMP HUMPHRIES
34492	R	08/27/18	CO E 121ST INF CAMP WHEELER GA
34554	W	02/09/18	317TH INF CAMP LEE
34609	W	10/04/18	CAMP HUMPHRIES
34677	W	10/04/18	CAMP HUMPHRIES
34705	W	02/09/18	317TH INF CAMP LEE

SER. NO.	MODEL	MM/DD/YY	USAGE
34708		07/23/18	CAMP DEVENS (STOCK TEST)
34781	W	10/04/18	CAMP HUMPHRIES
34806	W	10/04/10	CAMP HUMPHRIES
34860	W	10/04/18	CAMP HUMPHRIES
34863		12/03/17	CO B 117TH ENGR AEF
35062	W	10/04/18	CAMP HUMPHRIES
35099	E	02/25/18	349TH INF
35223	R	08/19/18	CO H 122ND INF
35291	W	10/04/18	CAMP HUMPHRIES
35292		09/22/18	CO C 27TH ENGR AEF
35375	W	08/12/18	CO I 22ND ENGR FT BENJ HARRISON
35467	E	01/17/18	CO H 350TH INF
35467	E	03/20/18	CO M 350TH INF
35475		09/22/18	CO C 27TH ENGR AEF
35541	E	02/25/18	349TH INF
35550	W	02/04/18	317TH INF CAMP LEE
35550	W	02/09/18	317TH INF CAMP LEE
35560	W	07/12/18	CO K 34TH ENGR FT BENJ HARRISON
35611	W	10/04/18	CAMP HUMPHRIES
35619		07/23/18	CAMP DEVENS (STOCK TEST)
35637	E	02/25/18	349TH INF
35674	E	02/19/18	349TH INF
35736	W	10/04/18	CAMP HUMPHRIES
35845	W	07/12/18	CO K 34TH ENGR FT BENJ HARRISON
35845	W	08/12/18	CO I 22ND ENGR FT BENJ HARRISON
35922	W	10/04/18	CAMP HUMPHRIES
35929	W	10/04/18	CAMP HUMPHRIES
35949		09/22/18	CO C 27TH ENGR AEF
35994		09/22/18	CO C 27TH ENGR AEF
36045	W	10/19/18	FAILED IN SERVICE
36045	W	10/28/18	DEFECTIVE BARRELS
36045	W	10/29/18	WATERTOWN ARS (EXAMINED)
36106	E	02/25/18	349TH INF
36139	W	10/04/18	CAMP HUMPHRIES
36179	W	08/12/18	CO I 22ND ENGR FT BENJ HARRISON
36215	E	02/25/18	349TH INF
36237	E	02/25/18	349TH INF
36273	R	04/12/18	CO F 353RD INF CAMP FUNSTON
36312	E	06/30/34	SHANGHAI VOLUNTEER CORPS
36336	R	08/12/18	CO I 22ND ENGR FT BENJ HARRISON
36447		09/22/18	CO C 27TH ENGR AEF
36461		09/22/18	CO C 27TH ENGR AEF
36532	W	10/04/18	CAMP HUMPHRIES
36533	W	10/19/18	FAILED IN SERVICE
36533	W	10/29/18	WATERTOWN ARS (EXAMINED)
36533	W	10/28/18	DEFECTIVE BARRELS
36591	E	02/19/18	349TH INF
36643	W	10/04/18	CAMP HUMPHRIES
36648		09/22/18	CO C 27TH ENGR AEF
36736	E	02/19/18	349TH INF
36737	E	12/02/18	CAMP LEWIS
36737		07/15/18	91ST DIV
36740		09/22/18	CO C 27TH ENGR AEF
36745	E	02/25/18	349TH INF
36752	E	02/25/18	349TH INF
36856	W	10/04/18	CAMP HUMPHRIES
36867	E	02/19/18	349TH INF
36880		09/22/18	CO C 27TH ENGR AEF
36887	E	02/25/18	349TH INF
36944	W	02/09/18	317TH INF CAMP LEE
36986	W	10/04/18	CAMP HUMPHRIES
37015	E	02/25/18	349TH INF
37045	E	01/17/18	CO H 350TH INF (DEFECTIVE)
37045	E	02/27/18	CO C 313TH AMMO TRAIN
37203		09/22/18	CO C 27TH ENGR AEF
37296	W	02/09/18	317TH INF CAMP LEE
37410		09/22/18	CO C 27TH ENGR AEF
37580	W	10/04/18	CAMP HUMPHRIES
37596		12/17/17	359TH INF
37596	E	05/01/18	WATERTOWN ARS (TESTED)
37648	W	10/04/18	CAMP HUMPHRIES
37650		09/22/18	CO C 27TH ENGR AEF
37682		09/22/18	CO C 27TH ENGR AEF
37757		12/03/17	CO B 117TH ENGR AEF
37774	W	10/04/18	CAMP HUMPHRIES
37892		09/22/18	CO C 27TH ENGR AEF
37898		09/22/18	CO C 27TH ENGR AEF

SER. NO.	MODEL	MM/DD/YY	USAGE
37941	E	02/19/18	349TH INF
37941	E	02/25/18	349TH INF
38090	W	10/04/18	CAMP HUMPHRIES
38263		09/22/18	CO C 27TH ENGR AEF
38324	E	02/19/18	349TH INF
38394	W	10/04/18	CAMP HUMPHRIES
38402		09/22/18	CO C 27TH ENGR AEF
38441	W	03/39	PAUL JAEGER (SOLD BY NJ)
38495	W	10/04/18	CAMP HUMPHRIES
38506	W	10/04/18	CAMP HUMPHRIES
38571	E	02/25/18	349TH INF
38597	W	10/04/18	CAMP HUMPHRIES
38693	W	10/04/18	CAMP HUMPHRIES
38707	W	07/12/18	CO K 34TH ENGR FT BENJ HARRISON
38717	R	05/02/18	SPECIAL FIRING PIN TEST
38762	E	04/17/18	CAMP JACKSON TO MIDVALE
38773	E	02/19/18	349TH INF
38776	W	10/04/18	CAMP HUMPHRIES
38786	W	10/04/18	CAMP HUMPHRIES
38799		09/22/18	CO C 27TH ENGR AEF
38874	W	10/04/18	CAMP HUMPHRIES
38962	W	10/04/18	CAMP HUMPHRIES
39034	W	10/04/18	CAMP HUMPHRIES
39056	E	02/25/18	349TH INF
39096	W	02/09/18	317TH INF CAMP LEE
39172	E	02/25/18	349TH INF
39184	E	01/31/18	CO F 313TH AMMO TRAIN CAMP DODGE
39184	E	01/31/18	313TH AMMO TRAIN
39184	E	03/20/18	CO M 350TH INF
39247	E	02/18/18	CO C 313TH AMMO TRAIN
39247	E	03/20/18	CO M 350TH INF
39261	W	02/09/18	317TH INF CAMP LEE
39330	W	07/12/18	CO K 34TH ENGR FT BENJ HARRISON
39348	W	10/04/18	CAMP HUMPHRIES
39374	E	02/19/18	349TH INF
39447	W	02/09/18	317TH INF CAMP LEE
39486	W	10/04/18	CAMP HUMPHRIES
39507		03/13/45	USNAS ST SIMONS ISLAND (RETD)
39563	W	10/04/18	CAMP HUMPHRIES
39569		09/22/18	CO C 27TH ENGR AEF
39621	R	04/12/18	HQ CO 353RD INF CAMP FUNSTON
39784	E	02/25/18	349TH INF
39861	E	02/25/18	349TH INF
39895	W	10/04/18	CAMP HUMPHRIES
39945	W	02/09/18	317TH INF CAMP LEE
39946	W	02/04/18	317TH INF CAMP LEE
39985	W	10/04/18	CAMP HUMPHRIES
40085	R	09/05/18	AMERICAN CAN CO
40275	W	10/04/18	CAMP HUMPHRIES
40307	E	02/19/18	349TH INF
40307	E	02/25/18	349TH INF
40307		09/22/18	CO C 27TH ENGR AEF
40422	E	01/25/18	CO F 313TH AMMO TRAIN
40422	E	03/20/18	CO M 350TH INF
40424	W	10/04/18	CAMP HUMPHRIES
40447		09/22/18	CO C 27TH ENGR AEF
40482	W	07/12/18	CO K 34TH ENGR FT BENJ HARRISON
40687	W	02/09/18	317TH INF CAMP LEE
40700		09/22/18	CO C 27TH ENGR AEF
40739	E	02/25/18	349TH INF
40764	W	10/04/18	CAMP HUMPHRIES
40786	E	02/25/18	349TH INF
40897	E	02/25/18	349TH INF
40921	R	05/02/18	SPECIAL FIRING PIN TEST
40959	W	10/04/18	CAMP HUMPHRIES
40982	E	01/31/18	CO A 313TH AMMO TRAIN
40982	E	03/20/18	CO M 350TH INF
41019	W	11/17/17	WRA CO (PACKED - CASE 5047)
41019	W	12/04/17	CO B 302ND ENGR AEF (DRAWN)
41030	E	02/25/18	349TH INF
41139		09/22/18	CO C 27TH ENGR AEF
41197	E	02/25/18	349TH INF
41260	E	02/19/18	349TH INF
41276	E	02/25/18	349TH INF
41295		09/22/18	CO C 27TH ENGR AEF
41344	W	08/12/18	CO I 22ND ENGR FT BENJ HARRISON
41374	E	02/25/18	349TH INF

SER. NO.	MODEL	MM/DD/YY	USAGE
41454	E	02/19/18	349TH INF
41459	E	02/25/18	349TH INF
41590	R	05/02/18	SPECIAL FIRING PIN TEST
41590	R	05/02/18	CAMP WADSWORTH (TEST)
41594	E	04/17/18	CAMP JACKSON TO MIDVALE
41621	E	01/23/18	CO D 313TH AMMO TRAIN CAMP DODGE
41621	E	03/20/18	CO M 350TH INF
41682	W	07/12/18	CO K 34TH ENGR FT BENJ HARRISON
41712	E	02/25/18	349TH INF
41735	E	02/25/18	349TH INF
42056		09/22/18	CO C 27TH ENGR AEF
42286	W	10/04/18	CAMP HUMPHRIES
42328	E	02/19/18	349TH INF
42425	E	02/25/18	349TH INF
42429	E	02/25/18	349TH INF
42596	R	05/02/18	CAMP WADSWORTH (TEST)
42596	R	05/02/18	SPECIAL FIRING PIN TEST
42611	R	04/11/18	CO A 353RD INF CAMP FUNSTON
42729	W	02/09/18	317TH INF CAMP LEE
42801	E	02/25/18	349TH INF
42814		09/22/18	CO C 27TH ENGR AEF
42956	E	02/25/18	349TH INF
43080	E	02/19/18	CAMP GORDON
43080	E	05/01/18	WATERTOWN ARS (TESTED)
43080		02/11/18	163RD INF
43133	W	10/04/18	CAMP HUMPHRIES
43219	W	08/12/18	CO I 22ND ENGR FT BENJ HARRISON
43253		09/22/18	CO C 27TH ENGR AEF
43259		09/22/18	CO C 27TH ENGR AEF
43287	E	02/25/18	349TH INF
43328		05/15/18	115TH INF (STRIKER TEST)
43418	E	02/25/18	349TH INF
43471	W	08/12/18	CO I 22ND ENGR FT BENJ HARRISON
43483	E	04/19/18	313TH AMMO TRAIN
43483	E	04/13/18	CO D 313RD AMMO TRAIN CAMP DODG
43550		09/22/18	CO C 27TH ENGR AEF
43656	W	10/04/18	CAMP HUMPHRIES
43818	W	10/04/18	CAMP HUMPHRIES
43865	W	08/12/18	CO I 22ND ENGR FT BENJ HARRISON
43907	W	10/04/18	CAMP HUMPHRIES
43909		09/22/18	CO C 27TH ENGR AEF
44035		09/22/18	CO C 27TH ENGR AEF
44141		09/22/18	CO C 27TH ENGR AEF
44164	R	04/11/18	CO H 353RD INF CAMP FUNSTON
44357		09/22/18	CO C 27TH ENGR AEF
44376	E	04/17/18	CAMP DIX TO MIDVALE
44495		09/22/18	CO C 27TH ENGR AEF
44530	E	05/01/18	WATERTOWN ARS (TESTED)
44530		12/17/17	360TH INF
44586	E	02/12/18	CO E 313TH AMMO TRAIN
44586	E	03/20/18	CO M 350TH INF
44599		09/22/18	CO C 27TH ENGR AEF
44623	E	02/19/18	349TH INF
44623	E	02/25/18	349TH INF
44772	W	02/09/18	317TH INF CAMP LEE
44801	E	02/19/18	349TH INF
44903	W	10/04/18	CAMP HUMPHRIES
44960	R	02/11/19	DOMINION ARSENAL QUEBEC
45160	W	10/04/18	CAMP HUMPHRIES
45291		09/22/18	CO C 27TH ENGR AEF
45430	R	04/24/18	CO G 353RD INF CAMP FUNSTON
45433	W	08/12/18	CO I 22ND ENGR FT BENJ HARRISON
45492	W	02/09/18	317TH INF CAMP LEE
45590		09/22/18	CO C 27TH ENGR AEF
45719	W	11/17/17	WRA CO (PACKED - CASE 5033)
45719	W	12/04/17	CO B 302ND ENGR AEF (DRAWN)
45758	W	10/04/18	CAMP HUMPHRIES
45802		09/22/18	CO C 27TH ENGR AEF
45825		09/22/18	CO C 27TH ENGR AEF
45861		09/22/18	CO C 27TH ENGR AEF
45871	W	02/04/18	317TH INF CAMP LEE
45927		09/22/18	CO C 27TH ENGR AEF
45928		07/23/18	CAMP DEVENS (STOCK TEST)
45963		09/22/18	CO C 27TH ENGR AEF
46077		09/22/18	CO C 27TH ENGR AEF
46251	E	04/17/18	CAMP JACKSON TO MIDVALE
46275		09/22/18	CO C 27TH ENGR AEF

SER. NO.	MODEL	MM/DD/YY	USAGE
46307	E	02/25/18	349TH INF
46496	W	10/04/18	CAMP HUMPHRIES
46582	W	02/09/18	317TH INF CAMP LEE
46593		09/22/18	CO C 27TH ENGR AEF
46605	W	10/04/18	CAMP HUMPHRIES
46611		09/22/18	CO C 27TH ENGR AEF
46666		09/22/18	CO C 27TH ENGR AEF
46788	R	05/18/18	CO C 353RD INF CAMP FUNSTON
46825		09/22/18	CO C 27TH ENGR AEF
46986		09/22/18	CO C 27TH ENGR AEF
47024		09/22/18	CO C 27TH ENGR AEF
47081		09/22/18	CO C 27TH ENGR AEF
47202	W	10/04/18	CAMP HUMPHRIES
47368	W	10/04/18	CAMP HUMPHRIES
47385		09/22/18	CO C 27TH ENGR AEF
47431	W	11/17/17	WRACO (PACKED - CASE 5033)
47431	W	12/04/17	CO B 302ND ENGR AEF (DRAWN)
47447	W	06 /99	CMP INVENTORY
47536		09/22/18	CO C 27TH ENGR AEF
47559	E	10/29/18	WATERTOWN ARS (EXAMINED)
47559	E	10/19/18	FAILED IN SERVICE
47559	E	04/23/18	REVIEWED BY ENGR BUREAU
47559		03/04/18	81ST DIV
47559	E	04/23/18	CAMP JACKSON
47574	W	10/04/18	CAMP HUMPHRIES
47702	R	05/18/18	CO C 353RD INF CAMP FUNSTON
47716		09/22/18	CO C 27TH ENGR AEF
47740		09/22/18	CO C 27TH ENGR AEF
47967	R	04/12/18	CO F 353RD INF CAMP FUNSTON
47973		09/22/18	CO C 27TH ENGR AEF
48151		09/22/18	CO C 27TH ENGR AEF
48155		07/23/18	CAMP DEVENS (STOCK TEST)
48178		09/22/18	CO C 27TH ENGR AEF
48258		09/22/18	CO C 27TH ENGR AEF
48264		09/22/18	CO C 27TH ENGR AEF
48286	R	08/12/18	CO I 22ND ENGR FT BENJ HARRISON
48418		09/22/18	CO C 27TH ENGR AEF
48471	W	11/17/17	WRA CO (PACKED - CASE 5033)
48471	W	12/04/17	CO B 302ND ENGR AEF (DRAWN)
48475		09/22/18	CO C 27TH ENGR AEF
48497		09/22/18	CO C 27TH ENGR AEF
48563		09/22/18	CO C 27TH ENGR AEF
48878		09/22/18	CO C 27TH ENGR AEF
48940		09/22/18	CO C 27TH ENGR AEF
48986		09/22/18	CO C 27TH ENGR AEF
49057		09/22/18	CO C 27TH ENGR AEF
49280	W	10/04/18	CAMP HUMPHRIES
49299		09/22/18	CO C 27TH ENGR AEF
49413	R	04/11/18	CO A 353RD INF CAMP FUNSTON
49444	W	07/12/18	CO K 34TH ENGR FT BENJ HARRISON
49556		09/22/18	CO C 27TH ENGR AEF
49621	W	10/04/18	CAMP HUMPHRIES
49661	W	10/04/18	CAMP HUMPHRIES
49721		09/22/18	CO C 27TH ENGR AEF
49728	W	06/30/43	HAWAII (IMPOUNDED ALIEN PROP)
49728	W	05/14/45	HAWAII MIL DEPT (RELEASED)
49780	E	06/26/18	CO E 361ST INF
49780	E	11/21/18	CAMP LEWIS
49798		09/22/18	CO C 27TH ENGR AEF
49835		09/22/18	CO C 27TH ENGR AEF
49839		09/22/18	CO C 27TH ENGR AEF
49858		09/22/18	CO C 27TH ENGR AEF
49928		09/22/18	CO C 27TH ENGR AEF
49992		09/22/18	CO C 27TH ENGR AEF
50018	R	04/25/18	55TH PIONEER INF
50050	E	04/11/18	328TH INF
50085		09/22/18	CO C 27TH ENGR AEF
50121		09/22/18	CO C 27TH ENGR AEF
50135		09/22/18	CO C 27TH ENGR AEF
50250	R	27	URBAN MA (ISSUED)
50296		09/22/18	CO C 27TH ENGR AEF
50426		09/22/18	CO C 27TH ENGR AEF
50430		07/23/18	CAMP DEVENS (STOCK TEST)
50456		09/22/18	CO C 27TH ENGR AEF
50552		09/22/18	CO C 27TH ENGR AEF
50629	W	10/04/18	CAMP HUMPHRIES
50666		09/22/18	CO C 27TH ENGR AEF

SER. NO.	MODEL	MM/DD/YY	USAGE
50796		09/22/18	CO C 27TH ENGR AEF
50880		09/22/18	CO C 27TH ENGR AEF
51058		09/22/18	CO C 27TH ENGR AEF
51064	W	10/04/18	CAMP HUMPHRIES
51086	W	12/04/17	CO B 302ND ENGR AEF (DRAWN)
51086	W	11/17/17	WRA CO (PACKED - CASE 5033)
51253	W	04/19/18	TRIGGER INTERCHANGEABILITY TEST
51258	R	04/11/18	CO A 353RD INF CAMP FUNSTON
51341	W	08/12/18	CO I 22ND ENGR FT BENJ HARRISON
51411		09/22/18	CO C 27TH ENGR AEF
51461		09/22/18	CO C 27TH ENGR AEF
51485		09/22/18	CO C 27TH ENGR AEF
51518	E	05/01/18	WATERTOWN ARS (TESTED)
51518	E	04/23/18	REVIEWED BY ENGR BUREAU
51518		12/17/17	328TH INF
51523		09/22/18	CO C 27TH ENGR AEF
51546	R	04/12/18	HQ CO 353RD INF CAMP FUNSTON
51548	R	04/12/18	CO F 353RD INF CAMP FUNSTON
51579		09/22/18	CO C 27TH ENGR AEF
51582		09/22/18	CO C 27TH ENGR AEF
51611		09/22/18	CO C 27TH ENGR AEF
51646	R	09/22/18	CO C 27TH ENGR AEF
51660	W	10/04/18	CAMP HUMPHRIES
51669	R	04/11/18	CO A 353RD INF CAMP FUNSTON
51968	W	11/17/17	WRACO (PACKED - CASE 5047)
51968	W	12/04/17	CO B 302ND ENGR AEF (DRAWN)
51969		09/22/18	CO C 27TH ENGR AEF
51973	W	11/17/17	WRA CO (PACKED - CASE 5047)
51973	W	12/04/17	CO B 302ND ENGR AEF (DRAWN)
52075	W	10/04/18	CAMP HUMPHRIES
52078		09/22/18	CO C 27TH ENGR AEF
52108		09/22/18	CO C 27TH ENGR AEF
52175		09/22/18	CO C 27TH ENGR AEF
52229	W	07/12/18	CO K 34TH ENGR FT BENJ HARRISON
52266		09/22/18	CO C 27TH ENGR AEF
52287	R	04/11/18	CO A 353RD INF CAMP FUNSTON
52326	R	04/11/18	CO H 353RD INF CAMP FUNSTON
52377	W	10/04/18	CAMP HUMPHRIES
52456	R	04/11/18	CO A 353RD INF CAMP FUNSTON
52466	W	10/04/18	CAMP HUMPHRIES
52475		09/22/18	CO C 27TH ENGR AEF
52521		09/22/18	CO C 27TH ENGR AEF
52592		09/22/18	CO C 27TH ENGR AEF
52661	R	04/12/18	HQ CO 353RD INF CAMP FUNSTON
52671	W	12/04/17	CO B 302ND ENGR AEF (DRAWN)
52671	W	11/17/17	WRA CO (PACKED - CASE 5033)
52691		09/22/18	CO C 27TH ENGR AEF
52697	R	04/12/18	CO F 353RD INF CAMP FUNSTON
52707	W	10/04/18	CAMP HUMPHRIES
52752	W	07/12/18	CO K 34TH ENGR FT BENJ HARRISON
52847	W	10/04/18	CAMP HUMPHRIES
52850		07/23/18	CAMP DEVENS (STOCK TEST)
52858	W	12/04/17	CO B 302ND ENGR AEF (DRAWN)
52858	W	11/17/17	WRA CO (PACKED - CASE 5033)
52896		09/22/18	CO C 27TH ENGR AEF
52922	R	04/12/18	CO F 353RD INF CAMP FUNSTON
52977	W	10/04/18	CAMP HUMPHRIES
53008	W	11/17/17	WRA CO (PACKED - CASE 5047)
53008	W	12/04/17	CO B 302ND ENGR AEF (DRAWN)
53017		09/22/18	CO C 27TH ENGR AEF
53088	E	02/09/18	BBL BURST CAMP GORDON
53088	E	05/01/18	WATERTOWN ARS (TESTED)
53088		09/17/18	FAILED IN SERVICE
53088	E	04/22/18	CAMP GORDON
53088	E	12/17/17	307TH AMMO TRAIN (SPLIT BBL)
53088	E	04/23/18	REVIEWED BY ENGR BUREAU
53099		09/22/18	CO C 27TH ENGR AEF
53176	W	10/04/18	CAMP HUMPHRIES
53186	W	07/12/18	CO K 34TH ENGR FT BENJ HARRISON
53205	E	04/18/18	307TH INF
53205	E	05/01/18	WATERTOWN ARS (TESTED)
53229		09/22/18	CO C 27TH ENGR AEF
53237		09/22/18	CO C 27TH ENGR AEF
53244	W	11/17/17	WRACO (PACKED - CASE 5033)
53244	W	12/04/17	CO B 302ND ENGR AEF (DRAWN)
53245	R	04/12/18	CO F 353RD INF CAMP FUNSTON
53289	W	10/04/18	CAMP HUMPHRIES

SER. NO.	MODEL	MM/DD/YY	USAGE
53295		09/22/18	CO C 27TH ENGR AEF
53303		09/22/18	CO C 27TH ENGR AEF
53317		09/22/18	CO C 27TH ENGR AEF
53324	W	08/12/18	CO I 22ND ENGR FT BENJ HARRISON
53371		09/22/18	CO C 27TH ENGR AEF
53413		09/22/18	CO C 27TH ENGR AEF
53478		09/22/18	CO C 27TH ENGR AEF
53492		09/22/18	CO C 27TH ENGR AEF
53527		09/22/18	CO C 27TH ENGR AEF
53574		09/22/18	CO C 27TH ENGR AEF
53587	W	10/04/18	CAMP HUMPHRIES
53591		09/22/18	CO C 27TH ENGR AEF
53643		09/22/18	CO C 27TH ENGR AEF
53669	R	08/12/18	CO I 22ND ENGR FT BENJ HARRISON
53679	R	05/02/18	CAMP WADSWORTH (TEST)
53679	R	05/02/18	SPECIAL FIRING PIN TEST
53686	W	10/04/18	CAMP HUMPHRIES
53688		09/22/18	CO C 27TH ENGR AEF
53701		09/22/18	CO C 27TH ENGR AEF
53742	R	05/18/18	CO C 353RD INF CAMP FUNSTON
53798		09/22/18	CO C 27TH ENGR AEF
53840	W	10/04/18	CAMP HUMPHRIES
53876	W	10/04/18	CAMP HUMPHRIES
53889		09/22/18	CO C 27TH ENGR AEF
53924		09/22/18	CO C 27TH ENGR AEF
54030	W	08/12/18	CO I 22ND ENGR FT BENJ HARRISON
54062	W	08/12/18	CO I 22ND ENGR FT BENJ HARRISON
54079	W	10/04/18	CAMP HUMPHRIES
54097		09/22/18	CO C 27TH ENGR AEF
54100	W	10/04/18	CAMP HUMPHRIES
54105		09/22/18	CO C 27TH ENGR AEF
54113	W	12/04/17	CO B 302ND ENGR AEF (DRAWN)
54113	W	11/17/17	WRACO (PACKED - CASE 5047)
54136	W	12/04/17	CO B 302ND ENGR AEF (DRAWN)
54136	W	11/17/17	WRACO (PACKED - CASE 5047)
54169		09/22/18	CO C 27TH ENGR AEF
54293		09/22/18	CO C 27TH ENGR AEF
54300		09/22/18	CO C 27TH ENGR AEF
54317		09/22/18	CO C 27TH ENGR AEF
54323	W	10/04/18	CAMP HUMPHRIES
54325		09/22/18	CO C 27TH ENGR AEF
54400		09/22/18	CO C 27TH ENGR AEF
54418	R	08/12/18	CO I 22ND ENGR FT BENJ HARRISON
54424		12/03/17	CO B 117TH ENGR AEF
54427		09/22/18	CO C 27TH ENGR AEF
54469	W	11/17/17	WRACO (PACKED - CASE 5047)
54469	W	12/04/17	CO B 302ND ENGR AEF (DRAWN)
54469	W	08/12/18	CO I 22ND ENGR FT BENJ HARRISON
54479		09/22/18	CO C 27TH ENGR AEF
54509	W	08/12/18	CO I 22ND ENGR FT BENJ HARRISON
54531		09/22/18	CO C 27TH ENGR AEF
54581		12/03/17	CO B 117TH ENGR AEF
54582	W	10/04/18	CAMP HUMPHRIES
54589		09/22/18	CO C 27TH ENGR AEF
54616	W	10/04/18	CAMP HUMPHRIES
54704		09/22/18	CO C 27TH ENGR AEF
54728		09/22/18	CO C 27TH ENGR AEF
54810	W	10/04/18	CAMP HUMPHRIES
54840	W	12/04/17	CO B 302ND ENGR AEF (DRAWN)
54840	W	11/17/17	WRACO (PACKED - CASE 5033)
54869	W	11/17/17	WRACO (PACKED - CASE 5047)
54869	W	12/04/17	CO B 302ND ENGR AEF (DRAWN)
54874	W	08/12/18	CO I 22ND ENGR FT BENJ HARRISON
54899	W	08/12/18	CO I 22ND ENGR FT BENJ HARRISON
54965	W	08/12/18	CO I 22ND ENGR FT BENJ HARRISON
55029		09/22/18	CO C 27TH ENGR AEF
55075		09/22/18	CO C 27TH ENGR AEF
55076		09/22/18	CO C 27TH ENGR AEF
55083	W	10/04/18	CAMP HUMPHRIES
55105		09/22/18	CO C 27TH ENGR AEF
55134		09/22/18	CO C 27TH ENGR AEF
55140	W	08/12/18	CO I 22ND ENGR FT BENJ HARRISON
55149		09/22/18	CO C 27TH ENGR AEF
55176	W	10/04/18	CAMP HUMPHRIES
55194		09/22/18	CO C 27TH ENGR AEF
55228	W	11/17/17	WRACO (PACKED - CASE 5033)
55228	W	12/04/17	CO B 302ND ENGR AEF (DRAWN)

SER. NO.	MODEL	MM/DD/YY	USAGE
55259	R	04/24/18	CO G 353RD INF CAMP FUNSTON
55279	W	10/04/18	CAMP HUMPHRIES
55302	W	10/04/18	CAMP HUMPHRIES
55411	W	08/12/18	CO I 22ND ENGR FT BENJ HARRISON
55507	W	10/04/18	CAMP HUMPHRIES
55548	W	11/17/17	WRACO (PACKED - CASE 5047)
55548	W	12/04/17	CO B 302ND ENGR AEF (DRAWN)
55656	W	08/12/18	CO I 22ND ENGR FT BENJ HARRISON
55721		09/22/18	CO C 27TH ENGR AEF
55721	W	10/04/18	CAMP HUMPHRIES
55732	W	08/12/18	CO I 22ND ENGR FT BENJ HARRISON
55739		09/22/18	CO C 27TH ENGR AEF
55750	W	12/04/17	CO B 302ND ENGR AEF (DRAWN)
55750	W	11/17/17	WRACO (PACKED - CASE 5033)
55823		09/22/18	CO C 27TH ENGR AEF
55831		09/22/18	CO C 27TH ENGR AEF
55833	W	10/04/18	CAMP HUMPHRIES
55849	E	01/03/19	130TH ENGR (FOUND)
55885	E	04/23/18	REVIEWED BY ENGR BUREAU
55885		09/17/18	FAILED IN SERVICE
55885	E	05/01/18	WATERTOWN ARS (TESTED)
56055	W	10/04/18	CAMP HUMPHRIES
56090		09/22/18	CO C 27TH ENGR AEF
56130		09/22/18	CO C 27TH ENGR AEF
56165		09/22/18	CO C 27TH ENGR AEF
56182	E	09/30/18	CAMP HANCOCK
56182	E	05/03/18	CAMP HANCOCK
56233	W	10/04/18	CAMP HUMPHRIES
56276	W	10/04/18	CAMP HUMPHRIES
56297		09/22/18	CO C 27TH ENGR AEF
56297		03/13/45	USNAS ST SIMONS ISLAND (RETD)
56329		12/17/17	336TH INF
56340	W	08/12/18	CO I 22ND ENGR FT BENJ HARRISON
56445	W	12/04/17	CO B 302ND ENGR AEF (DRAWN)
56445	W	11/17/17	WRACO (PACKED - CASE 5047)
56556		09/22/18	CO C 27TH ENGR AEF
56570	R	04/12/18	HQ CO 353RD INF CAMP FUNSTON
56583		09/22/18	CO C 27TH ENGR AEF
56606	R	02/28/18	NAT'L BRASS & COPPER TUBE CO
56616	W	10/04/18	CAMP HUMPHRIES
56735	W	10/04/18	CAMP HUMPHRIES
56799		09/22/18	CO C 27TH ENGR AEF
56834	W	08/12/18	CO I 22ND ENGR FT BENJ HARRISON
56844	W	10/04/18	CAMP HUMPHRIES
56890		09/22/18	CO C 27TH ENGR AEF
57127		09/22/18	CO C 27TH ENGR AEF
57144	W	07/12/18	CO K 34TH ENGR FT BENJ HARRISON
57234		09/22/18	CO C 27TH ENGR AEF
57293		09/22/18	CO C 27TH ENGR AEF
57355	W	10/04/18	CAMP HUMPHRIES
57366	W	10/04/18	CAMP HUMPHRIES
57430	W	10/04/18	CAMP HUMPHRIES
57589		09/22/18	CO C 27TH ENGR AEF
57612	W	10/04/18	CAMP HUMPHRIES
57637		09/22/18	CO C 27TH ENGR AEF
57660	W	10/04/18	CAMP HUMPHRIES
57683	W	10/04/18	CAMP HUMPHRIES
57690		12/03/17	CO B 117TH ENGR AEF
57697	W	10/04/18	CAMP HUMPHRIES
57727		09/22/18	CO C 27TH ENGR AEF
57744		09/22/18	CO C 27TH ENGR AEF
57749	W	10/04/18	CAMP HUMPHRIES
57760	W	10/04/18	CAMP HUMPHRIES
57836	W	10/04/18	CAMP HUMPHRIES
57843		09/22/18	CO C 27TH ENGR AEF
57916		09/22/18	CO C 27TH ENGR AEF
57940	W	10/04/18	CAMP HUMPHRIES
57958		09/22/18	CO C 27TH ENGR AEF
58049	W	10/04/18	CAMP HUMPHRIES
58143	W	10/04/18	CAMP HUMPHRIES
58185	W	08/12/18	CO I 22ND ENGR FT BENJ HARRISON
58252	W	10/04/18	CAMP HUMPHRIES
58378		09/22/18	CO C 27TH ENGR AEF
58393	W	08/12/18	CO I 22ND ENGR FT BENJ HARRISON
58551	W	08/12/18	CO I 22ND ENGR FT BENJ HARRISON
58587	W	10/04/18	CAMP HUMPHRIES
58608	W	10/04/18	CAMP HUMPHRIES

SER. NO.	MODEL	MM/DD/YY	USAGE
58680	W	10/04/18	CAMP HUMPHRIES
58795		09/22/18	CO C 27TH ENGR AEF
58816	E	02/25/18	349TH INF
58845		08/21/18	CO C 121ST INF
58991	W	10/04/18	CAMP HUMPHRIES
59055	E	02/25/18	349TH INF
59073	W	10/04/18	CAMP HUMPHRIES
59079	W	10/04/18	CAMP HUMPHRIES
59235	W	08/12/18	CO I 22ND ENGR FT BENJ HARRISON
59236	W	10/04/18	CAMP HUMPHRIES
59305	R	04/11/18	CO A 353RD INF CAMP FUNSTON
59630	W	08/12/18	CO I 22ND ENGR FT BENJ HARRISON
59632	E	02/25/18	349TH INF
59671		09/22/18	CO C 27TH ENGR AEF
59805		04/25/18	HQ 27TH ENGR CAMP MEADE (INQ)
59956	R	09/09/18	HDQTR TRAIN & MP CAMP WHEELER
60018	W	10/04/18	CAMP HUMPHRIES
60119		04/25/18	HQ 27TH ENGR CAMP MEADE (INQ)
60166	W	08/12/18	CO I 22ND ENGR FT BENJ HARRISON
60224	W	10/04/18	CAMP HUMPHRIES
60227		01/06/18	CO F 345TH INF CAMP PIKE
60231	W	08/12/18	CO I 22ND ENGR FT BENJ HARRISON
60232	W	10/04/18	CAMP HUMPHRIES
60275		04/25/18	HQ 27TH ENGR CAMP MEADE (INQ)
60320	R	27	URBAN MA (ISSUED)
60384	R	04/12/18	HQ CO 353RD INF CAMP FUNSTON
60391	W	10/04/18	CAMP HUMPHRIES
60578		09/22/18	CO C 27TH ENGR AEF
60637		04/25/18	HQ 27TH ENGR CAMP MEADE (INQ)
60695	E	05/01/18	WATERTOWN ARS (TESTED)
60714	W	10/04/18	CAMP HUMPHRIES
60727	W	10/04/18	CAMP HUMPHRIES
60747	W	08/12/18	CO I 22ND ENGR FT BENJ HARRISON
60795	E	04/23/18	REVIEWED BY ENGR BUREAU
60795		09/17/18	FAILED IN SERVICE
60795	E	05/01/18	WATERTOWN ARS (TESTED)
60795	E	02/09/18	BBL BURST CAMP GORDON
60795	E	12/17/17	328TH INF
60795	E	04/22/18	CAMP GORDON
60808	W	10/04/18	CAMP HUMPHRIES
60866	W	10/04/18	CAMP HUMPHRIES
60986	W	10/04/18	CAMP HUMPHRIES
61060	R	04/20/18	CO I 353RD INF CAMP FUNSTON
61244	W	10/04/18	CAMP HUMPHRIES
61255		12/03/17	CO B 117TH ENGR AEF
61367	W	10/04/18	CAMP HUMPHRIES
61481		09/22/18	CO C 27TH ENGR AEF
61504		12/03/17	CO B 117TH ENGR AEF
61550		09/17/18	FAILED IN SERVICE
61550	E	12/17/17	328TH INF
61550	E	02/09/18	BBL BURST CAMP GORDON
61550	E	04/23/18	REVIEWED BY ENGR BUREAU
61550	E	05/01/18	WATERTOWN ARS (TESTED)
61622	W	10/04/18	CAMP HUMPHRIES
61629	W	10/04/18	CAMP HUMPHRIES
61748	W	10/04/18	CAMP HUMPHRIES
61849	W	08/12/18	CO I 22ND ENGR FT BENJ HARRISON
61956		04/25/18	HQ 27TH ENGR CAMP MEADE (INQ)
62113	R	04/11/18	CO A 353RD INF CAMP FUNSTON
62140	W	10/04/18	CAMP HUMPHRIES
62149	R	02/28/18	NAT'L BRASS & COPPER TUBE CO
62176	R	04/12/18	CO F 353RD INF CAMP FUNSTON
62405		12/03/17	CO B 117TH ENGR AEF
62466	R	04/12/18	CO E 353RD INF CAMP FUNSTON
62467	R	04/12/18	CO F 353RD INF CAMP FUNSTON
62519	W	10/04/18	CAMP HUMPHRIES
62551	R	04/11/18	CO H 353RD INF CAMP FUNSTON
62562	R	04/12/18	CO F 353RD INF CAMP FUNSTON
62614	R	04/20/18	CO I 353RD INF CAMP FUNSTON
62623	E	01/24/18	CO C 322ND INF
62632	E	05/01/18	WATERTOWN ARS (TESTED)
62632	E	01/30/18	CAPT BIZZELL TO INSP DIV
62632	E	04/05/18	CAMP JACKSON
62647		12/03/17	CO B 117TH ENGR AEF
62674	W	10/04/18	CAMP HUMPHRIES
62745	E	05/01/18	WATERTOWN ARS (TESTED)
62745		09/17/18	FAILED IN SERVICE
63055	W	10/04/18	CAMP HUMPHRIES
63147	R	08/12/18	CO I 22ND ENGR FT BENJ HARRISON
63228	R	02/28/18	NAT'L BRASS & COPPER TUBE CO
63400	W	10/04/18	CAMP HUMPHRIES
63402	W	10/04/18	CAMP HUMPHRIES
63416	R	04/12/18	CO F 353RD INF CAMP FUNSTON
63589	R	04/11/18	CO H 353RD INF CAMP FUNSTON
63704		02/21/18	CO A 303RD ENGR CAMP DIX (ON HAND)
63715		12/03/17	CO B 117TH ENGR AEF
63804	R	09/21/18	CAMP FREMONT
63804	R	12/05/18	CAMP FREMONT
63911		12/03/17	CO B 117TH ENGR AEF
63929	W	07/01/43	USS SC-1005
64067		03 13/45	USNAS ST SIMONS ISLAND (RETD)
64214	R	04/12/18	CO F 353RD INF CAMP FUNSTON
64353	R	04/11/18	CO A 353RD INF CAMP FUNSTON
64426	W	08/12/18	CO I 22ND ENGR FT BENJ HARRISON
64430	W	10/04/18	CAMP HUMPHRIES
64440	W	10/04/18	CAMP HUMPHRIES
64582	W	10/04/18	CAMP HUMPHRIES
64633		12/03/17	CO B 117TH ENGR AEF
64682	W	10/04/18	CAMP HUMPHRIES
64819	R	04/24/18	CO G 353RD INF CAMP FUNSTON
65180	E	10/19/18	FAILED IN SERVICE
65180	E	10/29/18	WATERTOWN ARS (EXAMINED)
65180	E	11/02/18	BARREL FAILED
65180		03/04/18	81ST DIV
65180	E	04/23/18	CAMP JACKSON
65180	E	04/23/18	REVIEWED BY ENGR BUREAU
65187		12/03/17	CO B 117TH ENGR AEF
65226	W	10/04/18	CAMP HUMPHRIES
65442		12/03/17	CO B 117TH ENGR AEF
65445		12/03/17	CO B 117TH ENGR AEF
65796	R	10/23/18	CAMP GORDON
65843	W	02/01/18	26TH ENGR CAMP DIX
66088	R	06/11/18	337TH INF CAMP CUSTER
66422	R	02/28/18	NAT'L BRASS & COPPER TUBE CO
66718	W	02/01/18	26TH ENGR CAMP DIX
66884	R	02/28/18	NAT'L BRASS & COPPER TUBE CO
67754	W	10/04/18	CAMP HUMPHRIES
68129	W	02/09/18	317TH INF CAMP LEE
68536	W	10/04/18	CAMP HUMPHRIES
68670		12/03/17	CO B 117TH ENGR AEF
68675		12/03/17	CO B 117TH ENGR AEF
68679		12/03/17	CO B 117TH ENGR AEF
68681		01/23/18	CO B 303RD ENGR AEF (ON HAND)
68685		12/03/17	CO B 117TH ENGR AEF
68938	W	08/12/18	CO I 22ND ENGR FT BENJ HARRISON
69117		01/06/18	CO F 345TH INF CAMP PIKE
69237	E	06/16/19	EJ MUNCE
69273	E	06/16/19	EJ MUNCE (TURNED IN)
69460	W	02/09/18	317TH INF CAMP LEE
69533		02/07/19	RIMKUNAS REAR SIGHT
69541		12/03/17	CO B 117TH ENGR AEF
69549	R	02/28/18	NAT'L BRASS & COPPER TUBE CO
69616		12/03/17	CO B 117TH ENGR AEF
69800	E	07/01/18	324TH INF CAMP SEVIER
69806	E	06/19/18	HIGHEST NUMBER IN 321ST INF
69960	W	05/24/19	CO C 521ST ENGR (TURNED IN)
70116		12/03/17	CO B 117TH ENGR AEF
70217	W	02/08/18	26TH ENGR (BROKEN EXTRACTOR)
70416	E	04/17/18	INTERCHANGEABILITY TEST
70577		01/23/18	LYMAN GUN SIGHT CORP
70663	W	10/04/18	CAMP HUMPHRIES
70694		01/06/18	CO F 345TH INF CAMP PIKE
70721	R	02/28/18	NAT'L BRASS & COPPER TUBE CO
70758		09/22/18	CO C 27TH ENGR AEF
71123		12/03/17	CO B 117TH ENGR AEF
71355	E	04/19/18	313TH AMMO TRAIN
71355	E	04/13/18	CO D 313RD AMMO TRAIN CAMP DODG
71390	W	10/04/18	CAMP HUMPHRIES
71515		07/23/18	CAMP DEVENS (STOCK TEST)
71560	W	10/04/18	CAMP HUMPHRIES
71580		12/03/17	CO B 117TH ENGR AEF
71618		12/03/17	CO B 117TH ENGR AEF
71691		12/03/17	CO B 117TH ENGR AEF
71722		05/15/18	115TH INF (STRIKER TEST)

SER. NO.	MODEL	MM/DD/YY	USAGE
71817	W	10/04/18	CAMP HUMPHRIES
71919		02/21/18	CO A 303RD ENGR CAMP DIX (ON HAND)
72041		12/03/17	CO B 117TH ENGR AEF
72146	W	10/04/18	CAMP HUMPHRIES
72231		12/03/17	CO B 117TH ENGR AEF
72345	W	10/04/18	CAMP HUMPHRIES
72366		12/03/17	CO B 117TH ENGR AEF
72379		12/03/17	CO B 117TH ENGR AEF
72770		12/03/17	CO B 117TH ENGR AEF
72810	W	02/20/18	309TH FA (BROKEN XTRACTOR)
73263	W	10/04/18	CAMP HUMPHRIES
73329		12/03/17	CO B 117TH ENGR AEF
73501		12/03/17	CO B 117TH ENGR AEF
73711		02/21/18	CO A 303RD ENGR CAMP DIX (ON HAND)
73803	W	10/04/18	CAMP HUMPHRIES
73824	W	10/04/18	CAMP HUMPHRIES
73866	W	08/12/18	CO I 22ND ENGR FT BENJ HARRISON
73900	W	10/04/18	CAMP HUMPHRIES
74136	W	08/12/18	CO I 22ND ENGR FT BENJ HARRISON
74188	W	10/04/18	CAMP HUMPHRIES
74448		12/03/17	CO B 117TH ENGR AEF
74554	R	05/18/18	CO C 353RD INF CAMP FUNSTON
74744	W	08/21/35	US SW REFORMATRY EL RENO TO RIA
74773	W	10/04/18	CAMP HUMPHRIES
74792	W	08/12/18	CO I 22ND ENGR FT BENJ HARRISON
74820	W	10/04/18	CAMP HUMPHRIES
74987		12/03/17	CO B 117TH ENGR AEF
75074		12/03/17	CO B 117TH ENGR AEF
75261	W	10/04/18	CAMP HUMPHRIES
75297	E	05/26/19	EDDYSTONE TO OCO
75363	R	02/28/18	NAT'L BRASS & COPPER TUBE CO
75533	W	10/04/18	CAMP HUMPHRIES
75704	W	10/04/18	CAMP HUMPHRIES
75710	R	02/28/18	NAT'L BRASS & COPPER TUBE CO
75713		12/03/17	CO B 117TH ENGR AEF
75753	W	10/04/18	CAMP HUMPHRIES
75824		01/23/18	CO B 303RD ENGR AEF (ON HAND)
75873		07/23/18	CAMP DEVENS (STOCK TEST)
75938	W	08/12/18	CO I 22ND ENGR FT BENJ HARRISON
75957	W	10/04/18	CAMP HUMPHRIES
75959	E	06 /99	CMP INVENTORY
75961		02/21/18	CO A 303RD ENGR CAMP DIX (ON HAND)
76192	W	10/04/18	CAMP HUMPHRIES
76270		12/03/17	CO B 117TH ENGR AEF
76384	W	08/12/18	CO I 22ND ENGR FT BENJ HARRISON
76460		12/03/17	CO B 117TH ENGR AEF
76461		12/03/17	CO B 117TH ENGR AEF
76462		12/03/17	CO B 117TH ENGR AEF
76463		12/03/17	CO B 117TH ENGR AEF
76464		12/03/17	CO B 117TH ENGR AEF
76465		12/03/17	CO B 117TH ENGR AEF
76466		12/03/17	CO B 117TH ENGR AEF
76467		12/03/17	CO B 117TH ENGR AEF
76468		12/03/17	CO B 117TH ENGR AEF
76469		12/03/17	CO B 117TH ENGR AEF
76470		12/03/17	CO B 117TH ENGR AEF
76471		12/03/17	CO B 117TH ENGR AEF
76472		12/03/17	CO B 117TH ENGR AEF
76473		12/03/17	CO B 117TH ENGR AEF
76474		12/03/17	CO B 117TH ENGR AEF
76475		12/03/17	CO B 117TH ENGR AEF
76476		12/03/17	CO B 117TH ENGR AEF
76477		12/03/17	CO B 117TH ENGR AEF
76478		12/03/17	CO B 117TH ENGR AEF
76479		12/03/17	CO B 117TH ENGR AEF
76480		12/03/17	CO B 117TH ENGR AEF
76481		12/03/17	CO B 117TH ENGR AEF
76482		12/03/17	CO B 117TH ENGR AEF
76484		12/03/17	CO B 117TH ENGR AEF
76485		12/03/17	CO B 117TH ENGR AEF
76486		12/03/17	CO B 117TH ENGR AEF
76487		12/03/17	CO B 117TH ENGR AEF
76488		12/03/17	CO B 117TH ENGR AEF
76697		02/21/18	CO A 303RD ENGR CAMP DIX (ON HAND)
76857		12/03/17	CO B 117TH ENGR AEF
76881	W	08/12/18	CO I 22ND ENGR FT BENJ HARRISON
76894	W	10/04/18	CAMP HUMPHRIES
76906		12/03/17	CO B 117TH ENGR AEF
76967	W	10/04/18	CAMP HUMPHRIES
77036	E	07/29/18	CAMP MACARTHUR
77036	E	12/05/18	CAMP MACARTHUR
77096	W	10/04/18	CAMP HUMPHRIES
77179		12/03/17	CO B 117TH ENGR AEF
77180		12/03/17	CO B 117TH ENGR AEF
77186		12/03/17	CO B 117TH ENGR AEF
77187		12/03/17	CO B 117TH ENGR AEF
77199		12/03/17	CO B 117TH ENGR AEF
77200		12/03/17	CO B 117TH ENGR AEF
77408	W	08/12/18	CO I 22ND ENGR FT BENJ HARRISON
77513	W	08/12/18	CO I 22ND ENGR FT BENJ HARRISON
77552	E	04/17/18	INTERCHANGEABILITY TEST
77592		12/03/17	CO B 117TH ENGR AEF
77650		12/03/17	CO B 117TH ENGR AEF
77657	W	08/12/18	CO I 22ND ENGR FT BENJ HARRISON
77732		12/03/17	CO B 117TH ENGR AEF
77800	E	07/01/18	324TH INF CAMP SEVIER
77862	W	08/12/18	CO I 22ND ENGR FT BENJ HARRISON
78523	W	10/04/18	CAMP HUMPHRIES
78588	R	08/19/18	CO M 121ST INF
79241	W	08/12/18	CO I 22ND ENGR FT BENJ HARRISON
79308	W	08/12/18	CO I 22ND ENGR FT BENJ HARRISON
79467		05/15/18	115TH INF (STRIKER TEST)
79503	E	04/17/18	INTERCHANGEABILITY TEST
79753		02/21/18	CO A 303RD ENGR CAMP DIX (ON HAND)
79834	W	08/12/18	CO I 22ND ENGR FT BENJ HARRISON
80235	W	08/12/18	CO I 22ND ENGR FT BENJ HARRISON
80515	E	04/17/18	INTERCHANGEABILITY TEST
80648		03/13/45	USNAS ST SIMONS ISLAND (RETD)
80680	W	10/04/18	CAMP HUMPHRIES
80732		12/03/17	CO B 117TH ENGR AEF
80984	W	10/04/18	CAMP HUMPHRIES
80988		05/15/18	115TH INF (STRIKER TEST)
80999	E	12/05/18	CAMP MACARTHUR
80999	E	07/29/18	CAMP MACARTHUR
81054	W	08/12/18	CO I 22ND ENGR FT BENJ HARRISON
81090	W	08/12/18	CO I 22ND ENGR FT BENJ HARRISON
81232		12/03/17	CO B 117TH ENGR AEF
81473		12/03/17	CO B 117TH ENGR AEF
81603	W	10/04/18	CAMP HUMPHRIES
81680	W	08/12/18	CO I 22ND ENGR FT BENJ HARRISON
81691	W	02/08/18	26TH ENGR CAMP DIX (BROKEN EXTRACTOR)
81739	W	08/12/18	CO I 22ND ENGR FT BENJ HARRISON
81919	W	02/08/18	26TH ENGR CAMP DIX (BROKEN EXTRACTOR)
81933	W	08/12/18	CO I 22ND ENGR FT BENJ HARRISON
82060		02/21/18	CO A 303RD ENGR CAMP DIX (ON HAND)
82126	W	05/22/18	CAMP GRANT ILL
82126	W	05/22/18	CAMP DIX
82221	W	08/12/18	CO I 22ND ENGR FT BENJ HARRISON
82222		07/23/18	CAMP DEVENS (STOCK TEST)
82233	W	08/12/18	CO I 22ND ENGR FT BENJ HARRISON
82284	W	08/12/18	CO I 22ND ENGR FT BENJ HARRISON
82333		12/03/17	CO B 117TH ENGR AEF
82383	W	08/12/18	CO I 22ND ENGR FT BENJ HARRISON
82449	E	05/26/19	EDDYSTONE TO OCO
82552		12/03/17	CO B 117TH ENGR AEF
82593		12/03/17	CO B 117TH ENGR AEF
82666		12/03/17	CO B 117TH ENGR AEF
82715	R	06/07/18	330TH FA CAMP CUSTER
82784	W	08/12/18	CO I 22ND ENGR FT BENJ HARRISON
82787	W	10/04/18	CAMP HUMPHRIES
82835		12/03/17	CO B 117TH ENGR AEF
82854		12/03/17	CO B 117TH ENGR AEF
82878	E	08 05/18	CO B 343RD INF CAMP GRANT
83057		12/03/17	CO B 117TH ENGR AEF
83112		12/03/17	CO B 117TH ENGR AEF
83185	W	08/12/18	CO I 22ND ENGR FT BENJ HARRISON
83215	W	10/04/18	CAMP HUMPHRIES
83222	W	08/12/18	CO I 22ND ENGR FT BENJ HARRISON
83282	W	08/12/18	CO I 22ND ENGR FT BENJ HARRISON
83368	W	08/12/18	CO I 22ND ENGR FT BENJ HARRISON
83434	W	08/12/18	CO I 22ND ENGR FT BENJ HARRISON
83438		12/03/17	CO B 117TH ENGR AEF
83489	W	02/08/18	26TH ENGR CAMP DIX (BROKEN EXTRACTOR)
83517		12/03/17	CO B 117TH ENGR AEF

SER. NO.	MODEL	MM/DD/YY	USAGE
83533		12/03/17	CO B 117TH ENGR AEF
83585	W	10/04/18	CAMP HUMPHRIES
83623	W	08/12/18	CO I 22ND ENGR FT BENJ HARRISON
83629		12/03/17	CO B 117TH ENGR AEF
83758	W	02/12/18	309TH FA CAMP DIX
83770	W	02/08/18	26TH ENGR CAMP DIX (BROKEN EXTRACTOR)
83788	R	06/11/18	340TH INF CAMP CUSTER
83870	W	08/12/18	CO I 22ND ENGR FT BENJ HARRISON
83970	W	08/12/18	CO I 22ND ENGR FT BENJ HARRISON
83992	W	08/12/18	CO I 22ND ENGR FT BENJ HARRISON
84029	W	08/12/18	CO I 22ND ENGR FT BENJ HARRISON
84041	W	08/12/18	CO I 22ND ENGR FT BENJ HARRISON
84164	W	10/04/18	CAMP HUMPHRIES
84201		12/03/17	CO B 117TH ENGR AEF
84231	W	08/12/18	CO I 22ND ENGR FT BENJ HARRISON
84265	W	08/12/18	CO I 22ND ENGR FT BENJ HARRISON
84281		12/03/17	CO B 117TH ENGR AEF
84293	W	08/12/18	CO I 22ND ENGR FT BENJ HARRISON
84315	W	08/12/18	CO I 22ND ENGR FT BENJ HARRISON
84346	W	08/12/18	CO I 22ND ENGR FT BENJ HARRISON
84347	W	08/12/18	CO I 22ND ENGR FT BENJ HARRISON
84418		12/03/17	CO B 117TH ENGR AEF
84423	W	08/12/18	CO I 22ND ENGR FT BENJ HARRISON
84481	W	08/12/18	CO I 22ND ENGR FT BENJ HARRISON
84510	W	04/19/18	TRIGGER INTERCHANGEABILITY TEST
84512	W	08/12/18	CO I 22ND ENGR FT BENJ HARRISON
84575	W	08/12/18	CO I 22ND ENGR FT BENJ HARRISON
84580	W	02/12/18	309TH FA CAMP DIX
84634	W	08/12/18	CO I 22ND ENGR FT BENJ HARRISON
84816		01/23/18	CO B 303RD ENGR AEF (ON HAND)
84835	W	10/04/18	CAMP HUMPHRIES
84929	W	10/04/18	CAMP HUMPHRIES
85008	W	10/04/18	CAMP HUMPHRIES
85089	W	08/12/18	CO I 22ND ENGR FT BENJ HARRISON
85161		12/03/17	CO B 117TH ENGR AEF
85191	W	08/12/18	CO I 22ND ENGR FT BENJ HARRISON
85212	W	02/08/18	26TH ENGR CAMP DIX (BROKEN EXTRACTOR)
85257		01/06/18	CO F 345TH INF CAMP PIKE
85412	W	08/12/18	CO I 22ND ENGR FT BENJ HARRISON
85472	W	08/12/18	CO I 22ND ENGR FT BENJ HARRISON
85475	W	08/12/18	CO I 22ND ENGR FT BENJ HARRISON
85643	W	05/22/18	CAMP GRANT ILL
85643	W	05/22/18	CAMP DIX
85650	W	10/04/18	CAMP HUMPHRIES
85664	W	02/08/18	26TH ENGR (BROKEN EXTRACTOR)
85732	W	08/12/18	CO I 22ND ENGR FT BENJ HARRISON
85843	W	02/01/18	26TH ENGR CAMP DIX
85846	W	08/12/18	CO I 22ND ENGR FT BENJ HARRISON
85899	W	02/08/18	26TH ENGR (BROKEN EXTRACTOR)
86017	W	02/08/18	26TH ENGR (BROKEN EXTRACTOR)
86160	W	02/20/18	309TH FA (BROKEN XTRACTOR)
86438	W	08/12/18	CO I 22ND ENGR FT BENJ HARRISON
86462	W	08/12/18	CO I 22ND ENGR FT BENJ HARRISON
86581	W	10/04/18	CAMP HUMPHRIES
86612	W	10/04/18	CAMP HUMPHRIES
86636	W	02/07/18	26TH ENGR CAMP DIX
86636	W	02/08/18	26TH ENGR
86636	W	02/08/18	26TH ENGR (BROKEN EXTRACTOR)
86636	W	02/11/18	26TH ENGR CAMP DIX
86718	W	02/01/18	26TH ENGR CAMP DIX
86891	E	05/17/18	164TH DEPOT BRIG CAMP FUNSTON
87014	W	10/04/18	CAMP HUMPHRIES
87056	E	05/22/18	CAMP FUNSTON (RUSTY)
87066	W	10/04/18	CAMP HUMPHRIES
87078	W	08/12/18	CO I 22ND ENGR FT BENJ HARRISON
87202	W	08/12/18	CO I 22ND ENGR FT BENJ HARRISON
87367	W	10/04/18	CAMP HUMPHRIES
87459	W	02/07/18	26TH ENGR CAMP DIX
87459	W	02/11/18	26TH ENGR CAMP DIX
87459	W	02/08/18	26TH ENGR (BROKEN EXTRACTOR)
87459	W	02/08/18	26TH ENGR
87520	W	10/04/18	CAMP HUMPHRIES
87548	W	10/04/18	CAMP HUMPHRIES
87647	W	10/04/18	CAMP HUMPHRIES
87763	W	02/08/18	26TH ENGR CAMP DIX (BROKEN EXTRACTOR)
87833	W	10/04/18	CAMP HUMPHRIES
87853	W	08/12/18	CO I 22ND ENGR FT BENJ HARRISON
87863		05/15/18	115TH INF (STRIKER TEST)
87911		04/25/18	HQ 27TH ENGR CAMP MEADE (INQ)
88059		05/15/18	115TH INF (STRIKER TEST)
88254		05/15/18	115TH INF (STRIKER TEST)
88343	E	05/14/18	CO L 148TH INF
88369	R	08/12/18	CO I 22ND ENGR FT BENJ HARRISON
88387	R	04/15/18	146TH INF
88387	R	11/21/18	CAMP SHERIDAN
88414	W	02/20/18	309TH FA (BROKEN XTRACTOR)
88422		05/15/18	115TH INF (STRIKER TEST)
88504		05/15/18	115TH INF (STRIKER TEST)
88749	W	10/04/18	CAMP HUMPHRIES
88858		05/15/18	115TH INF (STRIKER TEST)
88910		02/21/18	CO A 303RD ENGR CAMP DIX (ON HAND)
89014	W	08/12/18	CO I 22ND ENGR FT BENJ HARRISON
89388		03/13/45	USNAS ST SIMONS ISLAND (RETD)
89700	E	07/01/18	324TH INF CAMP SEVIER
89708	R	06/25/18	SA FN TEST
89708	R	06/28/18	SA DAILY FUNCTION TEST
90591		05/15/18	115TH INF (STRIKER TEST)
91004		05/15/18	115TH INF (STRIKER TEST)
91358		01/23/18	CO B 303RD ENGR AEF (ON HAND)
91428		05/07/18	113TH AMMO TRAIN
91473	E	10/19/18	FAILED IN SERVICE
91473	E	10/29/18	WATERTOWN ARS (EXAMINED)
91473	E	04/23/18	REVIEWED BY ENGR BUREAU
91473	E	04/23/18	CAMP JACKSON
91473		03/04/18	81ST DIV
91473	E	02/08/18	306TH AMMO TRAIN CAMP JACKSON
91754		01/23/18	CO B 303RD ENGR AEF (ON HAND)
92160		02/21/18	CO A 303RD ENGR CAMP DIX (ON HAND)
92200	E	02/25/18	349TH INF
92218	W	10/04/18	CAMP HUMPHRIES
92271		02/21/18	CO A 303RD ENGR CAMP DIX (ON HAND)
92905		07/23/18	CAMP DEVENS (STOCK TEST)
93136		02/21/18	CO A 303RD ENGR CAMP DIX (ON HAND)
93151	E	01/27/34	GRIFFIN & HOWE (SALE)
93275		06/25/18	CO A 309TH ENGR CAMP SHERMAN
93338		02/21/18	CO A 303RD ENGR CAMP DIX (ON HAND)
93826		01/23/18	CO B 303RD ENGR AEF (ON HAND)
94006		06/25/18	CO A 309TH ENGR CAMP SHERMAN
94123	E	01/17/18	CO H 350TH INF (DEFECTIVE)
94123	E	02/12/18	CO E 313TH AMMO TRAIN
94128		02/21/18	CO A 303RD ENGR CAMP DIX (ON HAND)
94951		01/23/18	CO B 303RD ENGR AEF (ON HAND)
95890		02/21/18	CO A 303RD ENGR CAMP DIX (ON HAND)
95945		02/21/18	CO A 303RD ENGR CAMP DIX (ON HAND)
95967	W	10/04/18	CAMP HUMPHRIES
96227		06/25/18	CO A 309TH ENGR CAMP SHERMAN
96282		02/21/18	CO A 303RD ENGR CAMP DIX (ON HAND)
96386		01/23/18	CO B 303RD ENGR AEF (ON HAND)
96600		01/23/18	CO B 303RD ENGR AEF (ON HAND)
96605		02/21/18	CO A 303RD ENGR CAMP DIX (ON HAND)
96688	R	11/26/19	BUTTE & SUPERIOR MINING CO
96735		07/23/18	CAMP DEVENS (STOCK TEST)
96873		02/21/18	CO A 303RD ENGR CAMP DIX (ON HAND)
96942		02/21/18	CO A 303RD ENGR CAMP DIX (ON HAND)
96944		07/23/18	CAMP DEVENS (STOCK TEST)
97002		01/23/18	CO B 303RD ENGR AEF (ON HAND)
97014	W	05/24/19	CO C 521ST ENGR (TURNED IN)
97063		01/23/18	CO B 303RD ENGR AEF (ON HAND)
97159		05/15/18	115TH INF (STRIKER TEST)
97209	W	06 /99	CMP INVENTORY
97250	W	10/04/18	CAMP HUMPHRIES
97287		02/21/18	CO A 303RD ENGR CAMP DIX (ON HAND)
97413		07/23/18	CAMP DEVENS (STOCK TEST)
97475		01/23/18	CO B 303RD ENGR AEF (ON HAND)
97499	W	10/04/18	CAMP HUMPHRIES
97527		05/15/18	115TH INF (STRIKER TEST)
97579		02/21/18	CO A 303RD ENGR CAMP DIX (ON HAND)
97617		01/23/18	CO B 303RD ENGR AEF (ON HAND)
97663		02/21/18	CO A 303RD ENGR CAMP DIX (ON HAND)
97710	W	10/04/18	CAMP HUMPHRIES
97916		02/21/18	CO A 303RD ENGR CAMP DIX (ON HAND)
97936		02/21/18	CO A 303RD ENGR CAMP DIX (ON HAND)
97970		01/23/18	CO B 303RD ENGR AEF (ON HAND)
98147	W	10/04/18	CAMP HUMPHRIES

SER. NO.	MODEL	MM/DD/YY	USAGE
98153	R	06/11/18	337TH INF CAMP CUSTER
98459		02/21/18	CO A 303RD ENGR CAMP DIX (ON HAND)
98474		01/23/18	CO B 303RD ENGR AEF (ON HAND)
98516		07/23/18	CAMP DEVENS (STOCK TEST)
98537		01/23/18	CO B 303RD ENGR AEF (ON HAND)
98587		06/25/18	CO A 309TH ENGR CAMP SHERMAN
98667		07/23/18	CAMP DEVENS (STOCK TEST)
98742		01/23/18	CO B 303RD ENGR AEF (ON HAND)
98800		02/21/18	CO A 303RD ENGR CAMP DIX (ON HAND)
98825		02/21/18	CO A 303RD ENGR CAMP DIX (ON HAND)
98905		01/23/18	CO B 303RD ENGR AEF (ON HAND)
98986		02/21/18	CO A 303RD ENGR CAMP DIX (ON HAND)
99021		02/21/18	CO A 303RD ENGR CAMP DIX (ON HAND)
99041	W	10/04/18	CAMP HUMPHRIES
99068		01/23/18	CO B 303RD ENGR AEF (ON HAND)
99107	W	10/04/18	CAMP HUMPHRIES
99135	W	10/04/18	CAMP HUMPHRIES
99521		01/23/18	CO B 303RD ENGR AEF (ON HAND)
99591		05/15/18	115TH INF (STRIKER TEST)
99636		02/21/18	CO A 303RD ENGR CAMP DIX (ON HAND)
99654		02/21/18	CO A 303RD ENGR CAMP DIX (ON HAND)
99674		02/21/18	CO A 303RD ENGR CAMP DIX (ON HAND)
99726		01/23/18	CO B 303RD ENGR AEF (ON HAND)
99734		02/21/18	CO A 303RD ENGR CAMP DIX (ON HAND)
99775	W	01/16/18	301 AMMO TRAIN
99835		01/23/18	CO B 303RD ENGR AEF (ON HAND)
99868		01/23/18	CO B 303RD ENGR AEF (ON HAND)
99877		07/23/18	CAMP DEVENS (STOCK TEST)
99901		01/23/18	CO B 303RD ENGR AEF (ON HAND)
99953		02/21/18	CO A 303RD ENGR CAMP DIX (ON HAND)
100041		02/21/18	CO A 303RD ENGR CAMP DIX (ON HAND)
100125	W	01/16/18	301 AMMO TRAIN
100187	E	03/23/18	CO G 366TH INF CAMP DODGE
100187	E	05/01/18	WATERTOWN ARS (TESTED)
100187	E	05/28/18	CAMP DODGE (FAILED)
100187	E	03/14/18	BBL BURST CAMP DODGE
100195		01/23/18	CO B 303RD ENGR AEF (ON HAND)
100280		01/23/18	CO B 303RD ENGR AEF (ON HAND)
100283		02/21/18	CO A 303RD ENGR CAMP DIX (ON HAND)
100306		01/23/18	CO B 303RD ENGR AEF (ON HAND)
100314	W	10/04/18	CAMP HUMPHRIES
100329		02/21/18	CO A 303RD ENGR CAMP DIX (ON HAND)
100350		01/23/18	CO B 303RD ENGR AEF (ON HAND)
100408		01/23/18	CO B 303RD ENGR AEF (ON HAND)
100421		02/21/18	CO A 303RD ENGR CAMP DIX (ON HAND)
100442		01/23/18	CO B 303RD ENGR AEF (ON HAND)
100453		01/23/18	CO B 303RD ENGR AEF (ON HAND)
100456		02/21/18	CO A 303RD ENGR CAMP DIX (ON HAND)
100477		01/23/18	CO B 303RD ENGR AEF (ON HAND)
100490		02/21/18	CO A 303RD ENGR CAMP DIX (ON HAND)
100513		07/23/18	CAMP DEVENS (STOCK TEST)
100567		01/23/18	CO B 303RD ENGR AEF (ON HAND)
100590		01/23/18	CO B 303RD ENGR AEF (ON HAND)
100598		02/21/18	CO A 303RD ENGR CAMP DIX (ON HAND)
100638		02/21/18	CO A 303RD ENGR CAMP DIX (ON HAND)
100660		01/23/18	CO B 303RD ENGR AEF (ON HAND)
100725		02/21/18	CO A 303RD ENGR CAMP DIX (ON HAND)
100768		01/23/18	CO B 303RD ENGR AEF (ON HAND)
100854	E	05/15/18	CAMP HANCOCK STRIKER TEST
100886		02/21/18	CO A 303RD ENGR CAMP DIX (ON HAND)
100890		01/23/18	CO B 303RD ENGR AEF (ON HAND)
100948	E	08/17/18	CO K 133RD INF CAMP CODY
101017		01/23/18	CO B 303RD ENGR AEF (ON HAND)
101027	W	01/16/18	301 AMMO TRAIN
101071		02/21/18	CO A 303RD ENGR CAMP DIX (ON HAND)
101193		07/23/18	CAMP DEVENS (STOCK TEST)
101261		02/21/18	CO A 303RD ENGR CAMP DIX (ON HAND)
101293		02/21/18	CO A 303RD ENGR CAMP DIX (ON HAND)
101296	W	10/04/18	CAMP HUMPHRIES
101305	W	10/04/18	CAMP HUMPHRIES
101370		01/23/18	CO B 303RD ENGR AEF (ON HAND)
101374	E	01/25/18	CO G 345TH INF
101374		01/25/18	CO G 345TH INF CAMP PIKE
101419		07/23/18	CAMP DEVENS (STOCK TEST)
101422		02/21/18	CO A 303RD ENGR CAMP DIX (ON HAND)
101437		01/23/18	CO B 303RD ENGR AEF (ON HAND)
101452		02/21/18	CO A 303RD ENGR CAMP DIX (ON HAND)
101477		02/21/18	CO A 303RD ENGR CAMP DIX (ON HAND)
101502		02/21/18	CO A 303RD ENGR CAMP DIX (ON HAND)
101622		02/21/18	CO A 303RD ENGR CAMP DIX (ON HAND)
101656		07/23/18	CAMP DEVENS (STOCK TEST)
101687		01/23/18	CO B 303RD ENGR AEF (ON HAND)
101697		01/23/18	CO B 303RD ENGR AEF (ON HAND)
101711		01/23/18	CO B 303RD ENGR AEF (ON HAND)
101731		01/23/18	CO B 303RD ENGR AEF (ON HAND)
101734		01/23/18	CO B 303RD ENGR AEF (ON HAND)
101751	W	10/04/18	CAMP HUMPHRIES
101787		02/21/18	CO A 303RD ENGR CAMP DIX (ON HAND)
101806	E	02/25/18	349TH INF
101818		02/21/18	CO A 303RD ENGR CAMP DIX (ON HAND)
101837		02/21/18	CO A 303RD ENGR CAMP DIX (ON HAND)
101844		02/21/18	CO A 303RD ENGR CAMP DIX (ON HAND)
101877		01/23/18	CO B 303RD ENGR AEF (ON HAND)
101890		02/21/18	CO A 303RD ENGR CAMP DIX (ON HAND)
101897		02/21/18	CO A 303RD ENGR CAMP DIX (ON HAND)
101922		01/23/18	CO B 303RD ENGR AEF (ON HAND)
101930	W	10/04/18	CAMP HUMPHRIES
103205	E	01/25/18	CO G 345TH INF
103820	E	02/18/18	CO C 313TH AMMO TRAIN
103820		03/20/18	CO M 350TH INF
104125	E	01/25/18	CO G 345TH INF
105760	W	10/04/18	CAMP HUMPHRIES
106178	E	01/25/18	CO G 345TH INF
106195	W	10/04/18	CAMP HUMPHRIES
106248		05/15/18	115TH INF (STRIKER TEST)
106660	E	04/19/18	313TH AMMO TRAIN
107293		05/15/18	115TH INF (STRIKER TEST)
107758	R	10/23/18	CAMP GORDON
107827	W	10/04/18	CAMP HUMPHRIES
107848		05/15/18	115TH INF (STRIKER TEST)
107985	W	03 /39	PAUL JAEGER (SOLD BY NJ)
108205		08/21/18	HQ CO 124TH INF
108282	E	05/28/18	CAMP DODGE
108368	W	10/04/18	CAMP HUMPHRIES
108938	E	02/08/18	CO B 313TH AMMO TRAIN
108938		03/20/18	CO M 350TH INF
109341	W	10/04/18	CAMP HUMPHRIES
109559	W	10/04/18	CAMP HUMPHRIES
109682	W	04/04/18	308TH AMMO TRAIN
109682	W	05/23/18	ORD DEPT INSP DIV
109853	W	04/04/18	308TH AMMO TRAIN
109853		10/19/18	FAILED IN SERVICE
109853		10/29/18	WATERTOWN ARS (EXAMINED)
109888	W	04/04/18	308TH AMMO TRAIN
109888		10/19/18	FAILED IN SERVICE
109888		10/29/18	WATERTOWN ARS (EXAMINED)
110011	E	01/17/18	CO H 350TH INF (DEFECTIVE)
110011		02/28/18	CO C 313TH AMMO TRAIN
110349	E	04/19/18	313TH AMMO TRAIN
110779	E	02/18/18	CO C 313TH AMMO TRAIN
110779		03/20/18	CO M 350TH INF
111106		05/15/18	115TH INF (STRIKER TEST)
111221		08/21/18	CO E 123RD INF
111335		05/15/18	115TH INF (STRIKER TEST)
113420		05/15/18	115TH INF (STRIKER TEST)
111857	E	27	URBAN MA (ISSUED)
113243	W	04/04/18	308TH AMMO TRAIN
113243		10/19/18	FAILED IN SERVICE
113243		10/29/18	WATERTOWN ARS (EXAMINED)
113347	W	04/04/18	308TH AMMO TRAIN
113347		10/19/18	FAILED IN SERVICE
113347		10/29/18	WATERTOWN ARS (EXAMINED)
113420		05/15/18	115TH INF (STRIKER TEST)
113475	R	05/04/18	113TH INF
113574	E	04/20/18	314TH AMMO TRAIN
113577		05/15/18	115TH INF (STRIKER TEST)
113637	E	04/19/18	313TH AMMO TRAIN
114357		05/15/18	115TH INF (STRIKER TEST)
114375	R	12/02/18	CAMP MCCLELLAN
115049		05/15/18	115TH INF (STRIKER TEST)
115059		05/15/18	115TH INF (STRIKER TEST)
115235		05/15/18	115TH INF (STRIKER TEST)
116034	E	12/02/18	CAMP CODY
116117	R	06/11/18	340TH INF CAMP CUSTER

SER. NO.	MODEL	MM/DD/YY	USAGE
116209	R	04/28/18	116TH INF
116321		05/15/18	115TH INF (STRIKER TEST)
117894		05/15/18	115TH INF (STRIKER TEST)
118142		05/15/18	115TH INF (STRIKER TEST)
118190		09/17/18	FAILED IN SERVICE
118469		05/15/18	115TH INF (STRIKER TEST)
118482	R	10/23/18	CAMP GORDON
118637	R	04/22/18	CAMP GORDON
118637		10/19/18	FAILED IN SERVICE
118637		10/29/18	WATERTOWN ARS (EXAMINED)
119178	E	04/11/18	328TH INF
119691	R	10/18/18	CAMP GORDON
120611		09/04/18	CAMP WHEELER
120629		05/15/18	115TH INF (STRIKER TEST)
120735		09/17/18	FAILED IN SERVICE
121105	R	05/02/18	SPECIAL FIRING PIN TEST
121582		05/15/18	115TH INF (STRIKER TEST)
122127		05/15/18	115TH INF (STRIKER TEST)
122669	R	06/11/18	340TH INF CAMP CUSTER
122971		05/15/18	115TH INF (STRIKER TEST)
123228	R	04/24/18	114TH INF
123338	R	04/15/18	146TH INF
124058		05/15/18	115TH INF (STRIKER TEST)
124120		05/15/18	115TH INF (STRIKER TEST)
124244		05/15/18	115TH INF (STRIKER TEST)
124249		05/15/18	115TH INF (STRIKER TEST)
124399		05/15/18	115TH INF (STRIKER TEST)
124557	R	10/23/18	CAMP GORDON
124828		05/15/18	115TH INF (STRIKER TEST)
125092		05/15/18	115TH INF (STRIKER TEST)
125649	E	03/04/18	CO D 307TH ENGR
125649		09/17/18	FAILED IN SERVICE
125942	E	04/11/18	328TH INF
126011	R	09/14/18	124TH INF CAMP WHEELER
126408		05/15/18	115TH INF (STRIKER TEST)
126498		05/15/18	115TH INF (STRIKER TEST)
127960	E	03/04/18	CAMP GORDON
127985	E	04/18/18	307TH INF
128198		05/15/18	115TH INF (STRIKER TEST)
128290		05/15/18	115TH INF (STRIKER TEST)
128504		05/15/18	115TH INF (STRIKER TEST)
128549		05/15/18	115TH INF (STRIKER TEST)
128687		05/15/18	115TH INF (STRIKER TEST)
128885		05/15/18	115TH INF (STRIKER TEST)
128887		05/15/18	115TH INF (STRIKER TEST)
128905		05/15/18	115TH INF (STRIKER TEST)
129074		05/15/18	115TH INF (STRIKER TEST)
129082		05/15/18	115TH INF (STRIKER TEST)
129257		05/15/18	115TH INF (STRIKER TEST)
129268	R	04/27/18	114TH INF
129704		05/15/18	115TH INF (STRIKER TEST)
129946		05/15/18	115TH INF (STRIKER TEST)
130064	R	06/11/18	340TH INF CAMP CUSTER
130198		05/15/18	115TH INF (STRIKER TEST)
130212		05/15/18	115TH INF (STRIKER TEST)
130233	R	06/07/18	330TH FA CAMP CUSTER
130500		05/15/18	115TH INF (STRIKER TEST)
130506		05/15/18	115TH INF (STRIKER TEST)
130632		05/15/18	115TH INF (STRIKER TEST)
130649		05/15/18	115TH INF (STRIKER TEST)
130738		05/15/18	115TH INF (STRIKER TEST)
130747		05/15/18	115TH INF (STRIKER TEST)
131019	R	06/11/18	340TH INF CAMP CUSTER
131111		05/15/18	115TH INF (STRIKER TEST)
131120		05/15/18	115TH INF (STRIKER TEST)
131140		05/15/18	115TH INF (STRIKER TEST)
131663		05/15/18	115TH INF (STRIKER TEST)
131668		05/15/18	115TH INF (STRIKER TEST)
132071		05/15/18	115TH INF (STRIKER TEST)
132601	E	03/03/19	AEF
132880	R	05/02/18	SPECIAL FIRING PIN TEST
132880		05/23/18	ORD DEPT INSP DIV
132888		05/15/18	115TH INF (STRIKER TEST)
133031	W	04/22/18	324TH ART
133296		05/15/18	115TH INF (STRIKER TEST)
133362		05/15/18	115TH INF (STRIKER TEST)
133375		05/15/18	115TH INF (STRIKER TEST)

SER. NO.	MODEL	MM/DD/YY	USAGE
133383		05/15/18	115TH INF (STRIKER TEST)
133912		07/15/18	91ST DIV
134096		05/15/18	115TH INF (STRIKER TEST)
134171		05/15/18	115TH INF (STRIKER TEST)
134185		05/15/18	115TH INF (STRIKER TEST)
134232	W	07/12/18	CO K 34TH ENGR FT BENJ HARRISON
134277		07/23/18	CAMP DEVENS (STOCK TEST)
134358		05/15/18	115TH INF (STRIKER TEST)
134738		05/15/18	115TH INF (STRIKER TEST)
134923		05/15/18	115TH INF (STRIKER TEST)
135049	R	08/26/18	106TH AMMO TRAIN CAMP WHEELER
135367	R	06/11/18	340TH INF CAMP CUSTER
136443	R	06/11/18	340TH INF CAMP CUSTER
136480	R	06/11/18	340TH INF CAMP CUSTER
136614		07/23/18	CAMP DEVENS (STOCK TEST)
137155		06/25/18	CO A 309TH ENGR CAMP SHERMAN
137324	W	03/15/18	FA ACCURACY TEST
137427	W	05/20/18	CAMP SHERMAN (DEFECTIVE)
137467	W	04/04/18	308TH ENGR
137467		05/23/18	ORD DEPT INSP DIV
137575	R	11/10/22	AUGUSTA ARS (OVERHAULED)
137668	W	11/10/22	AUGUSTA ARS (OVERHAULED)
137977	R	04/24/18	116TH INF
137986		05/15/18	115TH INF (STRIKER TEST)
138126		05/15/18	115TH INF (STRIKER TEST)
138293		05/15/18	115TH INF (STRIKER TEST)
138411		05/15/18	115TH INF (STRIKER TEST)
138583	R	06/11/18	340TH INF CAMP CUSTER
138644	W	05/02/18	SPECIAL FIRING PIN TEST
138644	R	05/02/18	SPECIAL FIRING PIN TEST
138654	R	05/02/18	SPECIAL FIRING PIN TEST
139049		05/15/18	115TH INF (STRIKER TEST)
139142		05/15/18	115TH INF (STRIKER TEST)
139318	R	06/11/18	340TH INF CAMP CUSTER
139379	R	06/11/18	340TH INF CAMP CUSTER
139401	R	06/11/18	340TH INF CAMP CUSTER
139405	R	06/11/18	340TH INF CAMP CUSTER
139467		05/15/18	115TH INF (STRIKER TEST)
139475		05/15/18	115TH INF (STRIKER TEST)
139518	R	06/11/18	337TH INF CAMP CUSTER
139543	W	08/12/18	CO I 22ND ENGR FT BENJ HARRISON
139576	R	06/11/18	337TH INF CAMP CUSTER
139634		05/15/18	115TH INF (STRIKER TEST)
139895	R	05/02/18	CAMP WADSWORTH (TEST)
139895	R	05/02/18	SPECIAL FIRING PIN TEST
140009	R	06/11/18	340TH INF CAMP CUSTER
140436	W	08/12/18	CO I 22ND ENGR FT BENJ HARRISON
140727	R	08/15/18	INTERCHANGE OF MAY & JUNE PROD
140852	R	06/11/18	340TH INF CAMP CUSTER
141152	W	03/15/18	FA ACCURACY TEST
141464	W	08/12/18	CO I 22ND ENGR FT BENJ HARRISON
141479		00/00/18	CO A 33RD ENGR AEF SS DWINSK
141781	R	06/07/18	330TH FA CAMP CUSTER
141947		07/23/18	CAMP DEVENS (STOCK TEST)
142254	R	06/11/18	340TH INF CAMP CUSTER
142594	W	08/12/18	CO I 22ND ENGR FT BENJ HARRISON
142607	R	08/15/18	INTERCHANGE OF MAY & JUNE PROD
142607	R	09/16/18	INTERCHANGEABILITY TEST
142624	R	06/11/18	337TH INF CAMP CUSTER
142764	R	06/11/18	340TH INF CAMP CUSTER
142960		07/23/18	CAMP DEVENS (STOCK TEST)
143362		07/23/18	CAMP DEVENS (STOCK TEST)
143463	R	06/07/18	330TH FA CAMP CUSTER
143483	E	05/01/18	WATERTOWN ARS (TESTED)
143483		12/17/17	327TH INF
143542	E	04/17/18	CAMP GORDON TO MIDVALE
143606	R	08/19/18	CO I 121ST INF
143918	W	08/12/18	CO I 22ND ENGR FT BENJ HARRISON
144660	R	08/17/18	CAMP SEVIER
144868	R	06/11/18	337TH INF CAMP CUSTER
144913	W	10/24/18	17TH INF
145119	R	06/11/18	340TH INF CAMP CUSTER
145159		05/15/18	115TH INF (STRIKER TEST)
145342	W	08/12/18	CO I 22ND ENGR FT BENJ HARRISON
145353	R	06/11/18	337TH INF CAMP CUSTER
145417	W	08/12/18	CO I 22ND ENGR FT BENJ HARRISON
145481	E	05/02/18	CAMP WADSWORTH (TEST)

SER. NO.	MODEL	MM/DD/YY	USAGE
145481		05/02/18	SPECIAL FIRING PIN TEST
145658	W	08/12/18	CO I 22ND ENGR FT BENJ HARRISON
145674	R	06/07/18	330TH FA CAMP CUSTER
145769	R	06/11/18	340TH INF CAMP CUSTER
145998		08/21/18	CO D 121ST INF
146126		04/25/18	HQ 27TH ENGR CAMP MEADE (INQ)
146157		04/25/18	HQ 27TH ENGR CAMP MEADE (INQ)
146161	R	06/11/18	340TH INF CAMP CUSTER
147177		04/25/18	HQ 27TH ENGR CAMP MEADE (INQ)
147797	W	07/12/18	CO K 34TH ENGR FT BENJ HARRISON
148211		04/25/18	HQ 27TH ENGR CAMP MEADE (INQ)
148445		04/25/18	HQ 27TH ENGR CAMP MEADE (INQ)
148513		04/25/18	HQ 27TH ENGR CAMP MEADE (INQ)
148595	R	06/07/18	330TH FA CAMP CUSTER
148650	W	05/22/18	CAMP DIX
148650		05/22/18	CAMP GRANT ILL
148931	R	06/11/18	340TH INF CAMP CUSTER
149021	W	10/04/18	CAMP HUMPHRIES
149183	R	06/11/18	340TH INF CAMP CUSTER
149471	R	06/07/18	330TH FA CAMP CUSTER
149585	R	06/07/18	330TH FA CAMP CUSTER
149621	R	06/11/18	340TH INF CAMP CUSTER
149626	R	06/07/18	330TH FA CAMP CUSTER
149630		04/25/18	HQ 27TH ENGR CAMP MEADE (INQ)
149681		04/25/18	HQ 27TH ENGR CAMP MEADE (INQ)
149732	W	.27	URBAN MA (ISSUED)
149842		04/25/18	HQ 27TH ENGR CAMP MEADE (INQ)
149869	R	06/07/18	330TH FA CAMP CUSTER
150286		04/25/18	HQ 27TH ENGR CAMP MEADE (INQ)
150431	E	05/20/18	165TH DEPOT BRIG CAMP TRAVIS
150473	R	06/11/18	340TH INF CAMP CUSTER
150514	R	06/11/18	340TH INF CAMP CUSTER
150624		04/25/18	HQ 27TH ENGR CAMP MEADE (INQ)
150781	R	06/11/18	340TH INF CAMP CUSTER
151225		03/28/18	TO 40TH ENGR FT MEYER
151228	R	06/07/18	330TH FA CAMP CUSTER
151316	W	07/12/18	CO K 34TH ENGR FT BENJ HARRISON
151513	R	06/11/18	340TH INF CAMP CUSTER
151693		04/25/18	HQ 27TH ENGR CAMP MEADE (INQ)
151897	W	03/01/18	NAT'L BRASS & COPPER TUBE CO
151942	R	05/02/18	SPECIAL FIRING PIN TEST
151942	R	05/02/18	CAMP WADSWORTH (TEST)
152023	R	06/11/18	340TH INF CAMP CUSTER
152208	R	06/07/18	330TH FA CAMP CUSTER
152463	R	10/23/18	CAMP GORDON
152519		04/25/18	HQ 27TH ENGR CAMP MEADE (INQ)
152634	R	06/07/18	330TH FA CAMP CUSTER
153081	R	06/11/18	340TH INF CAMP CUSTER
153158	R	06/11/18	337TH INF CAMP CUSTER
153271	R	06/11/18	340TH INF CAMP CUSTER
153763		08/21/18	CO D 123RD INF
153815	R	06/11/18	340TH INF CAMP CUSTER
154071	W	10/04/18	CAMP HUMPHRIES
154423	R	08/17/18	CAMP SEVIER
154649	R	09/26/18	COO AEF (EXTRACTOR TEST)
154668		00/00/18	CO A 33RD ENGR AEF SS DWINSK
154745	W	03/01/18	NAT'L BRASS & COPPER TUBE CO
155418	R	06/11/18	340TH INF CAMP CUSTER
155440		00/00/18	CO A 33RD ENGR AEF SS DWINSK
155447		08/21/18	CO H 124TH INF
155627	R	06/11/18	339TH INF CAMP CUSTER
156035	R	06/11/18	340TH INF CAMP CUSTER
156081	R	06/07/18	330TH FA CAMP CUSTER
156143		000/0/18	CO A 33RD ENGR AEF SS DWINSK
156744		00/00/18	CO A 33RD ENGR AEF SS DWINSK
156774	R	06/07/18	330TH FA CAMP CUSTER
156912	R	06/11/18	340TH INF CAMP CUSTER
157057	R	06/11/18	340TH INF CAMP CUSTER
157604		00/00/18	CO A 33RD ENGR AEF SS DWINSK
157701	R	06/11/18	340TH INF CAMP CUSTER
157759		00/00/18	CO A 33RD ENGR AEF SS DWINSK
157811	R	06/11/18	340TH INF CAMP CUSTER
158032		07/23/18	CAMP DEVENS (STOCK TEST)
158303	R	06/11/18	340TH INF CAMP CUSTER
158443	R	10/04/18	CAMP HUMPHRIES
158474		07/23/18	CAMP DEVENS (STOCK TEST)
158659		00/00/18	CO A 33RD ENGR AEF SS DWINSK
158966	R	06/07/18	330TH FA CAMP CUSTER
159534		00/00/18	CO A 33RD ENGR AEF SS DWINSK
159659	R	06/11/18	340TH INF CAMP CUSTER
159895	R	06/11/18	340TH INF CAMP CUSTER
160279	R	06/11/18	340TH INF CAMP CUSTER
160291		07/15/18	91ST DIV
160291	E	12/02/18	CAMP LEWIS
160296	R	06/11/18	340TH INF CAMP CUSTER
160338	W	03/01/18	NAT'L BRASS & COPPER TUBE CO
160421	R	06/11/18	340TH INF CAMP CUSTER
160433	R	06/11/18	340TH INF CAMP CUSTER
160443	W	03/01/18	NAT'L BRASS & COPPER TUBE CO
160567		08/21/18	CO B 122ND INF
160772		00/00/18	CO A 33RD ENGR AEF SS DWINSK
160902	W	03/01/18	NAT'L BRASS & COPPER TUBE CO
161288	R	06/08/18	338TH INF CAMP CUSTER
161300	R	06/07/18	330TH FA CAMP CUSTER
161562	W	03/01/18	NAT'L BRASS & COPPER TUBE CO
161579	R	06/11/18	340TH INF CAMP CUSTER
161636	W	03/01/18	NAT'L BRASS & COPPER TUBE CO
161791	W	03/01/18	NAT'L BRASS & COPPER TUBE CO
161845	R	06/11/18	340TH INF CAMP CUSTER
161938	W	03/01/18	NAT'L BRASS & COPPER TUBE CO
162148	R	06/08/18	338TH INF CAMP CUSTER
162158	W	03/39	PAUL JAEGER (SOLD BY NJ)
162240	E	05/16/18	CAMP FUNSTON (RUSTY)
162472	R	06/07/18	330TH FA CAMP CUSTER
162573	E	06/08/18	CO A 1ST PIONEER INF
162590	R	06/11/18	340TH INF CAMP CUSTER
162749	R	06/22/48	VETERANS ADMIN (TO RIA)
162843		08/21/18	CO K 124TH INF
162860	R	06/07/18	330TH FA CAMP CUSTER
162967	R	06/11/18	337TH INF CAMP CUSTER
162981	W	03/01/18	NAT'L BRASS & COPPER TUBE CO
163014	R	08/28/18	CO K 122ND INF CAMP WHEELER GA
163040	R	08/19/18	COE122ND INF
163040	R	08/31/18	COE122ND INF CAMP GREENE NC
163344		08/21/18	SUPPLY CO 122ND INF
163370	R	06/08/18	338TH INF CAMP CUSTER
164226	W	03/07/23	APG
164698	R	06/11/18	40TH INF CAMP CUSTER
164815	R	06/11/18	340TH INF CAMP CUSTER
165126	R	06/11/18	340TH INF CAMP CUSTER
165301	R	10/23/18	CAMP GORDON
165369	R	06/11/18	340TH INF CAMP CUSTER
165448	R	06/11/18	340TH INF CAMP CUSTER
165885	R	06/11/18	337TH INF CAMP CUSTER
165966	R	06/11/18	337TH INF CAMP CUSTER
166032		08/21/18	CO C 121ST INF
166348	R	06/07/18	330TH FA CAMP CUSTER
166503		07/23/18	CAMP DEVENS (STOCK TEST)
166935		07/23/18	CAMP DEVENS (STOCK TEST)
166976	R	06/11/18	340TH INF CAMP CUSTER
166980	R	06/11/18	340TH INF CAMP CUSTER
167058	R	06/11/18	337TH INF CAMP CUSTER
167108	R	06/08/18	338TH INF CAMP CUSTER
167112	W	09/14/18	CO D 4TH PIONEER INF
167173	E	04/17/18	CAMP GORDON TO MIDVALE
167686	E	06/08/18	CO L 1ST PIONEER INF
167808	R	07/19/18	SA DAILY FUNCTION TEST
167962	E	04/11/18	328TH INF
168648	R	06/11/18	340TH INF CAMP CUSTER
168926		07/23/18	CAMP DEVENS (STOCK TEST)
169010	E	04/11/18	328TH INF
169459	R	06/11/18	337TH INF CAMP CUSTER
169498	R	10/04/18	CAMP HUMPHRIES
169522	R	06/11/18	340TH INF CAMP CUSTER
169560		07/23/18	CAMP DEVENS (STOCK TEST)
169615	R	06/11/18	340TH INF CAMP CUSTER
169700	R	06/11/18	340TH INF CAMP CUSTER
169781	R	06/11/18	340TH INF CAMP CUSTER
169788	R	08/15/18	INTERCHANGE OF MAY & JUNE PROD
169788	R	09/16/18	INTERCHANGEABILITY TEST
170059	R	06/11/18	340TH INF CAMP CUSTER
170178	R	06/11/18	340TH INF CAMP CUSTER
170221		07/23/18	CAMP DEVENS (STOCK TEST)
170364	R	09/26/18	COO AEF (EXTRACTOR TEST)

SER. NO.	MODEL	MM/DD/YY	USAGE
170712		07/23/18	CAMP DEVENS (STOCK TEST)
170942	R	08/27/18	CO G 121ST INF CAMP WHEELER GA
171156	E	03/04/18	CO D 328TH INF
171156	E	05/01/18	WATERTOWN ARS (TESTED)
171156		09/17/18	FAILED IN SERVICE
171280		07/23/18	CAMP DEVENS (STOCK TEST)
171400		07/23/18	CAMP DEVENS (STOCK TEST)
171598	R	06/11/18	340TH INF CAMP CUSTER
171650		07/23/18	CAMP DEVENS (STOCK TEST)
171869	R	06/11/18	337TH INF CAMP CUSTER
171917		07/23/18	CAMP DEVENS (STOCK TEST)
171964	R	08/15/18	INTERCHANGE OF MAY & JUNE PROD
172147		07/23/18	CAMP DEVENS (STOCK TEST)
172319		07/23/18	CAMP DEVENS (STOCK TEST)
172421	R	06/11/18	337TH INF CAMP CUSTER
172499		08/21/18	CO L 124TH INF
172616	R	08/12/18	CO I 22ND ENGR FT BENJ HARRISON
172688		07/23/18	CAMP DEVENS (STOCK TEST)
172776	R	06/08/18	338TH INF CAMP CUSTER
173114		08/21/18	CO F 123RD INF
173168		07/23/18	CAMP DEVENS (STOCK TEST)
173258	R	08/26/18	WAREHOUSE CAMP WHEELER GA
173542		07/15/18	91ST DIV
173542	E	11/21/18	CAMP LEWIS
173551	R	06/08/18	338TH INF CAMP CUSTER
173567		07/23/18	CAMP DEVENS (STOCK TEST)
173583	R	06/07/18	330TH FA CAMP CUSTER
173984		07/23/18	CAMP DEVENS (STOCK TEST)
173990	R	06/11/18	337TH INF CAMP CUSTER
174008		07/23/18	CAMP DEVENS (STOCK TEST)
174224	R	06/11/18	340TH INF CAMP CUSTER
174379	R	06/11/18	340TH INF CAMP CUSTER
174503	R	06/11/18	340TH INF CAMP CUSTER
174574	R	06/07/18	330TH FA CAMP CUSTER
174756	R	06/11/18	337TH INF CAMP CUSTER
175158	W	11/05/43	USAT STEPHEN CRANE X-93
175250	R	08/19/18	CO I 121ST INF
175310	R	08/31/18	CO G 122ND INF CAMP GREENE NC
175360	R	06/11/18	340TH INF CAMP CUSTER
175959	W	03/07/23	APG
176010	R	08/19/18	CO L 121ST INF
176090		06/25/18	CO A 309TH ENGR CAMP SHERMAN
176168	R	06/11/18	340TH INF CAMP CUSTER
176474	E	04/17/18	CAMP GORDON TO MIDVALE
176556	R	06/11/18	339TH INF CAMP CUSTER
176967	R	06/11/18	340TH INF CAMP CUSTER
177214	R	06/11/18	339TH INF CAMP CUSTER
177239	R	06/11/18	337TH INF CAMP CUSTER
177300	R	06/11/18	340TH INF CAMP CUSTER
177472	R	08/15/18	INTERCHANGE OF MAY & JUNE PROD
178057	E	04/17/18	CAMP GORDON TO MIDVALE
178095	W	10/06/20	USAT LAKE LOUISE (DAMAGED)
178108	W	03/07/23	APG
178216	R	06/11/18	340TH INF CAMP CUSTER
178233		08/21/18	CO A 121ST INF
178259	R	06/11/18	337TH INF CAMP CUSTER
178516	R	06/11/18	340TH INF CAMP CUSTER
178527		12/17/17	327TH INF
178527	E	04/23/18	REVIEWED BY ENGR BUREAU
178527	E	05/01/18	WATERTOWN ARS (TESTED)
178527		09/17/18	FAILED IN SERVICE
178531	R	06/07/18	330TH FA CAMP CUSTER
178630	E	04/18/18	307TH INF
178738	R	06/11/18	340TH INF CAMP CUSTER
178952	R	06/11/18	340TH INF CAMP CUSTER
179200	R	08/27/18	CO M 122ND INF CAMP WHEELER GA
179267	W	11/10/22	AUGUSTA ARS (OVERHAULED)
179387	R	06/11/18	340TH INF CAMP CUSTER
179392		05/11/31	APG
179393	R	08/19/18	CO M 121ST INF
179463	R	11/10/22	AUGUSTA ARS (OVERHAULED)
179560	R	06/07/18	330TH FA CAMP CUSTER
179599	R	06/07/18	330TH FA CAMP CUSTER
179739	R	06/08/18	338TH INF CAMP CUSTER
180117	R	08/19/18	COE122ND INF
180260	W	03/07/23	APG
180398		05/11/31	APG
180574		06/25/18	CO A 309TH ENGR CAMP SHERMAN
180593	R	06/11/18	340TH INF CAMP CUSTER
180603		08/21/18	COE123RD INF
180765	R	06/11/18	340TH INF CAMP CUSTER
180847	R	06/11/18	340TH INF CAMP CUSTER
180936	R	06/11/18	337TH INF CAMP CUSTER
181377	R	06/11/18	340TH INF CAMP CUSTER
181745		05/11/31	APG
181802	R	06/11/18	340TH INF CAMP CUSTER
182010	R	06/11/18	340TH INF CAMP CUSTER
182125	W	03/07/23	APG
182220	W	.27	URBAN MA (ISSUED)
182292	R	06/11/18	340TH INF CAMP CUSTER
182372	E	04/11/18	325TH INF
182372	E	04/11/18	328TH INF
182391	W	08/15/18	SMALL ARMS FIRING SCHOOL
182391	E	08/15/18	BATTLE SIGHT TEST
182442		05/11/31	APG
182492		06/25/18	CO A 309TH ENGR CAMP SHERMAN
182576	R	06/11/18	339TH INF CAMP CUSTER
182779	R	06/08/18	338TH INF CAMP CUSTER
182803	E	08/15/18	BATTLE SIGHT TEST
182803	W	08/15/18	SMALL ARMS FIRING SCHOOL
183276	R	06/07/18	330TH FA CAMP CUSTER
183286		09/22/18	CO C 27TH ENGR AEF
183464	R	06/11/18	340TH INF CAMP CUSTER
183529	W	08/21/35	US SW REFORMATRY EL RENO TO RIA
183586	W	08/15/50	VA
183675	E	04/17/18	INTERCHANGEABILITY TEST
183989	R	06/11/18	340TH INF CAMP CUSTER
184391	R	06/07/18	330TH FA CAMP CUSTER
184490	E	04/18/18	307TH INF
184523	R	06/11/18	340TH INF CAMP CUSTER
184588	R	06/11/18	340TH INF CAMP CUSTER
184648	R	06/11/18	340TH INF CAMP CUSTER
184791	R	06/11/18	340TH INF CAMP CUSTER
184908	W	08/21/35	US SW REFORMATRY EL RENO TO RIA
184914	R	06/11/18	340TH INF CAMP CUSTER
184952	R	06/11/18	340TH INF CAMP CUSTER
184999	R	06/11/18	340TH INF CAMP CUSTER
185054	E	01/19/18	REM ARMS TO EQUIP DIV
185205	R	06/11/18	340TH INF CAMP CUSTER
185257	R	06/08/18	338TH INF CAMP CUSTER
185318	W	08/21/35	US SW REFORMATRY EL RENO TO RIA
185378	R	06/11/18	337TH INF CAMP CUSTER
185391	R	06/11/18	340TH INF CAMP CUSTER
185649	R	06/11/18	340TH INF CAMP CUSTER
185677	R	06/11/18	339TH INF CAMP CUSTER
185684	W	04/19/18	TRIGGER INTERCHANGEABILITY TEST
185802	R	06/07/18	330TH FA CAMP CUSTER
185809	W	08/20/18	SMALL ARMS FIRING SCHOOL
185816	R	06/11/18	340TH INF CAMP CUSTER
185882	R	06/11/18	340TH INF CAMP CUSTER
185907	R	06/11/18	340TH INF CAMP CUSTER
185917	R	06/11/18	340TH INF CAMP CUSTER
185973	R	06/11/18	340TH INF CAMP CUSTER
185977	W	06/24/36	W MUKRZYCKI
186060	R	05/02/18	SPECIAL FIRING PIN TEST
186064	R	06/11/18	340TH INF CAMP CUSTER
186191	R	06/11/18	340TH INF CAMP CUSTER
186194		08/21/18	CO B 122ND INF
186268	R	06/11/18	340TH INF CAMP CUSTER
186593	R	06/11/18	340TH INF CAMP CUSTER
186711	R	06/11/18	340TH INF CAMP CUSTER
186803	R	06/07/18	330TH FA CAMP CUSTER
186844	W	04/19/18	TRIGGER INTERCHANGEABILITY TEST
187059	R	06/11/18	340TH INF CAMP CUSTER
187121	R	08/15/18	INTERCHANGE OF MAY & JUNE PROD
187246	R	08/15/18	INTERCHANGE OF MAY & JUNE PROD
187536	R	06/11/18	340TH INF CAMP CUSTER
187586	E	06/08/18	CO L 1ST PIONEER INF
187768	R	06/07/18	330TH FA CAMP CUSTER
187918	R	06/07/18	330TH FA CAMP CUSTER
188005	W	.27	URBAN MA (ISSUED)
188050	R	06/11/18	340TH INF CAMP CUSTER
188072	E	04/17/18	CAMP GORDON TO MIDVALE
188119	R	06/11/18	339TH INF CAMP CUSTER

SER. NO.	MODEL	MM/DD/YY	USAGE
188123	R	06/11/18	340TH INF CAMP CUSTER
188151	R	06/11/18	339TH INF CAMP CUSTER
188233	R	06/11/18	340TH INF CAMP CUSTER
188284	W	08/21/35	US SW REFORMATRY EL RENO TO RIA
188507	W	08/21/35	US SW REFORMATRY EL RENO TO RIA
188522	R	06/11/18	340TH INF CAMP CUSTER
188561	R	08/19/18	CO L 122ND INF
188610	R	06/11/18	340TH INF CAMP CUSTER
188621	R	06/11/18	340TH INF CAMP CUSTER
188772	W	08/21/35	US SW REFORMATRY EL RENO TO RIA
188787	R	06/11/18	340TH INF CAMP CUSTER
188806	R	06/11/18	340TH INF CAMP CUSTER
188856	R	06/07/18	330TH FA CAMP CUSTER
189126	W	08/20/18	SMALL ARMS FIRING SCHOOL
189170	R	06/11/18	339TH INF CAMP CUSTER
189272	E	04/17/18	INTERCHANGEABILITY TEST
189289	R	06/11/18	340TH INF CAMP CUSTER
189373	R	06/11/18	340TH INF CAMP CUSTER
189402	R	06/08/18	338TH INF CAMP CUSTER
189633	R	06/07/18	330TH FA CAMP CUSTER
189642	W	08/21/35	US SW REFORMATRY EL RENO TO RIA
189795	R	06/08/18	338TH INF CAMP CUSTER
189938	R	06/11/18	340TH INF CAMP CUSTER
189969	E	04/18/18	307TH INF
190132	R	06/11/18	337TH INF CAMP CUSTER
190172		04/15/19	V CZEGKA
190176	R	06/11/18	337TH INF CAMP CUSTER
190209	R	06 /99	CMP INVENTORY
190257	R	06/11/18	340TH INF CAMP CUSTER
190542	R	06/07/18	330TH FA CAMP CUSTER
190635	R	06/11/18	340TH INF CAMP CUSTER
190727	R	06/11/18	340TH INF CAMP CUSTER
190759	R	06/11/18	340TH INF CAMP CUSTER
191025	R	10/04/18	CAMP HUMPHRIES
191086	W	08/21/35	US SW REFORMATRY EL RENO TO RIA
191140	R	06/11/18	337TH INF CAMP CUSTER
191145	W	08/20/18	SMALL ARMS FIRING SCHOOL
191147	R	06/11/18	340TH INF CAMP CUSTER
191338	R	06/11/18	340TH INF CAMP CUSTER
191665	R	06/11/18	340TH INF CAMP CUSTER
191666	R	06/11/18	340TH INF CAMP CUSTER
191714	E	10/14/18	CAMP GORDON
191799		08/21/18	CO A 121ST INF
191857	R	06/11/18	340TH INF CAMP CUSTER
192260	R	06/11/18	337TH INF CAMP CUSTER
192269	R	06/11/18	340TH INF CAMP CUSTER
192284	R	06/11/18	340TH INF CAMP CUSTER
192335	R	06/11/18	340TH INF CAMP CUSTER
192442		08/21/18	CO D 121ST INF
192443	R	06/11/18	340TH INF CAMP CUSTER
192680	R	06/11/18	340TH INF CAMP CUSTER
192773	R	06/11/18	339TH INF CAMP CUSTER
192838	R	08/27/18	CO C 121ST INF CAMP WHEELER GA
193047	R	06/11/18	340TH INF CAMP CUSTER
193069	R	06/11/18	340TH INF CAMP CUSTER
193215	R	08/19/18	CO M 121ST INF
193228	R	06/11/18	340TH INF CAMP CUSTER
193417	R	08/12/18	CO I 22ND ENGR FT BENJ HARRISON
193529	R	08/26/18	WAREHOUSE CAMP WHEELER GA
193539	R	06/08/18	338TH INF CAMP CUSTER
193573	R	06/11/18	340TH INF CAMP CUSTER
193636	R	06/11/18	340TH INF CAMP CUSTER
193701	R	06/11/18	340TH INF CAMP CUSTER
193904		08/21/18	CO F 123RD INF
194120	R	06/11/18	340TH INF CAMP CUSTER
194344	R	06/11/18	340TH INF CAMP CUSTER
194437	R	11/10/22	AUGUSTA ARS (OVERHAULED)
194592		08/21/18	CO K 124TH INF
194615	R	06/11/18	340TH INF CAMP CUSTER
194664	R	08/26/18	WAREHOUSE CAMP WHEELER GA
194681	R	06/11/18	337TH INF CAMP CUSTER
194932	R	06/11/18	340TH INF CAMP CUSTER
195000	R	06/11/18	340TH INF CAMP CUSTER
195060	R	06/08/18	338TH INF CAMP CUSTER
195143	R	06/08/18	338TH INF CAMP CUSTER
195271	R	08/19/18	CO M 122ND INF
195378	R	08/26/18	WAREHOUSE 106TH AMMO TRAIN
195565		08/21/18	CO I 124TH INF
195603	R	08/19/18	CO H 122ND INF
195603	R	08/31/18	CO H 122ND INF CAMP GREENE NC
195634	R	08/26/18	WAREHOUSE SUP CO 121ST INF
195918	E	04/17/18	INTERCHANGEABILITY TEST
196145		08/21/18	CO C 122ND INF
196261	R	06/11/18	340TH INF CAMP CUSTER
196442	R	08/12/18	CO I 22ND ENGR FT BENJ HARRISON
196653	R	06/08/18	338TH INF CAMP CUSTER
196796	R	06/11/18	340TH INF CAMP CUSTER
196799	E	11/10/22	AUGUSTA ARS (OVERHAULED)
196915	R	06/11/18	337TH INF CAMP CUSTER
197055	R	06/11/18	339TH INF CAMP CUSTER
197122	R	08/26/18	WAREHOUSE CAMP WHEELER GA
197242	R	06/11/18	340TH INF CAMP CUSTER
197270		08/21/18	CO I 124TH INF
197279	R	06/11/18	340TH INF CAMP CUSTER
197397	R	11/10/22	AUGUSTA ARS (OVERHAULED)
197405	R	08/27/18	CO D 122ND INF CAMP WHEELER GA
197412	R	07/03/18	SA FN TEST
197534	R	08/27/18	CO C 122ND INF CAMP WHEELER GA
197778		08/21/18	CO A 122ND INF
197883	R	08/19/18	CO M 122ND INF
197996		08/21/18	CO C 121ST INF
198238		08/21/18	HQ CO 124TH INF
198247	R	06/11/18	340TH INF CAMP CUSTER
198293		03 13/45	USNAS ST SIMONS ISLAND (RETD)
198507	R	06/11/18	337TH INF CAMP CUSTER
198747	R	08/19/18	CO L 122ND INF
198801		08/21/18	CO C 122ND INF
198848	R	06/11/18	340TH INF CAMP CUSTER
198930	R	06/08/18	338TH INF CAMP CUSTER
198949	R	08/19/18	CO L 121ST INF
199260	R	06/11/18	339TH INF CAMP CUSTER
199734	R	08/15/18	INTERCHANGE OF MAY & JUNE PROD
199937	R	08/19/18	CO M 122ND INF
200309	R	06/07/18	330TH FA CAMP CUSTER
200317	R	06/11/18	337TH INF CAMP CUSTER
200336	R	08/27/18	CO C 122ND INF CAMP WHEELER GA
200433	E	05/22/18	CAMP FUNSTON (RUSTY)
200487	R	08/26/18	WAREHOUSE CAMP WHEELER GA
200598	R	06/11/18	340TH INF CAMP CUSTER
200656	E	08/15/18	BATTLE SIGHT TEST
200656	W	08/15/18	SMALL ARMS FIRING SCHOOL
200672	R	06/11/18	340TH INF CAMP CUSTER
201321	R	06/11/18	337TH INF CAMP CUSTER
201406	E	06/21/18	114TH ENGR CAMP BEAUREGARD
201420	R	10/18/18	CAMP GORDON
201425	R	08/26/18	WAREHOUSE CAMP WHEELER GA
201843	R	08/26/18	WAREHOUSE CAMP WHEELER GA
201915	R	06/11/18	340TH INF CAMP CUSTER
201935	R	12/02/18	CAMP WHEELER
201936	R	08 05/18	106TH ENGR AEF (DEF)
201977	R	08/19/18	CO L 122ND INF
202148		08/21/18	CO A 124TH INF
202156	R	08/19/18	CO I 122ND INF
202339		08/21/18	CO B 124TH INF
202681	E	05/20/18	165TH DEPOT BRIG CAMP TRAVIS
202754	R	09/14/18	124TH INF CAMP WHEELER
202762		12/01/25	LITTLE ROCK AIR INT DEP TO RIA
202919		08/21/18	CO A 123RD INF
202959	R	08/19/18	CO H 122ND INF
202959	R	08/31/18	CO H 122ND INF CAMP GREENE NC
202992	E	04/18/18	307TH INF
202992	E	05/01/18	WATERTOWN ARS (TESTED)
203021		08/21/18	COE124TH INF
203080	R	08/26/18	WAREHOUSE CAMP WHEELER GA
203108	R	06/11/18	340TH INF CAMP CUSTER
203116	R	08/30/18	WAREHOUSE CAMP WHEELER GA
203150	R	06/11/18	337TH INF CAMP CUSTER
203766		08/21/18	CO F 124TH INF
203795	R	08/14/18	122ND INF CAMP WHEELER
203795		08/21/18	SUPPLY CO 122ND INF
203969	R	10/04/18	CAMP HUMPHRIES
204150	R	08/19/18	CO L 122ND INF
204208		08/21/18	CO G 123RD INF
204367	R	06/11/18	337TH INF CAMP CUSTER

SER. NO.	MODEL	MM/DD/YY	USAGE
204403	R	08/26/18	WAREHOUSE 106TH AMMO TRAIN
204465		08/21/18	CO B 124TH INF
204515	R	09/26/18	COO AEF (EXTRACTOR TEST)
204553	W	03/25/18	INTERCHANGEABILITY TEST
204681	R	08/30/18	WAREHOUSE CAMP WHEELER GA
204710	R	06/11/18	337TH INF CAMP CUSTER
204875	R	08/19/18	CO G 122ND INF
204875	R	08/31/18	CO G 122ND INF CAMP GREENE NC
204900		08/21/18	HQ CO 124TH INF
205098	R	06/11/18	339TH INF CAMP CUSTER
205118		08/21/18	HQ CO 124TH INF
205189		04/09/18	156TH INF
205235	R	06/11/18	337TH INF CAMP CUSTER
205313	R	08/26/18	WAREHOUSE CAMP WHEELER GA
205611	R	04/26/19	VICTOR CZEGKA
205633		08/21/18	CO A 123RD INF
205639	R	06/11/18	337TH INF CAMP CUSTER
205807	R	06/11/18	339TH INF CAMP CUSTER
205849		08/21/18	CO B 121ST INF
205976		08/21/18	CO A 124TH INF
206025		08/21/18	CO A 121ST INF
206095	R	06/11/18	339TH INF CAMP CUSTER
206107	R	08/27/18	COE121ST INF CAMP WHEELER GA
206111	R	08/27/18	CO I 122ND INF CAMP WHEELER GA
206162		08/21/18	CO C 123RD INF
206178		08/21/18	CO C 123RD INF
206274	R	06/11/18	339TH INF CAMP CUSTER
206404	R	08/27/18	CO I 122ND INF CAMP WHEELER GA
206530	E	04/10/18	CO E 309TH INF
206731	R	08/19/18	CO I 122ND INF
206756	R	08/26/18	WAREHOUSE SUP CO 121ST INF
206797	E	04/13/18	CO D 313RD AMMO TRAIN CAMP DODG
206797	E	04/19/18	313TH AMMO TRAIN
206818	R	06/11/18	340TH INF CAMP CUSTER
207185	R	06/11/18	337TH INF CAMP CUSTER
207261		08/21/18	CO F 124TH INF
207273		08/21/18	CO D 122ND INF
207273	R	08/27/18	CO D 122ND INF CAMP WHEELER GA
207347	R	08/27/18	CO I 122ND INF CAMP WHEELER GA
207375	W	03/25/18	INTERCHANGEABILITY TEST
207390		08/21/18	CO G 123RD INF
207633	E	08/15/18	BATTLE SIGHT TEST
207633	W	08/15/18	SMALL ARMS FIRING SCHOOL
207673	W	05/02/18	CAMP WADSWORTH (TEST)
207772	R	08/19/18	CO L 121ST INF
207915		08/21/18	CO B 122ND INF
207950		08/21/18	CO L 124TH INF
207987		12/17/17	307TH ENGR (LOOSE BBL)
208186	E	04/17/18	INTERCHANGEABILITY TEST
208212		08/21/18	CO B 122ND INF
208489		08/21/18	CO A 123RD INF
208489	R	09/09/18	CO A 123RD INF
208555	R	06/11/18	337TH INF CAMP CUSTER
208641	R	06/11/18	339TH INF CAMP CUSTER
208789	R	06/11/18	340TH INF CAMP CUSTER
208896	E	05/10/18	CAMP HANCOCK (TEST)
208916	R	06/11/18	340TH INF CAMP CUSTER
208972	R	10/04/18	CAMP HUMPHRIES
209139	R	08/27/18	CO G 121ST INF CAMP WHEELER GA
209292	R	09/09/18	106TH ENGR
209367		08/21/18	CO D 123RD INF
209370		08/21/18	CO D 122ND INF
209394	R	06/11/18	337TH INF CAMP CUSTER
209444	R	08/30/18	WAREHOUSE CAMP WHEELER GA
209564	W	08/15/18	SMALL ARMS FIRING SCHOOL
209564	E	08/15/18	BATTLE SIGHT TEST
209643	R	08/19/18	CO K 121ST INF
209687	R	08/27/18	CO E 121ST INF CAMP WHEELER GA
209693	W	03/25/18	INTERCHANGEABILITY TEST
209702	R	08/19/18	CO M 121ST INF
209726	R	08/19/18	CO M 122ND INF
209869	R	06/07/18	330TH FA CAMP CUSTER
209984	R	10/18/18	CAMP GORDON
210215	R	06/11/18	337TH INF CAMP CUSTER
210219		12/17/17	326TH INF
210219	E	04/23/18	REVIEWED BY ENGR BUREAU
210219	E	05/01/18	WATERTOWN ARS (TESTED)
210240	R	08/19/18	CO L 122ND INF
210329	R	08/19/18	CO M 121ST INF
210469	R	08/26/18	WAREHOUSE CAMP WHEELER GA
210615	R	08/26/18	WAREHOUSE 106TH AMMO TRAIN
210616	R	08/26/18	WAREHOUSE 106TH AMMO TRAIN
210638	R	07/15/18	SA FN TEST
210680	R	06/11/18	337TH INF CAMP CUSTER
210749	R	06/11/18	339TH INF CAMP CUSTER
210753	R	08/31/18	CO G 122ND INF CAMP GREENE NC
210787		08/21/18	CO D 123RD INF
210888	R	06/07/18	330TH FA CAMP CUSTER
211062	E	04/17/18	INTERCHANGEABILITY TEST
211081	R	08/31/18	CO G 122ND INF CAMP GREENE NC
211171	R	08/30/18	WAREHOUSE CAMP WHEELER GA
211257		08/21/18	CO E 123RD INF
211268		08/21/18	CO F 123RD INF
211362	W	08/20/18	SMALL ARMS FIRING SCHOOL
211384	R	08/27/18	CO L 122ND INF CAMP WHEELER GA
211409	W	03/25/18	INTERCHANGEABILITY TEST
211525	R	08/27/18	CO L 122ND INF CAMP WHEELER GA
211561	W	03/25/18	INTERCHANGEABILITY TEST
211650	R	08/19/18	CO I 121ST INF
211787	R	06/11/18	337TH INF CAMP CUSTER
211890	R	08/19/18	CO I 122ND INF
211917	R	08/19/18	CO M 121ST INF
211947		08/21/18	HQ CO 124TH INF
212001		08/21/18	CO A 122ND INF
212008	R	08/26/18	WAREHOUSE CAMP WHEELER GA
212063	R	08/19/18	CO F 122ND INF
212102	R	06/11/18	337TH INF CAMP CUSTER
212167	R	08/19/18	CO L 122ND INF
212330	R	08/19/18	CO E 122ND INF
212335	E	04/17/18	INTERCHANGEABILITY TEST
212397	R	08/31/18	CO H 122ND INF CAMP GREENE NC
212397	R	08/19/18	CO H 122ND INF
212402	R	08/31/18	CO G 122ND INF CAMP GREENE NC
212403	R	08/26/18	WAREHOUSE SUP CO 121ST INF
212412	W	05/02/18	SPECIAL FIRING PIN TEST
212412	W	05/02/18	CAMP WADSWORTH (TEST)
212420	W	08/20/18	SMALL ARMS FIRING SCHOOL
212441	R	06/11/18	340TH INF CAMP CUSTER
212456		08/21/18	HQ CO 124TH INF
212459	R	08/19/18	CO E 122ND INF
212489	R	08/26/18	WAREHOUSE 121ST INF CAMP WHEELER
212589	R	08/15/50	VA
212593	R	08/19/18	CO K 122ND INF
212615		08/21/18	CO D 122ND INF
212666	R	08/26/18	WAREHOUSE CAMP WHEELER GA
212729	R	08/19/18	CO M 121ST INF
212777	R	06/11/18	337TH INF CAMP CUSTER
212783		08/21/18	CO D 122ND INF
212865	R	06/11/18	337TH INF CAMP CUSTER
212892	E	04/17/18	INTERCHANGEABILITY TEST
212896		08/21/18	CO L 124TH INF
213003	R	08/26/18	WAREHOUSE SUP CO 121ST INF
213004		08/21/18	CO H 123RD INF
213051	R	08/19/18	CO F 122ND INF
213067	R	08/26/18	WAREHOUSE CAMP WHEELER GA
213072	R	08/30/18	WAREHOUSE CAMP WHEELER GA
213148		08/21/18	CO C 124TH INF
213150		08/21/18	CO G 124TH INF
213159		08/21/18	CO A 123RD INF
213205		08/21/18	CO D 122ND INF
213294	R	08/19/18	CO M 121ST INF
213298	R	08/26/18	WAREHOUSE SUP CO 121ST INF
213358	R	08/26/18	WAREHOUSE CAMP WHEELER GA
213364	R	08/31/18	CO F 122ND INF CAMP GREENE NC
213472	R	08/31/18	CO G 122ND INF CAMP GREENE NC
213537	R	08/26/18	WAREHOUSE SUP CO 121ST INF
213577		08/21/18	CO A 124TH INF
213586		08/21/18	CO D 121ST INF
213609	R	08/19/18	CO M 121ST INF
213639		08/21/18	CO H 124TH INF
213699		08/21/18	CO D 123RD INF
213839	R	08/19/18	CO H 122ND INF
213914		08/21/18	CO H 124TH INF
213983	E	04/17/18	INTERCHANGEABILITY TEST

SER. NO.	MODEL	MM/DD/YY	USAGE
214002	R	08/19/18	CO L 122ND INF
214039	E	10/04/18	CAMP HUMPHRIES
214045	R	08/19/18	CO F 122ND INF
214048	R	08/31/18	CO F 122ND INF CAMP GREENE NC
214088		08/21/18	CO F 123RD INF
214101		08/21/18	CO A 122ND INF
214116	W	08/15/18	SMALL ARMS FIRING SCHOOL
214116	E	08/15/18	BATTLE SIGHT TEST
214119		08/21/18	CO C 124TH INF
214183	R	06/24/18	SA FN TEST
214183	R	06/28/18	SA DAILY FUNCTION TEST
214346	W	0424/19	BUREAU OF MINES
214429	R	08/19/18	CO L 122ND INF
214484	R	08/19/18	CO E 122ND INF
214592	R	08/19/18	CO L 121ST INF
214657	R	08/19/18	CO I 122ND INF
214678		08/21/18	CO B 124TH INF
214745	R	06/11/18	340TH INF CAMP CUSTER
214772	R	08/19/18	CO L 121ST INF
214947	R	08/19/18	CO E 122ND INF
214947	R	08/31/18	CO E 122ND INF CAMP GREENE NC
215077	E	08/15/18	BATTLE SIGHT TEST
215077	W	08/15/18	SMALL ARMS FIRING SCHOOL
215095	R	08/19/18	CO L 121ST INF
215121		08/21/18	CO C 121ST INF
215150	R	08/31/18	CO E 122ND INF CAMP GREENE NC
215166	R	08/27/18	CO F 121ST INF CAMP WHEELER GA
215192	E	08/15/18	BATTLE SIGHT TEST
215192	W	08/15/18	SMALL ARMS FIRING SCHOOL
215259		08/21/18	CO A 124TH INF
215346		08/21/18	CO M 124TH INF
215355	R	06/11/18	339TH INF CAMP CUSTER
215418	R	09/26/18	COO AEF (EXTRACTOR TEST)
215575		08/21/18	CO A 123RD INF
215625		08/21/18	CO B 122ND INF
215654		08/21/18	CO E 124TH INF
215788		08/21/18	CO D 121ST INF
215812	R	08/26/18	WAREHOUSE SUP CO 121ST INF
215877	R	06/11/18	340TH INF CAMP CUSTER
215882		08/21/18	CO M 124TH INF
215882	R	09/09/18	CO M 124TH INF
215887	R	06/11/18	337TH INF CAMP CUSTER
215904	R	08/30/18	WAREHOUSE CAMP WHEELER GA
215959	R	08/19/18	CO K 121ST INF
215965	R	08/19/18	CO K 121ST INF
216072		08/21/18	CO F 123RD INF
216089	R	08/19/18	CO K 121ST INF
216094	R	08/19/18	CO L 121ST INF
216181	R	08/30/18	WAREHOUSE CAMP WHEELER GA
216276		08/21/18	CO G 123RD INF
216297	R	08/27/18	CO B 122ND INF CAMP WHEELER GA
216363	R	08/26/18	WAREHOUSE CAMP WHEELER GA
216404	R	08/30/18	WAREHOUSE CAMP WHEELER GA
216427	R	08/26/18	WAREHOUSE SUP CO 121ST INF
216433		08/21/18	CO F 124TH INF
216460	E	08/15/18	BATTLE SIGHT TEST
216460	W	08/15/18	SMALL ARMS FIRING SCHOOL
216498	R	08/26/18	WAREHOUSE CAMP WHEELER GA
216508	R	08/26/18	WAREHOUSE 121ST INF CAMP WHEELER
216564	R	08/27/18	CO C 122ND INF CAMP WHEELER GA
216581	R	08/19/18	CO E 122ND INF
216581	R	08/31/18	CO E 122ND INF CAMP GREENE NC
216583	R	08/19/18	CO M 122ND INF
216603	R	08/30/18	WAREHOUSE CAMP WHEELER GA
216698		08/21/18	CO E 124TH INF
216699		08/21/18	CO F 124TH INF
216702		08/21/18	CO H 124TH INF
216712	R	08/15/18	INTERCHANGE OF MAY & JUNE PROD
216712	R	09/16/18	INTERCHANGEABILITY TEST
216737	R	08/19/18	CO E 122ND INF
216737	R	08/31/18	CO E 122ND INF CAMP GREENE NC
216740	R	08/19/18	CO H 122ND INF
216740	R	08/31/18	CO H 122ND INF CAMP GREENE NC
216745		08/21/18	CO D 122ND INF
216752	R	08/26/18	WAREHOUSE 121ST INF CAMP WHEELER
216773	R	08/26/18	WAREHOUSE 106TH AMMO TRAIN
216824		08/21/18	HQ CO 124TH INF
216825	E	05/17/18	164TH DEPOT BRIG CAMP FUNSTON
216892		08/21/18	CO G 124TH INF
217029		08/21/18	CO C 123RD INF
217033		08/21/18	CO G 123RD INF
217046	R	08/26/18	WAREHOUSE CAMP WHEELER GA
217075		08/21/18	CO F 123RD INF
217094	R	08/27/18	CO C 121ST INF CAMP WHEELER GA
217106		08/21/18	HQ CO 124TH INF
217139	R	08/19/18	CO K 121ST INF
217180	R	08/19/18	CO K 122ND INF
217207	R	08/19/18	CO F 122ND INF
217284		08/21/18	CO A 121ST INF
217305	E	08/15/18	BATTLE SIGHT TEST
217305	W	08/15/18	SMALL ARMS FIRING SCHOOL
217370		08/21/18	CO B 122ND INF
217380		08/21/18	CO D 121ST INF
217438	R	08/26/18	WAREHOUSE SUP CO 121ST INF
217454	R	08/30/18	WAREHOUSE CAMP WHEELER GA
217498	R	08/27/18	CO I 122ND INF CAMP WHEELER GA
217507	R	08/27/18	CO D 121ST INF CAMP WHEELER GA
217519		08/21/18	CO C 124TH INF
217553	R	08/30/18	WAREHOUSE CAMP WHEELER GA
217624	R	08/26/18	WAREHOUSE CAMP WHEELER GA
217668		08/21/18	CO I 124TH INF
217671	E	04/17/18	INTERCHANGEABILITY TEST
217730	R	08/26/18	WAREHOUSE SUP CO 121ST INF
217743	R	06/28/18	SA FN TEST
217788	R	09/09/18	106TH ENGR
217830		08/21/18	CO G 124TH INF
217892		08/21/18	CO G 123RD INF
217894	R	08/15/18	INTERCHANGE OF MAY & JUNE PROD
217916	R	08/19/18	CO K 121ST INF
218043		08/21/18	CO C 124TH INF
218059		08/21/18	CO B 124TH INF
218083		08/21/18	CO M 124TH INF
218084	R	08/19/18	CO L 122ND INF
218112	R	08/26/18	CO F 106TH AMMO TRAIN
218142	R	08/19/18	CO H 122ND INF
218186	R	08/27/18	CO C 122ND INF CAMP WHEELER GA
218214	R	08/19/18	CO K 121ST INF
218215	R	08/26/18	WAREHOUSE CAMP WHEELER GA
218231	R	08/26/18	WAREHOUSE 106TH AMMO TRAIN
218284	R	08/19/18	CO L 122ND INF
218286	R	08/19/18	CO E 122ND INF
218291		08/21/18	CO B 123RD INF
218302	R	08/27/18	CO D 122ND INF CAMP WHEELER GA
218323	R	08/19/18	CO L 121ST INF
218329	R	08/19/18	CO F 122ND INF
218329	R	08/31/18	CO F 122ND INF CAMP GREENE NC
218379		08/21/18	CO C 123RD INF
218391		08/21/18	CO C 122ND INF
218418	R	08/26/18	WAREHOUSE CAMP WHEELER GA
218448		08/21/18	CO E 123RD INF
218452	R	07/08/18	SA FN TEST
218456	R	08/27/18	CO F 121ST INF CAMP WHEELER GA
218462		08/21/18	CO B 122ND INF
218509	R	08/19/18	CO M 121ST INF
218547	R	08/26/18	WAREHOUSE 106TH AMMO TRAIN
218553	R	08/19/18	CO I 121ST INF
218677	R	08/19/18	CO K 121ST INF
218750	R	08/27/18	CO M 122ND INF CAMP WHEELER GA
218840	R	08/26/18	WAREHOUSE CAMP WHEELER GA
218841	R	08/30/18	WAREHOUSE CAMP WHEELER GA
218866		08/21/18	CO C 123RD INF
218876		08/21/18	CO D 124TH INF
218896	R	08/31/18	CO F 122ND INF CAMP GREENE NC
219006		08/21/18	CO E 124TH INF
219037		08/21/18	CO B 122ND INF
219042		08/21/18	CO G 123RD INF
219236	R	08/26/18	WAREHOUSE SUP CO 121ST INF
219251		08/21/18	CO E 123RD INF
219303	R	08/31/18	CO G 122ND INF CAMP GREENE NC
219304		08/21/18	SUPPLY CO 122ND INF
219331		08/21/18	CO D 122ND INF
219339	R	05/17/18	164TH DEPOT BRIG CAMP FUNSTON
219364	R	08/19/18	CO K 121ST INF
219461	R	08/19/18	CO K 122ND INF

SER. NO.	MODEL	MM/DD/YY	USAGE
219484		08/21/18	CO D 123RD INF
219514		08/21/18	CO A 122ND INF
219543	R	08/27/18	CO C 122ND INF CAMP WHEELER GA
219582		03/28/18	TO 40TH ENGR FT MEYER
219583	R	08/19/18	CO M 122ND INF
219630	R	08/19/18	CO L 122ND INF
219671		08/21/18	CO E 123RD INF
219685		08/21/18	CO I 124TH INF
219757		08/21/18	CO F 124TH INF
219775	R	08/26/18	CO A 122ND INF CAMP WHEELER GA
219805		08/21/18	CO A 124TH INF
219818	R	08/30/18	WAREHOUSE CAMP WHEELER GA
219889	R	08/30/18	WAREHOUSE CAMP WHEELER GA
219897		08/21/18	CO C 121ST INF
219950		06/24/18	MODIFIED STOCK
219950	E	08/17/18	STOCK BOLT TEST
219976	R	08/26/18	WAREHOUSE 121ST INF CAMP WHEELER
219990		08/21/18	CO D 123RD INF
219992		08/21/18	CO C 122ND INF
220001	R	08/19/18	CO L 121ST INF
220025	R	08/12/18	CO I 22ND ENGR FT BENJ HARRISON
220026	R	08/26/18	WAREHOUSE SUP CO 121ST INF
220030	R	08/19/18	CO L 121ST INF
220114	W	04/01/18	INTERCHANGEABILITY TEST
220118		08/21/18	CO E 123RD INF
220153	R	08/31/18	CO G 122ND INF CAMP GREENE NC
220182		08/21/18	CO D 123RD INF
220223		06/24/18	MODIFIED STOCK
220278	R	08/27/18	CO C 122ND INF CAMP WHEELER GA
220356	R	08/19/18	CO F 122ND INF
220399	R	08/19/18	CO H 122ND INF
220416		08/21/18	CO F 124TH INF
220439	R	08/26/18	WAREHOUSE SUP CO 121ST INF
220502	R	08/19/18	CO F 122ND INF
220504	R	08/26/18	WAREHOUSE 106TH AMMO TRAIN
220553		08/21/18	CO H 123RD INF
220620	R	10/04/18	CAMP HUMPHRIES
220629	R	08/26/18	WAREHOUSE SUP CO 121ST INF
220636		08/21/18	CO E 123RD INF
220661	R	08/27/18	CO C 122ND INF CAMP WHEELER GA
220696	R	08/30/18	WAREHOUSE CAMP WHEELER GA
220856	R	08/31/18	CO F 122ND INF CAMP GREENE NC
220869		08/21/18	CO F 123RD INF
220923	R	08/31/18	CO G 122ND INF CAMP GREENE NC
220933	R	09/09/18	106TH AMMUNITION TRAIN
220955	R	08/27/18	CO E 121ST INF CAMP WHEELER GA
220985	R	09/09/18	106TH ENGR
221045	R	09/09/18	106TH ENGR
221075	E	04/17/18	INTERCHANGEABILITY TEST
221122	R	09/09/18	106TH ENGR
221123		08/21/18	CO E 123RD INF
221135		08/21/18	CO L 124TH INF
221199		08/21/18	CO K 124TH INF
221229		08/21/18	CO C 123RD INF
221283	R	08/30/18	WAREHOUSE CAMP WHEELER GA
221289	R	08/19/18	CO I 121ST INF
221306		08/21/18	CO D 121ST INF
221313	R	08/31/18	CO E 122ND INF CAMP GREENE NC
221341		08/21/18	CO D 123RD INF
221343	R	08/19/18	CO M 121ST INF
221375	R	08/26/18	WAREHOUSE CAMP WHEELER GA
221385		08/21/18	CO B 121ST INF
221469	R	08/26/18	WAREHOUSE CAMP WHEELER GA
221539	R	08/19/18	CO G 122ND INF
221590	R	08/26/18	WAREHOUSE SUP CO 121ST INF
221632		08/21/18	CO I 124TH INF
221633		08/21/18	CO C 123RD INF
221654		08/21/18	CO D 121ST INF
221659	E	06/27/18	SA FN TEST
221660		08/21/18	CO A 123RD INF
221661		08/21/18	Q CO 124TH INF
221667		08/21/18	CO G 123RD INF
221669	R	08/27/18	CO A 121ST INF CAMP WHEELER GA
221679	R	08/26/18	WAREHOUSE 121ST INF CAMP WHEELER
221681		06/23/18	CAMP BOWIE (FOUND ON RANGE)
221690	R	08/19/18	CO E 122ND INF
221755	R	08/31/18	CO G 122ND INF CAMP GREENE NC
221771	R	08/19/18	CO I 122ND INF
221776		08/21/18	CO B 121ST INF
221803	R	01/09/43	WWII DONATION TO US NAVY
221872		08/21/18	CO A 122ND INF
221902		08/21/18	CO A 124TH INF
221973	R	08/19/18	CO M 122ND INF
221996	R	08/31/18	CO F 122ND INF CAMP GREENE NC
221999	E	04/17/18	INTERCHANGEABILITY TEST
222042		08/21/18	CO I 124TH INF
222161	R	08/31/18	CO E 122ND INF CAMP GREENE NC
222267		08/21/18	CO A 123RD INF
222270	R	08/30/18	WAREHOUSE CAMP WHEELER GA
222287	R	08/19/18	CO L 122ND INF
222390		08/21/18	CO M 124TH INF
222496		08/21/18	CO F 123RD INF
222510	R	08/26/18	WAREHOUSE CAMP WHEELER GA
222578	R	08/19/18	CO H 122ND INF
222578	R	08/31/18	CO H 122ND INF CAMP GREENE NC
222606	R	08/19/18	CO H 122ND INF
222618		08/21/18	CO A 123RD INF
222654	R	08/19/18	CO L 121ST INF
222692		08/21/18	CO A 121ST INF
222692		11/06/25	USMC
222833	R	08/19/18	CO H 122ND INF
222864	R	08/27/18	CO F 121ST INF CAMP WHEELER GA
222870	W	04/01/18	INTERCHANGEABILITY TEST
222894		08/21/18	CO G 123RD INF
222904	R	08/19/18	CO M 122ND INF
222926	R	08/26/18	WAREHOUSE 121ST INF CAMP WHEELER
222997		08/21/18	CO H 124TH INF
223050		08/21/18	CO D 124TH INF
223084	R	08/19/18	CO F 122ND INF
223111	W	04/01/18	INTERCHANGEABILITY TEST
223299		08/21/18	CO D 123RD INF
223388	R	08/19/18	CO L 121ST INF
223392	E	04/17/18	INTERCHANGEABILITY TEST
223436		08/21/18	CO C 124TH INF
223444	R	08/19/18	CO K 122ND INF
223661	R	08/31/18	CO F 122ND INF CAMP GREENE NC
223718		08/21/18	CO H 123RD INF
223787	R	09/09/18	106TH ENGR
223891	R	08/19/18	CO I 121ST INF
224022		08/21/18	CO I 124TH INF
224042	R	08/14/18	CAMP WHEELER
224042	R	08/14/18	122ND INF CAMP WHEELER
224093	R	08/31/18	CO G 122ND INF CAMP GREENE NC
224097	R	08/26/18	WAREHOUSE CAMP WHEELER GA
224108	R	08/26/18	WAREHOUSE 106TH AMMO TRAIN
224110	R	08/19/18	CO I 122ND INF
224128	E	04/17/18	INTERCHANGEABILITY TEST
224177	R	08/26/18	WAREHOUSE CAMP WHEELER GA
224220		08/21/18	CO E 123RD INF
224227	R	08/26/18	WAREHOUSE CAMP WHEELER GA
224229		08/21/18	CO G 124TH INF
224381		11/06/25	USMC
224452		08/21/18	HQ CO 124TH INF
224464		03/28/18	TO 40TH ENGR FT MEYER
224566	R	08/19/18	CO K 122ND INF
224579		08/21/18	CO B 123RD INF
224635		03 13/45	USNAS ST SIMONS ISLAND (RETD)
224708		08/21/18	CO A 123RD INF
224715	R	08/19/18	CO L 122ND INF
224716		08/21/18	CO F 123RD INF
224743	R	08/31/18	CO F 122ND INF CAMP GREENE NC
224758		08/21/18	HQ CO 124TH INF
224824		08/21/18	CO C 123RD INF
224968	W	06/22/48	VETERANS ADMIN (TO RIA)
225008	R	09/16/18	INTERCHANGEABILITY TEST
225008	R	08/15/18	INTERCHANGE OF MAY & JUNE PROD
225067		08/21/18	CO C 123RD INF
225075	R	08/19/18	CO I 122ND INF
225076	E	04/17/18	INTERCHANGEABILITY TEST
225087	R	08/31/18	CO F 122ND INF CAMP GREENE NC
225135	W	06/22/48	VETERANS ADMIN (TO RIA)
225142		08/21/18	CO D 121ST INF
225152	E	04/17/18	INTERCHANGEABILITY TEST
225196	R	08/26/18	WAREHOUSE SUP CO 121ST INF

SER. NO.	MODEL	MM/DD/YY	USAGE
225248	R	08/19/18	CO I 121ST INF
225270		08/21/18	CO F 123RD INF
225344	R	08/14/18	CAMP WHEELER
225344	R	08/14/18	122ND INF CAMP WHEELER
225344		08/21/18	SUPPLY CO 122ND INF
225374	R	06/22/48	VETERANS ADMIN (TO RIA)
225386		08/21/18	CO F 124TH INF
225396	R	08/19/18	CO F 122ND INF
225459	R	08/26/18	CO F 106TH AMMO TRAIN
225459	R	09/09/18	106TH AMMUNITION TRAIN
225494	R	07/03/18	SA FN TEST
225501		08/21/18	HQ CO 124TH INF
225549		08/21/18	CO C 122ND INF
225570	R	08/31/18	CO E 122ND INF CAMP GREENE NC
225570	R	08/19/18	CO E 122ND INF
225628		08/21/18	CO D 124TH INF
225629		08/21/18	CO C 123RD INF
225673		08/21/18	CO D 124TH INF
225698		08/21/18	CO A 124TH INF
225720	W	04/01/18	INTERCHANGEABILITY TEST
225720	R	08/19/18	CO E 122ND INF
225720	R	08/31/18	CO E 122ND INF CAMP GREENE NC
225741	W	04/01/18	INTERCHANGEABILITY TEST
225833		08/21/18	CO C 122ND INF
225851	R	08/26/18	WAREHOUSE 122ND INF CAMP WHEELER
225861	E	04/17/18	INTERCHANGEABILITY TEST
225921	R	08/27/18	CO C 121ST INF CAMP WHEELER GA
225922	R	08/30/18	WAREHOUSE CAMP WHEELER GA
225929	R	08/26/18	WAREHOUSE CAMP WHEELER GA
225942		08/21/18	CO F 123RD INF
225964	R	08/27/18	CO I 122ND INF CAMP WHEELER GA
226024	R	08/19/18	CO F 122ND INF
226145	R	08/19/18	CO M 121ST INF
226197		08/21/18	CO G 124TH INF
226328	R	08/26/18	CO F 106TH AMMO TRAIN
226344	R	08/28/18	CO K 122ND INF CAMP WHEELER GA
226357		11/06/25	USMC
226535		08/21/18	CO I 124TH INF
226536		08/21/18	CO B 122ND INF
226540	R	08/30/18	WAREHOUSE CAMP WHEELER GA
226752		06/23/18	CAMP BOWIE (FOUND ON RANGE)
226812	R	08/30/18	WAREHOUSE CAMP WHEELER GA
226883	R	08/30/18	WAREHOUSE CAMP WHEELER GA
226904		08/21/18	HQ CO 124TH INF
226962		08/21/18	CO B 123RD INF
226971	R	08/19/18	CO K 121ST INF
227182	R	08/26/18	WAREHOUSE CAMP WHEELER GA
227240	R	08/19/18	CO K 122ND INF
227252		08/21/18	CO B 122ND INF
227344	E	04/17/18	INTERCHANGEABILITY TEST
227355		08/21/18	CO B 122ND INF
227408	R	08/15/18	INTERCHANGE OF MAY & JUNE PROD
227409	R	10/19/18	CAMP GORDON
227527		08/21/18	CO A 122ND INF
227626	R	08/19/18	CO I 121ST INF
227627		08/21/18	CO F 123RD INF
227660	R	08/19/18	CO H 122ND INF
227689	R	08/14/18	CAMP WHEELER
227745	R	08/19/18	CO M 122ND INF
227761		08/21/18	CO F 124TH INF
227869	R	08/17/18	CAMP SHERIDAN
227896		08/21/18	CO F 124TH INF
227900		08/21/18	CO D 122ND INF
227991		08/21/18	CO A 122ND INF
228029		08/21/18	CO F 123RD INF
228093	R	08/26/18	WAREHOUSE 106TH AMMO TRAIN
228097	R	08/19/18	CO L 121ST INF
228189		08/21/18	CO D 123RD INF
228196	R	08/31/18	CO H 122ND INF CAMP GREENE NC
228210	R	08/28/18	CO K 122ND INF CAMP WHEELER GA
228210	E	04/17/18	INTERCHANGEABILITY TEST
228246		08/21/18	CO C 124TH INF
228247	R	08/26/18	WAREHOUSE CAMP WHEELER GA
228332		08/21/18	CO I 124TH INF
228487		08/21/18	CO A 121ST INF
228640		08/21/18	CO C 124TH INF
228687	R	08/19/18	CO I 121ST INF
228783	R	08/31/18	CO F 122ND INF CAMP GREENE NC
228799	R	08/19/18	CO H 122ND INF
228834	R	08/26/18	WAREHOUSE CAMP WHEELER GA
228858		08/21/18	CO K 124TH INF
228928	R	08/19/18	CO E 122ND INF
228986		08/21/18	CO D 124TH INF
229219	R	08/26/18	WAREHOUSE 106TH AMMO TRAIN
229389		08/21/18	CO A 122ND INF
229441		08/21/18	CO C 123RD INF
229583		08/21/18	CO K 124TH INF
229584		08/21/18	CO D 124TH INF
229630	E	07/07/18	164TH INF
229630	E	09/30/18	CAMP BOWIE
229717		08/21/18	CO G 123RD INF
229773	R	08/26/18	WAREHOUSE 121ST INF CAMP WHEELER
229786		08/21/18	CO C 123RD INF
229830		08/21/18	CO A 124TH INF
229868		08/21/18	CO A 122ND INF
229960	E	05/20/18	165TH DEPOT BRIG CAMP TRAVIS
229970		08/21/18	CO E 124TH INF
230041	E	04/17/18	INTERCHANGEABILITY TEST
230044	E	03/03/19	CAMP ZACHARY TAYLOR
230064		11/06/25	USMC
230065		08/21/18	CO D 122ND INF
230074		08/02/11	CO A 122ND INF
230391		08/21/18	CO D 124TH INF
230394	R	08/14/18	122ND INF CAMP WHEELER
230525	R	08/19/18	CO L 121ST INF
230581	R	09/27/18	CO E 13TH INF
230598	R	08/26/18	WAREHOUSE 121ST INF CAMP WHEELER
230600	R	08/19/18	CO M 121ST INF
230780	R	09/26/18	COO AEF (EXTRACTOR TEST)
231002	R	08/19/18	CO L 122ND INF
231070	R	08/27/18	CO B 121ST INF CAMP WHEELER GA
231097	R	08/30/18	WAREHOUSE CAMP WHEELER GA
231164		03/28/18	TO 40TH ENGR FT MEYER
231202		08/21/18	CO D 124TH INF
231206	R	08/27/18	CO A 121ST INF CAMP WHEELER GA
231235	R	08/26/18	WAREHOUSE 121ST INF CAMP WHEELER
231282		08/21/18	CO I 124TH INF
231492	R	08/31/18	CO F 122ND INF CAMP GREENE NC
231492	R	08/19/18	CO F 122ND INF
231524	R	08/19/18	CO F 122ND INF
231631	E	04/17/18	INTERCHANGEABILITY TEST
231666	E	07/07/18	164TH INF
231666	E	11/21/18	CAMP BOWIE
231670	R	08/19/18	CO M 122ND INF
231800		08/21/18	CO B 123RD INF
231973	R	08/19/18	CO K 121ST INF
231975	R	08/26/18	WAREHOUSE CAMP WHEELER GA
232022		08/21/18	CO A 123RD INF
232234	R	09/26/18	CAMP GORDON
232486		08/21/18	CO C 124TH INF
232551		08/21/18	CO D 122ND INF
232553		08/21/18	CO E 124TH INF
232625	R	08/19/18	CO L 122ND INF
232745	R	08/26/18	WAREHOUSE CAMP WHEELER GA
232842	R	08/19/18	CO L 122ND INF
232866	R	08/26/18	WAREHOUSE CAMP WHEELER GA
232890		08/21/18	HQ CO 124TH INF
233022	E	04/17/18	INTERCHANGEABILITY TEST
233041	R	06/06/18	SA FUNCTION TEST
233086		08/21/18	CO A 122ND INF
233350	R	08/26/18	WAREHOUSE SUP CO 121ST INF
233519		08/21/18	CO E 123RD INF
233523		08/21/18	CO C 121ST INF
233608	R	10/14/18	CAMP GORDON
233700		08/21/18	CO A 124TH INF
233773	R	08/26/18	WAREHOUSE CAMP WHEELER GA
233798	R	08/26/18	WAREHOUSE SUP CO 121ST INF
233870		11/06/25	USMC
234062	R	08/31/18	CO E 122ND INF CAMP GREENE NC
234241		08/21/18	CO G 123RD INF
234247	R	08/26/18	WAREHOUSE CAMP WHEELER GA
234252	R	08/15/18	INTERCHANGE OF MAY & JUNE PROD
234330	W	04/22/18	INTERCHANGEABILITY TEST
234493	R	08/12/18	CO I 22ND ENGR FT BENJ HARRISON

SER. NO.	MODEL	MM/DD/YY	USAGE
234503	R	11/10/22	AUGUSTA ARS (OVERHAULED)
234504		08/21/18	CO D 124TH INF
234545	R	08/19/18	CO H 122ND INF
234657		08/21/18	CO H 123RD INF
234756	R	08/19/18	CO K 122ND INF
234775		11/06/25	USMC
234794	R	08/19/18	CO E 122ND INF
234795		08/21/18	CO B 122ND INF
234862	R	08/19/18	CO E 122ND INF
234955		08/21/18	CO H 123RD INF
234965		08/21/18	CO A 124TH INF
235203	E	05/17/18	164TH DEPOT BRIG CAMP FUNSTON
235258	R	06/11/18	337TH INF CAMP CUSTER
235280		08/21/18	CO B 124TH INF
235348	R	08/15/18	INTERCHANGE OF MAY & JUNE PROD
235372		08/21/18	CO B 122ND INF
235514	R	10/14/18	CAMP GORDON
235678	R	08/19/18	CO L 121ST INF
235715		08/21/18	CO F 124TH INF
235939	R	08/31/18	CO H 122ND INF CAMP GREENE NC
235946		08/21/18	CO E 123RD INF
236028	R	08/27/18	CO E 121ST INF CAMP WHEELER GA
236139		08/21/18	CO A 123RD INF
236485	R	09/21/18	CAMP FREMONT
236485	R	11/15/18	CAMP FREMONT
236587	R	10/14/18	CAMP GORDON
237167	R	10/19/18	CAMP GORDON
237263		03/28/18	TO 40TH ENGR FT MEYER
237312	R	08/12/18	CO I 22ND ENGR FT BENJ HARRISON
237321	R	08/19/18	CO E 122ND INF
237430		08/21/18	CO F 124TH INF
237442	R	08/19/18	CO M 121ST INF
237473		08/21/18	CO C 122ND INF
237535	R	08/27/18	CO C 121ST INF CAMP WHEELER GA
237635	R	08/19/18	CO E 122ND INF
237635	R	08/31/18	CO E 122ND INF CAMP GREENE NC
237698		08/21/18	CO B 122ND INF
237722	E	04/17/18	INTERCHANGEABILITY TEST
237762		08/21/18	CO C 121ST INF
237868		08/21/18	CO A 123RD INF
237898	R	08/31/18	CO G 122ND INF CAMP GREENE NC
237898	R	08/19/18	CO G 122ND INF
237905		08/21/18	CO G 124TH INF
237942	R	08/26/18	WAREHOUSE CAMP WHEELER GA
238015	R	08/26/18	WAREHOUSE CAMP WHEELER GA
238092		08/21/18	CO A 124TH INF
238108	R	10/18/18	CAMP GORDON
238191		08/21/18	CO M 124TH INF
238219		08/21/18	CO B 122ND INF
238232	R	10/23/18	CAMP GORDON
238236	R	08/15/18	INTERCHANGE OF MAY & JUNE PROD
238268		08/21/18	CO C 122ND INF
238281	R	08/19/18	CO H 122ND INF
238318	W	06/22/48	VETERANS ADMIN (TO RIA)
238321		08/21/18	CO A 123RD INF
238498	R	08/31/18	CO E 122ND INF CAMP GREENE NC
238802		08/21/18	CO H 124TH INF
238856	R	08/27/18	CO G 121ST INF CAMP WHEELER GA
238930		08/21/18	CO C 121ST INF
239078	R	08/19/18	CO H 122ND INF
239137	E	10/18/18	CAMP GORDON
239155		08/21/18	CO A 123RD INF
239176	R	08/26/18	WAREHOUSE SUP CO 121ST INF
239256	R	08/27/18	CO F 121ST INF CAMP WHEELER GA
239259		08/21/18	CO C 123RD INF
239308	R	08/27/18	CO D 122ND INF CAMPWHEELER GA
239313	R	08/19/18	CO M 121ST INF
239346		08/21/18	CO A 122ND INF
239362		08/21/18	HQ CO 124TH INF
239554	R	08/27/18	CO A 121ST INF CAMPWHEELER GA
239680	R	08/27/18	CO E 121ST INF CAMPWHEELER GA
239776		08/21/18	CO A 122ND INF
239813	R	08/27/18	CO B 121ST INF CAMPWHEELER GA
239926		08/21/18	CO D 121ST INF
239964	R	09/09/18	106TH ENGR
240007		08/21/18	CO F 123RD INF
240062		08/21/18	CO E 124TH INF
240243	R	09/26/18	COO AEF (EXTRACTOR TEST)
240371	R	08/26/18	WAREHOUSE CAMP WHEELER GA
240488	R	08/19/18	CO H 122ND INF
240578	R	08/19/18	CO L 121ST INF
240827	W	04/19/18	TRIGGER INTERCHANGEABILITY TEST
240954		08/21/18	HQ CO 124TH INF
241083	R	08/31/18	CO G 122ND INF CAMP GREENE NC
241083	R	08/19/18	CO G 122ND INF
241299		08/21/18	CO M 124TH INF
241394	R	08/19/18	CO M 122ND INF
241418		12/17/17	327TH INF
241418	E	05/01/18	WATERTOWN ARS (TESTED)
241529	R	08/19/18	CO M 122ND INF
241700		08/21/18	CO C 124TH INF
241785		08/21/18	CO H 123RD INF
241786	R	08/19/18	CO M 121ST INF
241829		08/21/18	CO F 123RD INF
241934	R	08/27/18	CO M 122ND INF CAMP WHEELER GA
241942	R	08/19/18	CO E 122ND INF
241966	R	08/19/18	CO K 122ND INF
242189	R	08/19/18	CO E 122ND INF
242198		08/21/18	CO A 121ST INF
242272		08/21/18	CO A 122ND INF
242278	E	05/17/18	164TH DEPOT BRIG CAMP FUNSTON
242417	R	08/19/18	CO K 121ST INF
242544	R	08/26/18	WAREHOUSE CAMP WHEELER GA
242610	R	08/31/18	CO F 122ND INF CAMP GREENE NC
242618		08/21/18	CO H 123RD INF
242760	E	04/10/18	CO F 310TH INF
242809		08/21/18	CO A 124TH INF
242821		08/21/18	CO D 123RD INF
242853	R	08/19/18	CO E 122ND INF
242876	R	08/26/18	WAREHOUSE SUP CO 121ST INF
242975	R	08/26/18	WAREHOUSE CAMP WHEELER GA
243001	R	06/18/18	SA FN TEST
243008		08/21/18	CO D 121ST INF
243277	R	08/26/18	WAREHOUSE CAMP WHEELER GA
243333	R	08/27/18	CO G 121ST INF CAMP WHEELER GA
243452		08/21/18	CO K 124TH INF
243508	R	10/04/18	CAMP HUMPHRIES
243897		08/21/18	CO K 124TH INF
243915	R	08/12/18	CO I 22ND ENGR FT BENJ HARRISON
244031	R	08/31/18	CO E 122ND INF CAMP GREENE NC
244087	E	06/24/43	MIL GOVT HAWAII (RELEASED)
244751	E	06/21/18	114TH ENGR CAMP BEAUREGARD
244836	R	08/12/18	CO I 22ND ENGR FT BENJ HARRISON
244879	R	09/26/18	COO AEF (EXTRACTOR TEST)
245021		08/21/18	CO B 122ND INF
245853	R	08/27/18	CO D 121ST INF CAMP WHEELER GA
245878	R	08/27/18	CO M 122ND INF CAMP WHEELER GA
245994	R	06/18/18	SA FN TEST
246025	R	08/19/18	CO H 122ND INF
246025	R	08/31/18	CO H 122ND INF CAMP GREENE NC
246113	R	06/06/18	SA FUNCTION TEST
246238	W	04/19/18	TRIGGER INTERCHANGEABILITY TEST
246318	R	08/26/18	WAREHOUSE CAMP WHEELER GA
246872		08/21/18	CO D 124TH INF
247063	R	08/15/18	INTERCHANGE OF MAY & JUNE PROD
247063	R	09/16/18	INTERCHANGEABILITY TEST
247162	E	06/21/18	114TH ENGR CAMP BEAUREGARD
247347	W	06/99	CMP INVENTORY
247350	E	06/21/18	114TH ENGR CAMP BEAUREGARD
247374	W	06/22/48	VETERANS ADMIN (TO RIA)
247593		08/21/18	CO A 122ND INF
247628	R	08/19/18	CO M 122ND INF
247767		08/21/18	CO M 124TH INF
247771	R	10/18/18	CAMP GORDON
247868	W	04/19/18	TRIGGER INTERCHANGEABILITY TEST
248025		06/23/18	CAMP BOWIE (FOUND ON RANGE)
248071	W	04/22/18	INTERCHANGEABILITY TEST
248216	R	08/15/18	INTERCHANGE OF MAY & JUNE PROD
248333	R	06/11/18	337TH INF CAMP CUSTER
248571		08/21/18	CO A 121ST INF
248639		08/21/18	CO E 123RD INF
249017	R	08/27/18	CO G 121ST INF CAMP WHEELER GA
249036	R	06/11/18	337TH INF CAMP CUSTER
249158	R	08/27/18	CO C 121ST INF CAMP WHEELER GA

SER. NO.	MODEL	MM/DD/YY	USAGE
249254	W	04/19/18	TRIGGER INTERCHANGEABILITY TEST
249271	R	10/04/18	CAMP HUMPHRIES
249415		08/21/18	CO D 123RD INF
249483	R	09/21/18	CAMP FREMONT
249483	R	11/15/18	CAMP FREMONT
249729	R	08/26/18	WAREHOUSE 106TH AMMO TRAIN
249773	R	10/23/18	CAMP GORDON
249933	W	04/19/18	TRIGGER INTERCHANGEABILITY TEST
250128	W	04/19/18	TRIGGER INTERCHANGEABILITY TEST
250171	W	04/22/18	INTERCHANGEABILITY TEST
250184	W	04/19/18	TRIGGER INTERCHANGEABILITY TEST
250352	W	04/19/18	TRIGGER INTERCHANGEABILITY TEST
250472	R	08/30/18	WAREHOUSE CAMP WHEELER GA
250481		08/21/18	HQ CO 124TH INF
250840		08/21/18	CO B 123RD INF
250976	E	06/21/18	114TH ENGR CAMP BEAUREGARD
251499	R	08/19/18	CO M 122ND INF
251578	R	08/19/18	CO G 122ND INF
251578	R	08/31/18	CO G 122ND INF CAMP GREENE NC
251582	R	10/19/18	CAMP GORDON
251658		08/21/18	CO G 123RD INF
251923	W	04/19/18	TRIGGER INTERCHANGEABILITY TEST
252067	R	08/19/18	CO E 122ND INF
252394	R	08/26/18	WAREHOUSE CAMP WHEELER GA
252408		08/21/18	CO A 123RD INF
252481	W	04/19/18	TRIGGER INTERCHANGEABILITY TEST
252568		08/21/18	CO A 124TH INF
252860		06/23/18	CAMP BOWIE (FOUND ON RANGE)
252952	W	05/20/18	INTERCHANGEABILITY TEST
252954	R	08/15/18	INTERCHANGE OF MAY & JUNE PROD
253113	W	08/15/18	INTERCHANGE OF MAY & JUNE PROD
253416	W	04/22/18	INTERCHANGEABILITY TEST
253483	R	08/26/18	WAREHOUSE CAMP WHEELER GA
253549	W	04/22/18	INTERCHANGEABILITY TEST
253645	E	06/27/18	HIGHEST NUMBER IN 323RD INF
254216		08/21/18	CO F 123RD INF
254367		08/21/18	CO E 124TH INF
254594	R	08/26/18	WAREHOUSE 106TH AMMO TRAIN
254935	R	08/19/18	CO L 122ND INF
255021		08/21/18	CO F 123RD INF
255045	R	08/19/18	CO M 122ND INF
255052		08/21/18	CO H 123RD INF
255143	R	08/26/18	WAREHOUSE CAMP WHEELER GA
255237	R	08/26/18	WAREHOUSE 106TH AMMO TRAIN
255468	R	06/06/18	SA FUNCTION TEST
255483	W	11/10/22	AUGUSTA ARS (OVERHAULED)
255537	R	08/15/18	INTERCHANGE OF MAY & JUNE PROD
255597		08/21/18	CO D 124TH INF
255662	R	08/19/18	CO L 121ST INF
255766		08/21/18	CO L 124TH INF
255871	R	10/18/18	CAMP GORDON
255917		08/21/18	CO C 124TH INF
255949	R	08/26/18	WAREHOUSE CAMP WHEELER GA
255991	R	08/26/18	WAREHOUSE CAMP WHEELER GA
255997		08/21/18	CO F 123RD INF
255998	R	08/19/18	CO L 121ST INF
256115		08/21/18	CO E 123RD INF
256176		08/21/18	CO K 124TH INF
256471	R	08/15/18	INTERCHANGE OF MAY & JUNE PROD
256762		08/21/18	CO K 124TH INF
256764	R	08/26/18	WAREHOUSE CAMP WHEELER GA
256831	W	08/15/18	INTERCHANGE OF MAY & JUNE PROD
256942	R	06/06/18	SA FUNCTION TEST
257180	R	08/19/18	CO F 122ND INF
257290	E	05/17/18	164TH DEPOT BRIG CAMP FUNSTON
257307		08/21/18	CO C 121ST INF
257321	R	10/14/18	CAMP GORDON
257403		08/21/18	CO D 124TH INF
257406		08/21/18	CO A 121ST INF
257438		08/21/18	CO A 123RD INF
257504	R	09/26/18	COO AEF (EXTRACTOR TEST)
257542	R	08/19/18	CO M 122ND INF
257620	R	08/27/18	CO C 121ST INF CAMP WHEELER GA
257662	R	08/19/18	CO K 121ST INF
257664	R	08/19/18	CO M 122ND INF
257665	R	09/09/18	106TH ENGR
257698	R	08/26/18	WAREHOUSE 106TH AMMO TRAIN
257811	R	08/15/18	INTERCHANGE OF MAY & JUNE PROD
257820	R	09/09/18	106TH AMMUNITION TRAIN
257958	R	08/27/18	CO A 121ST INF CAMP WHEELER GA
258027	R	08/15/18	INTERCHANGE OF MAY & JUNE PROD
258133	R	08/19/18	CO L 121ST INF
258162		08/21/18	CO D 123RD INF
258221	R	08/26/18	WAREHOUSE SUP CO 121ST INF
258367	R	08/26/18	WAREHOUSE CAMP WHEELER GA
258538	R	10/04/18	CAMP HUMPHRIES
258693	R	08/19/18	CO I 122ND INF
259283	R	10/14/18	CAMP GORDON
259722	R	11/10/22	AUGUSTA ARS (OVERHAULED)
259742		08/21/18	CO G 124TH INF
259777		08/21/18	CO B 121ST INF
259795	R	08/30/18	WAREHOUSE CAMP WHEELER GA
259849		08/21/18	CO D 122ND INF
259861		08/21/18	CO D 122ND INF
260196	R	08/15/18	INTERCHANGE OF MAY & JUNE PROD
260215		08/21/18	CO G 124TH INF
261039	R	08/19/18	CO F 122ND INF
261039	R	08/31/18	CO F 122ND INF CAMP GREENE NC
261110		11/06/25	USMC
261127		08/21/18	COA 123RD INF
261224		11/21/23	APG (TRACER TEST)
261460	R	08/19/18	CO H 122ND INF
261688	R	08/31/18	CO H 122ND INF CAMP GREENE NC
261792	R	10/04/18	CAMP HUMPHRIES
261805	W	11/10/22	AUGUSTA ARS (OVERHAULED)
261945	R	09/26/18	COO AEF (EXTRACTOR TEST)
262160	E	05/16/18	CAMP FUNSTON (RUSTY)
262235	E	05/10/18	CAMP HANCOCK (TEST)
262248	R	08/19/18	CO H 122ND INF
262377	W	11/10/22	AUGUSTA ARS (OVERHAULED)
262502		08/21/18	CO E 123RD INF
262507	R	08/26/18	WAREHOUSE CAMP WHEELER GA
262518		08/21/18	CO H 123RD INF
262530	R	08/15/18	INTERCHANGE OF MAY & JUNE PROD
262563		11/06/25	USMC
262691	R	08/19/18	CO F 122ND INF
262706	R	08/19/18	CO F 122ND INF
262706	R	08/31/18	CO F 122ND INF CAMP GREENE NC
262737		08/21/18	CO H 123RD INF
262803	R	08/30/18	WAREHOUSE CAMP WHEELER GA
262830		08/21/18	CO L 124TH INF
262970	R	08/27/18	CO B 121ST INF CAMP WHEELER GA
263183		08/21/18	CO C 124TH INF
263236	R	08/12/18	CO I 22ND ENGR FT BENJ HARRISON
263362		03/05/23	USMC
263439	R	08/19/18	CO L 121ST INF
263470		08/21/18	CO B 123RD INF
263554		08/21/18	CO C 121ST INF
263647		08/21/18	CO A 124TH INF
263657	R	10/14/18	CAMP GORDON
263662		11/06/25	USMC
263956	R	08/15/18	INTERCHANGE OF MAY & JUNE PROD
263996		08/21/18	CO E 124TH INF
264078		08/21/18	CO B 122ND INF
264280	R	08/19/18	CO M 122ND INF
264299	R	08/27/18	CO E 121ST INF CAMP WHEELER GA
264561	R	08/19/18	CO H 122ND INF
264564	R	08/27/18	CO I 122ND INF CAMP WHEELER GA
264659	R	08/19/18	CO M 121ST INF
264717	R	08/26/18	WAREHOUSE SUP CO 121ST INF
264753	R	08/19/18	CO K 122ND INF
264781	R	08/19/18	CO L 121ST INF
264839		08/21/18	SUPPLY CO 122ND INF
264893		08/21/18	CO D 122ND INF
265043		08/21/18	CO D 124TH INF
265094	R	08/26/18	WAREHOUSE CAMP WHEELER GA
265174		08/21/18	CO E 124TH INF
265274	R	08/19/18	CO K 122ND INF
265425	R	08/15/18	INTERCHANGE OF MAY & JUNE PROD
265546		08/21/18	CO B 123RD INF
265585	R	08/27/18	CO C 121ST INF CAMP WHEELER GA
265647	R	08/27/18	CO H 121ST INF CAMP WHEELER GA
265692		08/21/18	CO C 121ST INF
265756		08/21/18	CO H 123RD INF

SER. NO.	MODEL	MM/DD/YY	USAGE
265784	R	08/31/18	CO H 122ND INF CAMP GREENE NC
265840	R	08/15/18	INTERCHANGE OF MAY & JUNE PROD
265867		03/06/23	USMC
265950	E	06/21/18	CO C 143RD INF CAMP BOWIE
265967	R	08/27/18	CO F 121ST INF CAMP WHEELER GA
266135	R	08/19/18	CO H 122ND INF
266135	R	08/31/18	CO H 122ND INF CAMP GREENE NC
266154		08/21/18	CO C 121ST INF
266252	R	08/19/18	CO K 121ST INF
266355		08/21/18	CO C 121ST INF
266488	R	08/19/18	CO K 121ST INF
266518		08/21/18	CO F 124TH INF
266553	R	08/26/18	WAREHOUSE CAMP WHEELER GA
266590		08/21/18	CO E 123RD INF
266616	R	08/27/18	CO G 121ST INF CAMP WHEELER GA
266629	R	08/19/18	CO E 122ND INF
266711		08/21/18	CO I 124TH INF
266721	R	08/19/18	CO I 122ND INF
266906		08/21/18	CO D 121ST INF
266954		08/21/18	CO C 123RD INF
266965	R	09/09/18	106TH ENGR
266983	R	08/19/18	CO H 122ND INF
267061	R	08/27/18	CO B 121ST INF CAMP WHEELER GA
267112	R	08/19/18	CO K 122ND INF
267113		08/21/18	CO A 121ST INF
267117	R	08/31/18	CO G 122ND INF CAMP GREENE NC
267132	R	08/26/18	WAREHOUSE 106TH AMMO TRAIN
267196	R	08/19/18	CO K 122ND INF
267219	R	08/19/18	CO L 122ND INF
267233		08/21/18	CO H 123RD INF
267244		08/21/18	CO H 123RD INF
267399	R	08/19/18	CO K 122ND INF
267404	R	08/19/18	CO M 122ND INF
267462		11/06/25	USMC
267607	E	05/22/18	CAMP FUNSTON (RUSTY)
267694		08/21/18	HQ CO 124TH INF
267756		08/21/18	CO B 122ND INF
267867	R	08/19/18	CO E 122ND INF
268141		08/21/18	CO F 124TH INF
268226	R	08/27/18	CO D 122ND INF CAMP WHEELER GA
268381		08/21/18	CO K 124TH INF
268558	R	08/19/18	CO K 121ST INF
268631		08/21/18	CO F 123RD INF
268668		08/21/18	CO M 124TH INF
268671	R	08/31/18	CO F 122ND INF CAMP GREENE NC
268830		08/21/18	CO A 123RD INF
268854		08/21/18	CO A 124TH INF
268856		08/21/18	HQ CO 124TH INF
268889	R	08/19/18	CO M 121ST INF
268960	R	08/19/18	CO L 121ST INF
268983		08/21/18	CO F 124TH INF
268987		08/21/18	CO M 124TH INF
269001	R	08/26/18	WAREHOUSE CAMP WHEELER GA
269067		08/21/18	CO E 123RD INF
269134	R	08/19/18	CO F 122ND INF
269220	E	05/16/18	CAMP FUNSTON (RUSTY)
269268		08/21/18	CO H 124TH INF
269293	R	10/04/18	CAMP HUMPHRIES
269366	R	08/15/18	INTERCHANGE OF MAY & JUNE PROD
269366	R	09/16/18	INTERCHANGEABILITY TEST
269433	R	08/15/18	INTERCHANGE OF MAY & JUNE PROD
269504	R	08/27/18	CO B 121ST INF CAMP WHEELER GA
269515		08/21/18	CO D 123RD INF
269533	R	10/04/18	CAMP HUMPHRIES
269657	R	08/19/18	CO I 122ND INF
269701	W	08/15/18	INTERCHANGE OF MAY & JUNE PROD
269718		08/21/18	CO B 124TH INF
269793	R	08/26/18	WAREHOUSE CAMP WHEELER GA
269798	R	08/19/18	CO F 122ND INF
269810	R	08/19/18	CO H 122ND INF
269835	R	08/19/18	CO L 122ND INF
269850		08/21/18	CO C 123RD INF
269867	R	08/26/18	WAREHOUSE 106TH AMMO TRAIN
269979		08/21/18	CO F 124TH INF
270034	R	08/30/18	WAREHOUSE CAMP WHEELER GA
270036	R	08/27/18	CO D 121ST INF CAMP WHEELER GA
270076	R	08/26/18	WAREHOUSE CAMP WHEELER GA
270130	R	08/26/18	WAREHOUSE 121ST INF CAMP WHEELER
270156		08/21/18	CO D 122ND INF
270186	R	08/19/18	CO K 122ND INF
270220		08/21/18	CO B 121ST INF
270246		08/21/18	CO C 122ND INF
270274		08/21/18	CO L 124TH INF
270291	R	08/19/18	CO H 122ND INF
270291	R	08/31/18	CO H 122ND INF CAMP GREENE NC
270419		08/21/18	CO E 124TH INF
270455		08/21/18	CO D 123RD INF
270469	R	08/26/18	WAREHOUSE CAMP WHEELER GA
270480	R	08/19/18	CO M 122ND INF
270518		08/21/18	HQ CO 124TH INF
270542	R	08/19/18	CO L 122ND INF
270553	R	08/26/18	WAREHOUSE 106TH AMMO TRAIN
270578	R	09/09/18	106TH ENGR
270588		08/21/18	CO B 121ST INF
270602	R	08/19/18	CO K 121ST INF
270664		08/21/18	CO A 122ND INF
270707	R	08/19/18	CO L 122ND INF
270771		08/21/18	CO B 124TH INF
270782	W	11/10/22	AUGUSTA ARS (OVERHAULED)
270849		08/21/18	CO E 124TH INF
270856	R	08/19/18	CO F 122ND INF
270944	R	08/19/18	CO M 122ND INF
270985	E	06/21/18	114TH ENGR CAMP BEAUREGARD
270989	R	08/19/18	CO F 122ND INF
271035	R	10/23/18	CAMP GORDON
271039	R	08/31/18	CO E 122ND INF CAMP GREENE NC
271113	R	08/27/18	CO B 122ND INF CAMP WHEELER GA
271185	R	09/09/18	106TH ENGR
271207	R	08/31/18	CO F 122ND INF CAMP GREENE NC
271213		08/21/18	CO C 122ND INF
271280	R	09/26/18	COO AEF (EXTRACTOR TEST)
271296	W	11/10/22	AUGUSTA ARS (OVERHAULED)
271323	R	08/19/18	CO M 122ND INF
271387	R	08/31/18	CO E 122ND INF CAMP GREENE NC
271513		08/21/18	CO A 122ND INF
271531		08/21/18	CO I 124TH INF
271555	R	08/19/18	CO H 122ND INF
271583	R	09/09/18	106TH AMMUNITION TRAIN
271597		08/21/18	HQ CO 124TH INF
271626	R	08/19/18	CO K 122ND INF
271638	R	08/19/18	CO F 122ND INF
271648		08/21/18	CO C 123RD INF
271689		08/21/18	CO C 122ND INF
271694		08/21/18	CO C 123RD INF
271712	R	08/31/18	CO G 122ND INF CAMP GREENE NC
271738	R	08/19/18	CO L 121ST INF
271739	R	08/31/18	CO F 122ND INF CAMP GREENE NC
271752	R	08/31/18	CO H 122ND INF CAMP GREENE NC
271779	R	08/19/18	CO K 121ST INF
271779		11/06/25	USMC
271780	R	08/19/18	CO F 122ND INF
271811		08/21/18	CO B 123RD INF
271861		08/21/18	CO E 123RD INF
271893	R	07/03/18	SA FN TEST
271929		08/21/18	CO F 124TH INF
271988	R	08/27/18	CO E 121ST INF CAMP WHEELER GA
272001	R	08/19/18	CO H 122ND INF
272027		08/21/18	CO D 124TH INF
272054		08/21/18	CO C 123RD INF
272106		08/21/18	CO D 123RD INF
272149	R	08/19/18	CO H 122ND INF
272149	R	08/31/18	CO H 122ND INF CAMP GREENE NC
272201	R	08/26/18	WAREHOUSE SUP CO 121ST INF
272220		08/21/18	CO F 123RD INF
272261	R	08/19/18	CO K 121ST INF
272304	R	08/19/18	CO I 122ND INF
272333		08/21/18	CO G 123RD INF
272351		08/21/18	HQ CO 124TH INF
272357		08/21/18	CO D 123RD INF
272388		08/21/18	CO E 124TH INF
272449		08/21/18	CO C 123RD INF
272473	R	09/09/18	106TH ENGR
272514	R	08/19/18	CO I 121ST INF
272517	R	08/19/18	CO I 122ND INF

SER. NO.	MODEL	MM/DD/YY	USAGE
272521		08/21/18	CO K 124TH INF
272526		08/21/18	CO C 122ND INF
272614	E	04/17/18	INTERCHANGEABILITY TEST
272619	R	08/19/18	CO K 121ST INF
272678		08/21/18	CO A 121ST INF
272685		08/21/18	CO B 123RD INF
272691	R	08/19/18	CO L 121ST INF
272724		08/21/18	CO C 121ST INF
272738		08/21/18	CO C 124TH INF
272749	R	08/27/18	CO C 121ST INF CAMP WHEELER GA
272763	R	08/19/18	CO E 122ND INF
272763	R	08/31/18	CO E 122ND INF CAMP GREENE NC
272777	W	05/13/18	INTERCHANGEABILITY TEST
272792		08/21/18	CO A 121ST INF
272815	R	08/26/18	WAREHOUSE CAMP WHEELER GA
272816		08/21/18	CO A 121ST INF
272820		08/21/18	CO H 123RD INF
272824		08/21/18	CO D 122ND INF
272827		08/21/18	CO B 123RD INF
272856		08/21/18	CO B 123RD INF
272864		08/21/18	CO H 124TH INF
272867	E	04/17/18	INTERCHANGEABILITY TEST
272895	R	08/19/18	CO F 122ND INF
272897		08/21/18	CO A 124TH INF
272907	R	08/19/18	CO E 122ND INF
272969	R	08/19/18	CO H 122ND INF
272983	R	08/19/18	CO E 122ND INF
272983	R	08/31/18	CO E 122ND INF CAMP GREENE NC
273015	E	07/07/18	164TH INF
273015	E	09/30/18	CAMP BOWIE
273032		08/21/18	CO G 124TH INF
273095	R	08/19/18	CO L 121ST INF
273104	R	08/30/18	WAREHOUSE CAMP WHEELER GA
273126	R	08/19/18	CO I 122ND INF
273134	R	08/30/18	WAREHOUSE CAMP WHEELER GA
273169	R	08/30/18	WAREHOUSE CAMP WHEELER GA
273182	R	08/19/18	CO L 121ST INF
273200	R	08/26/18	WAREHOUSE 121ST INF CAMP WHEELER
273216		08/21/18	CO C 124TH INF
273218	R	08/19/18	CO M 122ND INF
273246		08/21/18	CO F 123RD INF
273247		08/21/18	CO C 121ST INF
273274		08/21/18	CO B 122ND INF
273275	R	08/31/18	CO H 122ND INF CAMP GREENE NC
273275	R	08/19/18	CO H 122ND INF
273297	R	08/19/18	CO I 122ND INF
273335		11/06/25	USMC
273414	R	08/19/18	CO M 122ND INF
273430	R	08/26/18	WAREHOUSE 121ST INF CAMP WHEELER
273511		08/21/18	CO A 121ST INF
273520		08/21/18	CO A 122ND INF
273555	R	08/19/18	CO E 122ND INF
273579	R	08/19/18	CO G 122ND INF
273579	R	08/31/18	CO G 122ND INF CAMP GREENE NC
273583	R	08/19/18	CO H 122ND INF
273590		08/21/18	CO A 123RD INF
273628	R	11/01/18	REMINGTON (TARGETED)
273628	R	10/28/18	ENDURANCE TEST
273674		08/21/18	CO F 123RD INF
273700	R	09/09/18	CO M 123RD INF
273701	R	08/19/18	CO K 121ST INF
273714		08/21/18	CO I 124TH INF
273747	R	08/19/18	CO K 121ST INF
273778		08/21/18	HQ CO 124TH INF
273784		08/21/18	CO H 124TH INF
273786	R	08/31/18	CO E 122ND INF CAMP GREENE NC
273820		08/21/18	CO E 124TH INF
273848		08/21/18	CO C 122ND INF
273848	R	08/27/18	CO C 122ND INF CAMP WHEELER GA
273901	R	08/26/18	WAREHOUSE SUP CO 121ST INF
273925	R	08/19/18	CO K 122ND INF
273933	R	08/26/18	WAREHOUSE SUP CO 121ST INF
273965	R	09/09/18	106TH ENGR
274006	R	08/19/18	CO F 122ND INF
274010	E	05/03/18	CAPT BREESE (BATTLE SIGHT TEST)
274013	R	08/19/18	CO L 122ND INF
274044	E	05/10/18	CAMP HANCOCK (TEST)
274077		08/21/18	CO D 122ND INF
274093		08/21/18	CO F 123RD INF
274113	R	08/26/18	WAREHOUSE CAMP WHEELER GA
274114		08/21/18	CO B 121ST INF
274116		08/21/18	CO A 124TH INF
274188	R	08/19/18	CO L 121ST INF
274224		08/21/18	CO E 123RD INF
274236	R	08/19/18	CO I 121ST INF
274237		08/21/18	CO H 124TH INF
274339		08/21/18	CO D 123RD INF
274340	R	08/19/18	CO I 122ND INF
274410	R	08/26/18	WAREHOUSE CAMP WHEELER GA
274414		08/21/18	CO G 124TH INF
274428	R	08/27/18	CO C 121ST INF CAMP WHEELER GA
274443	R	08/26/18	WAREHOUSE 121ST INF CAMP WHEELER
274464		08/21/18	CO D 122ND INF
274519		08/21/18	CO A 121ST INF
274545	R	09/09/18	HDQTR TRAIN & MP CAMP WHEELER
274553		08/21/18	CO C 122ND INF
274559		08/21/18	CO H 123RD INF
274560	R	08/27/18	CO I 122ND INF CAMP WHEELER GA
274571	R	08/26/18	WAREHOUSE CAMP WHEELER GA
274586	R	08/1/918	CO K 121ST INF
274588	R	08/19/18	CO E 122ND INF
274588	R	08/31/18	CO E 122ND INF CAMP GREENE NC
274617	R	08/26/18	WAREHOUSE CAMP WHEELER GA
274624	R	08/31/18	CO E 122ND INF CAMP GREENE NC
274624	R	08/19/18	CO E 122ND INF
274633	R	08/19/18	CO F 122ND INF
274689	E	06/21/18	114TH ENGR CAMP BEAUREGARD
274697	R	08/31/18	CO H 122ND INF CAMP GREENE NC
274769	R	08/19/18	CO G 122ND INF
274856	R	08/26/18	WAREHOUSE CAMP WHEELER GA
274857	R	08/19/18	CO H 122ND INF
274863		08/21/18	CO A 124TH INF
274882	R	08/19/18	CO L 121ST INF
274906		08/21/18	CO B 121ST INF
274941	R	06/17/18	SA FUNCTION TEST
274959	R	08/26/18	WAREHOUSE CAMP WHEELER GA
274975		08/21/18	CO C 124TH INF
274992		08/21/18	CO C 122ND INF
275012		08/21/18	CO A 123RD INF
275012	R	09/09/18	CO A 123RD INF
275023		08/21/18	CO B 123RD INF
275036	R	08/31/18	CO F 122ND INF CAMP GREENE NC
275038		08/21/18	SUPPLY CO 122ND INF
275038	R	09/09/18	SUPPLY CO 122ND INF
275088	R	08/19/18	CO K 122ND INF
275103	W	05/13/18	INTERCHANGEABILITY TEST
275161	R	08/31/18	CO E 122ND INF CAMP GREENE NC
275161	R	08/19/18	CO E 122ND INF
275162		08/21/18	CO E 124TH INF
275166	R	08/26/18	WAREHOUSE CAMP WHEELER GA
275230		08/21/18	CO C 121ST INF
275286		08/21/18	CO C 122ND INF
275288		08/21/18	CO G 123RD INF
275295	R	08/31/18	CO G 122ND INF CAMP GREENE NC
275352	R	08/27/18	CO I 122ND INF CAMP WHEELER GA
275379		08/21/18	CO B 123RD INF
275421	W	05/13/18	INTERCHANGEABILITY TEST
275445		08/21/18	CO H 124TH INF
275449		08/21/18	HQ CO 124TH INF
275451		08/21/18	CO B 122ND INF
275480		08/21/18	CO A 121ST INF
275549		08/21/18	CO C 121ST INF
275565	R	08/26/18	WAREHOUSE CAMP WHEELER GA
275588	R	06/06/18	SA FUNCTION TEST
275596		08/21/18	CO D 123RD INF
275641	R	08/27/18	CO G 121ST INF CAMP WHEELER GA
275654		08/21/18	HQ CO 124TH INF
275699		08/21/18	CO C 123RD INF
275722	R	08/19/18	CO L 121ST INF
275741	R	09/09/18	106TH ENGR
275845	W	11/10/22	AUGUSTA ARS (OVERHAULED)
275894	R	08/19/18	CO L 122ND INF
275907	R	08/19/18	CO K 121ST INF
275921	R	08/27/18	CO G 121ST INF CAMP WHEELER GA

SER. NO.	MODEL	MM/DD/YY	USAGE
275936		08/21/18	CO E 123RD INF
276004		08/21/18	CO G 124TH INF
276033		08/21/18	CO B 121ST INF
276056	R	08/30/18	WAREHOUSE CAMP WHEELER GA
276081		08/21/18	CO H 123RD INF
276090		08/21/18	CO C 122ND INF
276117	R	08/26/18	WAREHOUSE SUP CO 121ST INF
276125	R	08/19/18	CO K 121ST INF
276185	R	08/19/18	CO M 122ND INF
276197	R	08/19/18	CO F 122ND INF
276212	R	08 05/18	106TH ENGR AEF (DEF)
276267	R	08/19/18	CO M 122ND INF
276286		08/21/18	CO A 122ND INF
276289		08/21/18	CO M 124TH INF
276294		08/21/18	CO C 122ND INF
276302	R	08/30/18	WAREHOUSE CAMP WHEELER GA
276308	R	08/26/18	WAREHOUSE SUP CO 121ST INF
276327	R	08/19/18	CO L 121ST INF
276337	R	08/26/18	WAREHOUSE CAMP WHEELER GA
276342		08/21/18	CO C 123RD INF
276351		08/21/18	SUPPLY CO 122ND INF
276404		08/21/18	CO B 122ND INF
276405	R	11/10/22	AUGUSTA ARS (OVERHAULED)
276432		08/21/18	CO F 124TH INF
276449		08/21/18	CO A 124TH INF
276455	R	08/26/18	WAREHOUSE CAMP WHEELER GA
276475		08/21/18	CO H 124TH INF
276493	R	08/30/18	WAREHOUSE CAMP WHEELER GA
276512	R	08/19/18	CO K 122ND INF
276527		08/21/18	CO H 123RD INF
276591		08/21/18	CO B 124TH INF
276598		08/21/18	CO A 121ST INF
276606		08/21/18	CO C 124TH INF
276666		08/21/18	CO K 124TH INF
276667	R	08/31/18	CO F 122ND INF CAMP GREENE NC
276710		08/21/18	CO A 124TH INF
276717		08/21/18	CO A 123RD INF
276742		08/21/18	CO L 124TH INF
276747	E	05/17/18	164TH DEPOT BRIG CAMP FUNSTON
276754	R	08/26/18	WAREHOUSE CAMP WHEELER GA
276760		08/21/18	CO K 124TH INF
276790	R	08/27/18	CO C 122ND INF CAMP WHEELER GA
276801		08/21/18	CO B 121ST INF
276807	R	08/19/18	CO H 122ND INF
276813		08/21/18	CO B 124TH INF
276847	E	09/30/18	CAMP BOWIE
276847	E	07/07/18	164TH INF
276873		08/21/18	CO F 123RD INF
276898	R	08/19/18	CO H 122ND INF
276916	R	08/31/18	CO E 122ND INF CAMP GREENE NC
276940		08/21/18	CO G 124TH INF
276942	W	11/10/22	AUGUSTA ARS (OVERHAULED)
276957	E	05/17/18	164TH DEPOT BRIG CAMP FUNSTON
276995	R	08/31/18	CO H 122ND INF CAMP GREENE NC
277019		08/21/18	CO G 123RD INF
277044	R	08/26/18	WAREHOUSE CAMP WHEELER GA
277056		08/21/18	CO A 121ST INF
277076	R	08/30/18	WAREHOUSE CAMP WHEELER GA
277091		08/21/18	CO H 124TH INF
277180	R	08/27/18	CO F 121ST INF CAMP WHEELER GA
277221	R	08/19/18	CO L 122ND INF
277284	R	08/19/18	CO I 122ND INF
277285	R	08/19/18	CO E 122ND INF
277292	R	08/27/18	CO I 122ND INF CAMP WHEELER GA
277318		08/21/18	CO G 123RD INF
277322	R	08/30/18	WAREHOUSE CAMP WHEELER GA
277330	R	08/30/18	WAREHOUSE CAMP WHEELER GA
277335	R	08/19/18	CO I 122ND INF
277373		08/21/18	CO C 123RD INF
277418	R	08/19/18	CO E 122ND INF
277434		08/21/18	CO G 123RD INF
277458		08/21/18	CO C 121ST INF
277460		08/21/18	CO G 124TH INF
277488	R	08/19/18	CO H 122ND INF
277498	R	08/31/18	CO H 122ND INF CAMP GREENE NC
277551	R	08/26/18	WAREHOUSE CAMP WHEELER GA
277556		08/21/18	CO H 123RD INF
277561	R	08/27/18	CO A 121ST INF CAMP WHEELER GA
277573	R	08/27/18	CO I 122ND INF CAMP WHEELER GA
277575		08/21/18	CO F 123RD INF
277595	E	05/20/18	165TH DEPOT BRIG CAMP TRAVIS
277602		08/21/18	CO E 123RD INF
277785	R	08/27/18	CO C 121ST INF CAMP WHEELER GA
277866		08/21/18	CO D 121ST INF
277882		08/21/18	CO H 123RD INF
277952	R	09/14/18	CAMP WHEELER
277952	R	09/14/18	124TH INF
277954		09/04/18	CAMP WHEELER
277954	R	11/21/18	CAMP WHEELER
277969		08/21/18	CO I 124TH INF
277988	R	08/15/18	INTERCHANGE OF MAY & JUNE PROD
278067	R	08/27/18	CO B 121ST INF CAMP WHEELER GA
278069	R	08/19/18	CO K 121ST INF
278150	R	10/14/18	CAMP GORDON
278218	R	08/31/18	CO E 122ND INF CAMP GREENE NC
278228		08/21/18	CO B 124TH INF
278321	R	08/15/18	INTERCHANGE OF MAY & JUNE PROD
278321	R	09/16/18	INTERCHANGEABILITY TEST
278322		08/21/18	CO C 122ND INF
278333		08/21/18	CO H 123RD INF
278347	W	03/39	PAUL JAEGER (SOLD BY NJ)
278352		08/21/18	CO D 123RD INF
278406	R	08/19/18	CO K 121ST INF
278446	R	08/31/18	CO G 122ND INF CAMP GREENE NC
278446	R	08/19/18	CO G 122ND INF
278463		08/21/18	CO D 124TH INF
278468		08/21/18	CO D 123RD INF
278483	R	08/26/18	WAREHOUSE CAMP WHEELER GA
278518		08/21/18	CO E 123RD INF
278525		08/21/18	CO L 124TH INF
278599	R	08/19/18	CO L 121ST INF
278655		08/21/18	CO B 124TH INF
278656	R	08/26/18	WAREHOUSE SUP CO 121ST INF
278669	R	09/09/18	CO I 124TH INF
278669		08/21/18	CO I 124TH INF
278681	R	08/27/18	CO B 122ND INF CAMP WHEELER GA
278689		08/21/18	CO A 122ND INF
278692	R	08/19/18	CO F 122ND INF
278719		08/21/18	CO H 124TH INF
278801		08/21/18	CO E 123RD INF
278821		08/21/18	CO D 122ND INF
278861	R	08/27/18	CO D 122ND INF CAMP WHEELER GA
278869	R	08/15/18	INTERCHANGE OF MAY & JUNE PROD
278927		08/21/18	CO A 123RD INF
278931	R	08/19/18	CO L 122ND INF
278951		08/21/18	CO B 122ND INF
279002	R	08/27/18	CO F 121ST INF CAMP WHEELER GA
279016	E	04/06/18	CO A 110TH INF CAMP HANCOCK
279017	R	08/26/18	WAREHOUSE CAMP WHEELER GA
279028		08/21/18	CO D 123RD INF
279083	R	08/31/18	CO H 122ND INF CAMP GREENE NC
279096		08/21/18	SUPPLY CO 122ND INF
279101	R	08/26/18	WAREHOUSE 106TH AMMO TRAIN
279146	R	08/26/18	WAREHOUSE 121ST INF CAMP WHEELER
279164	R	08/31/18	CO E 122ND INF CAMP GREENE NC
279164	R	08/19/18	CO E 122ND INF
279171	R	08/26/18	61ST BRIGADE HQ CAMP WHEELER GA
279175	E	06/21/18	114TH ENGR CAMP BEAUREGARD
279216		08/21/18	CO H 123RD INF
279227		08/21/18	CO A 122ND INF
279310	R	08/26/18	WAREHOUSE 106TH AMMO TRAIN
279343	R	08/19/18	CO K 122ND INF
279357	R	08/19/18	CO E 122ND INF
279359	R	08/19/18	CO M 121ST INF
279405		08/21/18	CO C 121ST INF
279446		08/21/18	CO G 124TH INF
279468		08/21/18	CO A 121ST INF
279474		08/21/18	CO C 123RD INF
279512	R	08/19/18	CO H 122ND INF
279512	R	08/31/18	CO H 122ND INF CAMP GREENE NC
279520		08/21/18	CO H 124TH INF
279549	R	08/19/18	CO K 122ND INF
279554		12/01/25	LITTLE ROCK AIR INT DEP TO RIA
279588		08/21/18	CO D 124TH INF

SER. NO.	MODEL	MM/DD/YY	USAGE
279616		08/21/18	CO E 124TH INF
279763	R	09/09/18	106TH AMMUNITION TRAIN
279767		08/21/18	CO B 121ST INF
279838		08/21/18	CO B 124TH INF
279848		08/21/18	CO G 123RD INF
279877	R	08/27/18	CO H 121ST INF CAMP WHEELER GA
279893		08/21/18	CO C 122ND INF
279895	R	08/26/18	WAREHOUSE CAMP WHEELER GA
279900	R	08/19/18	CO M 121ST INF
279913		08/21/18	SUPPLY CO 122ND INF
279922	R	08/19/18	CO K 122ND INF
279944		08/21/18	CO L 124TH INF
279955		08/21/18	CO A 121ST INF
279966	R	08/19/18	CO K 121ST INF
279974		08/21/18	HQ CO 124TH INF
280064		08/21/18	CO C 121ST INF
280100	R	08/19/18	CO F 122ND INF
280141	R	08/26/18	WAREHOUSE CAMP WHEELER GA
280204	R	08/26/18	WAREHOUSE CAMP WHEELER GA
280223	E	08/17/18	STOCK BOLT TEST
280242	R	08/19/18	CO F 122ND INF
280246	R	08/26/18	WAREHOUSE CAMP WHEELER GA
280253	R	08/19/18	CO L 122ND INF
280278	R	08/19/18	CO H 122ND INF
280278	R	08/31/18	CO H 122ND INF CAMP GREENE NC
280288		08/21/18	CO D 124TH INF
280304	R	08/26/18	WAREHOUSE CAMP WHEELER GA
280362		08/21/18	CO D 121ST INF
280366	R	08/26/18	WAREHOUSE CAMP WHEELER GA
280371	R	07/11/18	SA FN TEST
280447	R	08/26/18	WAREHOUSE SUP CO 121ST INF
280481		08/21/18	CO D 124TH INF
280483	R	08/26/18	CO A 122ND INF CAMP WHEELER GA
280513	R	08/19/18	CO I 121ST INF
280548		08/21/18	CO A 124TH INF
280570		08/21/18	CO A 121ST INF
280593	R	08/26/18	WAREHOUSE CAMP WHEELER GA
280655		08/21/18	CO F 123RD INF
280658		08/21/18	CO H 123RD INF
280670	R	09/09/18	CO M 124TH INF
280686	R	08/19/18	CO I 122ND INF
280713	R	11/21/18	CAMP WHEELER
280713		09/04/18	CAMP WHEELER
280713	R	09/14/18	CAMP WHEELER
280713	R	09/14/18	124TH INF
280735	R	09/09/18	106TH ENGR
280761	R	08/19/18	CO L 121ST INF
280769	R	08/19/18	CO E 122ND INF
280769	R	08/31/18	CO E 122ND INF CAMP GREENE NC
280879	R	09/09/18	SUPPLY CO 121ST INF
280899		08/21/18	CO E 123RD INF
280901	R	08/15/18	INTERCHANGE OF MAY & JUNE PROD
280919	R	08/19/18	CO F 122ND INF
280919	R	08/31/18	CO F 122ND INF CAMP GREENE NC
280955		08/21/18	CO K 124TH INF
280965	R	08/19/18	CO L 121ST INF
280992	R	08/19/18	CO K 121ST INF
281004		08/21/18	CO A 121ST INF
281016	R	08/27/18	CO C 121ST INF CAMP WHEELER GA
281032	R	08/19/18	CO L 122ND INF
281047	R	08/31/18	CO G 122ND INF CAMP GREENE NC
281071		08/21/18	CO A 121ST INF
281107	R	08/31/18	CO G 122ND INF CAMP GREENE NC
281134		08/21/18	CO G 124TH INF
281147	R	08/15/18	INTERCHANGE OF MAY & JUNE PROD
281176	R	08/19/18	CO H 122ND INF
281181	R	08/26/18	WAREHOUSE CAMP WHEELER GA
281187	R	08/19/18	CO G 122ND INF
281189	R	08/27/18	CO E 121ST INF CAMP WHEELER GA
281190		08/21/18	CO E 124TH INF
281198	W	05/13/18	INTERCHANGEABILITY TEST
281210	R	08/19/18	CO K 121ST INF
281227	R	08/26/18	WAREHOUSE CAMP WHEELER GA
281243	R	08/19/18	CO M 121ST INF
281245		08/21/18	CO K 124TH INF
281260	R	08/26/18	WAREHOUSE CAMP WHEELER GA
281261	R	08/19/18	CO I 122ND INF
281265		08/21/18	HQ CO 124TH INF
281292	R	08/19/18	CO I 122ND INF
281338	R	08/31/18	CO H 122ND INF CAMP GREENE NC
281379		08/21/18	CO G 124TH INF
281381		08/21/18	CO C 123RD INF
281402	R	08/26/18	WAREHOUSE CAMP WHEELER GA
281439		08/21/18	CO F 123RD INF
281484		08/21/18	CO B 122ND INF
281528	R	08/26/18	CO F 106TH AMMO TRAIN
281528	R	09/09/18	106TH AMMUNITION TRAIN
281557		08/21/18	CO B 122ND INF
281564	R	09/09/18	CO M 123RD INF
281615		08/21/18	CO H 123RD INF
281631	R	08/27/18	CO M 122ND INF CAMP WHEELER GA
281687	R	08/31/18	CO E 122ND INF CAMP GREENE NC
281702	R	09/09/18	106TH ENGR
281709		08/21/18	CO A 124TH INF
281733		08/21/18	CO C 123RD INF
281741	R	08/26/18	61ST BRIGADE HQ CAMP WHEELER GA
281794	E	05/20/18	165TH DEPOT BRIG CAMP TRAVIS
281825	R	08/19/18	CO K 121ST INF
281833		08/21/18	CO E 124TH INF
281840		08/21/18	CO E 124TH INF
281843		08/21/18	CO A 122ND INF
281853	R	08/26/18	WAREHOUSE 121ST INF CAMP WHEELER
281860	R	08/26/18	WAREHOUSE CAMP WHEELER GA
281871	R	08/19/18	CO H 122ND INF
281873	R	08/19/18	CO M 121ST INF
281925	R	08/19/18	CO E 122ND INF
281959	R	08/19/18	CO F 122ND INF
281990		08/21/18	CO C 123RD INF
282004		08/21/18	CO A 121ST INF
282031	R	08/26/18	WAREHOUSE 121ST INF CAMP WHEELER
282043	R	08/27/18	CO B 121ST INF CAMP WHEELER GA
282046	R	08/19/18	CO H 122ND INF
282053	R	08/27/18	CO B 121ST INF CAMP WHEELER GA
282072	R	09/09/18	106TH AMMUNITION TRAIN
282112		08/21/18	CO A 123RD INF
282113		08/21/18	CO G 124TH INF
282127	R	08/19/18	CO L 122ND INF
282144	R	08/19/18	CO K 122ND INF
282171		08/21/18	HQ CO 124TH INF
282173		08/21/18	CO H 123RD INF
282213	R	08/27/18	CO B 122ND INF CAMP WHEELER GA
282214	R	08/19/18	CO H 122ND INF
282214	R	08/31/18	CO H 122ND INF CAMP GREENE NC
282246		08/21/18	CO B 124TH INF
282278	R	08/19/18	CO L 121ST INF
282279		08/21/18	CO A 123RD INF
282294		08/21/18	CO A 124TH INF
282309	R	08/26/18	62ND BRIGADE HQ CAMP WHEELER GA
282316	E	05/16/18	CAMP FUNSTON (RUSTY)
282325	R	08/19/18	CO G 122ND INF
282325	R	08/31/18	CO G 122ND INF CAMP GREENE NC
282398	R	08/27/18	CO D 121ST INF CAMP WHEELER GA
282417	R	08/19/18	CO M 122ND INF
282451	R	08/19/18	CO L 121ST INF
282455	R	08/19/18	CO M 121ST INF
282465		08/21/18	CO A 124TH INF
282503	R	08/26/18	WAREHOUSE CAMP WHEELER GA
282530	R	08/31/18	CO F 122ND INF CAMP GREENE NC
282531	R	08/19/18	CO F 122ND INF
282564	R	08/19/18	CO K 121ST INF
282587		08/21/18	CO G 124TH INF
282591		08/21/18	CO H 123RD INF
282609		08/21/18	CO C 122ND INF
282612	R	08/19/18	CO K 121ST INF
282624	R	08/26/18	WAREHOUSE SUP CO 121ST INF
282626		08/21/18	CO A 124TH INF
282638	R	08/26/18	WAREHOUSE CAMP WHEELER GA
282646		08/21/18	CO H 123RD INF
282647	R	08/19/18	CO L 121ST INF
282649		08/21/18	CO D 121ST INF
282675		08/21/18	CO H 123RD INF
282698	R	08/19/18	CO K 121ST INF
282736	R	08/19/18	CO L 121ST INF
282739		08/21/18	CO C 123RD INF

SER. NO.	MODEL	MM/DD/YY	USAGE
282747	R	08/31/18	CO G 122ND INF CAMP GREENE NC
282747	R	08/19/18	CO G 122ND INF
282772		08/21/18	CO A 123RD INF
282782		08/21/18	CO C 123RD INF
282796	R	08/19/18	CO K 122ND INF
282806		08/21/18	CO E 123RD INF
282808		08/21/18	CO C 121ST INF
282832	R	08/19/18	CO G 122ND INF
282890	R	08/31/18	CO F 122ND INF CAMP GREENE NC
282911		08/21/18	CO E 123RD INF
282934	R	08/19/18	CO H 122ND INF
282942	R	08/26/18	61ST BRIGADE HQ CAMP WHEELER GA
282955	R	08/27/18	CO F 121ST INF CAMP WHEELER GA
282984		08/21/18	CO F 124TH INF
283044		08/21/18	CO B 123RD INF
283079		08/21/18	CO G 123RD INF
283112		08/21/18	CO K 124TH INF
283120	R	08/19/18	CO K 121ST INF
283121		08/21/18	CO B 122ND INF
283127		08/21/18	CO E 123RD INF
283177	E	11/21/18	CAMP BOWIE
283177	E	06/21/18	CO C 143RD INF CAMP BOWIE
283224	R	08/30/18	WAREHOUSE CAMP WHEELER GA
283235	R	08/19/18	CO I 122ND INF
283266	R	08/30/18	WAREHOUSE CAMP WHEELER GA
283270		08/21/18	CO I 124TH INF
283273	R	08/30/18	WAREHOUSE CAMP WHEELER GA
283305		08/21/18	CO B 123RD INF
283324		08/21/18	CO C 122ND INF
283362	R	06/21/18	SA FN TEST
283409		08/21/18	CO H 123RD INF
283440	R	08/26/18	WAREHOUSE CAMP WHEELER GA
283459		08/21/18	CO B 123RD INF
283494	R	08/19/18	CO H 122ND INF
283536		08/21/18	CO E 124TH INF
283574	R	08/31/18	CO E 122ND INF CAMP GREENE NC
283622		08/21/18	CO A 122ND INF
283626	R	08/26/18	WAREHOUSE CAMP WHEELER GA
283709	R	08/19/18	CO H 122ND INF
283751	R	08/14/18	CAMP WHEELER
283751	R	08/14/18	122ND INF CAMP WHEELER
283751	R	08/21/18	SUPPLY CO 122ND INF
283754		08/21/18	CO F 124TH INF
283764	R	08/26/18	WAREHOUSE CAMP WHEELER GA
283847		08/21/18	CO A 122ND INF
283847	R	08/26/18	CO A 122ND INF CAMP WHEELER GA
283854	R	08/19/18	CO K 121ST INF
283864		08/21/18	CO D 123RD INF
283919	R	11/10/22	AUGUSTA ARS (OVERHAULED)
283927		08/21/18	CO K 124TH INF
283939	R	08/27/18	CO G 121ST INF CAMP WHEELER GA
283966		08/21/18	CO H 124TH INF
283971	R	08/19/18	CO H 122ND INF
283976		08/21/18	CO H 124TH INF
283999		08/21/18	CO C 123RD INF
284027		08/21/18	CO A 121ST INF
284042	R	08/26/18	WAREHOUSE CAMP WHEELER GA
284060	R	08/26/18	WAREHOUSE 106TH AMMO TRAIN
284071		08/21/18	CO G 123RD INF
284080	R	08/26/18	CO A 122ND INF CAMP WHEELER GA
284080	R	08/19/18	CO H 122ND INF
284106	R	08/27/18	CO H 121ST INF CAMP WHEELER GA
284133	R	08/31/18	CO G 122ND INF CAMP GREENE NC
284149		08/21/18	CO A 123RD INF
284156		08/21/18	CO D 122ND INF
284169	R	08/19/18	CO I 122ND INF
284175	R	08/19/18	CO H 122ND INF
284183	R	08/26/18	WAREHOUSE CAMP WHEELER GA
284191	R	08/19/18	CO E 122ND INF
284220		08/21/18	CO G 123RD INF
284238		08/21/18	CO B 124TH INF
284243	R	08/19/18	CO I 121ST INF
284249		08/21/18	CO H 124TH INF
284270		08/21/18	CO C 123RD INF
284273		08/21/18	CO C 121ST INF
284285		08/21/18	CO C 121ST INF
284290	R	09/09/18	106TH ENGR
284306	R	08/19/18	CO L 122ND INF
284404	R	08/19/18	CO L 121ST INF
284454	R	08/26/18	WAREHOUSE CAMP WHEELER GA
284484		08/21/18	CO H 124TH INF
284499		08/21/18	CO B 123RD INF
284507	R	08/30/18	WAREHOUSE CAMP WHEELER GA
284518	R	08/19/18	CO K 121ST INF
284544	R	08/19/18	CO K 121ST INF
284568		08/21/18	CO A 122ND INF
284571		08/21/18	CO G 124TH INF
284577	R	08/19/18	CO K 121ST INF
284583	R	08/19/18	CO L 122ND INF
284605	R	08/19/18	CO I 122ND INF
284656	R	08/28/18	CO K 122ND INF CAMP WHEELER GA
284666	R	08/27/18	CO I 122ND INF CAMP WHEELER GA
284685		08/21/18	HQ CO 124TH INF
284765	R	08/19/18	CO M 122ND INF
284778	R	08/26/18	WAREHOUSE CAMP WHEELER GA
284785		08/21/18	CO M 124TH INF
284842	R	08/26/18	WAREHOUSE 106TH AMMO TRAIN
284852	W	05/13/18	INTERCHANGEABILITY TEST
284894		08/21/18	CO D 124TH INF
284902	R	08/19/18	CO L 121ST INF
284918	R	08/26/18	WAREHOUSE CAMP WHEELER GA
284984		08/21/18	CO D 123RD INF
285009		08/21/18	CO C 121ST INF
285072	R	08/19/18	CO M 121ST INF
285081	R	08/19/18	CO K 121ST INF
285105		08/21/18	CO F 123RD INF
285186		08/21/18	CO B 124TH INF
285222	R	08/19/18	CO F 122ND INF
285223		08/21/18	CO F 123RD INF
285247	R	08/19/18	CO M 122ND INF
285250	R	08/19/18	CO M 122ND INF
285304	R	08/19/18	CO L 121ST INF
285306		08/21/18	CO A 122ND INF
285307	E	06/21/18	114TH ENGR CAMP BEAUREGARD
285307		08/21/18	CO F 123RD INF
285308	R	08/19/18	CO F 122ND INF
285336		08/21/18	CO A 121ST INF
285337		08/21/18	CO F 123RD INF
285390	R	08/26/18	WAREHOUSE CAMP WHEELER GA
285404		08/21/18	CO C 121ST INF
285419		08/21/18	CO D 123RD INF
285428		08/21/18	CO B 122ND INF
285431	R	08/19/18	CO H 122ND INF
285449	R	08/27/18	CO D 122ND INF CAMP WHEELER GA
285456	R	08/19/18	CO F 122ND INF
285456	R	08/31/18	CO F 122ND INF CAMP GREENE NC
285462	R	08/26/18	WAREHOUSE CAMP WHEELER GA
285472		08/21/18	CO E 123RD INF
285498		08/21/18	CO B 124TH INF
285523		08/21/18	CO C 122ND INF
285531	R	08/19/18	CO I 121ST INF
285567	R	08/19/18	CO M 122ND INF
285572	R	08/27/18	CO B 122ND INF CAMP WHEELER GA
285578	R	08/19/18	CO E 122ND INF
285601	R	06/17/18	SA FUNCTION TEST
285667		08/21/18	CO E 124TH INF
285669	E	01/09/43	WWII DONATION TO US NAVY
285674		08/21/18	CO L 124TH INF
285694		08/21/18	CO D 122ND INF
285700	R	08/19/18	CO K 122ND INF
285701		08/21/18	CO A 123RD INF
285730	R	08/31/18	CO H 122ND INF CAMP GREENE NC
285745	R	08/31/18	CO E 122ND INF CAMP GREENE NC
285752		08/21/18	CO D 121ST INF
285790		08/21/18	CO B 122ND INF
285802	R	09/09/18	106TH ENGR
285822		08/21/18	HQ CO 124TH INF
285825		08/21/18	CO M 124TH INF
285840	R	08/26/18	WAREHOUSE CAMP WHEELER GA
285859	R	08/31/18	CO E 122ND INF CAMP GREENE NC
285906		08/21/18	CO A 123RD INF
285915		08/21/18	CO F 123RD INF
285962	R	08/26/18	WAREHOUSE CAMP WHEELER GA
286002		08/21/18	CO D 122ND INF

SER. NO.	MODEL	MM/DD/YY	USAGE
286041	E	05/16/18	CAMP FUNSTON (RUSTY)
286052	R	08/19/18	CO L 121ST INF
286091		08/21/18	CO C 123RD INF
286128	W	03/39	PAUL JAEGER (SOLD BY NJ)
286138	R	08/19/18	CO I 122ND INF
286154	R	08/19/18	CO I 121ST INF
286172		08/21/18	CO C 121ST INF
286181		08/21/18	CO G 124TH INF
286200	R	09/09/18	106TH ENGR
286263	W	03/39	PAUL JAEGER (SOLD BY NJ)
286293		08/21/18	HQ CO 124TH INF
286301	R	08/26/18	WAREHOUSE 121ST INF CAMP WHEELER
286304	R	08/19/18	CO H 122ND INF
286305	R	08/30/18	WAREHOUSE CAMP WHEELER GA
286306	R	08/26/18	WAREHOUSE CAMP WHEELER GA
286324	R	08/19/18	CO I 121ST INF
286325		08/21/18	CO G 124TH INF
286381		08/21/18	CO H 123RD INF
286451	W	03/39	PAUL JAEGER (SOLD BY NJ)
286452		08/21/18	CO B 122ND INF
286467	R	08/19/18	CO K 122ND INF
286474	W	05/20/18	INTERCHANGEABILITY TEST
286484		08/21/18	CO H 123RD INF
286526	R	08/19/18	CO E 122ND INF
286552	R	08/19/18	CO M 121ST INF
286635	R	08/19/18	CO M 122ND INF
286643	R	09/09/18	106TH ENGR
286663		08/21/18	CO A 122ND INF
286670	R	08/26/18	CO A 122ND INF CAMP WHEELER GA
286693		08/21/18	CO A 124TH INF
286725		08/21/18	CO A 121ST INF
286727	R	08/31/18	CO F 122ND INF CAMP GREENE NC
286771		08/21/18	CO C 121ST INF
286812	R	08/26/18	WAREHOUSE 106TH AMMO TRAIN
286825	R	08/19/18	CO H 122ND INF
286864		08/21/18	CO C 124TH INF
286902	R	08/19/18	CO K 122ND INF
286921	R	08/26/18	WAREHOUSE 106TH AMMO TRAIN
286931		08/21/18	CO H 123RD INF
286949		08/21/18	CO D 121ST INF
286956	R	08/26/18	CO A 122ND INF CAMP WHEELER GA
286983	R	08/19/18	CO M 121ST INF
287029	R	08/19/18	CO K 122ND INF
287043		08/21/18	SUPPLY CO 122ND INF
287055	R	08/19/18	CO G 122ND INF
287116	R	08/19/18	CO L 121ST INF
287147	R	08/19/18	CO K 122ND INF
287151		08/21/18	CO A 122ND INF
287198	R	08/26/18	WAREHOUSE CAMP WHEELER GA
287218	R	08/19/18	CO H 122ND INF
287223	R	08/19/18	CO K 121ST INF
287251	R	08/19/18	CO E 122ND INF
287255	R	08/19/18	CO E 122ND INF
287274		08/21/18	CO A 121ST INF
287296		08/21/18	CO A 124TH INF
287297	R	09/09/18	SUPPLY CO 122ND INF
287310	R	08/26/18	WAREHOUSE CAMP WHEELER GA
287316	R	08/19/18	CO I 122ND INF
287327		08/21/18	CO I 124TH INF
287355	R	08/26/18	WAREHOUSE CAMP WHEELER GA
287378	R	08/19/18	CO L 121ST INF
287407	R	09/09/18	106TH ENGR
287410		08/21/18	CO G 124TH INF
287433		08/21/18	CO C 123RD INF
287439	R	08/27/18	CO C 122ND INF CAMP WHEELER GA
287470		08/21/18	CO B 122ND INF
287473	W	05/20/18	INTERCHANGEABILITY TEST
287478	R	08/19/18	CO L 121ST INF
287479		08/21/18	CO C 121ST INF
287523		08/21/18	CO C 122ND INF
287526		08/21/18	CO F 123RD INF
287529	R	08/19/18	CO M 121ST INF
287544	R	06/21/18	SA FN TEST
287544	R	06/28/18	SA DAILY FUNCTION TEST
287551	R	08/19/18	CO I 121ST INF
287560		08/21/18	CO E 123RD INF
287617		08/21/18	CO C 121ST INF
287633		08/21/18	CO D 121ST INF
287682		08/21/18	CO C 121ST INF
287684	R	08/26/18	WAREHOUSE CAMP WHEELER GA
287689	R	08/14/18	122ND INF CAMP WHEELER
287690		08/21/18	CO C 123RD INF
287707	R	08/19/18	CO E 122ND INF
287733	R	08/30/18	WAREHOUSE CAMP WHEELER GA
287734	R	08/19/18	CO L 121ST INF
287768	R	08/30/18	WAREHOUSE CAMP WHEELER GA
287772		08/21/18	CO B 121ST INF
287773	R	08/31/18	CO F 122ND INF CAMP GREENE NC
287782	R	08/26/18	WAREHOUSE 121ST INF CAMP WHEELER
287784	R	08/26/18	WAREHOUSE SUP CO 121ST INF
287821	R	08/19/18	CO E 122ND INF
287821	R	08/31/18	CO E 122ND INF CAMP GREENE NC
287832	R	08/31/18	CO F 122ND INF CAMP GREENE NC
287883		08/21/18	CO F 124TH INF
287949	R	08/19/18	CO G 122ND INF
287949	R	08/31/18	CO G 122ND INF CAMP GREENE NC
287958	R	08/19/18	CO L 121ST INF
287991	R	08/19/18	CO K 121ST INF
287992		08/21/18	CO D 122ND INF
288020	R	08/15/18	INTERCHANGE OF MAY & JUNE PROD
288043		08/21/18	CO B 123RD INF
288064	R	08/26/18	WAREHOUSE CAMP WHEELER GA
288072		08/21/18	CO H 123RD INF
288111	R	09/09/18	106TH ENGR
288117	R	08/27/18	CO G 121ST INF CAMP WHEELER GA
288121	R	08/31/18	CO H 122ND INF CAMP GREENE NC
288126	R	08/26/18	WAREHOUSE CAMP WHEELER GA
288153	R	08/19/18	CO K 121ST INF
288194	R	08/19/18	CO H 122ND INF
288227		08/21/18	CO B 121ST INF
288246	E	05/20/18	165TH DEPOT BRIG CAMP TRAVIS
288292	R	08/26/18	CO A 122ND INF CAMP WHEELER GA
288300	R	08/31/18	CO E 122ND INF CAMP GREENE NC
288326		08/21/18	CO I 124TH INF
288346		08/21/18	CO C 122ND INF
288363	R	08/26/18	WAREHOUSE 121ST INF CAMP WHEELER
288371	R	08/26/18	WAREHOUSE SUP CO 121ST INF
288415		08/21/18	CO G 124TH INF
288421	R	09/09/18	106TH ENGR
288484		08/21/18	CO H 123RD INF
288489	R	08/19/18	CO K 121ST INF
288492	R	08/19/18	CO L 121ST INF
288496		08/21/18	CO B 122ND INF
288500	R	08/19/18	CO I 121ST INF
288533	R	08/31/18	CO E 122ND INF CAMP GREENE NC
288536		08/21/18	CO C 122ND INF
288554		08/21/18	CO C 123RD INF
288558	R	08/26/18	WAREHOUSE CAMP WHEELER GA
288570	R	08/30/18	WAREHOUSE CAMP WHEELER GA
288605		08/21/18	CO L 124TH INF
288619		08/21/18	CO A 124TH INF
288620	R	09/09/18	SUPPLY CO 121ST INF
288667	R	08/26/18	WAREHOUSE CAMP WHEELER GA
288684	R	08/26/18	WAREHOUSE CAMP WHEELER GA
288739		08/21/18	HQ CO 124TH INF
288740	R	08/26/18	WAREHOUSE CAMP WHEELER GA
288741		08/21/18	CO D 124TH INF
288765	R	08/19/18	CO E 122ND INF
288765	R	08/31/18	CO E 122ND INF CAMP GREENE NC
288793	R	08/27/18	CO B 121ST INF CAMP WHEELER GA
288812	R	08/26/18	WAREHOUSE CAMP WHEELER GA
288826	R	08/26/18	WAREHOUSE 121ST INF CAMP WHEELER
288834	R	08/19/18	CO I 121ST INF
288870		08/21/18	CO E 124TH INF
288871	R	08/19/18	CO H 122ND INF
288874		08/21/18	CO I 124TH INF
288878	R	08/30/18	WAREHOUSE CAMP WHEELER GA
288904		08/21/18	CO A 124TH INF
288934	R	08/27/18	CO L 122ND INF CAMP WHEELER GA
288950	E	04/17/18	INTERCHANGEABILITY TEST
288964	R	08/26/18	WAREHOUSE CAMP WHEELER GA
288965		08/21/18	CO A 121ST INF
288996		08/21/18	CO C 121ST INF
289005		08/21/18	CO D 124TH INF

SER. NO.	MODEL	MM/DD/YY	USAGE
289019		08/21/18	CO F 124TH INF
289020	R	08/26/18	WAREHOUSE CAMP WHEELER GA
289052	R	08/19/18	CO K 121ST INF
289052		08/21/18	CO C 121ST INF
289057		08/21/18	CO A 121ST INF
289072	R	08/19/18	CO F 122ND INF
289076	R	08/19/18	CO G 122ND INF
289077	R	08/19/18	CO F 122ND INF
289099	W	05/20/18	INTERCHANGEABILITY TEST
289111	R	08/26/18	61ST BRIGADE HQ CAMP WHEELER GA
289171	R	08/19/18	CO F 122ND INF
289194	R	08/30/18	WAREHOUSE CAMP WHEELER GA
289195		08/21/18	CO A 124TH INF
289214		08/21/18	CO I 124TH INF
289252	R	08/19/18	CO H 122ND INF
289276		08/21/18	CO A 122ND INF
289305	R	08/19/18	CO L 121ST INF
289369	R	08/26/18	WAREHOUSE CAMP WHEELER GA
289371	R	08/26/18	WAREHOUSE 121ST INF CAMP WHEELER
289382	R	08/19/18	CO M 122ND INF
289390		08/21/18	CO C 124TH INF
289411		08/21/18	CO G 123RD INF
289420		08/21/18	CO A 123RD INF
289458		08/21/18	CO B 122ND INF
289466		08/21/18	CO B 124TH INF
289493	R	09/09/18	106TH ENGR
289495		08/21/18	CO H 124TH INF
289505		08/21/18	CO A 124TH INF
289511	R	08/19/18	CO F 122ND INF
289515	R	08/19/18	CO I 122ND INF
289557	R	08/26/18	WAREHOUSE 121ST INF CAMP WHEELER
289565	R	08/19/18	CO I 122ND INF
289625	R	08/19/18	CO F 122ND INF
289647		08/21/18	CO B 121ST INF
289682	R	08/26/18	WAREHOUSE 106TH AMMO TRAIN
289688		08/21/18	CO D 121ST INF
289691	R	08/26/18	WAREHOUSE CAMP WHEELER GA
289732		08/21/18	CO H 124TH INF
289734	R	08/19/18	CO L 122ND INF
289745	R	08/26/18	WAREHOUSE 121ST INF CAMP WHEELER
289757	R	08/26/18	WAREHOUSE 121ST INF CAMP WHEELER
289827	R	08/27/18	CO C 122ND INF CAMP WHEELER GA
289831		08/21/18	CO H 123RD INF
289836	R	08/19/18	CO H 122ND INF
289869		08/21/18	CO K 124TH INF
289877		08/21/18	CO D 122ND INF
289883		08/21/18	CO C 122ND INF
289928		08/21/18	CO C 123RD INF
289947	R	08/26/18	WAREHOUSE CAMP WHEELER GA
289947	W	07/10/18	SA FN TEST
289950		08/21/18	CO A 124TH INF
289970	R	08/19/18	CO L 121ST INF
289970	E	11/06/18	CAMP FOREST
289972		08/21/18	CO H 124TH INF
289998	W	06/17/18	SA FUNCTION TEST
290009	R	08/26/18	WAREHOUSE 106TH AMMO TRAIN
290020	R	08/26/18	WAREHOUSE CAMP WHEELER GA
290054	R	08/19/18	CO G 122ND INF
290109		08/21/18	CO A 123RD INF
290137		08/21/18	CO I 124TH INF
290138	R	09/09/18	106TH ENGR
290138	R	06/99	CMP INVENTORY
290143	E	11/10/22	AUGUSTA ARS (OVERHAULED)
290145		08/21/18	CO C 122ND INF
290152		08/21/18	CO A 123RD INF
290166	R	08/19/18	CO I 122ND INF
290205	R	08/31/18	CO F 122ND INF CAMP GREENE NC
290240	R	08/19/18	CO E 122ND INF
290241	R	08/19/18	CO I 121ST INF
290246	R	08/19/18	CO K 122ND INF
290253		08/21/18	CO M 124TH INF
290288	R	08/27/18	CO M 122ND INF CAMP WHEELER GA
290299	R	08/19/18	CO H 122ND INF
290299	R	08/31/18	CO H 122ND INF CAMP GREENE NC
290332	R	08/28/18	CO K 122ND INF CAMP WHEELER GA
290352	R	08/27/18	CO C 122ND INF CAMP WHEELER GA
290363	R	08/19/18	CO L 121ST INF
290371	R	08/19/18	CO H 122ND INF
290409	R	08/31/18	CO G 122ND INF CAMP GREENE NC
290434		08/21/18	CO B 121ST INF
290444		08/21/18	CO K 124TH INF
290449		08/21/18	CO G 124TH INF
290453	R	08/31/18	CO G 122ND INF CAMP GREENE NC
290498		08/21/18	CO C 124TH INF
290518		08/21/18	CO I 124TH INF
290519	R	08/19/18	CO K 122ND INF
290540		08/21/18	CO A 121ST INF
290551	R	08/14/18	122ND INF CAMP WHEELER
290551		08/21/18	SUPPLY CO 122ND INF
290554	R	08/15/18	INTERCHANGE OF MAY & JUNE PROD
290579		08/21/18	CO B 124TH INF
290711		08/21/18	SUPPLY CO 122ND INF
290728	R	08/31/18	CO F 122ND INF CAMP GREENE NC
290753	R	08/19/18	CO E 122ND INF
290754	R	08/26/18	WAREHOUSE CAMP WHEELER GA
290758		08/21/18	CO A 121ST INF
290789	R	08/19/18	CO H 122ND INF
290798		08/21/18	CO E 123RD INF
290814		08/21/18	CO K 124TH INF
290817	R	08/31/18	CO H 122ND INF CAMP GREENE NC
290834	R	08/26/18	WAREHOUSE CAMP WHEELER GA
290865	W	03/39	PAUL JAEGER (SOLD BY NJ)
290879	R	08/19/18	CO H 122ND INF
290887		08/21/18	CO C 124TH INF
290907		08/21/18	CO C 124TH INF
290960	R	08/31/18	CO E 122ND INF CAMP GREENE NC
290981		03/17/18	111TH ENGR AEF
291006		08/21/18	CO H 124TH INF
291008	R	08/26/18	WAREHOUSE CAMP WHEELER GA
291022	R	08/27/18	CO E 121ST INF CAMP WHEELER GA
291044		03/17/18	111TH ENGR AEF
291079	R	08/26/18	WAREHOUSE 121ST INF CAMP WHEELER
291087		08/21/18	CO F 124TH INF
291092	R	08/30/18	WAREHOUSE CAMP WHEELER GA
291141		08/21/18	CO D 121ST INF
291164		03/17/18	111TH ENGR AEF
291166	R	08/26/18	WAREHOUSE CAMP WHEELER GA
291171	R	08/31/18	CO E 122ND INF CAMP GREENE NC
291187	R	08/31/18	CO H 122ND INF CAMP GREENE NC
291193	R	08/26/18	WAREHOUSE CAMP WHEELER GA
291215	R	08/26/18	WAREHOUSE 121ST INF CAMP WHEELER
291253	R	08/14/18	122ND INF CAMP WHEELER
291278		08/21/18	CO A 123RD INF
291326		08/21/18	CO G 124TH INF
291328		08/21/18	CO G 124TH INF
291335		08/21/18	CO C 122ND INF
291350		08/21/18	CO G 123RD INF
291354	R	08/19/18	CO M 121ST INF
291357		08/21/18	CO B 121ST INF
291364	R	08/28/18	CO K 122ND INF CAMP WHEELER GA
291379		08/21/18	CO M 124TH INF
291407		08/21/18	CO G 124TH INF
291427	R	08/27/18	CO C 121ST INF CAMP WHEELER GA
291448	R	08/26/18	WAREHOUSE CAMP WHEELER GA
291458	R	08/27/18	CO C 122ND INF CAMP WHEELER GA
291467		08/21/18	CO F 123RD INF
291477	R	08/28/18	CO K 122ND INF CAMP WHEELER GA
291482	R	08/15/18	INTERCHANGE OF MAY & JUNE PROD
291484		08/21/18	CO H 123RD INF
291532	R	08/27/18	CO L 122ND INF CAMP WHEELER GA
291575	R	08/26/18	WAREHOUSE 121ST INF CAMP WHEELER
291576	R	08/19/18	CO H 122ND INF
291600	R	08/19/18	CO I 122ND INF
291636		03/17/18	111TH ENGR AEF
291640		08/21/18	CO D 123RD INF
291647		08/21/18	CO D 123RD INF
291677	R	08/27/18	CO L 122ND INF CAMP WHEELER GA
291681	R	08/26/18	WAREHOUSE SUP CO 121ST INF
291688	R	08/19/18	CO H 122ND INF
291717	R	08/26/18	WAREHOUSE 106TH AMMO TRAIN
291736	R	08/31/18	CO G 122ND INF CAMP GREENE NC
291736	R	08/19/18	CO G 122ND INF
291749	R	08/26/18	WAREHOUSE CAMP WHEELER GA
291750	R	08/19/18	CO E 122ND INF

SER. NO.	MODEL	MM/DD/YY	USAGE
291750	R	08/31/18	CO E 122ND INF CAMP GREENE NC
291767	R	08/31/18	CO H 122ND INF CAMP GREENE NC
291813	R	08/19/18	CO M 122ND INF
291825	R	08/26/18	WAREHOUSE CAMP WHEELER GA
291826	R	08/19/18	CO K 121ST INF
291833	R	08/26/18	WAREHOUSE CAMP WHEELER GA
291848	R	08/31/18	CO G 122ND INF CAMP GREENE NC
291853	R	08/27/18	CO M 122ND INF CAMP WHEELER GA
291873	R	08/19/18	CO G 122ND INF
291873	R	08/31/18	CO G 122ND INF CAMP GREENE NC
291914		08/21/18	CO F 123RD INF
291916		08/21/18	CO G 124TH INF
291926	R	08/26/18	WAREHOUSE CAMP WHEELER GA
291948	R	08/27/18	CO M 122ND INF CAMP WHEELER GA
291951		08/21/18	CO G 123RD INF
291961	R	08/19/18	CO L 121ST INF
291963	R	08/27/18	CO M 122ND INF CAMP WHEELER GA
291995	W	03/39	PAUL JAEGER (SOLD BY NJ)
292019		08/21/18	CO B 122ND INF
292050		08/21/18	CO K 124TH INF
292058		08/21/18	CO C 122ND INF
292086		08/21/18	CO C 121ST INF
292103	W	05/20/18	INTERCHANGEABILITY TEST
292122	R	08/26/18	WAREHOUSE CAMP WHEELER GA
292130	R	08/26/18	WAREHOUSE SUP CO 121ST INF
292139		09/02/43	USAT MADLO (TL-1) S-218
292144		08/21/18	CO C 121ST INF
292148	R	08/26/18	WAREHOUSE CAMP WHEELER GA
292157		08/21/18	HQ CO 124TH INF
292168	R	08/19/18	CO F 122ND INF
292212		08/21/18	CO K 124TH INF
292217	E	01/29/34	L F HANCE
292230	R	08/19/18	CO M 121ST INF
292268		03/17/18	111TH ENGR AEF
292270	R	08/26/18	WAREHOUSE 106TH AMMO TRAIN
292313	R	08/19/18	CO I 122ND INF
292331		08/21/18	CO B 123RD INF
292346	R	08/31/18	CO H 122ND INF CAMP GREENE NC
292426	W	05/20/18	INTERCHANGEABILITY TEST
292431	R	05/17/18	164TH DEPOT BRIG CAMP FUNSTON
292443	R	08/27/18	CO G 121ST INF CAMP WHEELER GA
292452	R	08/26/18	WAREHOUSE CAMP WHEELER GA
292471		08/21/18	CO G 123RD INF
292526		08/21/18	CO H 123RD INF
292542	R	08/27/18	CO E 121ST INF CAMP WHEELER GA
292546	R	09/09/18	CO M 123RD INF
292550	R	08/26/18	WAREHOUSE SUP CO 121ST INF
292555		08/21/18	CO M 124TH INF
292571		08/21/18	CO M 124TH INF
292602	R	08/19/18	CO H 122ND INF
292617		08/21/18	HQ CO 124TH INF
292618		08/21/18	CO K 124TH INF
292680		03/17/18	111TH ENGR AEF
292705	R	08/30/18	WAREHOUSE CAMP WHEELER GA
292706		08/21/18	CO E 124TH INF
292719		03/17/18	111TH ENGR AEF
292740		08/21/18	CO L 124TH INF
292756	R	08/30/18	WAREHOUSE CAMP WHEELER GA
292808	R	08/27/18	CO C 121ST INF CAMP WHEELER GA
292848		08/21/18	HQ CO 124TH INF
292854	R	08/26/18	WAREHOUSE CAMP WHEELER GA
292902		08/21/18	CO I 124TH INF
292961	R	09/09/18	CO M 123RD INF
292981		03/17/18	111TH ENGR AEF
293004	R	08/19/18	CO M 122ND INF
293015	R	08/27/18	CO H 121ST INF CAMP WHEELER GA
293085		08/21/18	CO I 124TH INF
293110	R	08/19/18	CO H 122ND INF
293116		08/21/18	CO G 123RD INF
293128	R	08/19/18	CO M 122ND INF
293129	R	06/28/18	SA DAILY FUNCTION TEST
293129	R	06/25/18	SA FN TEST
293145	R	08/26/18	WAREHOUSE CAMP WHEELER GA
293163	R	08/19/18	CO F 122ND INF
293171	R	08/26/18	WAREHOUSE 121ST INF CAMP WHEELER
293213	R	08/26/18	WAREHOUSE CAMP WHEELER GA
293260		08/21/18	CO F 124TH INF
293262	W	05/20/18	INTERCHANGEABILITY TEST
293267	R	08/19/18	CO I 122ND INF
293279		08/21/18	CO K 124TH INF
293285		08/21/18	CO F 123RD INF
293318	R	08/15/18	INTERCHANGE OF MAY & JUNE PROD
293322	W	03 /39	PAUL JAEGER (SOLD BY NJ)
293362	R	06/28/18	SA DAILY FUNCTION TEST
293363	R	08/19/18	CO F 122ND INF
293389	R	08/26/18	WAREHOUSE CAMP WHEELER GA
293479	R	08/19/18	CO I 122ND INF
293486		08/21/18	SUPPLY CO 122ND INF
293591	R	08/27/18	CO G 121ST INF CAMP WHEELER GA
293631	R	08/19/18	CO K 121ST INF
293647		08/21/18	CO G 123RD INF
293656		08/21/18	CO E 124TH INF
293665		03/17/18	111TH ENGR AEF
293727	R	06/11/18	340TH INF CAMP CUSTER
293841	R	08/27/18	CO B 122ND INF CAMP WHEELER GA
293849		08/21/18	CO D 121ST INF
293850	R	07/15/18	SA FN TEST
293863		08/21/18	HQ CO 124TH INF
293877		08/21/18	CO A 124TH INF
293909		08/21/18	CO D 123RD INF
293928	R	09/09/18	CO M 123RD INF
293944		08/21/18	CO E 124TH INF
293947	R	07/08/18	SA FN TEST
293986	R	08/19/18	CO G 122ND INF
293986	R	08/31/18	CO G 122ND INF CAMP GREENE NC
293994	R	08/19/18	CO G 122ND INF
293995		08/21/18	CO C 123RD INF
293997	R	08/19/18	CO M 122ND INF
294007	R	08/28/18	CO K 122ND INF CAMP WHEELER GA
294059	R	08/19/18	CO K 122ND INF
294140	R	08/19/18	CO E 122ND INF
294156	R	08/15/18	INTERCHANGE OF MAY & JUNE PROD
294171		08/21/18	CO G 123RD INF
294190	R	08/31/18	CO G 122ND INF CAMP GREENE NC
294254		03/17/18	111TH ENGR AEF
294256		08/21/18	CO B 122ND INF
294257	R	08/31/18	CO E 122ND INF CAMP GREENE NC
294274		08/21/18	HQ CO 124TH INF
294301		08/21/18	CO A 124TH INF
294317		08/21/18	CO M 124TH INF
294323		08/21/18	CO C 123RD INF
294349	R	08/27/18	CO D 122ND INF CAMP WHEELER GA
294377		08/21/18	CO E 124TH INF
294384	R	08/31/18	CO G 122ND INF CAMP GREENE NC
294387		08/21/18	CO D 121ST INF
294403	R	08/31/18	CO H 122ND INF CAMP GREENE NC
294410	R	08/19/18	CO H 122ND INF
294447		08/21/18	CO F 123RD INF
294465	R	08/30/18	WAREHOUSE CAMP WHEELER GA
294470	R	08/27/18	CO I 122ND INF CAMP WHEELER GA
294503	R	08/19/18	CO L 122ND INF
294527	R	06/17/18	SA FUNCTION TEST
294534		03/17/18	111TH ENGR AEF
294551	R	08/19/18	CO M 122ND INF
294584		03/17/18	111TH ENGR AEF
294592		08/21/18	CO A 122ND INF
294596	R	08/31/18	CO G 122ND INF CAMP GREENE NC
294607	R	08/27/18	CO F 121ST INF CAMP WHEELER GA
294656	R	08/26/18	WAREHOUSE SUP CO 121ST INF
294662	E	02/15/18	PACKED
294702	R	08/26/18	WAREHOUSE CAMP WHEELER GA
294704		08/21/18	CO B 123RD INF
294727	R	08/19/18	CO H 122ND INF
294751	R	08/19/18	CO H 122ND INF
294751	R	08/31/18	CO H 122ND INF CAMP GREENE NC
294768	R	08/26/18	WAREHOUSE CAMP WHEELER GA
294785		08/21/18	CO A 123RD INF
294790		06/23/18	CAMP BOWIE (FOUND ON RANGE)
294792	R	08/26/18	WAREHOUSE 121ST INF CAMP WHEELER
294840	R	08/19/18	CO F 122ND INF
294840	R	08/31/18	CO F 122ND INF CAMP GREENE NC
294848	R	08/19/18	CO K 122ND INF
294856		08/21/18	CO B 122ND INF
294908	R	08/26/18	WAREHOUSE CAMP WHEELER GA

SER. NO.	MODEL	MM/DD/YY	USAGE
294914	R	09/09/18	106TH AMMUNITION TRAIN
295012	R	08/31/18	CO F 122ND INF CAMP GREENE NC
295014		08/21/18	CO B 124TH INF
295025		08/21/18	CO E 124TH INF
295034		08/21/18	CO A 123RD INF
295036	R	08/27/18	CO E 121ST INF CAMP WHEELER GA
295046	R	09/09/18	106TH AMMUNITION TRAIN
295090	R	09/09/18	HDQTR TRAIN & MP CAMP WHEELER
295096	R	07/23/18	SA FN TEST
295099		08/21/18	CO A 122ND INF
295101	R	08/31/18	CO H 122ND INF CAMP GREENE NC
295134	W	05/20/18	INTERCHANGEABILITY TEST
295140		08/21/18	CO A 123RD INF
295175	R	08/19/18	CO K 121ST INF
295192	R	09/09/18	106TH ENGR
295203		08/21/18	CO H 123RD INF
295266		08/21/18	CO H 124TH INF
295276	R	08/27/18	CO H 121ST INF CAMP WHEELER GA
295330	W	27	URBAN MA (ISSUED)
295333	R	08/19/18	CO F 122ND INF
295349	R	08/19/18	CO K 122ND INF
295350		08/21/18	CO C 122ND INF
295359	R	08/27/18	CO D 121ST INF CAMP WHEELER GA
295360		08/21/18	CO F 124TH INF
295361	R	08/26/18	WAREHOUSE CAMP WHEELER GA
295404		08/21/18	CO G 123RD INF
295419	R	08/30/18	WAREHOUSE CAMP WHEELER GA
295454		08/21/18	CO B 124TH INF
295475	R	08/26/18	CO F 106TH AMMO TRAIN
295478	R	08/31/18	CO E 122ND INF CAMP GREENE NC
295493		08/21/18	CO B 121ST INF
295503	R	08/19/18	CO H 122ND INF
295523	R	08/19/18	CO I 122ND INF
295526	R	08/26/18	WAREHOUSE CAMP WHEELER GA
295534		08/21/18	CO D 123RD INF
295541	R	08/19/18	CO K 121ST INF
295602	R	08/14/18	122ND INF CAMP WHEELER
295608	R	08/26/18	WAREHOUSE 106TH AMMO TRAIN
295667	R	08/27/18	CO F 121ST INF CAMP WHEELER GA
295684	R	08/19/18	CO E 122ND INF
295697	R	08/31/18	CO E 122ND INF CAMP GREENE NC
295713	R	08/19/18	CO M 121ST INF
295738		08/21/18	CO B 121ST INF
295754		08/21/18	CO M 124TH INF
295758	R	08/27/18	CO C 122ND INF CAMP WHEELER GA
295820		08/21/18	CO G 123RD INF
295821	R	08/26/18	WAREHOUSE CAMP WHEELER GA
295827	R	08/26/18	WAREHOUSE 106TH AMMO TRAIN
295849		08/21/18	CO D 124TH INF
295858		08/21/18	CO F 124TH INF
295884		08/21/18	CO G 124TH INF
295887	R	08/26/18	WAREHOUSE CAMP WHEELER GA
295895	R	08/26/18	WAREHOUSE CAMP WHEELER GA
295907		08/21/18	HQ CO 124TH INF
295942		08/21/18	CO B 121ST INF
295996	R	08/26/18	WAREHOUSE CAMP WHEELER GA
296018		08/21/18	CO E 124TH INF
296051		08/21/18	CO H 123RD INF
296063	R	08/26/18	WAREHOUSE CAMP WHEELER GA
296066	R	08/19/18	CO H 122ND INF
296066	R	08/31/18	CO H 122ND INF CAMP GREENE NC
296089	R	08/26/18	WAREHOUSE 106TH AMMO TRAIN
296118		03/17/18	111TH ENGR AEF
296119	R	08/30/18	WAREHOUSE CAMP WHEELER GA
296129		08/21/18	CO D 124TH INF
296145	R	08/26/18	WAREHOUSE SUP CO 121ST INF
296163	R	08/30/18	WAREHOUSE CAMP WHEELER GA
296200	R	08/31/18	CO G 122ND INF CAMP GREENE NC
296222		08/21/18	CO C 123RD INF
296236	R	09/09/18	106TH AMMUNITION TRAIN
296260		08/21/18	CO L 124TH INF
296287	R	08/30/18	WAREHOUSE CAMP WHEELER GA
296301		08/21/18	CO B 122ND INF
296406		08/21/18	CO B 124TH INF
296455		08/21/18	HQ CO 124TH INF
296494	R	08/19/18	CO E 122ND INF
296510		08/21/18	CO B 123RD INF
296523	R	08/26/18	WAREHOUSE 106TH AMMO TRAIN
296527		08/21/18	CO C 123RD INF
296551	R	08/26/18	WAREHOUSE CAMP WHEELER GA
296567	R	08/26/18	WAREHOUSE SUP CO 121ST INF
296582		08/21/18	CO A 122ND INF
296603	R	08/19/18	CO H 122ND INF
296605		08/21/18	CO H 123RD INF
296623	R	08/19/18	CO E 122ND INF
296631	R	08/26/18	WAREHOUSE 121ST INF CAMP WHEELER
296689	R	06/26/18	SA FN TEST
296703		08/21/18	CO B 123RD INF
296709	R	08/30/18	WAREHOUSE CAMP WHEELER GA
296721	R	08/26/18	WAREHOUSE 106TH AMMO TRAIN
296770	R	08/19/18	CO H 122ND INF
296811	R	08/26/18	WAREHOUSE CAMP WHEELER GA
296834		08/21/18	CO D 123RD INF
296857	R	08/31/18	CO G 122ND INF CAMP GREENE NC
296863		03/17/18	111TH ENGR AEF
296867		08/21/18	CO D 124TH INF
296872		08/21/18	CO M 124TH INF
296881	R	08/19/18	CO I 122ND INF
296884	R	08/30/18	WAREHOUSE CAMP WHEELER GA
296914	R	08/19/18	CO K 122ND INF
296925	R	08/19/18	CO L 121ST INF
296928	R	08/26/18	WAREHOUSE 121ST INF CAMP WHEELER
296949		08/21/18	CO H 124TH INF
296984	R	27	URBAN MA (ISSUED)
296987	R	08/30/18	WAREHOUSE CAMP WHEELER GA
297014		08/21/18	CO D 123RD INF
297085		08/21/18	CO C 124TH INF
297114	R	08/26/18	WAREHOUSE CAMP WHEELER GA
297158	R	08/27/18	CO H 121ST INF CAMP WHEELER GA
297161	R	08/26/18	WAREHOUSE SUP CO 121ST INF
297164	R	08/27/18	CO F 121ST INF CAMP WHEELER GA
297185		08/21/18	CO B 123RD INF
297187		08/21/18	CO A 123RD INF
297211		08/21/18	CO F 124TH INF
297216		08/21/18	CO M 124TH INF
297233		08/21/18	CO H 123RD INF
297243	R	08/27/18	CO C 122ND INF CAMP WHEELER GA
297244	R	08/27/18	CO A 121ST INF CAMP WHEELER GA
297271		08/21/18	CO H 123RD INF
297286	R	09/09/18	106TH AMMUNITION TRAIN
297292		08/21/18	CO B 124TH INF
297295	R	06/12/18	SA FUNCTION TEST
297303	R	08/26/18	WAREHOUSE SUP CO 121ST INF
297305		08/21/18	HQ CO 124TH INF
297349	R	08/19/18	CO M 122ND INF
297352		08/21/18	CO I 124TH INF
297357		08/21/18	CO B 123RD INF
297360	R	08/26/18	WAREHOUSE CAMP WHEELER GA
297364	R	08/19/18	CO H 122ND INF
297372	R	08/26/18	WAREHOUSE SUP CO 121ST INF
297378	R	08/31/18	CO F 122ND INF CAMP GREENE NC
297401	R	08/26/18	WAREHOUSE CAMP WHEELER GA
297436	R	08/26/18	WAREHOUSE CAMP WHEELER GA
297474		08/21/18	CO D 122ND INF
297519	R	08/26/18	WAREHOUSE 106TH AMMO TRAIN
297532		08/21/18	CO G 123RD INF
297533		03/17/18	111TH ENGR AEF
297564		08/19/18	CO G 121ST INF
297584		08/19/18	CO L 122ND INF
297664		08/21/18	CO F 124TH INF
297709		08/19/18	CO E 122ND INF
297722		08/21/18	CO E 123RD INF
297738		08/19/18	CO H 121ST INF
297783		08/19/18	CO E 122ND INF
297787		08/21/18	CO C 122ND INF
297810		08/21/18	CO B 122ND INF
297887		08/21/18	CO B 123RD INF
297906		08/21/18	CO C 124TH INF
297930		08/21/18	CO A 121ST INF
297953		08/21/18	CO D 122ND INF
297962		08/21/18	CO C 123RD INF
297971	W	27	URBAN MA (ISSUED)
297973		08/21/18	CO C 124TH INF
297996	R	09/09/18	106TH ENGR

SER. NO.	MODEL	MM/DD/YY	USAGE
298080	E	05/10/18	CAMP HANCOCK (TEST)
298105		08/19/18	CO F 122ND INF
298108		08/21/18	CO A 123RD INF
298131		08/18/18	CO I 121ST INF
298229		08/19/18	CO G 122ND INF
298240		08/21/18	CO C 122ND INF
298267		08/21/18	CO A 121ST INF
298291		08/21/18	CO G 123RD INF
298315		08/21/18	CO H 123RD INF
298337		08/21/18	CO B 122ND INF
298366		08/21/18	CO B 122ND INF
298393		08/21/18	CO A 124TH INF
298450		08/21/18	CO E 124TH INF
298459		08/21/18	CO A 121ST INF
298465		08/19/18	CO K 121ST INF
298524		08/19/18	CO L 122ND INF
298615		08/19/18	CO F 122ND INF
298620		08/21/18	CO E 123RD INF
298649		08/21/18	CO I 124TH INF
298667		08/21/18	CO C 123RD INF
298679		08/21/18	HQ CO 124TH INF
298690		12/01/25	LITTLE ROCK AIR INT DEP TO RIA
298714		12/01/25	LITTLE ROCK AIR INT DEP TO RIA
298745		08/21/18	CO B 121ST INF
298813		08/19/18	CO H 122ND INF
298910		08/21/18	CO M 124TH INF
298928		08/19/18	CO K 121ST INF
298942		08/21/18	CO D 121ST INF
298955		08/19/18	CO F 121ST INF
298961		08/19/18	CO L 122ND INF
298978		08/19/18	CO K 121ST INF
299030		08/21/18	CO E 123RD INF
299032		08/19/18	CO E 121ST INF
299145		08/21/18	CO C 122ND INF
299186		08/21/18	CO B 124TH INF
299232		08/21/18	CO B 123RD INF
299237		08/21/18	CO A 122ND INF
299246		08/21/18	CO D 124TH INF
299256		08/21/18	CO F 124TH INF
299289		08/19/18	CO L 121ST INF
299329		08/19/18	CO E 122ND INF
299333		08/21/18	CO L 124TH INF
299344		08/21/18	CO F 123RD INF
299350		08/21/18	CO F 124TH INF
299385		08/21/18	CO B 124TH INF
299395		08/21/18	CO B 124TH INF
299447		08/21/18	CO A 122ND INF
299464		08/19/18	CO M 122ND INF
299515		08/19/18	CO E 122ND INF
299627		08/21/18	CO B 123RD INF
299635		08/19/18	CO H 122ND INF
299650		08/19/18	CO K 121ST INF
299654		08/21/18	CO E 124TH INF
299786		08/19/18	CO L 122ND INF
299834		08/21/18	CO D 121ST INF
299859		08/19/18	CO I 121ST INF
299883		08/19/18	CO G 122ND INF
299899		08/19/18	CO M 121ST INF
299921		08/21/18	CO D 123RD INF
300000	R	07/26/18	SA FN TEST
300011		08/21/18	CO C 123RD INF
300025		08/21/18	CO L 124TH INF
300029		08/21/18	CO M 124TH INF
300168		08/21/18	CO C 124TH INF
300185		08/19/18	CO H 122ND INF
300217		08/19/18	CO K 122ND INF
300263		08/21/18	CO L 124TH INF
300268		08/21/18	CO I 124TH INF
300270		08/21/18	CO E 124TH INF
300303		08/21/18	CO A 121ST INF
300304		08/21/18	CO D 121ST INF
300308	R	09/09/18	106TH AMMUNITION TRAIN
300323		08/19/18	CO L 122ND INF
300402		08/19/18	CO F 122ND INF
300452		08/21/18	CO F 123RD INF
300453		08/21/18	CO M 124TH INF
300513		08/19/18	CO K 122ND INF
300537		08/21/18	CO H 124TH INF
300579		08/21/18	CO L 124TH INF
300594		08/21/18	CO A 123RD INF
300628		08/21/18	CO F 124TH INF
300641		08/19/18	CO H 122ND INF
300668		08/21/18	CO F 124TH INF
300681		08/21/18	CO D 123RD INF
300689		08/21/18	CO E 123RD INF
300696		08/21/18	CO C 124TH INF
300703		08/21/18	CO B 123RD INF
300758		08/19/18	CO E 122ND INF
300765		08/21/18	CO C 123RD INF
300876		08/19/18	CO M 121ST INF
300887		08/21/18	CO C 123RD INF
300897		08/21/18	CO A 124TH INF
300898		08/21/18	CO E 123RD INF
300911		08/21/18	CO E 124TH INF
300921		08/21/18	SUPPLY CO 122ND INF
300966	E	03/28/18	302ND ENGR
300994		08/21/18	CO G 124TH INF
301015		08/21/18	CO B 122ND INF
301097		08/19/18	CO M 121ST INF
301127		08/19/18	CO H 121ST INF
301142		08/19/18	CO H 122ND INF
301144		08/21/18	CO I 124TH INF
301182		08/21/18	CO G 123RD INF
301184		08/21/18	CO C 123RD INF
301222		08/21/18	CO C 121ST INF
301291	R	07/13/18	SA FN TEST
301315		08/21/18	CO C 124TH INF
301403		08/21/18	CO H 122ND INF
301414		09/09/18	106TH ENGR
301422		08/19/18	CO H 121ST INF
301457		08/21/18	CO C 122ND INF
301458		08/21/18	CO M 122ND INF
301459		08/21/18	CO A 122ND INF
301460		08/19/18	CO E 121ST INF
301467		08/19/18	CO I 121ST INF
301487		08/21/18	CO I 124TH INF
301492		08/21/18	CO D 124TH INF
301546		08/21/18	CO A 121ST INF
301633		08/21/18	CO F 124TH INF
301707		08/19/18	CO I 122ND INF
301709		08/21/18	CO B 122ND INF
301729		08/21/18	CO D 124TH INF
301762		08/19/18	CO I 122ND INF
301766		08/21/18	CO D 122ND INF
301819		08/19/18	CO L 121ST INF
301841		08/21/18	CO K 124TH INF
301863		08/21/18	CO A 123RD INF
301887		08/21/18	CO B 121ST INF
301906		08/19/18	CO L 122ND INF
301962		08/19/18	CO E 121ST INF
301970		08/21/18	CO D 121ST INF
301996	R	07/16/18	SA FN TEST
302154		08/19/18	CO M 122ND INF
302277		08/21/18	HQ CO 124TH INF
302282		08/19/18	CO F 122ND INF
302296		08/21/18	CO A 121ST INF
302297		08/21/18	HQ CO 124TH INF
302311		08/21/18	HQ CO 124TH INF
302328		08/19/18	CO G 122ND INF
302373		08/19/18	CO F 121ST INF
302379		08/21/18	CO C 123RD INF
302469		08/21/18	CO C 122ND INF
302498	W	07/03/18	SA FN TEST
302550		08/19/18	CO M 121ST INF
302559		08/21/18	CO F 123RD INF
302589		08/21/18	CO L 124TH INF
302590		08/19/18	CO L 121ST INF
302592	R	11/21/18	CAMP WHEELER
302598		08/21/18	CO A 124TH INF
302620		12/01/25	LITTLE ROCK AIR INT DEP TO RIA
302638	R	09/09/18	106TH ENGR
302690		08/19/18	CO H 122ND INF
302716	R	09/09/18	CO I 124TH INF
302736		08/21/18	CO H 124TH INF

SER. NO.	MODEL	MM/DD/YY	USAGE
302743		08/21/18	CO B 124TH INF
302750		08/21/18	CO E 123RD INF
302773		08/19/18	CO F 121ST INF
302815		08/19/18	CO I 121ST INF
302818		08/21/18	HQ CO 124TH INF
302829		08/21/18	CO G 124TH INF
302861		08/21/18	CO H 124TH INF
302869		08/21/18	CO H 123RD INF
302904		08/21/18	CO M 124TH INF
302922		08/21/18	CO D 121ST INF
302945		08/21/18	CO D 123RD INF
303002		08/19/18	CO I 121ST INF
303041		08/21/18	CO D 121ST INF
303046		08/19/18	CO L 121ST INF
303124		08/19/18	CO L 122ND INF
303160		08/21/18	CO A 124TH INF
303209		12/01/25	LITTL ROCK AIR INT DEP TO RIA
303264		08/21/18	CO D 122ND INF
303303		08/19/18	CO K 122ND INF
303350		08/21/18	CO D 121ST INF
303354		08/19/18	CO H 122ND INF
303359		08/19/18	CO M 122ND INF
303363		08/19/18	CO K 122ND INF
303366		08/19/18	CO F 121ST INF
303419		08/19/18	CO K 122ND INF
303442		08/21/18	CO D 124TH INF
303473		08/19/18	CO H 122ND INF
303540		08/21/18	CO M 124TH INF
303542		08/21/18	CO E 123RD INF
303589	R	09/09/18	CO A123RD INF
303636		08/21/18	CO C 124TH INF
303641		08/21/18	CO C 122ND INF
303644		08/21/18	CO A 122ND INF
303651		12/01/25	LITTLE ROCK AIR INT DEP TO RIA
303664		08/21/18	CO A 122ND INF
303667		08/21/18	CO B 124TH INF
303689		08/21/18	CO A 123RD INF
303701		08/21/18	CO L 124TH INF
303758		08/21/18	HQ CO 124TH INF
303762		08/21/18	CO K 124TH INF
303808		08/21/18	CO H 124TH INF
303849		08/21/18	CO B 124TH INF
303854		08/21/18	CO M 124TH INF
303894		08/21/18	CO F 124TH INF
303902	R	07/19/18	SA DAILY FUNCTION TEST
303971		08/21/18	CO B 124TH INF
304021	R	08/30/18	WAREHOUSE CAMP WHEELER GA
304025		08/21/18	CO E 123RD INF
304099	R	09/09/18	106TH AMMUNITION TRAIN
304109	R	08/19/18	CO I 122ND INF
304117	R	08/26/18	WAREHOUSE CAMP WHEELER GA
304122		08/21/18	CO H 123RD INF
304133		08/21/18	CO K 124TH INF
304138		08/21/18	CO B 122ND INF
304141	R	08/19/18	CO I 122ND INF
304151	R	08/28/18	CO K 122ND INF CAMP WHEELER GA
304167	R	08/19/18	CO M 121ST INF
304192		08/21/18	CO D 123RD INF
304264		08/21/18	CO G 123RD INF
304287		08/21/18	CO M 124TH INF
304350		08/21/18	CO G 123RD INF
304383	R	08/26/18	WAREHOUSE SUP CO 121ST INF
304388	R	08/19/18	CO E 122ND INF
304399	R	08/26/18	WAREHOUSE CAMP WHEELER GA
304407		08/21/18	CO E 124TH INF
304412	R	08/26/18	WAREHOUSE CAMP WHEELER GA
304449	R	08/26/18	CO F 106TH AMMO TRAIN
304449	R	09/09/18	106TH AMMUNITION TRAIN
304456	R	08/27/18	CO C 122ND INF CAMP WHEELER GA
304469	R	08/19/18	CO I 122ND INF
304487	E	01/27/34	GRIFFIN & HOWE (SALE)
304494	R	08/19/18	CO L 122ND INF
304502		08/21/18	CO A 123RD INF
304511		08/21/18	CO C 124TH INF
304516	R	08/27/18	CO C 121ST INF CAMP WHEELER GA
304541		08/21/18	CO H 124TH INF
304562		08/21/18	CO F 124TH INF

SER. NO.	MODEL	MM/DD/YY	USAGE
304580		08/21/18	CO D 121ST INF
304672	R	08/19/18	CO M 121ST INF
304696		08/21/18	CO E 123RD INF
304710	R	08/19/18	CO L 122ND INF
304722	R	08/26/18	WAREHOUSE 121ST INF CAMP WHEELER
304725	R	08/26/18	WAREHOUSE 106TH AMMO TRAIN
304747	R	08/27/18	CO C 122ND INF CAMP WHEELER GA
304764	R	02/05/18	REMINGTON TO TREASURY DEPT
304770		08/21/18	C O E 124TH INF
304859	R	08/19/18	CO F 122ND INF
304859	R	08/31/18	CO F 122ND INF CAMP GREENE NC
304917	R	08/31/18	CO E 122ND INF CAMP GREENE NC
304928	E	09/04/18	HQ CO 9TH INF CAMP FREMONT
304928	E	12/02/18	CAMP FREMONT
304929		08/21/18	CO B 123RD INF
304944	R	08/27/18	CO B 122ND INF CAMP WHEELER GA
304945	R	08/26/18	WAREHOUSE 106TH AMMO TRAIN
304961	E	02/15/18	PACKED
305000		08/21/18	CO B 123RD INF
305021	R	08/19/18	CO K 121ST INF
305032		08/21/18	CO A 123RD INF
305061	R	08/19/18	CO K 122ND INF
305070		08/21/18	CO A 121ST INF
305100	R	08/30/18	WAREHOUSE CAMP WHEELER GA
305128	R	08/26/18	WAREHOUSE 106TH AMMO TRAIN
305138	R	08/26/18	WAREHOUSE CAMP WHEELER GA
305144		08/21/18	CO E 123RD INF
305150	R	08/19/18	CO H 122ND INF
305196		08/21/18	CO G 123RD INF
305220		08/21/18	CO F 124TH INF
305225		08/21/18	CO B 123RD INF
305241	R	08/19/18	CO L 121ST INF
305244		08/21/18	CO D 121ST INF
305303	R	08/30/18	WAREHOUSE CAMP WHEELER GA
305317		08/21/18	HQ CO 124TH INF
305352	R	08/27/18	CO D 121ST INF CAMP WHEELER GA
305362		02/17/44	USAT DEUTGAN S-108
305368	E	05/17/18	164TH DEPOT BRIG CAMP FUNSTON
305409	R	08/19/18	CO F 122ND INF
305413		08/21/18	CO H 124TH INF
305426		08/21/18	CO B 123RD INF
305446	R	08/31/18	CO G 122ND INF CAMP GREENE NC
305452	R	08/26/18	WAREHOUSE 106TH AMMO TRAIN
305453	R	08/26/18	WAREHOUSE SUP CO 121ST INF
305466	R	08/26/18	WAREHOUSE 121ST INF CAMP WHEELER
305481		08/21/18	CO B 123RD INF
305485	R	08/19/18	CO E 122ND INF
305500	R	08/30/18	WAREHOUSE CAMP WHEELER GA
305508	E	05/20/18	165TH DEPOT BRIG CAMP TRAVIS
305517		08/21/18	CO G 123RD INF
305527		08/21/18	CO A 123RD INF
305573	E	02/15/18	PACKED
305579	R	08/19/18	CO E 122ND INF
305580	R	08/27/18	CO C 122ND INF CAMP WHEELER GA
305589	R	08/30/18	WAREHOUSE CAMP WHEELER GA
305595		08/21/18	CO D 123RD INF
305638	R	08/26/18	WAREHOUSE 106TH AMMO TRAIN
305705	W	.27	URBAN MA (ISSUED)
305751	R	08/31/18	CO E 122ND INF CAMP GREENE NC
305761	R	08/19/18	CO E 122ND INF
305793	E	07/31/18	47TH CO 164TH DEPOT BRIG
305808		08/21/18	CO D 124TH INF
305842		12/01/25	LITTLE ROCK AIR INT DEP TO RIA
305845		08/21/18	CO L 124TH INF
305872		08/21/18	CO H 124TH INF
305878	R	08/30/18	WAREHOUSE CAMP WHEELER GA
305945		08/21/18	CO H 123RD INF
305964		08/21/18	CO H 123RD INF
305986	R	08/26/18	WAREHOUSE CAMP WHEELER GA
306020		08/21/18	CO A 124TH INF
306076		08/21/18	CO K 124TH INF
306166		12/01/25	LITTLE ROCK AIR INT DEP TO RIA
306172	R	08/26/18	WAREHOUSE SUP CO 121ST INF
306182		08/21/18	CO B 123RD INF
306192	R	08/19/18	CO H 122ND INF
306207		08/21/18	CO B 123RD INF
306211		08/21/18	CO K 124TH INF

SER. NO.	MODEL	MM/DD/YY	USAGE
306226		08/21/18	CO G 123RD INF
306234	R	08/26/18	WAREHOUSE 106TH AMMO TRAIN
306242		08/21/18	SUPPLY CO 122ND INF
306243	R	08/27/18	CO I 122ND INF CAMP WHEELER GA
306280	R	08/31/18	CO F 122ND INF CAMP GREENE NC
306315	R	09/09/18	106TH AMMUNITION TRAIN
306336	R	08/27/18	CO D 122ND INF CAMP WHEELER GA
306372	R	08/31/18	CO H 122ND INF CAMP GREENE NC
306385		08/21/18	HQ CO 124TH INF
306392		08/21/18	CO D 122ND INF
306418		08/21/18	CO E 124TH INF
306437	R	08/27/18	CO L 122ND INF CAMP WHEELER GA
306442		08/21/18	CO G 123RD INF
306457	R	08/19/18	CO E 122ND INF
306457	R	08/31/18	CO E 122ND INF CAMP GREENE NC
306464		08/21/18	CO C 122ND INF
306476		08/02/11	8CO D 124TH INF
306489	R	08/31/18	CO H 122ND INF CAMP GREENE NC
306495		08/21/18	HQ CO 124TH INF
306499		08/21/18	CO I 124TH INF
306499	R	09/09/18	CO I 124TH INF
306517		08/21/18	CO B 124TH INF
306526	R	08/31/18	CO H 122ND INF CAMP GREENE NC
306526	R	08/19/18	CO H 122ND INF
306529	R	08/27/18	CO D 122ND INF CAMP WHEELER GA
306530		08/21/18	CO C 124TH INF
306532		08/21/18	CO A 121ST INF
306538	R	08/26/18	WAREHOUSE SUP CO 121ST INF
306555		08/21/18	CO C 122ND INF
306588	R	09/09/18	HDQTR TRAIN & MP CAMP WHEELER
306599		08/21/18	CO H 123RD INF
306617	R	08/19/18	CO I 122ND INF
306620	R	08/27/18	CO H 121ST INF CAMP WHEELER GA
306640		08/21/18	CO C 122ND INF
306642	E	02/15/18	PACKED
306709		08/21/18	CO E 123RD INF
306734	R	08/19/18	CO M 122ND INF
306761		08/21/18	CO I 124TH INF
306812		08/21/18	CO B 121ST INF
306825	R	08/19/18	CO M 121ST INF
306866	R	09/09/18	106TH ENGR
306872		08/21/18	CO G 124TH INF
306879		08/21/18	CO H 123RD INF
306885		08/21/18	CO D 121ST INF
306886		08/21/18	CO D 124TH INF
306891	R	08/27/18	CO B 122ND INF CAMP WHEELER GA
306898	R	08/31/18	CO G 122ND INF CAMP GREENE NC
306906	R	09/09/18	106TH ENGR
306924	R	08/19/18	CO F 122ND INF
306936	R	08/26/18	WAREHOUSE 106TH AMMO TRAIN
306937		08/21/18	CO E 123RD INF
306961		08/21/18	CO D 123RD INF
307011		08/21/18	CO B 123RD INF
307017	R	08/26/18	WAREHOUSE SUP CO 121ST INF
307020		08/21/18	CO H 124TH INF
307034	R	08/27/18	CO H 121ST INF CAMP WHEELER GA
307069	R	08/27/18	CO C 122ND INF CAMP WHEELER GA
307069		08/21/18	CO C 122ND INF
307086		08/21/18	CO K 124TH INF
307117	R	08/19/18	CO M 122ND INF
307140		12/01/25	LITTLE ROCK AIR INT DEP TO RIA
307156	R	08/19/18	CO F 122ND INF
307163	R	08/19/18	CO H 122ND INF
307175	R	08/19/18	CO M 121ST INF
307217		08/21/18	CO E 123RD INF
307250	R	08/19/18	CO I 122ND INF
307252	R	08/27/18	CO D 121ST INF CAMP WHEELER GA
307272		08/21/18	CO H 123RD INF
307275	R	08/31/18	CO F 122ND INF CAMP GREENE NC
307325	R	08/30/18	WAREHOUSE CAMP WHEELER GA
307326		08/21/18	CO C 123RD INF
307379	R	08/26/18	WAREHOUSE SUP CO 121ST INF
307389	R	08/19/18	CO F 122ND INF
307409	R	08/19/18	CO I 122ND INF
307474		08/21/18	CO L 124TH INF
307475	R	08/31/18	CO F 122ND INF CAMP GREENE NC
307508	R	08/31/18	CO G 122ND INF CAMP GREENE NC
307544	R	08/26/18	WAREHOUSE CAMP WHEELER GA
307566	W	07/07/18	SA FN TEST
307582	R	08/19/18	CO H 122ND INF
307594	R	08/26/18	WAREHOUSE 106TH AMMO TRAIN
307615	R	08/26/18	WAREHOUSE CAMP WHEELER GA
307618		08/21/18	CO B 123RD INF
307644	R	08/19/18	CO F 122ND INF
307644	R	08/31/18	CO F 122ND INF CAMP GREENE NC
307666		08/21/18	CO C 124TH INF
307676	R	08/26/18	WAREHOUSE SUP CO 121ST INF
307679		08/21/18	CO K 124TH INF
307696	R	08/26/18	WAREHOUSE 106TH AMMO TRAIN
307700	R	08/19/18	CO I 122ND INF
307725	R	08/19/18	CO F 122ND INF
307744	R	08/30/18	WAREHOUSE CAMP WHEELER GA
307760	R	08/31/18	CO F 122ND INF CAMP GREENE NC
307760	R	08/19/18	CO F 122ND INF
307814	R	08/26/18	WAREHOUSE SUP CO 121ST INF
307815		08/21/18	CO G 124TH INF
307816		08/21/18	CO M 124TH INF
307897		08/21/18	CO D 122ND INF
307947		08/21/18	CO F 123RD INF
307952		08/21/18	CO E 123RD INF
308017	R	08/26/18	WAREHOUSE CAMP WHEELER GA
308061		08/21/18	CO B 124TH INF
308090	R	08/31/18	CO G 122ND INF CAMP GREENE NC
308096	R	08/27/18	CO L 122ND INF CAMP WHEELER GA
308106	R	08/19/18	CO H 122ND INF
308148	R	08/26/18	WAREHOUSE 121ST INF CAMP WHEELER
308190		08/21/18	CO B 122ND INF
308218	R	08/26/18	WAREHOUSE SUP CO 121ST INF
308233		08/21/18	CO C 121ST INF
308243	R	08/19/18	CO I 122ND INF
308289	R	08/19/18	CO L 122ND INF
308338	R	08/19/18	CO H 122ND INF
308353		08/21/18	CO C 123RD INF
308358		08/21/18	HQ CO 124TH INF
308362	R	08/19/18	CO F 122ND INF
308365	R	08/31/18	CO F 122ND INF CAMP GREENE NC
308385		08/21/18	CO B 124TH INF
308416		08/21/18	CO A 123RD INF
308434		08/21/18	CO C 123RD INF
308447	W	06/12/18	SA FUNCTION TEST
308453		08/21/18	CO H 123RD INF
308459		08/21/18	CO B 123RD INF
308473		08/21/18	CO C 122ND INF
308554		08/21/18	CO I 124TH INF
308571	E	02/15/18	302ND ENGR (INSP BY OAKES)
308571	E	04/24/18	CAMP UPTON
308571	E	04/24/18	REVIEWED BY ENGR BUREAU
308571	E	05/01/18	WATERTOWN ARS (TESTED)
308571		09/17/18	FAILED IN SERVICE
308576	R	08/19/18	CO I 122ND INF
308603	R	08/26/18	WAREHOUSE CAMP WHEELER GA
308623		08/21/18	CO F 124TH INF
308655	R	08/19/18	CO M 122ND INF
308664		08/21/18	CO F 124TH INF
308665		08/21/18	HQ CO 124TH INF
308670		07/24/18	48TH CO 164TH DEPOT BRIG
308679	R	08/19/18	CO F 122ND INF
308716	R	08/19/18	CO F 122ND INF
308716	R	08/31/18	CO F 122ND INF CAMP GREENE NC
308735		08/21/18	CO L 124TH INF
308794		08/21/18	CO F 124TH INF
308799	R	08/27/18	CO F 121ST INF CAMP WHEELER GA
308802	R	08/19/18	CO E 122ND INF
308810		08/21/18	CO C 122ND INF
308857		08/21/18	CO D 122ND INF
308890	R	09/09/18	106TH AMMUNITION TRAIN
308962		08/21/18	CO B 124TH INF
308963		08/21/18	CO B 121ST INF
308974	E	06/19/18	CAMP FUNSTON (RUSTY)
308974		08/21/18	CO A 122ND INF
308976	R	08/31/18	CO H 122ND INF CAMP GREENE NC
308977		08/21/18	CO B 121ST INF
309038		08/21/18	CO B 124TH INF
309045	R	08/19/18	CO H 122ND INF

SER. NO.	MODEL	MM/DD/YY	USAGE
309049	R	08/19/18	CO F 122ND INF
309058	R	08/26/18	WAREHOUSE CAMP WHEELER GA
309063	R	08/26/18	WAREHOUSE 121ST INF CAMP WHEELER
309149		08/21/18	HQ CO 124TH INF
309170	R	08/31/18	CO E 122ND INF CAMP GREENE NC
309226		08/21/18	CO D 123RD INF
309233		08/21/18	CO C 124TH INF
309258	R	08/26/18	WAREHOUSE 106TH AMMO TRAIN
309274		08/21/18	CO B 123RD INF
309277	E	05/16/18	CAMP FUNSTON (RUSTY)
309307	R	08/26/18	WAREHOUSE CAMP WHEELER GA
309315		08/21/18	CO E 123RD INF
309329	R	08/26/18	WAREHOUSE 121ST INF CAMP WHEELER
309351	R	08/19/18	CO M 122ND INF
309372		08/21/18	CO D 123RD INF
309380		08/21/18	CO C 122ND INF
309381		08/21/18	HQ CO 124TH INF
309388		08/21/18	CO C 123RD INF
309401	R	08/27/18	CO F 121ST INF CAMP WHEELER GA
309410	R	08/30/18	WAREHOUSE CAMP WHEELER GA
309437	R	08/26/18	WAREHOUSE 106TH AMMO TRAIN
309445		08/21/18	CO G 123RD INF
309506	R	08/26/18	WAREHOUSE CAMP WHEELER GA
309522	R	08/28/18	CO K 122ND INF CAMP WHEELER GA
309524		08/21/18	CO D 122ND INF
309525	W	27	URBAN MA (ISSUED)
309531		08/21/18	CO B 122ND INF
309550	R	08/31/18	CO G 122ND INF CAMP GREENE NC
309551	R	08/19/18	CO E 122ND INF
309581	R	08/19/18	CO F 122ND INF
309597		08/21/18	CO C 123RD INF
309603	E	05/10/18	CAMP HANCOCK (TEST)
309616	R	08/26/18	WAREHOUSE CAMP WHEELER GA
309631	R	08/26/18	WAREHOUSE SUP CO 121ST INF
309638		08/21/18	CO G 123RD INF
309645	R	09/09/18	CO M 124TH INF
309672		08/21/18	CO M 124TH INF
309686	R	08/19/18	CO E 122ND INF
309686	R	08/31/18	CO E 122ND INF CAMP GREENE NC
309732	W	27	URBAN MA (ISSUED)
309756	R	08/31/18	CO H 122ND INF CAMP GREENE NC
309759		08/21/18	CO D 123RD INF
309808	E	06/24/18	HIGHEST NUMBER IN 322ND INF
309810	R	09/09/18	CO A 123RD INF
309832		08/21/18	CO E 123RD INF
309835		08/21/18	CO A 122ND INF
309867		08/21/18	CO B 121ST INF
309869	R	08/27/18	CO E 121ST INF CAMP WHEELER GA
309882		08/21/18	CO B 123RD INF
309943		08/21/18	CO A 124TH INF
309965		08/21/18	CO A 122ND INF
310043	R	08/27/18	CO B 122ND INF CAMP WHEELER GA
310054		08/21/18	CO G 123RD INF
310058	R	08/27/18	CO E 121ST INF CAMP WHEELER GA
310073	R	08/19/18	CO I 121ST INF
310159	R	06/17/18	SA FUNCTION TEST
310187	W	05/24/33	W B ANDREWS
310188		08/21/18	CO C 122ND INF
310261		08/21/18	CO E 124TH INF
310286	R	08/27/18	CO A 121ST INF CAMP WHEELER GA
310343	W	10/24/18	CO B 36TH INF CAMP DEVENS
310357		08/21/18	CO H 123RD INF
310384	R	08/19/18	CO F 122ND INF
310408	R	08/26/18	WAREHOUSE CAMP WHEELER GA
310430	R	08/19/18	CO E 122ND INF
310454	W	06/13/18	SA DAILY FUNCTION TEST
310522		08/21/18	CO A 124TH INF
310531	R	09/09/18	106TH AMMUNITION TRAIN
310594	R	08/26/18	WAREHOUSE 106TH AMMO TRAIN
310599		08/21/18	CO H 124TH INF
310603	R	08/19/18	CO K 121ST INF
310647		08/21/18	CO B 122ND INF
310665	R	08/27/18	CO G 121ST INF CAMP WHEELER GA
310667	R	08/19/18	CO E 122ND INF
310670	R	08/19/18	CO K 121ST INF
310692	R	08/19/18	CO H 122ND INF
310706	R	08/26/18	WAREHOUSE 121ST INF CAMP WHEELER
310725		09/04/18	CAMP WHEELER
310725	R	09/14/18	124TH INF CAMP WHEELER
310725	R	11/21/18	CAMP WHEELER
310739	R	08/26/18	WAREHOUSE SUP CO 121ST INF
310745	R	08/19/18	CO I 121ST INF
310765		08/21/18	CO B 123RD INF
310805	R	07/02/18	SA FN TEST
310859	R	09/09/18	106TH ENGR
310913	R	08/26/18	WAREHOUSE CAMP WHEELER GA
310952	R	08/26/18	WAREHOUSE CAMP WHEELER GA
310982	R	08/19/18	CO F 122ND INF
310982	R	08/31/18	CO F 122ND INF CAMP GREENE NC
311018	R	08/27/18	CO E 121ST INF CAMP WHEELER GA
311041		08/21/18	CO B 123RD INF
311079	R	08/26/18	WAREHOUSE 106TH AMMO TRAIN
311106	R	08/26/18	WAREHOUSE CAMP WHEELER GA
311141	E	07 28/20	INQUIRY
311152	R	08/19/18	CO H 122ND INF
311161	R	08/19/18	CO I 122ND INF
311229		08/21/18	CO D 122ND INF
311298	R	08/30/18	WAREHOUSE CAMP WHEELER GA
311312		08/21/18	CO D 122ND INF
311466	R	08/19/18	CO M 121ST INF
311470	R	08/19/18	CO H 122ND INF
311518		08/21/18	CO B 122ND INF
311534	W	06/13/18	SA DAILY FUNCTION TEST
311617		08/21/18	CO F 123RD INF
311627		08/21/18	CO H 123RD INF
311646		08/21/18	CO M 124TH INF
311647	R	08/27/18	CO D 122ND INF CAMP WHEELER GA
311674	R	08/19/18	CO I 122ND INF
311704		08/21/18	CO A 122ND INF
311733		08/21/18	CO B 122ND INF
311740		08/21/18	CO H 123RD INF
311758		08/21/18	CO A 124TH INF
311778		08/21/18	CO A 124TH INF
311833		08/21/18	CO I 124TH INF
311837	R	08/19/18	CO E 122ND INF
311838		08/21/18	CO A 124TH INF
311860	R	08/26/18	WAREHOUSE SUP CO 121ST INF
311888	R	08/19/18	CO E 122ND INF
311898	R	08/26/18	WAREHOUSE 106TH AMMO TRAIN
311913	R	08/19/18	CO K 122ND INF
311940	R	08/26/18	WAREHOUSE CAMP WHEELER GA
311943	R	08/31/18	CO F 122ND INF CAMP GREENE NC
311943	R	08/19/18	CO F 122ND INF
311947		08/21/18	CO C 124TH INF
311979		08/21/18	CO B 121ST INF
311982	R	08/26/18	WAREHOUSE CAMP WHEELER GA
312007	R	08/19/18	CO I 122ND INF
312024	W	27	URBAN MA (ISSUED)
312057		08/21/18	HQ CO 124TH INF
312202	R	08/27/18	CO E 121ST INF CAMP WHEELER GA
312216	R	08/27/18	CO F 121ST INF CAMP WHEELER GA
312221	R	08/19/18	CO M 122ND INF
312260		08/21/18	CO A 123RD INF
312265	R	08/26/18	CO A 122ND INF CAMP WHEELER GA
312268	R	08/19/18	CO H 122ND INF
312268	R	08/31/18	CO H 122ND INF CAMP GREENE NC
312330	R	09/09/18	106TH ENGR
312384	R	08/27/18	CO F 121ST INF CAMP WHEELER GA
312402		08/21/18	CO D 123RD INF
312533	R	11/10/22	AUGUSTA ARS (OVERHAULED)
312533		08/21/18	CO C 121ST INF
312575		08/21/18	CO C 124TH INF
312577		08/21/18	CO C 124TH INF
312592		08/21/18	CO K 124TH INF
312632	R	08/19/18	CO I 122ND INF
312685	R	08/19/18	CO H 122ND INF
312709	R	08/19/18	CO M 122ND INF
312752	W	06/13/18	SA DAILY FUNCTION TEST
312798		08/21/18	CO H 124TH INF
312800		08/21/18	CO K 124TH INF
312804	R	08/26/18	WAREHOUSE 106TH AMMO TRAIN
312805	R	08/19/18	CO L 122ND INF
312922	R	08/19/18	CO M 121ST INF
312933	R	09/09/18	106TH ENGR

SER. NO.	MODEL	MM/DD/YY	USAGE
312953		08/21/18	CO K 124TH INF
312981	R	08/26/18	WAREHOUSE 122ND INF CAMP WHEELER
312990		08/21/18	HQ CO 124TH INF
313106		08/21/18	CO A 123RD INF
313123		08/21/18	CO A 123RD INF
313125	R	08/19/18	CO E 122ND INF
313125	R	08/31/18	CO E 122ND INF CAMP GREENE NC
313129	R	08/26/18	WAREHOUSE 106TH AMMO TRAIN
313141	R	08/27/18	CO C 121ST INF CAMP WHEELER GA
313207		08/21/18	CO C 123RD INF
313210	R	08/31/18	CO F 122ND INF CAMP GREENE NC
313210	R	08/19/18	CO F 122ND INF
313220		08/21/18	CO A 122ND INF
313228	R	08/19/18	CO I 122ND INF
313280		08/21/18	CO A 124TH INF
313302	R	08/19/18	CO L 121ST INF
313306	R	08/19/18	CO L 121ST INF
313367		08/21/18	CO F 123RD INF
313496	R	08/26/18	WAREHOUSE 106TH AMMO TRAIN
313514	R	08/19/18	CO K 122ND INF
313534		08/21/18	CO G 123RD INF
313539	R	08/26/18	WAREHOUSE CAMP WHEELER GA
313542	W	06/13/18	SA DAILY FUNCTION TEST
313584	R	08/26/18	WAREHOUSE CAMP WHEELER GA
313680	R	08/26/18	WAREHOUSE SUP CO 121ST INF
313703	R	08/30/18	WAREHOUSE CAMP WHEELER GA
313791	R	08/27/18	CO C 122ND INF CAMP WHEELER GA
313803	R	08/26/18	WAREHOUSE CAMP WHEELER GA
313841	R	08/14/18	122ND INF CAMP WHEELER
313885	R	08/27/18	CO I 122ND INF CAMP WHEELER GA
313907	R	08/19/18	CO E 122ND INF
313973		08/21/18	CO C 122ND INF
314156		08/21/18	CO D 123RD INF
314206	R	08/19/18	CO L 122ND INF
314294		08/21/18	CO E 124TH INF
314359	R	08/26/18	WAREHOUSE 106TH AMMO TRAIN
314429	R	09/09/18	CO I 124TH INF
314575	R	08/26/18	WAREHOUSE SUP CO 121ST INF
314629		08/21/18	CO H 123RD INF
314776		08/21/18	CO C 124TH INF
314951	R	08/19/18	CO G 122ND INF
314990		08/21/18	CO A 122ND INF
315048	R	08/26/18	WAREHOUSE 106TH AMMO TRAIN
315059		08/21/18	CO D 124TH INF
315077	R	08/26/18	WAREHOUSE SUP CO 121ST INF
315079	E	05/10/18	CAMP HANCOCK (TEST)
315096		08/21/18	CO E 123RD INF
315136		08/21/18	CO B 124TH INF
315332	R	08/19/18	CO K 121ST INF
315362	E	02/15/18	PACKED
315377	R	09/09/18	SUPPLY CO 121ST INF
315403	R	09/09/18	106TH ENGR
315404		08/21/18	CO G 123RD INF
315451	R	08/27/18	CO B 122ND INF CAMP WHEELER GA
315452	R	08/19/18	CO K 121ST INF
315725		08/21/18	CO C 121ST INF
315873	R	08/26/18	WAREHOUSE SUP CO 121ST INF
315912	R	09/09/18	SUPPLY CO 121ST INF
315938		08/21/18	CO E 124TH INF
315978		08/21/18	CO E 124TH INF
315979	R	08/26/18	WAREHOUSE CAMP WHEELER GA
316149	R	08/27/18	CO M 122ND INF CAMP WHEELER GA
316178		08/21/18	CO D 124TH INF
316230	R	08/27/18	CO F 121ST INF CAMP WHEELER GA
316315		08/21/18	CO D 121ST INF
316334	R	08/19/18	CO M 122ND INF
316344	E	05/10/18	CAMP HANCOCK (TEST)
316417		08/21/18	CO I 124TH INF
316472	R	08/19/18	CO I 121ST INF
316499	R	08/26/18	WAREHOUSE SUP CO 121ST INF
316813	R	08/27/18	CO F 121ST INF CAMP WHEELER GA
316851	R	08/19/18	CO M 121ST INF
316973	R	08/31/18	CO E 122ND INF CAMP GREENE NC
317011	E	06/19/18	CAMP FUNSTON (RUSTY)
317018	R	08/26/18	WAREHOUSE SUP CO 121ST INF
317019	W	.27	URBAN MA (ISSUED)
317074		08/21/18	CO G 124TH INF
317076	W	06/13/18	SA DAILY FUNCTION TEST
317130	R	08/19/18	CO H 122ND INF
317130	R	08/31/18	CO H 122ND INF CAMP GREENE NC
317144	R	11/04/18	CAMP MCCLELLAN
317144	R	12/03/18	CO D 9TH AMMO TRAIN
317259		08/21/18	CO L 124TH INF
317303	R	08/31/18	CO G 122ND INF CAMP GREENE NC
317312		08/21/18	CO B 124TH INF
317383		08/21/18	CO D 123RD INF
317444		08/21/18	CO I 124TH INF
317503	R	07/15/18	SA FN TEST
317735	R	08/19/18	CO M 121ST INF
317768	R	08/30/18	WAREHOUSE CAMP WHEELER GA
317891		08/21/18	CO A 123RD INF
317909		08/21/18	CO H 124TH INF
317912		08/21/18	CO D 123RD INF
317947	R	08/26/18	WAREHOUSE CAMP WHEELER GA
317979	W	01/27/34	GRIFFIN & HOWE (SALE)
318025	W	08/15/18	INTERCHANGE OF MAY & JUNE PROD
318139		08/21/18	CO G 123RD INF
318233	E	05/03/18	112TH INF
318233	E	09/30/18	CAMP HANCOCK
318238	W	08/15/18	INTERCHANGE OF MAY & JUNE PROD
318314	R	08/17/18	CAMP SHERIDAN
318320	R	08/30/18	WAREHOUSE CAMP WHEELER GA
318423		08/21/18	CO I 124TH INF
318495		08/21/18	CO C 124TH INF
318499	R	08/27/18	CO B 122ND INF CAMP WHEELER GA
318533	R	08/26/18	WAREHOUSE 106TH AMMO TRAIN
318553	R	09/29/18	CO G 45TH INF
318755	W	08/15/18	INTERCHANGE OF MAY & JUNE PROD
319070		08/21/18	CO B 124TH INF
319143	W	08/15/18	INTERCHANGE OF MAY & JUNE PROD
319149		08/21/18	CO B 123RD INF
319185	R	07/24/18	SA FN TEST
319247	R	03/08/18	306TH MG BN
319333	W	.27	URBAN MA (ISSUED)
319427	E	04/24/18	REVIEWED BY ENGR BUREAU
319427	E	05/01/18	WATERTOWN ARS (TESTED)
319427	E	04/24/18	CAMP UPTON
319427	E	03/28/18	302ND ENGR
319427	E	03/06/18	306TH MG BN
319427		09/17/18	FAILED IN SERVICE
319661	R	08/19/18	CO L 122ND INF
319700		08/21/18	HQ CO 124TH INF
319722	W	08/15/18	INTERCHANGE OF MAY & JUNE PROD
319760		08/21/18	CO A 122ND INF
319792	W	06/17/18	SA FUNCTION TEST
319934	R	09/09/18	106TH ENGR
320229	W	06/17/18	SA FN TEST
320345	R	08/27/18	CO B 122ND INF CAMP WHEELER GA
320401	W	.27	URBAN MA (ISSUED)
320492	R	08/31/18	CO H 122ND INF CAMP GREENE NC
320494	R	08/19/18	CO H 122ND INF
320553		08/21/18	CO D 123RD INF
320574	E	05/22/18	CAMP FUNSTON (RUSTY)
320583	W	09/16/18	INTERCHANGEABILITY TEST
320583	W	08/15/18	INTERCHANGE OF MAY & JUNE PROD
320852	W	.27	URBAN MA (ISSUED)
320896	E	05/22/18	CAMP FUNSTON (RUSTY)
321029	E	07/31/18	47TH CO 164TH DEPOT BRIG
321050	R	08/26/18	WAREHOUSE 106TH AMMO TRAIN
321051	R	08/30/18	WAREHOUSE CAMP WHEELER GA
321064	W	08/15/18	INTERCHANGE OF MAY & JUNE PROD
321064	W	09/16/18	INTERCHANGEABILITY TEST
321098	W	08/15/18	INTERCHANGE OF MAY & JUNE PROD
321140	W	08/15/18	INTERCHANGE OF MAY & JUNE PROD
321212	E	03/12/18	FA VELOCITY TEST
321294	R	08/30/18	WAREHOUSE CAMP WHEELER GA
321348	R	08/27/18	CO D 122ND INF CAMP WHEELER GA
321459	W	08/15/18	INTERCHANGE OF MAY & JUNE PROD
321459	W	09/16/18	INTERCHANGEABILITY TEST
321469	R	06/99	CMP INVENTORY
321529		08/21/18	CO G 123RD INF
321604	W	08/15/18	INTERCHANGE OF MAY & JUNE PROD
321677		08/21/18	CO A 122ND INF
321691	R	08/26/18	WAREHOUSE 121ST INF CAMP WHEELER

SER. NO.	MODEL	MM/DD/YY	USAGE
321923		08/21/18	CO C 123RD INF
321962		04/27/44	USAT MARGARET TWAITS S-52
322038		11/20/44	USAT GEN MACARTHUR S-44
322329	R	09/09/18	CO A 123RD INF
322351	W	08/15/18	INTERCHANGE OF MAY & JUNE PROD
322392		08/21/18	CO A 123RD INF
322497	W	08/15/18	INTERCHANGE OF MAY & JUNE PROD
322505	R	08/26/18	WAREHOUSE 106TH AMMO TRAIN
322506	R	08/19/18	CO K 121ST INF
322610		08/21/18	CO C 124TH INF
322641	R	08/30/18	WAREHOUSE CAMP WHEELER GA
322671	E	03/15/18	FA ACCURACY TEST
322671	E	03/12/18	FA VELOCITY TEST
322717	W	08/15/18	INTERCHANGE OF MAY & JUNE PROD
322718		08/21/18	CO A 124TH INF
322735		08/21/18	CO F 124TH INF
322752	R	11/10/22	AUGUSTA ARS (OVERHAULED)
322927		08/21/18	CO C 123RD INF
322938	W	08/15/18	INTERCHANGE OF MAY & JUNE PROD
322974	W	09/16/18	INTERCHANGEABILITY TEST
322995	R	08/26/18	WAREHOUSE 106TH AMMO TRAIN
323037	W	06/17/18	SA FUNCTION TEST
323041	W	08/15/18	INTERCHANGE OF MAY & JUNE PROD
323088	W	08/15/18	INTERCHANGE OF MAY & JUNE PROD
323088	W	09/16/18	INTERCHANGEABILITY TEST
323130	W	08/15/18	INTERCHANGE OF MAY & JUNE PROD
323134	R	08/27/18	CO E 121ST INF CAMP WHEELER GA
323201		08/21/18	CO H 123RD INF
323231		08/21/18	HQ CO 124TH INF
323238	W	08/15/18	INTERCHANGE OF MAY & JUNE PROD
323238	W	09/16/18	INTERCHANGEABILITY TEST
323305	E	05/16/18	CAMP FUNSTON (RUSTY)
323314	R	08/26/18	WAREHOUSE SUP CO 121ST INF
323351	W	08/15/18	INTERCHANGE OF MAY & JUNE PROD
323351	W	09/16/18	INTERCHANGEABILITY TEST
323403	R	08/26/18	WAREHOUSE SUP CO 121ST INF
323459	E	05/16/18	CAMP FUNSTON (RUSTY)
323478	R	08/30/18	WAREHOUSE CAMP WHEELER GA
323503	R	09/09/18	106TH ENGR
323613		08/21/18	CO A 122ND INF
323617	R	07/07/18	SA FN TEST
323663	E	05/17/18	164TH DEPOT BRIG CAMP FUNSTON
323726	R	08/27/18	CO B 122ND INF CAMP WHEELER GA
323811		08/21/18	CO M 124TH INF
323887		08/21/18	CO K 124TH INF
323959		08/21/18	CO G 123RD INF
323974	W	09/16/18	INTERCHANGEABILITY TEST
323974	W	08/15/18	INTERCHANGE OF MAY & JUNE PROD
323985		08/21/18	CO I 124TH INF
324027	W	08/15/18	INTERCHANGE OF MAY & JUNE PROD
324027	W	09/16/18	INTERCHANGEABILITY TEST
324089	E	06/22/48	VETERANS ADMIN (TO RIA)
324089		08/21/18	CO L 124TH INF
324220	R	08/30/18	WAREHOUSE CAMP WHEELER GA
324235	R	08/26/18	WAREHOUSE 106TH AMMO TRAIN
324245		08/21/18	HQ CO 124TH INF
324251	W	08/15/18	INTERCHANGE OF MAY & JUNE PROD
324251	W	09/16/18	INTERCHANGEABILITY TEST
324380	R	08/19/18	CO M 121ST INF
324448	E	05/15/18	CAMP HANCOCK STRIKER TEST
324459	R	08/19/18	CO F 122ND INF
324501		08/21/18	CO H 123RD INF
324521	W	08/15/18	INTERCHANGE OF MAY & JUNE PROD
324648		08/21/18	CO F 124TH INF
324660	R	08/30/18	WAREHOUSE CAMP WHEELER GA
324662		08/21/18	CO D 123RD INF
324732	E	05/16/18	CAMP FUNSTON (RUSTY)
324859	R	08/31/18	CO F 122ND INF CAMP GREENE NC
324900	W	06/19/18	SA FN TEST
324965	W	08/15/18	INTERCHANGE OF MAY & JUNE PROD
325034	W	08/15/18	INTERCHANGE OF MAY & JUNE PROD
325034	W	09/16/18	INTERCHANGEABILITY TEST
325037	E	02/15/18	PACKED
325040	W	08/15/18	INTERCHANGE OF MAY & JUNE PROD
325040	W	09/16/18	INTERCHANGEABILITY TEST
325105	R	08/30/18	WAREHOUSE CAMP WHEELER GA
325296	W	08/15/18	INTERCHANGE OF MAY & JUNE PROD
325296	W	09/16/18	INTERCHANGEABILITY TEST
325394	W	08/15/18	INTERCHANGE OF MAY & JUNE PROD
325411	W	08/15/18	INTERCHANGE OF MAY & JUNE PROD
325411	W	09/16/18	INTERCHANGEABILITY TEST
325518	E	07/31/18	47TH CO 164TH DEPOT BRIG
325563	R	08/19/18	CO H 122ND INF
325563	R	08/31/18	CO H 122ND INF CAMP GREENE NC
325623		02/19/45	USS IY-7
325626	W	08/15/18	INTERCHANGE OF MAY & JUNE PROD
325777		08/21/18	HQ CO 124TH INF
326112		08/21/18	CO C 123RD INF
326128	W	08/15/18	INTERCHANGE OF MAY & JUNE PROD
326171	W	08/15/18	INTERCHANGE OF MAY & JUNE PROD
326269	W	08/15/18	INTERCHANGE OF MAY & JUNE PROD
326397	W	08/15/18	INTERCHANGE OF MAY & JUNE PROD
326460	R	08/30/18	WAREHOUSE CAMP WHEELER GA
326749	W	06/20/18	SA FN TEST
326787	R	08/30/18	WAREHOUSE CAMP WHEELER GA
326829	R	08/26/18	WAREHOUSE 106TH AMMO TRAIN
326852		08/21/18	CO B 122ND INF
326868		08/21/18	CO C 124TH INF
326971		08/21/18	CO I 124TH INF
327021		08/21/18	CO G 123RD INF
327124	R	08/26/18	WAREHOUSE 106TH AMMO TRAIN
327191		11/14/18	CO L 2ND INF
327201		10/15/18	CO L 2ND INF
327297	W	06/20/18	SA FN TEST
327398		08/21/18	CO B 122ND INF
327465	R	08/26/18	WAREHOUSE 121ST INF CAMP WHEELER
327507	R	08/19/18	CO K 122ND INF
327591		08/21/18	CO I 124TH INF
327626	W	06/24/18	SA FN TEST
327719	R	08/27/18	CO B 122ND INF CAMP WHEELER GA
327889		08/21/18	CO E 123RD INF
327948	R	08/19/18	CO K 121ST INF
328032	W	06/25/18	SA FN TEST
328065	E	05/17/18	164TH DEPOT BRIG CAMP FUNSTON
328091	E	11/10/22	AUGUSTA ARS (OVERHAULED)
328184	R	08/26/18	WAREHOUSE SUP CO 121ST INF
328282		07/29/18	CO L 2ND INF
328473	W	08/15/18	INTERCHANGE OF MAY & JUNE PROD
328784	R	08/01/18	SA FN TEST
328838		08/21/18	CO F 124TH INF
328900	W	07/08/18	INTERCHANGEABILITY TEST
328913	R	08/26/18	WAREHOUSE 106TH AMMO TRAIN
329054	R	08/26/18	WAREHOUSE CAMP WHEELER GA
329132	R	08/26/18	CO A 122ND INF CAMP WHEELER GA
329132	R	08/19/18	CO H 122ND INF
329227		08/21/18	CO C 124TH INF
329258		08/21/18	CO C 122ND INF
329320		09/06/18	CO L 2ND INF
329333	R	11/23/18	CAMP MCCLELLAN
329587	R	08/27/18	CO D 121ST INF CAMP WHEELER GA
329850	R	08/26/18	WAREHOUSE 106TH AMMO TRAIN
329867		08/21/18	SUPPLY CO 122ND INF
329876	W	06/25/18	SA FN TEST
329900	W	07/02/18	SA FN TEST
330034		08/21/18	CO A 122ND INF
330147		11/14/18	CO L 2ND INF
330270	W	07/01/18	SA FN TEST
330349	R	08/26/18	WAREHOUSE 106TH AMMO TRAIN
330378		08/21/18	CO C 121ST INF
330616		08/21/18	HQ CO 124TH INF
330697	R	05/17/18	164TH DEPOT BRIG CAMP FUNSTON
330761		08/21/18	CO G 123RD INF
330799	R	08/30/18	WAREHOUSE CAMP WHEELER GA
330974	R	09/09/18	106TH AMMUNITION TRAIN
331086		08/21/18	CO D 123RD INF
331149	E	05/01/18	WATERTOWN ARS (TESTED)
331156		08/21/18	CO H 123RD INF
331285	E	04/17/18	INTERCHANGEABILITY TEST
331520	E	05/16/18	CAMP FUNSTON (RUSTY)
331593	R	08/26/18	WAREHOUSE 121ST INF CAMP WHEELER
331619	R	08/30/18	WAREHOUSE CAMP WHEELER GA
331660	R	08/26/18	WAREHOUSE 106TH AMMO TRAIN
331683	E	05/01/18	WATERTOWN ARS (TESTED)
331742	R	08/30/18	WAREHOUSE CAMP WHEELER GA

SER. NO.	MODEL	MM/DD/YY	USAGE
331863	E	04/01/18	CO I 112TH INF CAMP HANCOCK
331863	E	04/23/18	REVIEWED BY ENGR BUREAU
331863		09/17/18	FAILED IN SERVICE
331891	R	08/27/18	CO M 122ND INF CAMP WHEELER GA
332574	R	08/26/18	WAREHOUSE CAMP WHEELER GA
332601	R	08/30/18	WAREHOUSE CAMP WHEELER GA
332758		05/11/31	APG
332859	R	07/19/18	SA FN TEST
332956	E	05/16/18	CAMP FUNSTON (RUSTY)
333008		08/21/18	CO I 124TH INF
333012	R	08/17/18	CAMP SHERIDAN
333149	E	04/24/18	REVIEWED BY ENGR BUREAU
333149	E	05/01/18	WATERTOWN ARS (TESTED)
333149		09/17/18	FAILED IN SERVICE
333149	E	03/12/18	306TH FA
333149	E	03/28/18	302ND ENGR
333149	E	04/24/18	CAMP UPTON
333204	E	05/20/18	165TH DEPOT BRIG CAMP TRAVIS
333270	R	08/26/18	WAREHOUSE SUP CO 121ST INF
333539		08/21/18	CO C 124TH INF
333860		08/21/18	CO A 122ND INF
334147		08/21/18	CO B 123RD INF
334182		08/21/18	CO A 123RD INF
334296	E	05/01/18	WATERTOWN ARS (TESTED)
334378		08/21/18	CO E 123RD INF
334393	E	02/15/18	PACKED
334450		07/30/18	SA FN TEST
334538	W	06/25/18	SA FN TEST
334602	R	08/26/18	WAREHOUSE 106TH AMMO TRAIN
334721		08/31/18	CO L 2ND INF
334790		08/21/18	SUPPLY CO 122ND INF
334900	W	06/28/18	SA FN TEST
334935	R	08/19/18	CO G 122ND INF
334935	R	08/31/18	CO G 122ND INF CAMP GREENE NC
335041	R	08/19/18	CO E 122ND INF
335041	R	08/31/18	CO E 122ND INF CAMP GREENE NC
335471	R	08/26/18	WAREHOUSE 106TH AMMO TRAIN
335574	W	06/27/18	SA FN TEST
335615	E	04/17/18	INTERCHANGEABILITY TEST
335616	R	05/17/18	164TH DEPOT BRIG CAMP FUNSTON
335815	W	07/08/18	INTERCHANGEABILITY TEST
335967		07/24/18	48TH CO 164TH DEPOT BRIG
335971		02/17/44	USAT DEUTGAN S-108
336352	W	07/08/18	INTERCHANGEABILITY TEST
336489	E	02/15/18	PACKED
336546	E	03/28/18	302ND ENGR CAMP UPTON (INSP BYRF NASH)
336546	E	04/24/18	REVIEWED BY ENGR BUREAU
336546	E	05/01/18	WATERTOWN ARS (TESTED)
336546		09/17/18	FAILED IN SERVICE
336732	R	08/26/18	WAREHOUSE CAMP WHEELER GA
336826	E	02/15/18	PACKED
337021	E	05/16/18	CAMP FUNSTON (RUSTY)
337069		08/21/18	CO H 123RD INF
337071	R	11/23/18	CAMP MCCLELLAN
337188	R	08/26/18	62ND BRIGADE HQ CAMP WHEELER GA
337204	W	07/08/18	INTERCHANGEABILITY TEST
337445	R	06/99	CMP INVENTORY
338182	E	05/16/18	CAMP FUNSTON (RUSTY)
338250	R	08/26/18	WAREHOUSE SUP CO 121ST INF
338445	R	08/26/18	106TH AMMO TRAIN CAMP WHEELER
338473	W	07/08/18	SA FN TEST
338508		08/21/18	CO G 124TH INF
338925	E	04/10/18	CO K 110TH INF CAMP HANCOCK
339528	E	04/17/18	INTERCHANGEABILITY TEST
339926	R	08/19/18	CO E 122ND INF
339968		08/21/18	CO A 124TH INF
340224		08/21/18	CO A 122ND INF
340224	R	08/26/18	CO A 122ND INF CAMP WHEELER GA
340301	R	09/09/18	4TH TRENCH MORTAR BATTY
340915	R	07/09/18	SA FN TEST
341261	E	05/16/18	CAMP FUNSTON (RUSTY)
341321		08/21/18	CO I 124TH INF
341334	W	07/08/18	SA FN TEST
341339	E	06/08/18	CO L 1ST PIONEER INF
341375	R	08/31/18	CO G 122ND INF CAMP GREENE NC
341417	E	01/27/34	GRIFFIN & HOWE (SALE)
341635	E	05/22/18	CAMP FUNSTON (RUSTY)

SER. NO.	MODEL	MM/DD/YY	USAGE
341726	W	07/08/18	INTERCHANGEABILITY TEST
341927	R	08/27/18	CO D 122ND INF CAMP WHEELER GA
341985	R	09/19/18	46TH INF CAMP SHERIDAN
342049	R	07/12/18	SA FN TEST
342247	E	03/12/18	FA VELOCITY TEST
342270	E	07/17/18	SA FN TEST
342315	W	07/08/18	SA FN TEST
342604	R	07/30/18	SA FN TEST
342665	E	06/19/18	CAMP FUNSTON (RUSTY)
342715	W	07/08/18	INTERCHANGEABILITY TEST
342811	R	05/17/18	164TH DEPOT BRIG CAMP FUNSTON
342831		08/21/18	CO H 124TH INF
343000	R	05/17/18	164TH DEPOT BRIG CAMP FUNSTON
343481	E	05/17/18	164TH DEPOT BRIG CAMP FUNSTON
343605	E	04/17/18	INTERCHANGEABILITY TEST
343740	R	08/27/18	CO C 121ST INF CAMP WHEELER GA
343767	R	08/26/18	WAREHOUSE 106TH AMMO TRAIN
343816	E	05/22/18	CAMP FUNSTON (RUSTY)
344015	E	04/17/18	INTERCHANGEABILITY TEST
344116	E	05/17/18	164TH DEPOT BRIG CAMP FUNSTON
344296	E	04/04/18	CO H 110TH INF CAMP HANCOCK
344296	E	04/23/18	REVIEWED BY ENGR BUREAU
344296	E	05/01/18	WATERTOWN ARS (TESTED)
344296		09/17/18	FAILED IN SERVICE
344486	E	07/31/18	47TH CO 164TH DEPOT BRIG
344520	R	05/17/18	164TH DEPOT BRIG CAMP FUNSTON
344723	R	10/26/42	MIL GOVT HAWAII (RELEASED)
344786	W	09/26/18	COO AEF (EXTRACTOR TEST)
344885	E	06/19/18	CAMP FUNSTON (RUSTY)
344989		08/21/18	CO A 124TH INF
345116	W	07/08/18	INTERCHANGEABILITY TEST
345175	W	07/08/18	INTERCHANGEABILITY TEST
345562	E	07/31/18	47TH CO 164TH DEPOT BRIG
345597	W	07/08/18	INTERCHANGEABILITY TEST
345707	R	08/15/50	VA
345958	E	07/31/18	47TH CO 164TH DEPOT BRIG
345984	E	05/17/18	164TH DEPOT BRIG CAMP FUNSTON
346122	W	07/08/18	INTERCHANGEABILITY TEST
346206	E	05/16/18	CAMP FUNSTON (RUSTY)
346243	E	04/17/18	INTERCHANGEABILITY TEST
346518	E	05/16/18	CAMP FUNSTON (RUSTY)
346660	W	09/26/18	COO AEF (EXTRACTOR TEST)
346684	E	05/17/18	164TH DEPOT BRIG CAMP FUNSTON
346747	R	05/17/18	164TH DEPOT BRIG CAMP FUNSTON
346810	E	04/17/18	INTERCHANGEABILITY TEST
346823		06/27/18	COL 2ND INF
346866	W	07/11/18	SA FN TEST
346989	W	09/26/18	COO AEF (EXTRACTOR TEST)
347297	E	05/20/18	165TH DEPOT BRIG CAMP TRAVIS
347333	E	07/31/18	47TH CO 164TH DEPOT BRIG
347780	E	04/17/18	INTERCHANGEABILITY TEST
347986	E	06/19/18	CAMP FUNSTON (RUSTY)
348025	W	09/26/18	COO AEF (EXTRACTOR TEST)
348200	E	05/16/18	CAMP FUNSTON (RUSTY)
348270	R	08/17/18	CAMP SHERIDAN
348315	W	02/10/19	NEW ENGLAND WESTINGHOUSE
348315	W	07/12/18	SA FN TEST
348835	E	06/19/18	CAMP FUNSTON (RUSTY)
348938	W	09/26/18	COO AEF (EXTRACTOR TEST)
349432	W	09/26/18	COO AEF (EXTRACTOR TEST)
349531	R	08/17/18	CAMP SHERIDAN
349540		09/01/18	CO L 2ND INF
349585	W	09/26/18	COO AEF (EXTRACTOR TEST)
349588	E	05/17/18	164TH DEPOT BRIG CAMP FUNSTON
349794	E	05/16/18	CAMP FUNSTON (RUSTY)
349886	E	05/17/18	164TH DEPOT BRIG CAMP FUNSTON
350523	E	05/16/18	CAMP FUNSTON (RUSTY)
350812	W	09/26/18	COO AEF (EXTRACTOR TEST)
351431	W	07/15/18	SA FN TEST
351533	W	09/26/18	COO AEF (EXTRACTOR TEST)
351625	W	07/15/18	SA FN TEST
351709	W	09/26/18	COO AEF (EXTRACTOR TEST)
351807	E	04/17/18	INTERCHANGEABILITY TEST
351988	W	07/16/18	SA FN TEST
352467		05/04/18	CO L 2ND INF
352467		08/01/18	CO L 2ND INF
352956	R	27	URBAN MA (ISSUED)

SER. NO.	MODEL	MM/DD/YY	USAGE
353098	E	05/16/18	CAMP FUNSTON (RUSTY)
353195	W	07/24/18	INTERCHANGEABILITY TEST
353381	R	05/17/18	164TH DEPOT BRIG CAMP FUNSTON
353459	W	04/17/18	SA FN TEST
354001	W	07/24/18	INTERCHANGEABILITY TEST
354646	W	10/24/18	CO B 36TH INF CAMP DEVENS
354988	E	05/17/18	164TH DEPOT BRIG CAMP FUNSTON
355129	E	06/19/18	CAMP FUNSTON (RUSTY)
355420		09/17/18	FAILED IN SERVICE
355420	E	05/01/18	WATERTOWN ARS (TESTED)
355420	E	04/23/18	REVIEWED BY ENGR BUREAU
355420	E	04/02/18	CO M 109TH INF CAMP HANCOCK
355466	W	07/24/18	INTERCHANGEABILITY TEST
355532	W	10/24/18	CO B 36TH INF CAMP DEVENS
355608	W	07/24/18	INTERCHANGEABILITY TEST
355800	W	10/24/18	CO B 36TH INF CAMP DEVENS
355867	E	05/17/18	164TH DEPOT BRIG CAMP FUNSTON
356315	W	07/17/18	SA FN TEST
356523	W	10/24/18	CO B 36TH INF CAMP DEVENS
356600	E	06/19/18	CAMP FUNSTON (RUSTY)
356630	E	05/17/18	164TH DEPOT BRIG CAMP FUNSTON
356651	W	10/24/18	CO B 36TH INF CAMP DEVENS
356997	W	10/24/18	CO B 36TH INF CAMP DEVENS
357020	W	10/24/18	CO B 36TH INF CAMP DEVENS
357398	E	11/10/22	AUGUSTA ARS (OVERHAULED)
357429	W	08/08/18	INTERCHANGEABILITY TEST
357526	W	07/19/18	SA FN TEST
357539	W	10/24/18	CO B 36TH INF CAMP DEVENS
357558	E	05/16/18	CAMP FUNSTON (RUSTY)
357567	W	07/20/18	SA FN TEST
357605	W	10/24/18	CO B 36TH INF CAMP DEVENS
357842	E	05/16/18	CAMP FUNSTON (RUSTY)
357897	E	05/16/18	CAMP FUNSTON (RUSTY)
357997	R	08/17/18	CAMP SHERIDAN
358007	E	05/17/18	164TH DEPOT BRIG CAMP FUNSTON
358140	W	10/24/18	CO B 36TH INF CAMP DEVENS
358575	W	10/24/18	CO B 36TH INF CAMP DEVENS
358661	W	02/03/19	36TH INF CAMP DEVENS (MISSING)
358724	W	10/24/18	CO B 36TH INF CAMP DEVENS
358835	E	04/17/18	INTERCHANGEABILITY TEST
359100	E	05/20/18	165TH DEPOT BRIG CAMP TRAVIS
359114		12/17/43	USAT JANE MOORHEAD S063
359234	W	10/24/18	CO B 36TH INF CAMP DEVENS
359376	W	10/24/18	CO B 36TH INF CAMP DEVENS
359406	R	07/30/18	SA FN TEST
359424	E	05/17/18	164TH DEPOT BRIG CAMP FUNSTON
359472	W	10/24/18	CO B 36TH INF CAMP DEVENS
359848	E	05/03/18	112TH INF
359848	E	11/21/18	CAMP HANCOCK
359868	E	04/17/18	INTERCHANGEABILITY TEST
359974	E	05/17/18	164TH DEPOT BRIG CAMP FUNSTON
360019	W	10/24/18	CO B 36TH INF CAMP DEVENS
360093	E	06/19/18	CAMP FUNSTON (RUSTY)
360157	W	10/24/18	CO B 36TH INF CAMP DEVENS
360310	W	10/24/18	CO B 36TH INF CAMP DEVENS
360459	W	10/24/18	CO B 36TH INF CAMP DEVENS
360659	W	10/24/18	CO B 36TH INF CAMP DEVENS
360769	W	08/01/18	SA DAILY TEST
360769	W	07/24/18	SA FN TEST
360769	W	02/10/19	NEW ENGLAND WESTINGHOUSE
360889	W	07/22/18	SA FN TEST
360892	W	10/24/18	CO B 36TH INF CAMP DEVENS
360975	E	05/17/18	164TH DEPOT BRIG CAMP FUNSTON
361049	E	05/17/18	164TH DEPOT BRIG CAMP FUNSTON
361104	W	10/24/18	CO B 36TH INF CAMP DEVENS
361151	E	05/16/18	CAMP FUNSTON (RUSTY)
361213	W	10/24/18	CO B 36TH INF CAMP DEVENS
361327	E	06/08/18	CO L 1ST PIONEER INF
361381	W	10/24/18	CO B 36TH INF CAMP DEVENS
361382	E	05/17/18	164TH DEPOT BRIG CAMP FUNSTON
361529	W	07/24/18	INTERCHANGEABILITY TEST
361550	E	05/16/18	CAMP FUNSTON (RUSTY)
361906	W	10/24/18	CO B 36TH INF CAMP DEVENS
361916	W	10/24/18	CO B 36TH INF CAMP DEVENS
361937	W	10/24/18	CO B 36TH INF CAMP DEVENS
361945	W	10/24/18	CO B 36TH INF CAMP DEVENS
362189	W	10/24/18	CO B 36TH INF CAMP DEVENS
362320	W	10/24/18	CO B 36TH INF CAMP DEVENS
362627	W	07/24/18	SA FN TEST
362699	R	11/23/18	CAMP MCCLELLAN
363095	W	10/24/18	CO B 36TH INF CAMP DEVENS
363258	W	10/24/18	CO B 36TH INF CAMP DEVENS
363326	W	10/24/18	CO B 36TH INF CAMP DEVENS
363445	E	06/19/18	CAMP FUNSTON (RUSTY)
363554	W	07/25/18	SA FN TEST
363554	W	08/01/18	SA DAILY TEST
363674	W	10/24/18	CO B 36TH INF CAMP DEVENS
363731	W	10/24/18	CO B 36TH INF CAMP DEVENS
363870	W	10/24/18	CO B 36TH INF CAMP DEVENS
363993	W	10/24/18	CO B 36TH INF CAMP DEVENS
364105	W	07/24/18	INTERCHANGEABILITY TEST
364181	E	01/27/34	GRIFFIN & HOWE (SALE)
364306	W	07/24/18	INTERCHANGEABILITY TEST
364557	W	07/24/18	INTERCHANGEABILITY TEST
364568	W	10/24/18	CO B 36TH INF CAMP DEVENS
364798	W	10/24/18	CO B 36TH INF CAMP DEVENS
364876	E	05/20/18	165TH DEPOT BRIG CAMP TRAVIS
364910	W	11/23/18	CAMP MCCLELLAN
364997	R	07/29/18	SA FN TEST
365021	W	10/24/18	CO B 36TH INF CAMP DEVENS
365044	W	10/24/18	CO B 36TH INF CAMP DEVENS
365080	E	04/18/18	CO C 309TH INF CAMP DIX
365099	W	10/24/18	CO B 36TH INF CAMP DEVENS
365136	W	10/24/18	CO B 36TH INF CAMP DEVENS
365155	R	08/17/18	CAMP SHERIDAN
365330	W	10/24/18	CO B 36TH INF CAMP DEVENS
365362	W	10/24/18	CO B 36TH INF CAMP DEVENS
365371	W	10/24/18	CO B 36TH INF CAMP DEVENS
365439	W	10/24/18	CO B 36TH INF CAMP DEVENS
365666	W	10/24/18	CO B 36TH INF CAMP DEVENS
365879	W	03/03/44	USS CARMICK (DD493)
365985	W	07/24/18	INTERCHANGEABILITY TEST
366040	E	05/22/18	CAMP FUNSTON (RUSTY)
366046	E	05/17/18	164TH DEPOT BRIG CAMP FUNSTON
366085	E	05/16/18	CAMP FUNSTON (RUSTY)
366138	W	10/24/18	CO B 36TH INF CAMP DEVENS
366412	W	10/24/18	CO B 36TH INF CAMP DEVENS
366554	E	05/20/18	165TH DEPOT BRIG CAMP TRAVIS
366728	W	08/08/18	INTERCHANGEABILITY TEST
366734	R	08/17/18	CAMP SHERIDAN
366892	E	05/20/18	165TH DEPOT BRIG CAMP TRAVIS
367253	W	11/23/18	CAMP MCCLELLAN
367391	W	11/23/18	CAMP MCCLELLAN
367545	R	07/30/18	SA FN TEST
367551	R	07/30/18	SA FN TEST
367906	W	08/08/18	INTERCHANGEABILITY TEST
368075	E	05/20/18	165TH DEPOT BRIG CAMP TRAVIS
368185	E	05/20/18	165TH DEPOT BRIG CAMP TRAVIS
368517	E	04/05/18	CO M 111TH INF CAMP HANCOCK
368517		09/17/18	FAILED IN SERVICE
368517	E	04/23/18	REVIEWED BY ENGR BUREAU
368517	E	05/01/18	WATERTOWN ARS (TESTED)
368582	W	08/08/18	INTERCHANGEABILITY TEST
369136	W	08/08/18	INTERCHANGEABILITY TEST
369448	R	07/30/18	SA FN TEST
369780	17W	11/23/18	CAMP MCCLELLAN
369900	17W	11/23/18	CAMP MCCLELLAN
370287	E	06/08/18	CO L 1ST PIONEER INF
370577		07/20/43	USAT RACHEL JACKSON X-64
370845	17W	11/23/18	CAMP MCCLELLAN
371742	E	05/20/18	165TH DEPOT BRIG CAMP TRAVIS
372067	E	05/20/18	165TH DEPOT BRIG CAMP TRAVIS
372418	E	05/20/18	165TH DEPOT BRIG CAMP TRAVIS
372465	R	08/19/18	CO H 122ND INF
372817	E	04/10/18	CO F 309TH INF
373124	R	11/10/22	AUGUSTA ARS (OVERHAULED)
374354	E	05/20/18	165TH DEPOT BRIG CAMP TRAVIS
374537	R	08/01/18	SA FN TEST
375084	E	05/20/18	165TH DEPOT BRIG CAMP TRAVIS
376175	R	07/26/18	SA FN TEST
376376	R	11/10/22	AUGUSTA ARS (OVERHAULED)
376389		08/21/18	CO H 123RD INF
376901	R	08/19/18	CO L 122ND INF
377138	R	11/10/22	AUGUSTA ARS (OVERHAULED)

SER. NO.	MODEL	MM/DD/YY	USAGE
377264	E	04/10/10	CO L 110TH INF CAMP HANCOCK
377264	E	04/23/18	REVIEWED BY ENGR BUREAU
377264	E	05/01/18	WATERTOWN ARS (TESTED)
377427	E	05/03/18	112TH INF
377427	E	09/30/18	CAMP HANCOCK
377630	E	06/30/18	HIGHEST NO. IN PROCESS
377925		08/21/18	CO L 2ND INF
378458	R	08/17/18	CAMP SHERIDAN
378599	R	08/01/18	SA FN TEST
379016	E	04/23/18	REVIEWED BY ENGR BUREAU
379016	E	05/01/18	WATERTOWN ARS (TESTED)
379016		09/17/18	FAILED IN SERVICE
379150		08/30/18	CO L 2ND INF
379301		09/17/18	FAILED IN SERVICE
379301	E	04/23/18	REVIEWED BY ENGR BUREAU
379301	E	05/01/18	WATERTOWN ARS (TESTED)
380250	R	08/01/18	SA DAILY TEST
380250	R	07/25/18	SA FN TEST
380855		08/21/18	CO A 122ND INF
381345	E	04/02/18	CO L 110TH INF CAMP HANCOCK
381345	E	04/23/18	REVIEWED BY ENGR BUREAU
381345	E	05/01/18	WATERTOWN ARS (TESTED)
381345		09/17/18	FAILED IN SERVICE
381451	E	06/08/18	CO I 1ST PIONEER INF
381987	R	06/99	CMP INVENTORY
381990	E	05/03/18	112TH INF
382487	R	09/03/18	MICH MOTOR SPECIALTIES
382629	R	08/19/18	CO L 122ND INF
385153		08/30/18	CO L 2ND INF
386147	E	05/01/18	WATERTOWN ARS (TESTED)
386147	E	04/10/18	CO I 112TH INF CAMP HANCOCK
386147	E	04/23/18	REVIEWED BY ENGR BUREAU
387058	W	05/22/18	CAMP DIX
387058	E	05/22/18	CAMP GRANT ILL
387058	E	05/22/18	CAMP DIX
388129	E	04/18/18	CO C 309TH INF CAMP DIX
388270	E	04/18/18	CO C 309TH INF
389138		02/14/19	CO L 2ND INF
389349	E	04/18/18	CO B 312TH INF CAMP DIX
389493	E	05/03/18	112TH INF
389493	E	12/02/18	CAMP HANCOCK
391109	E	11/21/18	CAMP HANCOCK
391309	E	05/03/18	112TH INF
391310	E	12/02/18	CAMP HANCOCK
391310	E	04/02/18	CO H 109TH INF CAMP HANCOCK
391310	E	04/23/18	REVIEWED BY ENGR BUREAU
391727	E	05/01/18	WATERTOWN ARS (TESTED)
391727	E	04/23/18	REVIEWED BY ENGR BUREAU
392449	E	04/10/18	CO F 310TH INF
392536	E	06/08/18	CO G 1ST PIONEER INF
393096	E	04/06/18	CO C 110TH INF CAMP HANCOCK
393096	E	04/23/18	REVIEWED BY ENGR BUREAU
393096	E	11/21/18	CAMP HANCOCK
393469	R	11/10/22	AUGUSTA ARS (OVERHAULED)
393498	R	08/19/18	CO I 122ND INF
393803	E	05/01/18	WATERTOWN ARS (TESTED)
393803	E	04/23/18	REVIEWED BY ENGR BUREAU
394642	E	06/08/18	CO G 1ST PIONEER INF
394760		03/01/18	CO L 2ND INF
395363	E	04/10/18	CO F 310TH INF
396818	E	04/10/18	CO E 312TH INF
397069		11/14/18	CO L 2ND INF
397388	E	04/10/18	CO F 310TH INF
398018	E	04/03/18	78TH DIV CAMP DIX (RUSTY)
398283	W	06/99	CMP INVENTORY
398704	E	04/29/18	MP CAMP DIX
398704	E	11/21/18	CAMP DIX
399436	E	05/13/18	312TH INF
399588	E	04/18/18	CO B 312TH INF CAMP DIX
399606	R	06/99	CMP INVENTORY
399754	E	06/08/18	CO L 1ST PIONEER INF
400584	E	04/17/18	CAMP DIX TO MIDVALE
401361	E	06/08/18	CO H 1ST PIONEER INF
401582	W	06/99	CMP INVENTORY
401741	E	04/10/18	CO F 309TH INF
403447	E	11/21/18	CAMP DIX
403447	E	05/13/18	303RD INF CAMP DIX
403493	R	06/99	CMP INVENTORY
404213	E	04/18/18	CO B 312TH INF CAMP DIX
404236		09/09/18	CO L 2ND INF
404323	E	04/10/18	CO D 309TH INF
404737	E	04/10/18	CO F 310TH INF
405571	E	06/08/18	CO B 1ST PIONEER INF
405879	E	04/03/18	78TH DIV (RUSTY)
405879	E	04/03/18	CAMP DIX (RUSTY)
406093	E	04/03/18	78TH DIV CAMP DIX (RUSTY)
407076	E	04/10/18	CO F 310TH INF
407448	R	01/10/19	CAMP LEE
408128	E	04/17/18	CAMP DIX TO MIDVALE
408550	R	27	URBAN MA (ISSUED)
409292	E	04/03/18	78TH DIV CAMP DIX (RUSTY)
409365	R	09/21/18	INTERCHANGEABILITY TEST
409978		07/20/43	USAT RACHEL JACKSON X-64
410074	E	06/08/18	CO L 1ST PIONEER INF
411395	R	27	URBAN MA (ISSUED)
411424	E	04/29/18	311TH INF
411466	E	04/03/18	78TH DIV CAMP DIX (RUSTY)
412284	E	04/10/18	CO F 310TH INF
412380	E	04/03/18	CAMP DIX (RUSTY)
412380	E	04/03/18	78TH DIV (RUSTY)
412517	E	04/18/18	CO B 312TH INF CAMP DIX
415227	E	06/08/18	CO L 1ST PIONEER INF
416800	E	04/03/18	78TH DIV CAMP DIX (RUSTY)
417402	E	04/29/18	312TH INF
417620	E	04/03/18	78TH DIV (RUSTY)
417620	E	04/22/18	312TH INF CAMP DIX (RUSTY)
417625	E	04/03/18	78TH DIV (RUSTY)
417977	E	04/29/18	311TH INF
418063	E	04/29/18	311TH INF
419108	E	04/29/18	311TH INF
419667	E	04/03/18	78TH DIV CAMP DIX (RUSTY)
420839	E	04/18/18	CO B 312TH INF CAMP DIX
420839	E	04/18/18	312TH INF
420981	R	27	URBAN MA (ISSUED)
421098	E	04/29/18	312TH INF
421657	E	04/29/18	311TH INF
421942	E	04/29/18	312TH INF
421942	E	04/03/18	78TH DIV CAMP DIX (RUSTY)
422126	E	04/03/18	CAMP DIX (RUSTY)
422126	E	04/03/18	78TH DIV (RUSTY)
422126	E	04/03/18	CAMP DIX
422634	E	04/22/18	312TH INF
423734	E	04/03/18	78TH DIV CAMP DIX (RUSTY)
423832	E	04/22/18	312TH INF
423832	E	08/07/18	312TH INF
424827	E	04/29/18	311TH INF
425172	E	04/29/18	311TH INF
426131	E	04/03/18	CAMP DIX
426131	E	04/03/18	78TH DIV (RUSTY)
426131	E	04/03/18	CAMP DIX (RUSTY)
426428	E	09/30/18	CAMP HANCOCK
426428	E	05/03/18	112TH INF
427172	E	04/17/18	CAMP DIX TO MIDVALE
427185	E	06/08/18	CO L 1ST PIONEER INF
427185	W	11/10/22	AUGUSTA ARS (OVERHAULED)
427612	E	04/17/18	CAMP DIX TO MIDVALE
427744	E	04/17/18	CAMP DIX TO MIDVALE
428216	E	06/08/18	CO I 1ST PIONEER INF
428268	E	04/22/18	312TH INF
428288	E	08/07/18	312TH INF
428984	E	06/10/18	CO K 1ST PIONEER INF
430517	E	04/03/18	78TH DIV (RUSTY)
430517	E	04/03/18	CAMP DIX (RUSTY)
431588	E	04/17/18	CAMP DIX TO MIDVALE
432816	E	04/17/18	CAMP DIX TO MIDVALE
432883	E	06/08/18	CO H 1ST PIONEER INF
434618	E	07/22/43	USS SC1009
434718	E	05/02/18	SPECIAL FIRING PIN TEST
434718	E	05/02/18	CAMP WADSWORTH (TEST)
434945	R	02/10/19	NEW ENGLAND WESTINGHOUSE
435191	E	06/08/18	CO I 1ST PIONEER INF
435210	E	06/10/18	CO K 1ST PIONEER INF
435358	E	06/08/18	CO A 1ST PIONEER INF
435699	E	04/17/18	CAMP DIX TO MIDVALE

SER. NO.	MODEL	MM/DD/YY	USAGE
435966	E	04/17/18	CAMP DIX TO MIDVALE
435974	E	05/13/18	312TH INF
436144	E	06/08/18	CO L 1ST PIONEER INF
436520	E	05/13/18	312TH INF CAMP DIX
436520	E	08/07/18	312TH INF
436520	E	11/21/18	CAMP DIX
436739	R	02/10/19	NEW ENGLAND WESTINGHOUSE
437400	E	06/08/18	CO G 1ST PIONEER INF
437600	E	04/17/18	CAMP DIX TO MIDVALE
437618	E	04/17/18	CAMP DIX TO MIDVALE
438380	E	06/08/18	CO G 1ST PIONEER INF
438817	E	06/08/18	CO I 1ST PIONEER INF
439720	E	06/08/18	CO B 1ST PIONEER INF
440176	E	06/10/18	CO K 1ST PIONEER INF
440523	E	06/11/18	CO M 1ST PIONEER INF
440710	E	06/08/18	CO H 1ST PIONEER INF
440736	E	04/17/18	CAMP DIX TO MIDVALE
441201	E	04/17/18	CAMP DIX TO MIDVALE
441271	E	06/08/18	CO G 1ST PIONEER INF
441414	E	06/08/18	CO L 1ST PIONEER INF
441507	E	06/08/18	CO B 1ST PIONEER INF
441962	E	06/10/18	CO K 1ST PIONEER INF
441963	E	06/08/18	338TH INF CAMP CUSTER
441963	E	06/08/18	CO A 1ST PIONEER INF
442058	E	06/08/18	CO H 1ST PIONEER INF
442065	E	06/11/18	CO M 1ST PIONEER INF
442376	E	06/08/18	CO I 1ST PIONEER INF
442529	E	04/17/18	CAMP DIX TO MIDVALE
443265	E	06/08/18	CO B 1ST PIONEER INF
443592	E	04/17/18	CAMP DIX TO MIDVALE
443985	E	06/08/18	CO B 1ST PIONEER INF
444031	W	01/09/43	WWII DONATION TO US NAVY
445087	W	27	URBAN MA (ISSUED)
445709	R	06/22/48	VETERANS ADMIN (TO RIA)
445822	E	05/02/18	SPECIAL FIRING PIN TEST
445922	E	05/02/18	CAMP WADSWORTH (TEST)
445977	R	06/99	CMP INVENTORY
445984	E	06/08/18	CO L 1ST PIONEER INF
446459	E	04/17/18	CAMP DIX TO MIDVALE
446627	E	06/08/18	CO H 1ST PIONEER INF
446833	E	06/08/18	CO I 1ST PIONEER INF
446984	E	06/08/18	CO L 1ST PIONEER INF
447095	E	06/08/18	CO B 1ST PIONEER INF
447738		11/10/04	USAT MULCRA S-141
448307	E	06/08/18	CO H 1ST PIONEER INF
452941	E	04/17/18	CAMP DIX TO MIDVALE
453715	E	04/17/18	CAMP DIX TO MIDVALE
453922	E	06/08/18	CO L 1ST PIONEER INF
454061	E	06/08/18	CO G 1ST PIONEER INF
454553	E	06/08/18	CO A 1ST PIONEER INF
455385	E	06/08/18	CO G 1ST PIONEER INF
455650	E	04/17/18	CAMP DIX TO MIDVALE
456897	E	06/08/18	CO A 1ST PIONEER INF
457041	E	06/11/18	CO M 1ST PIONEER INF
457169	R	09/21/18	INTERCHANGEABILITY TEST
457873	E	06/08/18	CO I 1ST PIONEER INF
457893	E	06/08/18	CO B 1ST PIONEER INF
458169	E	06/11/18	CO M 1ST PIONEER INF
458176	E	06/08/18	CO L 1ST PIONEER INF
460085	E	06/08/18	CO H 1ST PIONEER INF
460507	E	06/08/18	CO H 1ST PIONEER INF
460595	R	04/05/19	PERISCOPIC ATTACHMENT
464017	E	06/11/18	CO M 1ST PIONEER INF
464429	E	05/02/18	CAMP WADSWORTH SPECIAL FIRING PIN TEST
465752	E	06/11/18	CO M 1ST PIONEER INF
467143	W	05/24/38	HOSP FOR DEF DELINQUENTS TO RIA
471777	E	07/01/43	USS SC-1005
475857	E	09/05/23	W E BLACKSHEAR
478027	E	07/22/43	USS SC1009
478368	E	11/10/22	AUGUSTA ARS (OVERHAULED)
478557	R	09/24/35	US NW PENITENTIARY LEWISBURG
479182	W	06/99	CMP INVENTORY
479978	W	06/30/34	SHANGHAI VOLUNTEER CORPS
483021	E	01/27/34	GRIFFIN & HOWE (SALE)
483909	W	06/30/34	SHANGHAI VOLUNTEER CORPS
484605	R	02/11/19	DOMINION ARSENAL QUEBEC
484800	E	06/99	CMP INVENTORY
485145	E	09/26/18	COO AEF (EXTRACTOR TEST)
489289		08/21/18	CO K 124TH INF
490159	R	02/11/19	DOMINION ARSENAL QUEBEC
490763		05/11/31	APG
490788	W	06/30/34	SHANGHAI VOLUNTEER CORPS
491405	W	08/21/35	US SW REFORMATRY EL RENO TO RIA
491773	W	06/30/34	SHANGHAI VOLUNTEER CORPS
492012	W	06/30/34	SHANGHAI VOLUNTEER CORPS
492307	W	06/30/34	SHANGHAI VOLUNTEER CORPS
492831	W	06/30/34	SHANGHAI VOLUNTEER CORPS
493295	W	06/30/34	SHANGHAI VOLUNTEER CORPS
493595	W	03/18/42	HAWAII MIL DEPT (RELEASED)
493652	W	08/21/35	US SW REFORMATRY EL RENO TO RIA
494066	W	06/30/34	SHANGHAI VOLUNTEER CORPS
49419Z	W	06/30/34	SHANGHAI VOLUNTEER CORPS
494328	W	06/30/34	SHANGHAI VOLUNTEER CORPS
494563	W	06/30/34	SHANGHAI VOLUNTEER CORPS
494693	W	06/30/34	SHANGHAI VOLUNTEER CORPS
494846	W	06/30/34	SHANGHAI VOLUNTEER CORPS
495708	E	06/06/18	SA FUNCTION TEST
496139	W	06/30/34	SHANGHAI VOLUNTEER CORPS
496511	W	06/30/34	SHANGHAI VOLUNTEER CORPS
496951	W	06/30/34	SHANGHAI VOLUNTEER CORPS
498113	W	06/30/34	SHANGHAI VOLUNTEER CORPS
499232	E	12/05/18	CAMP MACARTHUR
499232	E	08/02/18	CAMP MACARTHUR
499814		11/21/23	APG (TRACER TEST)
501381	R	06/99	CMP INVENTORY
501719		09/17/18	FAILED IN SERVICE
502670	R	27	URBAN MA (ISSUED)
503567		05/11/31	APG
505008		05/11/31	APG
507118	E	07/06/42	FT MCCLELLAN (ACCIDENT)
507674	R	08/26/18	WAREHOUSE SUP CO 121ST INF
507841	W	12/30/18	INTERCHANGEABILITY TEST
507936	W	05/24/38	HOSP FOR DEF DELINQUENTS TO RIA
509308		11/21/23	APG (TRACER TEST)
509925	W	05/24/38	HOSP FOR DEF DELINQUENTS TO RIA
512353	W	05/24/38	HOSP FOR DEF DELINQUENTS TO RIA
512879	W	08/19/35	US DETENTION FARM MILAN TO RIA
512926		05/11/31	APG
513964	E	08/10/18	CO G 152ND INF CAMP SHELBY
513964	E	12/02/18	CAMP SHELBY
514443	W	05/24/38	HOSP FOR DEF DELINQUENTS TO RIA
515402		05/11/31	APG
515501	E	06/18/18	REVISED BATTLE SIGHT TEST
515589	W	12/30/18	INTERCHANGEABILITY TEST
515943	W	08/19/35	US DETENTION FARM MILAN TO RIA
516109	W	12/30/18	INTERCHANGEABILITY TEST
516795	W	08/19/35	US DETENTION FARM MILAN TO RIA
517594	W	08/19/35	US DETENTION FARM MILAN TO RIA
517669	W	12/30/18	INTERCHANGEABILITY TEST
518555	W	05/24/38	HOSP FOR DEF DELINQUENTS TO RIA
518770	W	12/30/18	INTERCHANGEABILITY TEST
519192	W	08/19/35	US DETENTION FARM MILAN TO RIA
519540	R	08/26/18	WAREHOUSE SUP CO 121ST INF
519798	W	05/24/38	HOSP FOR DEF DELINQUENTS TO RIA
520246	W	08/19/35	US DETENTION FARM MILAN TO RIA
521275	W	08/19/35	US DETENTION FARM MILAN TO RIA
522852	R	06/99	CMP INVENTORY
522946	W	08/19/35	US DETENTION FARM MILAN TO RIA
524565	W	08/19/35	US DETENTION FARM MILAN TO RIA
524805	W	08/19/35	US DETENTION FARM MILAN TO RIA
525118	W	05/24/38	HOSP FOR DEF DELINQUENTS TO RIA
526723	W	08/19/35	US DETENTION FARM MILAN TO RIA
527692	W	05/24/38	HOSP FOR DEF DELINQUENTS TO RIA
527703	R	06/99	CMP INVENTORY
531933	E	06/30/34	SHANGHAI VOLUNTEER CORPS
533633	W	05/24/38	HOSP FOR DEF DELINQUENTS TO RIA
533685	E	04/26/34	GRIFFIN & HOWE (SALE)
535073	W	05/24/38	HOSP FOR DEF DELINQUENTS TO RIA
539974	E	08/09/18	EDDYSTONE FROM CAMP SHERIDAN
539974	E	07/17/18	CAMP SHERIDAN
542072	E	05/24/18	SA FUNCTION TEST
543065	W	08/19/35	US DETENTION FARM MILAN TO RIA
543526	W	05/24/38	HOSP FOR DEF DELINQUENTS TO RIA
545084	E	05/20/18	SA (FN TEST)

SER. NO.	MODEL	MM/DD/YY	USAGE
548332	E	06/06/18	SA FUNCTION TEST
548781	E	07/22/18	REVISED BATTLE SIGHT TEST
550705	E	08/23/18	CO C 151ST INF
550705	E	12/02/18	CAMP SHELBY
555110	E	05/24/18	SA FUNCTION TEST
556914	E	06/06/18	SA FUNCTION TEST
558219	E	07/29/18	CAMP MACARTHUR
558219	E	12/05/18	CAMP MACARTHUR
558747	E	10/24/18	17TH INF
561398	E	05/20/18	SA (FN TEST)
561688	E	03/03/19	AEF
563021		12/30/43	USAT SAN NICHOLAS S-163
564348	E	05/24/18	SA FUNCTION TEST
564864	E	08/15/18	INTERCHANGE OF MAY & JUNE PROD
564864	E	09/16/18	INTERCHANGEABILITY TEST
565204	E	03/06/19	316TH INF (DAM)
569699	E	06/06/18	SA FUNCTION TEST
570121	E	06/28/18	SA DAILY FUNCTION TEST
574637	R	06/99	CMP INVENTORY
575745	E	10/24/18	17TH INF
576354	E	07/08/18	SA FN TEST
576386	E	07/22/18	REVISED BATTLE SIGHT TEST
577139	E	06/17/18	SA FUNCTION TEST
577358	R	09/23/18	25TH MG BN
577457	E	11/10/22	AUGUSTA ARS (OVERHAULED)
577956	E	06/28/18	SA DAILY FUNCTION TEST
578688	E	06/18/18	REVISED BATTLE SIGHT TEST
579600	E	08/15/18	INTERCHANGE OF MAY & JUNE PROD
580719	E	10/04/18	CAMP HUMPHRIES
581152	E	07/22/18	REVISED BATTLE SIGHT TEST
581185	E	06/18/18	REVISED BATTLE SIGHT TEST
581538	E	12/02/18	CAMP SHERIDAN
582629	E	06/12/18	SA FUNCTION TEST
582980	E	09/07/18	113TH TRENCH MORTAR BATTY
583357	E	10/04/18	CAMP HUMPHRIES
585420	E	10/24/18	17TH INF
587265	E	10/19/18	CAMP GORDON
590718	E	06/06/18	SA FUNCTION TEST
590778	E	05/20/18	SA (FN TEST)
590967	E	10/04/18	CAMP HUMPHRIES
591550	E	08/02/18	CAMP MACARTHUR
591550	E	12/05/18	CAMP MACARTHUR
593013	E	10/04/18	CAMP HUMPHRIES
594743	E	06/25/18	CO H 151ST INF (DAMAGED)
594743	E	08/13/18	151ST INF
594764	R	08/15/50	VA
596223	E	07/29/18	CAMP MACARTHUR
596223	E	12/05/18	CAMP MACARTHUR
596606	E	10/04/18	CAMP HUMPHRIES
598889	R	08/15/50	VA
599077	E	06/12/18	SA FUNCTION TEST
599085	E	10/04/18	CAMP HUMPHRIES
599348	E	07/09/18	SA FN TEST
599366	E	10/04/18	CAMP HUMPHRIES
600959	E	10/04/18	CAMP HUMPHRIES
601362	E	10/04/18	CAMP HUMPHRIES
603307	E	10/04/18	CAMP HUMPHRIES
603398	E	08/15/18	INTERCHANGE OF MAY & JUNE PROD
603436	E	10/18/18	CAMP GORDON
605831	E	06/18/18	REVISED BATTLE SIGHT TEST
606543	E	06/18/18	REVISED BATTLE SIGHT TEST
607011	E	09/09/18	CAMP SHELBY
607011	E	12/12/18	150TH INF
607825	E	08/15/18	INTERCHANGE OF MAY & JUNE PROD
608358	E	09/26/18	COO AEF (EXTRACTOR TEST)
609928	E	10/04/18	CAMP HUMPHRIES
610300	E	06/06/18	SA FUNCTION TEST
610331	E	08/29/18	CO L 152ND INF CAMP SHELBY
610331	E	12/02/18	CAMP SHELBY
610933	E	10/04/18	CAMP HUMPHRIES
611404	E	09/26/18	COO AEF (EXTRACTOR TEST)
617172	E	06/06/18	SA FUNCTION TEST
617856	E	10/04/18	CAMP HUMPHRIES
618468	E	10/04/18	CAMP HUMPHRIES
619561		07/20/43	USAT RACHEL JACKSON X-64
620497	E	10/04/18	CAMP HUMPHRIES
621419	E	06/12/18	SA FUNCTION TEST
621558	E	10/04/18	CAMP HUMPHRIES
622328	E	07/29/18	CAMP MACARTHUR
623934	E	07/29/18	CAMP MACARTHUR
625887	E	10/04/18	CAMP HUMPHRIES
626620	E	10/04/18	CAMP HUMPHRIES
627054	E	12/05/18	CAMP MACARTHUR
627054	E	07/29/18	CAMP MACARTHUR
629038	E	07/17/18	CAMP SHERIDAN
629996	E	07/22/18	REVISED BATTLE SIGHT TEST
631190	E	08/09/18	EDDYSTONE FROM CAMP SHERIDAN
631190	E	07/17/18	CAMP SHERIDAN
631601	E	07/08/18	SA FN TEST
631960	E	07/29/18	CAMP MACARTHUR
631960	E	12/05/18	CAMP MACARTHUR
632095	E	08/15/18	INTERCHANGE OF MAY & JUNE PROD
634061	E	10/24/18	17TH INF
636871	E	06/18/18	SA FN TEST
637122	E	10/04/18	CAMP HUMPHRIES
637433	E	08/16/18	152ND INF
638444		09/29/44	USAT MAIWARA S-146
639824	E	10/04/18	CAMP HUMPHRIES
639955	E	01/10/19	CAMP LEE
640820	E	06/17/18	CO C 22ND ENGR (DAMAGED)
640820	E	09/30/18	CAMP SHERIDAN
641368	E	08/15/18	INTERCHANGE OF MAY & JUNE PROD
641456	E	09/16/18	INTERCHANGEABILITY TEST
641456	E	08/15/18	INTERCHANGE OF MAY & JUNE PROD
642357	E	08/02/18	CAMP MACARTHUR
644632	E	06/17/18	SA FUNCTION TEST
645313	E	06/18/18	REVISED BATTLE SIGHT TEST
645313	E	08/20/18	SMALL ARMS FIRING SCHOOL
645670	E	12/02/18	CAMP SHELBY
645889	E	07/29/18	CAMP MACARTHUR
647358	R	06/02/20	HIGHEST S. N.
648175	E	10/04/18	CAMP HUMPHRIES
649161	E	08/15/18	INTERCHANGE OF MAY & JUNE PROD
649233	E	08/15/18	INTERCHANGE OF MAY & JUNE PROD
650452	E	08/15/18	INTERCHANGE OF MAY & JUNE PROD
652708	E	10/23/18	CAMP GORDON
652910	E	06/22/48	VETERANS ADMIN (TO RIA)
654620	E	06/28/18	SA DAILY FUNCTION TEST
655479	E	08/15/18	INTERCHANGE OF MAY & JUNE PROD
656775	E	06/12/18	SA FUNCTION TEST
660337	E	04/13/33	C I BRASELTON
662037	E	07/22/18	REVISED BATTLE SIGHT TEST
663722	E	08/15/18	INTERCHANGE OF MAY & JUNE PROD
663757	E	08/15/18	INTERCHANGE OF MAY & JUNE PROD
666751	E	07/22/18	REVISED BATTLE SIGHT TEST
666962	E	06/18/18	REVISED BATTLE SIGHT TEST
668419	E	08/15/18	INTERCHANGE OF MAY & JUNE PROD
670064	E	08/15/18	INTERCHANGE OF MAY & JUNE PROD
670620	E	08/15/18	INTERCHANGE OF MAY & JUNE PROD
670706	E	07/22/18	REVISED BATTLE SIGHT TEST
672990	E	08/15/18	INTERCHANGE OF MAY & JUNE PROD
679277	E	08/15/18	INTERCHANGE OF MAY & JUNE PROD
679318	E	07/18/18	SA FN TEST
679576	E	10/24/18	17TH INF
680650	E	06/22/48	VETERANS ADMIN (TO RIA)
680725	E	07/22/18	REVISED BATTLE SIGHT TEST
680773	E	07/22/18	REVISED BATTLE SIGHT TEST
681031	E	08/15/18	INTERCHANGE OF MAY & JUNE PROD
681903	E	06/18/18	REVISED BATTLE SIGHT TEST
682349	E	07/09/18	SA FN TEST
682694	E	10/04/18	CAMP HUMPHRIES
682826	E	08/21/18	CAMP SHELBY
682826	E	12/02/18	CAMP SHELBY
685650	E	06/28/18	SA DAILY FUNCTION TEST
688145	E	06/28/18	SA DAILY FUNCTION TEST
688960	E	06/28/18	SA DAILY FUNCTION TEST
689168	E	08/13/18	113TH TRENCH MORTAR BATTY
689168	E	09/30/18	CAMP SHELBY
690078	E	09/26/18	COO AEF (EXTRACTOR TEST)
690228	E	10/24/18	17TH INF
691147	E	07/22/18	REVISED BATTLE SIGHT TEST
691446	E	06/18/18	REVISED BATTLE SIGHT TEST
693486	E	07/13/18	CAMP SHELBY
693486	E	12/12/18	150TH INF

SER. NO.	MODEL	MM/DD/YY	USAGE
695424	E	06/18/18	REVISED BATTLE SIGHT TEST
695550	W	05/24/19	CO C 521ST ENGR (TURNED IN)
695678	E	06/28/18	SA DAILY FUNCTION TEST
695839	E	06/28/18	SA FN TEST
696528	E	08/15/18	INTERCHANGE OF MAY & JUNE PROD
697213	E	07/27/18	SA FN TEST
698164	E	08/15/18	INTERCHANGE OF MAY & JUNE PROD
698164	E	09/16/18	INTERCHANGEABILITY TEST
698394	E	08/15/18	INTERCHANGE OF MAY & JUNE PROD
699693		05/10/43	USAT JOELLEN S-184
700466	E	06/29/18	SA DAILY FUNCTION TEST
700542	E	08/15/18	INTERCHANGE OF MAY & JUNE PROD
700763	E	01/16/19	COE59TH PIONEER INF (DAM)
701773	E	10/23/18	CAMP GORDON
703650	E	07/22/18	REVISED BATTLE SIGHT TEST
704910	E	07/01/18	SA FN TEST
705298	E	07/22/18	REVISED BATTLE SIGHT TEST
705729	E	12/02/18	CAMP SHELBY
705729	E	08/21/18	CAMP SHELBY
706657	E	07/22/18	REVISED BATTLE SIGHT TEST
707687	E	12/12/18	150TH INF
707687	E	09/09/18	CAMP SHELBY
708145	E	07/02/18	SA FN TEST
710208	E	07/22/18	REVISED BATTLE SIGHT TEST
711059	E	09/26/18	COO AEF (EXTRACTOR TEST)
711098	E	08/15/18	INTERCHANGE OF MAY & JUNE PROD
712979	E	10/23/18	CAMP GORDON
714406	E	06/28/18	SA DAILY FUNCTION TEST
719404	E	07/22/18	REVISED BATTLE SIGHT TEST
720533	E	07/22/18	REVISED BATTLE SIGHT TEST
721258	E	07/01/18	SA FN TEST
721968	E	09/26/18	COO AEF (EXTRACTOR TEST)
722763	E	07/12/18	SA FN TEST
722763	E	07/22/18	SA (MALFUNCTION)
723453	E	10/18/18	CAMP GORDON
723731	E	07/22/18	REVISED BATTLE SIGHT TEST
725794	E	08/15/18	INTERCHANGE OF MAY & JUNE PROD
727434	E	07/22/18	REVISED BATTLE SIGHT TEST
727650	E	07/22/18	REVISED BATTLE SIGHT TEST
729069	E	07/03/18	SA FN TEST
732395	E	11/10/22	AUGUSTA ARS (OVERHAULED)
733433	E	06/22/48	VETERANS ADMIN (TO RIA)
733665	E	03/01/19	CAMP SHELBY
733798	E	08/15/18	INTERCHANGE OF MAY & JUNE PROD
733798	E	09/16/18	INTERCHANGEABILITY TEST
734051	E	08/15/18	INTERCHANGE OF MAY & JUNE PROD
734716	E	08/15/18	INTERCHANGE OF MAY & JUNE PROD
734825	E	10/04/18	CAMP HUMPHRIES
735209	E	07/16/18	SA FN TEST
737688	E	07/22/18	REVISED BATTLE SIGHT TEST
738561	E	08/15/18	INTERCHANGE OF MAY & JUNE PROD
738849	E	08/15/18	INTERCHANGE OF MAY & JUNE PROD
738880	E	07/22/18	REVISED BATTLE SIGHT TEST
739680	E	08/15/18	INTERCHANGE OF MAY & JUNE PROD
740636	E	07/22/18	REVISED BATTLE SIGHT TEST
741928	E	08/21/18	CAMP SHELBY
741928	E	12/02/18	CAMP SHELBY
742163	E	07/08/18	SA FN TEST
743355	E	08/15/18	INTERCHANGE OF MAY & JUNE PROD
743600	E	07/22/18	REVISED BATTLE SIGHT TEST
744513	E	10/19/18	CAMP GORDON
746043	E	07/22/18	REVISED BATTLE SIGHT TEST
746315	E	08/15/18	INTERCHANGE OF MAY & JUNE PROD
746774	E	10/23/18	CAMP GORDON
747109	E	08/15/18	INTERCHANGE OF MAY & JUNE PROD
748005	E	08/15/18	INTERCHANGE OF MAY & JUNE PROD
748005	E	09/16/18	INTERCHANGEABILITY TEST
748327	E	08/15/18	INTERCHANGE OF MAY & JUNE PROD
748561	E	08/15/18	INTERCHANGE OF MAY & JUNE PROD
748601	E	07/22/18	REVISED BATTLE SIGHT TEST
748878	E	07/23/18	SA FN TEST
749477	E	07/22/18	REVISED BATTLE SIGHT TEST
749728	E	07/22/18	REVISED BATTLE SIGHT TEST
749895	E	06/99	CMP INVENTORY
750400	E	08/15/18	INTERCHANGE OF MAY & JUNE PROD
752740	E	08/15/18	INTERCHANGE OF MAY & JUNE PROD
753346	E	07/22/18	REVISED BATTLE SIGHT TEST
753858	E	07/08/18	SA FN TEST
754654	E	10/04/18	CAMP HUMPHRIES
754719	E	08/15/18	INTERCHANGE OF MAY & JUNE PROD
755085	E	10/18/18	CAMP GORDON
755449	E	07/26/18	SAFN TEST
762785	E	07/27/18	SA FN TEST
762978	E	07/22/18	REVISED BATTLE SIGHT TEST
764152	E	07/22/18	REVISED BATTLE SIGHT TEST
765091	E	10/19/18	CAMP GORDON
765121	E	07/16/18	SA FN TEST
766329	E	07/22/18	REVISED BATTLE SIGHT TEST
767843	E	07/22/18	REVISED BATTLE SIGHT TEST
770934	E	07/12/18	SA FN TEST
770934	E	07/22/18	SA (MALFUNCTION)
771750	E	10/23/18	CAMP GORDON
773037	E	09/26/18	COO AEF (EXTRACTOR TEST)
773334		07/23/18	CAMP DEVENS (STOCK TEST)
774431	E	10/14/18	CAMP GORDON
775918	E	07/22/18	REVISED BATTLE SIGHT TEST
776276	E	10/04/18	CAMP HUMPHRIES
776980	E	09/26/18	COO AEF (EXTRACTOR TEST)
778479	E	07/22/18	REVISED BATTLE SIGHT TEST
781756	E	07/22/18	REVISED BATTLE SIGHT TEST
782579	E	10/14/18	CAMP GORDON
782652		07/20/43	USAT RACHEL JACKSON X-64
782835	E	10/04/18	CAMP HUMPHRIES
784276	R	10/14/18	CAMP GORDON
784354	E	10/04/18	CAMP HUMPHRIES
785732		08/21/18	CO D 122ND INF
785931	E	07/27/18	SA FN TEST
786096	E	10/04/18	CAMP HUMPHRIES
786595	E	10/19/18	CAMP GORDON
786685	E	07/19/18	SA FN TEST
787140		03/31/44	USAT WILLWATCH S-116
788297	E	10/04/18	CAMP HUMPHRIES
789481	E	10/04/18	CAMP HUMPHRIES
790276	E	10/14/18	CAMP GORDON
790964	E	10/14/18	CAMP GORDON
791725	E	10/19/18	CAMP GORDON
792177	E	07/22/18	REVISED BATTLE SIGHT TEST
792189	E	10/19/18	CAMP GORDON
792216	E	06/22/48	VETERANS ADMIN (TO RIA)
793374	E	06/30/18	HIGHEST NO. IN PROCESS
794010	E	07/22/18	REVISED BATTLE SIGHT TEST
794089	E	10/14/18	CAMP GORDON
794362	E	10/14/18	CAMP GORDON
795126	E	07/22/18	REVISED BATTLE SIGHT TEST
796644	E	09/26/18	COO AEF (EXTRACTOR TEST)
796828	E	09/26/18	COO AEF (EXTRACTOR TEST)
798569	E	10/04/18	CAMP HUMPHRIES
798757	E	10/04/18	CAMP HUMPHRIES
799673	E	07/22/18	REVISED BATTLE SIGHT TEST
800809	E	07/24/18	SA FN TEST
801404	E	10/04/18	CAMP HUMPHRIES
802944	E	10/04/18	CAMP HUMPHRIES
805488	E	07/22/18	REVISED BATTLE SIGHT TEST
806351	E	10/04/18	CAMP HUMPHRIES
806392	E	10/04/18	CAMP HUMPHRIES
808241	E	10/04/18	CAMP HUMPHRIES
810981	E	07/23/18	SA FN TEST
811827	E	10/04/18	CAMP HUMPHRIES
812526	E	10/04/18	CAMP HUMPHRIES
815720	E	10/04/18	CAMP HUMPHRIES
816652	E	10/04/18	CAMP HUMPHRIES
820398	E	06/17/18	CO C 22ND ENGR (DAMAGED)
821293	E	10/04/18	CAMP HUMPHRIES
824511	E	05/16/40	VETERANS ADMIN
825282	E	07/30/18	SA FN TEST
826062	E	06/99	CMP INVENTORY
828026	E	10/14/18	CAMP GORDON
828169	E	10/04/18	CAMP HUMPHRIES
829038	E	08/09/18	EDDYSTONE FROM CAMP SHERIDAN
829822	E	09/26/18	CAMP GORDON
832414	E	10/04/18	CAMP HUMPHRIES
832751	E	07/29/18	SA FN TEST
838896	E	06/99	CMP INVENTORY
841626	E	07/30/18	SA FN TEST

SER. NO.	MODEL	MM/DD/YY	USAGE
846063	E	08/29/18	MG-SMALL ARMS SECT
847499	E	10/23/18	CAMP GORDON
858220	E	10/23/18	CAMP GORDON
859046	E	08/05/18	BLEW UP WHILE TARGETING
861002	E	10/19/18	CAMP GORDON
867349	E	09/11/18	REJECTED BY INSPECTOR
869263	E	09/11/18	REJECTED BY INSPECTOR
871239	E	11/06/18	CAMP FOREST
871253	E	07/16/18	SA TO OO
871917	E	07/16/18	SA TO OO
872252		05/15/18	115TH INF (STRIKER TEST)
876905	E	06/99	CMP INVENTORY
876927	E	01/27/34	GRIFFIN & HOWE (SALE)
890943	E	11/06/18	CAMP FOREST
891594	E	11/06/18	CAMP FOREST
892574	E	08/29/18	MG-SMALL ARMS SECT
894411	E	11/06/18	CAMP FOREST
894953	E	08/29/18	MG-SMALL ARMS SECT
897609	E	09/15/18	214TH ENGR (LOST)
898316	E	10/26/18	CO B 209TH ENGR
899289	E	10/14/18	CAMP SERVIER
900327	E	11/06/18	CAMP FOREST
905202		04/04/43	USAT MIALA S-174
905644		07/20/43	USAT RACHEL JACKSON X-64
905835	E	05/16/40	VETERANS ADMIN
906246	E	11/06/18	CAMP FOREST
908107	E	08/29/18	MG-SMALL ARMS SECT
912191		02/20/44	USAT JOHNREID CAM S-95
912731	E	11/06/18	CAMP FOREST
912975	E	09/14/18	CO B 209TH ENGR
915591	E	04/26/34	GRIFFIN & HOWE (SALE)
917326	E	11/06/18	CAMP FOREST
918333	E	01/14/19	CO E 214TH ENGR
921445	E	11/06/18	CAMP FOREST
933054	E	11/10/22	AUGUSTA ARS (OVERHAULED)
933501	E	04/26/34	GRIFFIN & HOWE (SALE)
936930		03/31/44	USAT WILLWATCH S-116
938036		10/27/44	USAT WORTANNA S-145
938889	E	11/10/22	AUGUSTA ARS (OVERHAULED)
941522	E	11/10/22	AUGUSTA ARS (OVERHAULED)
944471	E	11/10/22	AUGUSTA ARS (OVERHAULED)
946466	E	11/10/22	AUGUSTA ARS (OVERHAULED)
951553		03/22/44	USAT VALKYRIE S-133
955446	E	11/10/22	AUGUSTA ARS (OVERHAULED)
960206		07/20/43	USAT RACHEL JACKSON X-64
960461	E	11/10/22	AUGUSTA ARS (OVERHAULED)
961579	E	11/10/22	AUGUSTA ARS (OVERHAULED)
962650	E	06/99	CMP INVENTORY
963516	E	1110//22	AUGUSTA ARS (OVERHAULED)
973722		05/10/43	USAT JOELLEN S-184
976601	E	11/10/22	AUGUSTA ARS (OVERHAULED)
977167	E	11/10/22	AUGUSTA ARS (OVERHAULED)
985002		02/20/44	USAT JOHNREID CAM S-95
986401	E	02/10/19	NEW ENGLAND WESTINGHOUSE
987468	E	06/22/48	VETERANS ADMIN (TO RIA)
998436		09/14/44	USAT WOMBAT S-79
1007433	E	06/22/48	VETERANS ADMIN (TO RIA)
1025402		10/27/44	USAT WORTANNA S-145
1028620		10/27/44	USAT WORTANNA S-145
1032738	E	06/22/48	VETERANS ADMIN (TO RIA)
1033192	E	06/22/48	VETERANS ADMIN (TO RIA)
1034301	E	06/22/48	VETERANS ADMIN (TO RIA)
1034680		11/10/44	USAT MULCRA S-141
1038261		07/20/43	USAT RACHEL JACKSON X-64
1042895	E	11/10/22	AUGUSTA ARS (OVERHAULED)
1043276		03/22/44	USAT POTTS S-123
1048065		11/10/44	USAT MULCRA S-141
1049210	E	10/22/42	CO C 3RD BIRTC
1050485		03/23/44	USAT MICHALE S-169
1051898	E	10/30/42	CO A 4TH BIRTC
1061325		12/30/43	USAT SAN NICHOLAS S-163
1062081	E	06/30/34	SHANGHAI VOLUNTEER CORPS
1063419	E	11/10/22	AUGUSTA ARS (OVERHAULED)
1063933	E	06/99	CMP INVENTORY
1064533	E	06/30/34	SHANGHAI VOLUNTEER CORPS
1064783	E	06/30/34	SHANGHAI VOLUNTEER CORPS
1067512		04/04/43	USAT MIALA S-174
1070240	E	06/99	CMP INVENTORY
1070677		03/22/44	USAT BELLBIRD S-120
1072297	E	06/22/48	VETERANS ADMIN (TO RIA)
1072323	E	06/30/34	SHANGHAI VOLUNTEER CORPS
1074309	E	06/22/48	VETERANS ADMIN (TO RIA)
1074461		05/10/43	USAT KIAORA S-185
1074981	E	06/30/34	SHANGHAI VOLUNTEER CORPS
1075033	E	06/22/48	VETERANS ADMIN (TO RIA)
1075528	E	06/30/34	SHANGHAI VOLUNTEER CORPS
1075828	E	06/30/35	SHANGHAI VOLUNTEER CORPS
1076541		07/20/43	USAT RACHEL JACKSON X-64
1083893	E	06/30/34	SHANGHAI VOLUNTEER CORPS
1087899	E	06/30/34	SHANGHAI VOLUNTEER CORPS
1092209	E	06/30/34	SHANGHAI VOLUNTEER CORPS
1093814		03/22/44	USAT BELLBIRD S-120
1097021		07/20/43	USAT RACHEL JACKSON X-64
1097900	E	06/30/34	SHANGHAI VOLUNTEER CORPS
1098913		07/20/43	USAT RACHEL JACKSON X-64
1099271		07/20/43	USAT RACHEL JACKSON X-64
1102352		12/28/43	USAT MELCUSTOMS S-100
1103185	E	06/99	CMP INVENTORY
1105484	E	02/27/42	CO D 1ST BIRTC
1105890		09/29/44	USAT MAIWARA S-146
1106674		05/10/43	USAT KIAORA S-185
1106738	E	06/30/34	SHANGHAI VOLUNTEER CORPS
1107300		09/14/44	USAT WOMBAT S-79
1107316	E	06/30/34	SHANGHAI VOLUNTEER CORPS
1108245	E	06/30/34	SHANGHAI VOLUNTEER CORPS
1108496		11/10/44	USAT MULCRA S-141
1108992	E	06/30/34	SHANGHAI VOLUNTEER CORPS
1109484		12/28/43	USAT MELCUSTOMS S-100
1110356	E	06/30/34	SHANGHAI VOLUNTEER CORPS
1114588	E	06/30/35	SHANGHAI VOLUNTEER CORPS
1115125	E	06/99	CMP INVENTORY
1116103	E	06/30/34	SHANGHAI VOLUNTEER CORPS
1118356	E	06/30/35	SHANGHAI VOLUNTEER CORPS
1119931	E	06/30/34	SHANGHAI VOLUNTEER CORPS
1120663		09/14/44	USAT WOMBAT S-79
1123077	E	06/30/34	SHANGHAI VOLUNTEER CORPS
1124588	E	06/30/34	SHANGHAI VOLUNTEER CORPS
1128899		03/22/44	USAT POTTS S-123
1130078	E	06/30/34	SHANGHAI VOLUNTEER CORPS
1131853	E	06/30/34	SHANGHAI VOLUNTEER CORPS
1132764	E	06/30/34	SHANGHAI VOLUNTEER CORPS
1133906	E	06/30/34	SHANGHAI VOLUNTEER CORPS
1134451	E	06/30/34	SHANGHAI VOLUNTEER CORPS
1135660	E	06/30/34	SHANGHAI VOLUNTEER CORPS
1138849	E	06/99	CMP INVENTORY
1154387	E	06/30/34	SHANGHAI VOLUNTEER CORPS
1155319	E	11/10/22	AUGUSTA ARS (OVERHAULED)
1156147		44	USAT TUHOES-132
1156531	E	06/30/34	SHANGHAI VOLUNTEER CORPS
1162243	E	06/30/34	SHANGHAI VOLUNTEER CORPS
1163225		07/20/43	USAT RACHEL JACKSON X-64
1163794	E	11/10/22	AUGUSTA ARS (OVERHAULED)
1166095	E	06/30/34	SHANGHAI VOLUNTEER CORPS
1166786	E	06/30/34	SHANGHAI VOLUNTEER CORPS
1174309	E	06/30/34	SHANGHAI VOLUNTEER CORPS
1174651	E	06/30/34	SHANGHAI VOLUNTEER CORPS
1176494	E	06/30/34	SHANGHAI VOLUNTEER CORPS
1177771	E	06/30/34	SHANGHAI VOLUNTEER CORPS
1179512	E	06/30/34	SHANGHAI VOLUNTEER CORPS
1180574	E	06/30/34	SHANGHAI VOLUNTEER CORPS
1181835	E	06/30/35	SHANGHAI VOLUNTEER CORPS
1182551	E	06/30/34	SHANGHAI VOLUNTEER CORPS
1182742	E	06/30/34	SHANGHAI VOLUNTEER CORPS
1183477	E	06/30/34	SHANGHAI VOLUNTEER CORPS
1183528		03/22/44	USAT VALKYRIES-133
1184140	E	06/30/34	SHANGHAI VOLUNTEER CORPS
1188531	E	06/30/35	SHANGHAI VOLUNTEER CORPS
1189186		07/29/43	USAT JESSIE
1191555	E	06/30/34	SHANGHAI VOLUNTEER CORPS
1192575	E	06/30/34	SHANGHAI VOLUNTEER CORPS
1193748	E	06/30/34	SHANGHAI VOLUNTEER CORPS
1194120	E	06/30/34	SHANGHAI VOLUNTEER CORPS
1194769	E	06/30/34	SHANGHAI VOLUNTEER CORPS
1195232	E	06/30/35	SHANGHAI VOLUNTEER CORPS

SER. NO.	MODEL	MM/DD/YY	USAGE
1196232	E	06/30/34	SHANGHAI VOLUNTEER CORPS
1196329	E	06/30/34	SHANGHAI VOLUNTEER CORPS
1197572	E	06/30/34	SHANGHAI VOLUNTEER CORPS
1199594		11/10/44	USAT MULCRAS-141
1200419	E	06/30/34	SHANGHAI VOLUNTEER CORPS
1200743	E	06/30/34	SHANGHAI VOLUNTEER CORPS
1200954	E	06/30/34	SHANGHAI VOLUNTEER CORPS
1201036		04/27/44	USAT MARGARET TWAITSS-52
1201953	E	06/30/34	SHANGHAI VOLUNTEER CORPS
1202297	E	06/30/34	SHANGHAI VOLUNTEER CORPS
1202859	E	06/30/34	SHANGHAI VOLUNTEER CORPS
1202938	E	06/30/35	SHANGHAI VOLUNTEER CORPS
1202988	E	06/30/34	SHANGHAI VOLUNTEER CORPS
1203547	E	06/30/34	SHANGHAI VOLUNTEER CORPS
1204550	E	06/30/34	SHANGHAI VOLUNTEER CORPS
1204623	E	06/30/35	SHANGHAI VOLUNTEER CORPS
1205018	E	06/30/34	SHANGHAI VOLUNTEER CORPS
1205525	E	06/30/34	SHANGHAI VOLUNTEER CORPS
1206420	E	06/30/34	SHANGHAI VOLUNTEER CORPS
1206747	E	06/30/34	SHANGHAI VOLUNTEER CORPS
1206978	E	06/30/34	SHANGHAI VOLUNTEER CORPS
1207446	E	06/30/34	SHANGHAI VOLUNTEER CORPS
1208573	E	06/30/34	SHANGHAI VOLUNTEER CORPS
1209312	E	06/30/35	SHANGHAI VOLUNTEER CORPS
1209512	E	06/30/34	SHANGHAI VOLUNTEER CORPS
1210138	E	06/30/35	SHANGHAI VOLUNTEER CORPS
1210159	E	06/30/34	SHANGHAI VOLUNTEER CORPS
1212771	E	06/30/34	SHANGHAI VOLUNTEER CORPS
1213431	E	06/30/34	SHANGHAI VOLUNTEER CORPS
1214749	E	06/30/34	SHANGHAI VOLUNTEER CORPS
1214817	E	06/30/34	SHANGHAI VOLUNTEER CORPS
1215844	E	06/30/34	SHANGHAI VOLUNTEER CORPS
1216032	E	06/30/34	SHANGHAI VOLUNTEER CORPS
1216817	E	06/30/34	SHANGHAI VOLUNTEER CORPS
1216912		11/10/44	USAT MULCRAS-141
1217494	E	06/30/34	SHANGHAI VOLUNTEER CORPS
1217817	E	06/30/34	SHANGHAI VOLUNTEER CORPS
1220202	E	06/30/34	SHANGHAI VOLUNTEER CORPS
1220328	E	06/30/35	SHANGHAI VOLUNTEER CORPS
1221238	E	06/30/34	SHANGHAI VOLUNTEER CORPS
1221622	E	06/30/35	SHANGHAI VOLUNTEER CORPS
1221632	E	06/30/34	SHANGHAI VOLUNTEER CORPS
1222248	E	06/30/34	SHANGHAI VOLUNTEER CORPS
1223166	E	06/30/34	SHANGHAI VOLUNTEER CORPS
1223550	E	06/30/34	SHANGHAI VOLUNTEER CORPS
1223974	E	06/30/34	SHANGHAI VOLUNTEER CORPS
1224110	E	06/30/34	SHANGHAI VOLUNTEER CORPS
1224552	E	06/30/35	SHANGHAI VOLUNTEER CORPS
1224632	E	06/30/34	SHANGHAI VOLUNTEER CORPS
1225042	E	06/30/34	SHANGHAI VOLUNTEER CORPS
1225187	E	06/99	CMP INVENTORY
1225802	E	06/30/35	SHANGHAI VOLUNTEER CORPS
1225862	E	06/30/34	SHANGHAI VOLUNTEER CORPS
1227190	E	06/30/34	SHANGHAI VOLUNTEER CORPS
1227198	E	06/30/35	SHANGHAI VOLUNTEER CORPS
1227300	E	06/30/34	SHANGHAI VOLUNTEER CORPS
1227458	E	06/02/42	CO C 13TH BN BRANCH IR TRAINING CENTER
1227515	E	06/30/34	SHANGHAI VOLUNTEER CORPS
1227808	E	06/30/34	SHANGHAI VOLUNTEER CORPS
1227922	E	06/30/34	SHANGHAI VOLUNTEER CORPS
1227963	E	06/30/34	SHANGHAI VOLUNTEER CORPS
1228858	E	06/30/34	SHANGHAI VOLUNTEER CORPS
1228991	E	06/30/34	SHANGHAI VOLUNTEER CORPS
1229613	E	06/30/34	SHANGHAI VOLUNTEER CORPS
1229924	E	06/30/34	SHANGHAI VOLUNTEER CORPS
1230218	E	06/30/34	SHANGHAI VOLUNTEER CORPS
1230961	E	06/30/34	SHANGHAI VOLUNTEER CORPS
1231074	E	06/30/34	SHANGHAI VOLUNTEER CORPS
1231222	E	06/30/35	SHANGHAI VOLUNTEER CORPS
1231296	E	06/30/34	SHANGHAI VOLUNTEER CORPS
1232098	E	06/30/34	SHANGHAI VOLUNTEER CORPS
1232148	E	06/30/34	SHANGHAI VOLUNTEER CORPS
1232173	E	06/30/34	SHANGHAI VOLUNTEER CORPS
1233096	E	06/30/34	SHANGHAI VOLUNTEER CORPS
1233678	E	06/30/34	SHANGHAI VOLUNTEER CORPS
1234471	E	06/30/34	SHANGHAI VOLUNTEER CORPS
1235069	E	06/30/34	SHANGHAI VOLUNTEER CORPS
1236069	E	06/30/35	SHANGHAI VOLUNTEER CORPS
1236176	E	06/30/34	SHANGHAI VOLUNTEER CORPS
1236336	E	06/30/34	SHANGHAI VOLUNTEER CORPS
1236407	E	06/30/34	SHANGHAI VOLUNTEER CORPS
1236536	E	06/30/35	SHANGHAI VOLUNTEER CORPS
1237397	E	06/30/34	SHANGHAI VOLUNTEER CORPS
1238123	E	06/30/34	SHANGHAI VOLUNTEER CORPS
1239184	E	06/30/34	SHANGHAI VOLUNTEER CORPS
1239325	E	06/30/34	SHANGHAI VOLUNTEER CORPS
1240758	E	06/30/34	SHANGHAI VOLUNTEER CORPS
1241279	E	06/30/34	SHANGHAI VOLUNTEER CORPS
1241608	E	06/30/34	SHANGHAI VOLUNTEER CORPS
1242960	E	06/30/34	SHANGHAI VOLUNTEER CORPS
1243555	E	06/30/34	SHANGHAI VOLUNTEER CORPS
1243960	E	06/30/35	SHANGHAI VOLUNTEER CORPS
1244099	E	06/30/34	SHANGHAI VOLUNTEER CORPS
1244385	E	06/30/34	SHANGHAI VOLUNTEER CORPS
1244644	E	02/06/43	USS J FRANKLIN BELL
1245214	E	06/30/34	SHANGHAI VOLUNTEER CORPS
1245419	E	06/30/34	SHANGHAI VOLUNTEER CORPS
1245690	E	06/30/34	SHANGHAI VOLUNTEER CORPS
1248758	E	06/30/35	SHANGHAI VOLUNTEER CORPS
1250202	E	06/30/35	SHANGHAI VOLUNTEER CORPS
1250452	E	06/30/34	SHANGHAI VOLUNTEER CORPS
1250951	E	06/30/34	SHANGHAI VOLUNTEER CORPS
1251980	E	06/30/34	SHANGHAI VOLUNTEER CORPS
1252173	E	06/30/35	SHANGHAI VOLUNTEER CORPS
1252430	E	06/30/34	SHANGHAI VOLUNTEER CORPS
1252495		07/20/43	USAT RACHEL JACKSON X-64
1254662	E	06/30/34	SHANGHAI VOLUNTEER CORPS
1255257	E	06/30/34	SHANGHAI VOLUNTEER CORPS
1255455	E	06/30/34	SHANGHAI VOLUNTEER CORPS
1256366	E	04/11/42	CO D 1ST BIRTC
1256663	E	06/30/34	SHANGHAI VOLUNTEER CORPS
1257860	E	06/30/35	SHANGHAI VOLUNTEER CORPS
1258375	E	06/30/34	SHANGHAI VOLUNTEER CORPS
1258526	E	06/30/35	SHANGHAI VOLUNTEER CORPS
1258586	E	06/30/34	SHANGHAI VOLUNTEER CORPS
1258832	E	06/30/34	SHANGHAI VOLUNTEER CORPS
1259058	E	03/31/39	F DRIGGERS
1259119	E	06/30/34	SHANGHAI VOLUNTEER CORPS
1259935	E	06/30/34	SHANGHAI VOLUNTEER CORPS
1260346	E	06/30/35	SHANGHAI VOLUNTEER CORPS
1261330	E	06/30/34	SHANGHAI VOLUNTEER CORPS
1261550	E	06/30/34	SHANGHAI VOLUNTEER CORPS
1261600	E	06/30/35	SHANGHAI VOLUNTEER CORPS
1262117	E	06/30/34	SHANGHAI VOLUNTEER CORPS
1262302	E	06/99	CMP INVENTORY
1262482	E	06/30/34	SHANGHAI VOLUNTEER CORPS
1262490	E	06/30/34	SHANGHAI VOLUNTEER CORPS
1263186	E	06/30/35	SHANGHAI VOLUNTEER CORPS
1263230	E	06/30/35	SHANGHAI VOLUNTEER CORPS
1263300	E	06/30/34	SHANGHAI VOLUNTEER CORPS
1263910	E	06/30/34	SHANGHAI VOLUNTEER CORPS
1263942	E	06/30/34	SHANGHAI VOLUNTEER CORPS
1263993	E	06/30/34	SHANGHAI VOLUNTEER CORPS
1264583	E	06/30/35	SHANGHAI VOLUNTEER CORPS
1264585	E	06/30/34	SHANGHAI VOLUNTEER CORPS
1264750	E	06/30/34	SHANGHAI VOLUNTEER CORPS
1265257	E	06/30/35	SHANGHAI VOLUNTEER CORPS
1265370	E	06/30/34	SHANGHAI VOLUNTEER CORPS
1266300	E	06/30/34	SHANGHAI VOLUNTEER CORPS
1266370	E	06/30/35	SHANGHAI VOLUNTEER CORPS
1266470	E	06/30/34	SHANGHAI VOLUNTEER CORPS
1266600	E	06/30/35	SHANGHAI VOLUNTEER CORPS
1266628	E	07/04/43	USSSC728
1266665	E	06/30/35	SHANGHAI VOLUNTEER CORPS
1266873	E	06/30/34	SHANGHAI VOLUNTEER CORPS
1267210	E	06/30/34	SHANGHAI VOLUNTEER CORPS
1267480	E	06/30/34	SHANGHAI VOLUNTEER CORPS
1268093	E	06/30/35	SHANGHAI VOLUNTEER CORPS
1268164	E	06/30/34	SHANGHAI VOLUNTEER CORPS
1268580	E	06/30/35	SHANGHAI VOLUNTEER CORPS
1268626	E	06/30/34	SHANGHAI VOLUNTEER CORPS
1268832	E	06/30/35	SHANGHAI VOLUNTEER CORPS
1268880	E	06/30/34	SHANGHAI VOLUNTEER CORPS
1269325	E	06/30/35	SHANGHAI VOLUNTEER CORPS

SER. NO.	MODEL	MM/DD/YY	USAGE
1269330	E	06/30/34	SHANGHAI VOLUNTEER CORPS
1269346	E	06/30/34	SHANGHAI VOLUNTEER CORPS
1269406	E	06/30/35	SHANGHAI VOLUNTEER CORPS
1270067	E	06/30/34	SHANGHAI VOLUNTEER CORPS
1270686	E	06/30/34	SHANGHAI VOLUNTEER CORPS
1270879	E	06/30/34	SHANGHAI VOLUNTEER CORPS
1271269	E	06/30/35	SHANGHAI VOLUNTEER CORPS
1271369	E	06/30/34	SHANGHAI VOLUNTEER CORPS
1271735	E	06/30/34	SHANGHAI VOLUNTEER CORPS
1272532	E	06/30/35	SHANGHAI VOLUNTEER CORPS
1272792	E	06/30/34	SHANGHAI VOLUNTEER CORPS
1272949	E	06/30/34	SHANGHAI VOLUNTEER CORPS
1273313	E	06/30/34	SHANGHAI VOLUNTEER CORPS
1273355	E	06/30/35	SHANGHAI VOLUNTEER CORPS
1273530	E	06/30/35	SHANGHAI VOLUNTEER CORPS
1273556	E	06/30/34	SHANGHAI VOLUNTEER CORPS
1273578	E	06/30/35	SHANGHAI VOLUNTEER CORPS
1273878	E	06/30/34	SHANGHAI VOLUNTEER CORPS
1273925	E	06/30/34	SHANGHAI VOLUNTEER CORPS
1274233	E	06/30/34	SHANGHAI VOLUNTEER CORPS
1274544	E	06/30/34	SHANGHAI VOLUNTEER CORPS
1274644	E	06/30/35	SHANGHAI VOLUNTEER CORPS
1274651	E	06/30/35	SHANGHAI VOLUNTEER CORPS
1274673	E	06/30/34	SHANGHAI VOLUNTEER CORPS
1274935	E	06/30/34	SHANGHAI VOLUNTEER CORPS
1274975	E	06/30/34	SHANGHAI VOLUNTEER CORPS
1275422	E	06/30/34	SHANGHAI VOLUNTEER CORPS
1275482	E	06/30/35	SHANGHAI VOLUNTEER CORPS
1275590	E	06/30/34	SHANGHAI VOLUNTEER CORPS
1275955	E	06/30/34	SHANGHAI VOLUNTEER CORPS
1276054	E	06/30/35	SHANGHAI VOLUNTEER CORPS
1276572	E	06/30/34	SHANGHAI VOLUNTEER CORPS
1276713	E	06/30/34	SHANGHAI VOLUNTEER CORPS
1277164	E	06/30/34	SHANGHAI VOLUNTEER CORPS
1277530	E	06/30/34	SHANGHAI VOLUNTEER CORPS
1277763	E	06/30/34	SHANGHAI VOLUNTEER CORPS
1278214	E	06/30/34	SHANGHAI VOLUNTEER CORPS
1278313	E	06/30/35	SHANGHAI VOLUNTEER CORPS
1278641	E	06/30/34	SHANGHAI VOLUNTEER CORPS
1278721	E	06/30/34	SHANGHAI VOLUNTEER CORPS
1279909	E	06/30/34	SHANGHAI VOLUNTEER CORPS
1280030	E	06/30/34	SHANGHAI VOLUNTEER CORPS
1280230	E	04/10/42	CO A 7TH BIRTC
1280967	E	06/30/34	SHANGHAI VOLUNTEER CORPS
1281643	E	06/30/34	SHANGHAI VOLUNTEER CORPS
1281644	E	06/30/34	SHANGHAI VOLUNTEER CORPS
1281831	E	06/30/35	SHANGHAI VOLUNTEER CORPS
1281844	E	06/30/35	SHANGHAI VOLUNTEER CORPS
1281967	E	06/30/34	SHANGHAI VOLUNTEER CORPS
1282094	E	06/30/34	SHANGHAI VOLUNTEER CORPS
1282096	E	06/30/35	SHANGHAI VOLUNTEER CORPS
1282205	E	06/30/34	SHANGHAI VOLUNTEER CORPS
1282358	E	06/30/34	SHANGHAI VOLUNTEER CORPS
1282406	E	06/30/34	SHANGHAI VOLUNTEER CORPS
1282579	E	06/30/35	SHANGHAI VOLUNTEER CORPS
1282831	E	06/30/34	SHANGHAI VOLUNTEER CORPS
1283005	E	06/30/34	SHANGHAI VOLUNTEER CORPS
1283194	E	06/30/34	SHANGHAI VOLUNTEER CORPS
1283601	E	06/30/34	SHANGHAI VOLUNTEER CORPS
1283780	E	06/30/34	SHANGHAI VOLUNTEER CORPS
1283920	E	06/30/34	SHANGHAI VOLUNTEER CORPS
1284031	E	06/30/34	SHANGHAI VOLUNTEER CORPS
1284089	E	06/30/34	SHANGHAI VOLUNTEER CORPS
1284289	E	06/30/35	SHANGHAI VOLUNTEER CORPS
1284291	E	06/30/34	SHANGHAI VOLUNTEER CORPS
1284533	E	06/30/34	SHANGHAI VOLUNTEER CORPS
1284673	E	06/30/34	SHANGHAI VOLUNTEER CORPS
1284895	E	06/30/34	SHANGHAI VOLUNTEER CORPS
1284917	E	06/30/34	SHANGHAI VOLUNTEER CORPS
1284949	E	06/30/34	SHANGHAI VOLUNTEER CORPS
1285369	E	06/30/35	SHANGHAI VOLUNTEER CORPS
1285629	E	06/30/34	SHANGHAI VOLUNTEER CORPS
1287494	E	06/30/35	SHANGHAI VOLUNTEER CORPS
1287963	E	06/30/35	SHANGHAI VOLUNTEER CORPS
1288020	E	06/30/34	SHANGHAI VOLUNTEER CORPS
1288094	E	06/30/35	SHANGHAI VOLUNTEER CORPS
1288330	E	06/30/35	SHANGHAI VOLUNTEER CORPS
1288415	E	06/30/34	SHANGHAI VOLUNTEER CORPS
1288520	E	06/30/34	SHANGHAI VOLUNTEER CORPS
1288690	E	06/30/34	SHANGHAI VOLUNTEER CORPS
1288928	E	06/30/34	SHANGHAI VOLUNTEER CORPS
1288967	E	06/30/34	SHANGHAI VOLUNTEER CORPS
1289149	E	06/30/34	SHANGHAI VOLUNTEER CORPS
1289376	E	06/30/34	SHANGHAI VOLUNTEER CORPS
1289378	E	06/30/35	SHANGHAI VOLUNTEER CORPS
1289583	E	06/30/34	SHANGHAI VOLUNTEER CORPS
1289828	E	06/30/34	SHANGHAI VOLUNTEER CORPS
1289949	E	06/30/34	SHANGHAI VOLUNTEER CORPS
1289957	E	06/30/34	SHANGHAI VOLUNTEER CORPS
1290063	E	06/30/34	SHANGHAI VOLUNTEER CORPS
1290225	E	06/30/34	SHANGHAI VOLUNTEER CORPS
1290735	E	06/30/34	SHANGHAI VOLUNTEER CORPS
1290970	E	06/30/34	SHANGHAI VOLUNTEER CORPS
1291029	E	06/30/34	SHANGHAI VOLUNTEER CORPS
1291052	E	06/30/34	SHANGHAI VOLUNTEER CORPS
1291296	E	06/30/35	SHANGHAI VOLUNTEER CORPS
1291620	E	06/30/34	SHANGHAI VOLUNTEER CORPS
1291904	E	06/30/34	SHANGHAI VOLUNTEER CORPS
1292104	E	06/30/34	SHANGHAI VOLUNTEER CORPS
1292350	E	06/30/35	SHANGHAI VOLUNTEER CORPS
1292479	E	06/30/34	SHANGHAI VOLUNTEER CORPS
1292712	E	06/30/34	SHANGHAI VOLUNTEER CORPS
1292922		03/13/45	USNASSTSIMONS ISLAND (RETD)
1293035	E	06/30/34	SHANGHAI VOLUNTEER CORPS
1293066	E	06/30/34	SHANGHAI VOLUNTEER CORPS
1293073	E	06/30/34	SHANGHAI VOLUNTEER CORPS
1293146	E	06/30/35	SHANGHAI VOLUNTEER CORPS
1293224	E	06/30/34	SHANGHAI VOLUNTEER CORPS
1293287	E	06/30/34	SHANGHAI VOLUNTEER CORPS
1293289	E	06/30/35	SHANGHAI VOLUNTEER CORPS
1293338	E	06/30/35	SHANGHAI VOLUNTEER CORPS
1293339	E	06/30/34	SHANGHAI VOLUNTEER CORPS
1293347	E	06/30/34	SHANGHAI VOLUNTEER CORPS
1293350	E	06/30/34	SHANGHAI VOLUNTEER CORPS
1293442	E	06/30/34	SHANGHAI VOLUNTEER CORPS
1293570	E	06/30/34	SHANGHAI VOLUNTEER CORPS
1293650	E	06/30/34	SHANGHAI VOLUNTEER CORPS
1293672	E	06/30/34	SHANGHAI VOLUNTEER CORPS
1293701	E	06/30/34	SHANGHAI VOLUNTEER CORPS
1293712	E	06/30/35	SHANGHAI VOLUNTEER CORPS
1293728	E	06/30/34	SHANGHAI VOLUNTEER CORPS
1293892	E	06/30/34	SHANGHAI VOLUNTEER CORPS
1293997	E	06/30/34	SHANGHAI VOLUNTEER CORPS
1294146	E	06/30/34	SHANGHAI VOLUNTEER CORPS
1294231	E	06/30/34	SHANGHAI VOLUNTEER CORPS
1294438	E	06/30/34	SHANGHAI VOLUNTEER CORPS
1294706	E	06/30/34	SHANGHAI VOLUNTEER CORPS
1294756	E	06/30/34	SHANGHAI VOLUNTEER CORPS
1294895	E	06/30/35	SHANGHAI VOLUNTEER CORPS
1295081	E	06/30/34	SHANGHAI VOLUNTEER CORPS
1295221	E	06/30/34	SHANGHAI VOLUNTEER CORPS
1295381	E	06/30/34	SHANGHAI VOLUNTEER CORPS
1295677	E	06/30/34	SHANGHAI VOLUNTEER CORPS
1295691	E	06/30/34	SHANGHAI VOLUNTEER CORPS
1296314	E	06/30/34	SHANGHAI VOLUNTEER CORPS
1296530	E	06/30/35	SHANGHAI VOLUNTEER CORPS
1296536	E	06/30/34	SHANGHAI VOLUNTEER CORPS
1296646	E	06/30/35	SHANGHAI VOLUNTEER CORPS
1296674	E	06/30/34	SHANGHAI VOLUNTEER CORPS
1296675	E	06/30/34	SHANGHAI VOLUNTEER CORPS
1296741	E	06/30/34	SHANGHAI VOLUNTEER CORPS
1296771	E	06/30/34	SHANGHAI VOLUNTEER CORPS
1296840	E	06/30/34	SHANGHAI VOLUNTEER CORPS
1296961	E	06/30/35	SHANGHAI VOLUNTEER CORPS
1296972	E	06/30/34	SHANGHAI VOLUNTEER CORPS
1296982	E	06/30/34	SHANGHAI VOLUNTEER CORPS
1296983	E	06/30/35	SHANGHAI VOLUNTEER CORPS
1296993	E	06/30/34	SHANGHAI VOLUNTEER CORPS
1297337	E	06/30/34	SHANGHAI VOLUNTEER CORPS
1297555	E	06/30/34	SHANGHAI VOLUNTEER CORPS
1297616	E	06/30/34	SHANGHAI VOLUNTEER CORPS
1297683	E	06/30/34	SHANGHAI VOLUNTEER CORPS
1297741	E	06/30/34	SHANGHAI VOLUNTEER CORPS
1297871	E	06/30/34	SHANGHAI VOLUNTEER CORPS

SER. NO.	MODEL	MM/DD/YY	USAGE
1297967	E	06/30/34	SHANGHAI VOLUNTEER CORPS
1298055	E	06/30/35	SHANGHAI VOLUNTEER CORPS
1298169	E	06/30/34	SHANGHAI VOLUNTEER CORPS
1298225	E	06/30/34	SHANGHAI VOLUNTEER CORPS
1298255	E	06/30/34	SHANGHAI VOLUNTEER CORPS
1298259	E	06/30/35	SHANGHAI VOLUNTEER CORPS
1298268	E	06/30/35	SHANGHAI VOLUNTEER CORPS
1298282	E	06/30/35	SHANGHAI VOLUNTEER CORPS
1298287	E	06/30/34	SHANGHAI VOLUNTEER CORPS
1298288	E	06/30/34	SHANGHAI VOLUNTEER CORPS
1298331	E	06/30/35	SHANGHAI VOLUNTEER CORPS
1298364	E	06/30/34	SHANGHAI VOLUNTEER CORPS
1298412	E	06/30/35	SHANGHAI VOLUNTEER CORPS
1298544	E	06/30/34	SHANGHAI VOLUNTEER CORPS
1298613	E	06/30/34	SHANGHAI VOLUNTEER CORPS
1298626	E	06/30/34	SHANGHAI VOLUNTEER CORPS
1298638	E	06/30/35	SHANGHAI VOLUNTEER CORPS
1298646	E	06/30/34	SHANGHAI VOLUNTEER CORPS
1298836	E	06/30/35	SHANGHAI VOLUNTEER CORPS
1298882	E	06/30/34	SHANGHAI VOLUNTEER CORPS
1298925	E	06/30/34	SHANGHAI VOLUNTEER CORPS
1298972	E	06/30/35	SHANGHAI VOLUNTEER CORPS
1298982	E	06/30/35	SHANGHAI VOLUNTEER CORPS
1299856	E	06/30/34	SHANGHAI VOLUNTEER CORPS
1299924	E	06/30/34	SHANGHAI VOLUNTEER CORPS
1299934	E	06/30/35	SHANGHAI VOLUNTEER CORPS
1299967	E	06/30/35	SHANGHAI VOLUNTEER CORPS
1300025	E	06/30/34	SHANGHAI VOLUNTEER CORPS
1300040	E	06/30/34	SHANGHAI VOLUNTEER CORPS
1300127	E	06/30/34	SHANGHAI VOLUNTEER CORPS
1300294	E	06/30/34	SHANGHAI VOLUNTEER CORPS
1300394	E	06/30/34	SHANGHAI VOLUNTEER CORPS
1300418	E	06/30/35	SHANGHAI VOLUNTEER CORPS
1300453	E	06/30/34	SHANGHAI VOLUNTEER CORPS
1300487	E	06/30/35	SHANGHAI VOLUNTEER CORPS
1300503	E	06/30/34	SHANGHAI VOLUNTEER CORPS
1300536	E	06/30/34	SHANGHAI VOLUNTEER CORPS
1300617	E	06/30/34	SHANGHAI VOLUNTEER CORPS
1300766	E	06/30/34	SHANGHAI VOLUNTEER CORPS
1300864	E	06/30/35	SHANGHAI VOLUNTEER CORPS
1301024	E	06/30/34	SHANGHAI VOLUNTEER CORPS
1301085	E	06/30/35	SHANGHAI VOLUNTEER CORPS
1301206	E	06/30/34	SHANGHAI VOLUNTEER CORPS
1301380	E	06/30/34	SHANGHAI VOLUNTEER CORPS
1301584	E	06/30/34	SHANGHAI VOLUNTEER CORPS
1301953	E	06/30/35	SHANGHAI VOLUNTEER CORPS
1302301	E	06/30/35	SHANGHAI VOLUNTEER CORPS
1302559	E	06/30/35	SHANGHAI VOLUNTEER CORPS
1302588	E	06/30/34	SHANGHAI VOLUNTEER CORPS
1302751	E	06/30/34	SHANGHAI VOLUNTEER CORPS
1302783	E	06/30/34	SHANGHAI VOLUNTEER CORPS
1302977	E	06/30/34	SHANGHAI VOLUNTEER CORPS
1303008	E	06/30/34	SHANGHAI VOLUNTEER CORPS
1303366	E	06/30/34	SHANGHAI VOLUNTEER CORPS
1303499	E	06/30/34	SHANGHAI VOLUNTEER CORPS
1303601	E	06/30/34	SHANGHAI VOLUNTEER CORPS
1303842	E	06/30/34	SHANGHAI VOLUNTEER CORPS
1303871	E	06/30/35	SHANGHAI VOLUNTEER CORPS
1303881	E	06/30/34	SHANGHAI VOLUNTEER CORPS
1304030	E	06/30/34	SHANGHAI VOLUNTEER CORPS
1304105	E	06/30/35	SHANGHAI VOLUNTEER CORPS
1304221	E	06/30/34	SHANGHAI VOLUNTEER CORPS
1304466	E	06/30/35	SHANGHAI VOLUNTEER CORPS
1304639	E	06/30/34	SHANGHAI VOLUNTEER CORPS
1304702	E	06/30/34	SHANGHAI VOLUNTEER CORPS
1304769	E	06/30/34	SHANGHAI VOLUNTEER CORPS
1304914	E	06/30/34	SHANGHAI VOLUNTEER CORPS
1305018	E	06/30/35	SHANGHAI VOLUNTEER CORPS
1305038	E	06/30/34	SHANGHAI VOLUNTEER CORPS
1305054	E	06/30/34	SHANGHAI VOLUNTEER CORPS
1305159	E	06/30/34	SHANGHAI VOLUNTEER CORPS
1305161	E	06/30/34	SHANGHAI VOLUNTEER CORPS
1305216	E	06/30/34	SHANGHAI VOLUNTEER CORPS
1305245	E	06/30/34	SHANGHAI VOLUNTEER CORPS
1305247	E	06/30/34	SHANGHAI VOLUNTEER CORPS
1305379	E	06/30/34	SHANGHAI VOLUNTEER CORPS
1305466	E	06/30/35	SHANGHAI VOLUNTEER CORPS
1305500	E	06/30/34	SHANGHAI VOLUNTEER CORPS
1305604	E	06/30/35	SHANGHAI VOLUNTEER CORPS
1305627	E	06/30/34	SHANGHAI VOLUNTEER CORPS
1305879	E	06/30/35	SHANGHAI VOLUNTEER CORPS
1305890	E	06/30/34	SHANGHAI VOLUNTEER CORPS
1305927	E	06/30/34	SHANGHAI VOLUNTEER CORPS
1305970	E	06/30/35	SHANGHAI VOLUNTEER CORPS
1305985	E	06/30/35	SHANGHAI VOLUNTEER CORPS
1306284	E	06/30/35	SHANGHAI VOLUNTEER CORPS
1306378	E	06/30/34	SHANGHAI VOLUNTEER CORPS
1306418	E	06/30/35	SHANGHAI VOLUNTEER CORPS
1306456	E	06/30/34	SHANGHAI VOLUNTEER CORPS
1306570	E	06/30/34	SHANGHAI VOLUNTEER CORPS
1306630	E	06/30/34	SHANGHAI VOLUNTEER CORPS
1306675	E	06/30/35	SHANGHAI VOLUNTEER CORPS
1306925	E	06/30/34	SHANGHAI VOLUNTEER CORPS
1307507	E	06/30/34	SHANGHAI VOLUNTEER CORPS
1307651	E	06/30/34	SHANGHAI VOLUNTEER CORPS
1307749	E	06/30/35	SHANGHAI VOLUNTEER CORPS
1308208	E	06/30/35	SHANGHAI VOLUNTEER CORPS
1308214	E	06/30/34	SHANGHAI VOLUNTEER CORPS
1308244	E	06/30/34	SHANGHAI VOLUNTEER CORPS
1308348	E	06/30/34	SHANGHAI VOLUNTEER CORPS
1308407	E	06/30/35	SHANGHAI VOLUNTEER CORPS
1308512	E	06/30/35	SHANGHAI VOLUNTEER CORPS
1308577	E	06/30/34	SHANGHAI VOLUNTEER CORPS
1308626	E	06/30/35	SHANGHAI VOLUNTEER CORPS
1308728	E	06/30/34	SHANGHAI VOLUNTEER CORPS
1308750	E	06/30/34	SHANGHAI VOLUNTEER CORPS
1308794	E	06/30/34	SHANGHAI VOLUNTEER CORPS
1308941	E	06/30/34	SHANGHAI VOLUNTEER CORPS
1309016	E	06/30/34	SHANGHAI VOLUNTEER CORPS
1309614	E	06/30/34	SHANGHAI VOLUNTEER CORPS
1309731	E	06/30/34	SHANGHAI VOLUNTEER CORPS
1309799	E	06/30/34	SHANGHAI VOLUNTEER CORPS
1309864	E	06/30/34	SHANGHAI VOLUNTEER CORPS
1309908	E	06/30/34	SHANGHAI VOLUNTEER CORPS
1309962	E	06/30/34	SHANGHAI VOLUNTEER CORPS
1310341	E	06/30/34	SHANGHAI VOLUNTEER CORPS
1310436	E	06/30/34	SHANGHAI VOLUNTEER CORPS
1310797	E	06/30/34	SHANGHAI VOLUNTEER CORPS
1310799	E	06/30/34	SHANGHAI VOLUNTEER CORPS
1310982	E	06/30/34	SHANGHAI VOLUNTEER CORPS
1311204	E	06/30/35	SHANGHAI VOLUNTEER CORPS
1311209	E	06/30/34	SHANGHAI VOLUNTEER CORPS
1311451	E	06/30/34	SHANGHAI VOLUNTEER CORPS
1311502	E	06/30/34	SHANGHAI VOLUNTEER CORPS
1311509	E	06/30/34	SHANGHAI VOLUNTEER CORPS
1311699	E	06/30/34	SHANGHAI VOLUNTEER CORPS
1311766	E	06/30/34	SHANGHAI VOLUNTEER CORPS
1311968	E	06/30/34	SHANGHAI VOLUNTEER CORPS
1312053	E	06/30/34	SHANGHAI VOLUNTEER CORPS
1312116	E	06/30/34	SHANGHAI VOLUNTEER CORPS
1312244	E	06/30/34	SHANGHAI VOLUNTEER CORPS
1312365	E	06/30/34	SHANGHAI VOLUNTEER CORPS
1312568	E	06/30/35	SHANGHAI VOLUNTEER CORPS
1312577	E	06/30/34	SHANGHAI VOLUNTEER CORPS
1312805	E	06/30/34	SHANGHAI VOLUNTEER CORPS
1312835	E	06/30/34	SHANGHAI VOLUNTEER CORPS
1313047	E	06/30/34	SHANGHAI VOLUNTEER CORPS
1313051	E	06/30/34	SHANGHAI VOLUNTEER CORPS
1313125	E	06/30/34	SHANGHAI VOLUNTEER CORPS
1313537	E	06/30/34	SHANGHAI VOLUNTEER CORPS
1313764	E	06/30/34	SHANGHAI VOLUNTEER CORPS
1313858	E	06/30/34	SHANGHAI VOLUNTEER CORPS
1314348	E	06/30/35	SHANGHAI VOLUNTEER CORPS
1314458	E	06/30/34	SHANGHAI VOLUNTEER CORPS
1314712	E	06/30/34	SHANGHAI VOLUNTEER CORPS
1314881	E	06/30/35	SHANGHAI VOLUNTEER CORPS
1314890	E	06/30/35	SHANGHAI VOLUNTEER CORPS
1314987	E	06/30/34	SHANGHAI VOLUNTEER CORPS
1315030	E	06/30/34	SHANGHAI VOLUNTEER CORPS
1315091	E	06/30/35	SHANGHAI VOLUNTEER CORPS
1315101	E	06/30/34	SHANGHAI VOLUNTEER CORPS
1315107	E	06/30/34	SHANGHAI VOLUNTEER CORPS
1315295	E	06/30/34	SHANGHAI VOLUNTEER CORPS
1315317	E	06/30/34	SHANGHAI VOLUNTEER CORPS

SER. NO.	MODEL	MM/DD/YY	USAGE
1315325	E	06/30/35	SHANGHAI VOLUNTEER CORPS
1315487	E	06/30/34	SHANGHAI VOLUNTEER CORPS
1315491	E	06/30/34	SHANGHAI VOLUNTEER CORPS
1315531	E	06/30/35	SHANGHAI VOLUNTEER CORPS
1315634	E	06/30/34	SHANGHAI VOLUNTEER CORPS
1315723	E	06/30/34	SHANGHAI VOLUNTEER CORPS
1315732	E	06/30/34	SHANGHAI VOLUNTEER CORPS
1315863	E	06/30/34	SHANGHAI VOLUNTEER CORPS
1315911	E	06/30/34	SHANGHAI VOLUNTEER CORPS
1315916	E	06/30/34	SHANGHAI VOLUNTEER CORPS
1315922	E	06/30/35	SHANGHAI VOLUNTEER CORPS
1316032	E	06/30/35	SHANGHAI VOLUNTEER CORPS
1316046	E	06/30/34	SHANGHAI VOLUNTEER CORPS
1316056	E	06/30/35	SHANGHAI VOLUNTEER CORPS
1316098	E	06/30/34	SHANGHAI VOLUNTEER CORPS
1316180	E	06/30/34	SHANGHAI VOLUNTEER CORPS
1316305	E	06/30/34	SHANGHAI VOLUNTEER CORPS
1316517	E	06/30/34	SHANGHAI VOLUNTEER CORPS
1316635	E	06/30/34	SHANGHAI VOLUNTEER CORPS
1316643	E	06/30/34	SHANGHAI VOLUNTEER CORPS
1316655	E	06/30/34	SHANGHAI VOLUNTEER CORPS
1316659	E	06/30/34	SHANGHAI VOLUNTEER CORPS
1316745	E	06/30/34	SHANGHAI VOLUNTEER CORPS
1317002	E	06/30/34	SHANGHAI VOLUNTEER CORPS
1317122	E	06/30/34	SHANGHAI VOLUNTEER CORPS
1317144	E	06/30/35	SHANGHAI VOLUNTEER CORPS
1317180	E	06/30/34	SHANGHAI VOLUNTEER CORPS
1317326	E	06/30/34	SHANGHAI VOLUNTEER CORPS
1317332	E	06/30/34	SHANGHAI VOLUNTEER CORPS
1317420	E	06/30/34	SHANGHAI VOLUNTEER CORPS
1317511	E	06/30/34	SHANGHAI VOLUNTEER CORPS
1317530	E	06/30/34	SHANGHAI VOLUNTEER CORPS
1317579	E	06/30/34	SHANGHAI VOLUNTEER CORPS
1317605	E	06/30/34	SHANGHAI VOLUNTEER CORPS
1317665	E	06/30/34	SHANGHAI VOLUNTEER CORPS
1317686	E	06/30/34	SHANGHAI VOLUNTEER CORPS
1317862	E	06/30/34	SHANGHAI VOLUNTEER CORPS
1317942	E	06/30/35	SHANGHAI VOLUNTEER CORPS
1318184	E	06/30/35	SHANGHAI VOLUNTEER CORPS
1318305	E	06/30/34	SHANGHAI VOLUNTEER CORPS
1318326	E	06/30/35	SHANGHAI VOLUNTEER CORPS
1318342	E	06/30/35	SHANGHAI VOLUNTEER CORPS
1318413	E	06/30/34	SHANGHAI VOLUNTEER CORPS
1318450	E	06/30/34	SHANGHAI VOLUNTEER CORPS
1318453	E	06/30/34	SHANGHAI VOLUNTEER CORPS
1318519	E	06/30/34	SHANGHAI VOLUNTEER CORPS
1318527	E	06/30/35	SHANGHAI VOLUNTEER CORPS
1318650	E	06/30/34	SHANGHAI VOLUNTEER CORPS
1318690	E	06/30/34	SHANGHAI VOLUNTEER CORPS
1318731	E	06/30/34	SHANGHAI VOLUNTEER CORPS
1319080	E	06/30/34	SHANGHAI VOLUNTEER CORPS
1319113	E	06/30/34	SHANGHAI VOLUNTEER CORPS
1319114	E	06/30/34	SHANGHAI VOLUNTEER CORPS
1319120	E	06/30/34	SHANGHAI VOLUNTEER CORPS
1319139	E	06/30/35	SHANGHAI VOLUNTEER CORPS
1319157	E	06/30/34	SHANGHAI VOLUNTEER CORPS
1319173	E	06/30/34	SHANGHAI VOLUNTEER CORPS
1319230	E	06/30/35	SHANGHAI VOLUNTEER CORPS
1319271	E	06/30/34	SHANGHAI VOLUNTEER CORPS
1319576	E	06/30/35	SHANGHAI VOLUNTEER CORPS
1319676	E	06/30/34	SHANGHAI VOLUNTEER CORPS
1319710	E	06/30/34	SHANGHAI VOLUNTEER CORPS
1319711	E	06/30/34	SHANGHAI VOLUNTEER CORPS
1320080	E	06/30/34	SHANGHAI VOLUNTEER CORPS
1320091	E	06/30/34	SHANGHAI VOLUNTEER CORPS
1320156	E	06/30/34	SHANGHAI VOLUNTEER CORPS
1320316	E	06/30/34	SHANGHAI VOLUNTEER CORPS
1320343	E	06/30/34	SHANGHAI VOLUNTEER CORPS
1320369	E	06/30/34	SHANGHAI VOLUNTEER CORPS
1320407	E	06/30/34	SHANGHAI VOLUNTEER CORPS
1320451	E	06/30/34	SHANGHAI VOLUNTEER CORPS
1320735	E	06/30/35	SHANGHAI VOLUNTEER CORPS
1320877	E	06/30/34	SHANGHAI VOLUNTEER CORPS
1320899	E	06/30/34	SHANGHAI VOLUNTEER CORPS
1321055	E	06/30/34	SHANGHAI VOLUNTEER CORPS
1321058	E	06/30/34	SHANGHAI VOLUNTEER CORPS
1321095	E	06/30/34	SHANGHAI VOLUNTEER CORPS

SER. NO.	MODEL	MM/DD/YY	USAGE
1321130	E	06/30/34	SHANGHAI VOLUNTEER CORPS
1321157	E	06/30/34	SHANGHAI VOLUNTEER CORPS
1321168	E	06/30/34	SHANGHAI VOLUNTEER CORPS
1321202	E	06/30/34	SHANGHAI VOLUNTEER CORPS
1321220	E	06/30/35	SHANGHAI VOLUNTEER CORPS
1321235	E	06/30/35	SHANGHAI VOLUNTEER CORPS
1321276	E	06/30/34	SHANGHAI VOLUNTEER CORPS
1321487	E	06/30/34	SHANGHAI VOLUNTEER CORPS
1321547	E	06/30/35	SHANGHAI VOLUNTEER CORPS
1321579	E	06/30/34	SHANGHAI VOLUNTEER CORPS
1321639	E	06/30/34	SHANGHAI VOLUNTEER CORPS
1321788	E	06/30/34	SHANGHAI VOLUNTEER CORPS
1321854	E	06/30/34	SHANGHAI VOLUNTEER CORPS
1321966	E	06/30/35	SHANGHAI VOLUNTEER CORPS
1322030	E	06/30/34	SHANGHAI VOLUNTEER CORPS
1322166	E	06/30/34	SHANGHAI VOLUNTEER CORPS
1322216	E	06/30/34	SHANGHAI VOLUNTEER CORPS
1322359	E	06/30/34	SHANGHAI VOLUNTEER CORPS
1322396	E	06/30/34	SHANGHAI VOLUNTEER CORPS
1322425	E	06/30/34	SHANGHAI VOLUNTEER CORPS
1322522	E	06/30/34	SHANGHAI VOLUNTEER CORPS
1322595	E	06/30/34	SHANGHAI VOLUNTEER CORPS
1322606	E	06/30/34	SHANGHAI VOLUNTEER CORPS
1322614	E	06/30/34	SHANGHAI VOLUNTEER CORPS
1322622	E	06/30/35	SHANGHAI VOLUNTEER CORPS
1322631	E	06/30/34	SHANGHAI VOLUNTEER CORPS
1322696	E	06/30/34	SHANGHAI VOLUNTEER CORPS
1322759	E	06/30/34	SHANGHAI VOLUNTEER CORPS
1322926	E	06/30/34	SHANGHAI VOLUNTEER CORPS
1322998	E	06/30/34	SHANGHAI VOLUNTEER CORPS
1323036	E	06/30/35	SHANGHAI VOLUNTEER CORPS
1323194	E	06/30/34	SHANGHAI VOLUNTEER CORPS
1323389		09/29/44	USAT MAIWARAS-146
1323454	E	06/30/34	SHANGHAI VOLUNTEER CORPS
1323559	E	06/30/34	SHANGHAI VOLUNTEER CORPS
1324348	E	06/30/34	SHANGHAI VOLUNTEER CORPS
1324355	E	06/30/34	SHANGHAI VOLUNTEER CORPS
1324389	E	06/30/35	SHANGHAI VOLUNTEER CORPS
1324444	E	06/30/34	SHANGHAI VOLUNTEER CORPS
1324538	E	06/30/35	SHANGHAI VOLUNTEER CORPS
1324573	E	06/30/34	SHANGHAI VOLUNTEER CORPS
1324586	E	06/30/34	SHANGHAI VOLUNTEER CORPS
1325421	E	06/30/34	SHANGHAI VOLUNTEER CORPS
1325449	E	06/30/34	SHANGHAI VOLUNTEER CORPS
1325559	E	06/30/34	SHANGHAI VOLUNTEER CORPS
1325586	E	06/30/35	SHANGHAI VOLUNTEER CORPS
1325605	E	06/30/34	SHANGHAI VOLUNTEER CORPS
1325743	E	06/30/34	SHANGHAI VOLUNTEER CORPS
1325824	E	06/30/34	SHANGHAI VOLUNTEER CORPS
1325977	E	06/30/34	SHANGHAI VOLUNTEER CORPS
1326031	E	06/30/35	SHANGHAI VOLUNTEER CORPS
1326097	E	06/30/34	SHANGHAI VOLUNTEER CORPS
1326146	E	06/30/35	SHANGHAI VOLUNTEER CORPS
1326216	E	06/30/34	SHANGHAI VOLUNTEER CORPS
1326339	E	06/30/34	SHANGHAI VOLUNTEER CORPS
1326594	E	06/30/34	SHANGHAI VOLUNTEER CORPS
1326659	E	06/30/34	SHANGHAI VOLUNTEER CORPS
1326667	E	06/30/34	SHANGHAI VOLUNTEER CORPS
1326684	E	06/30/34	SHANGHAI VOLUNTEER CORPS
1326710	E	06/30/34	SHANGHAI VOLUNTEER CORPS
1326726	E	06/30/35	SHANGHAI VOLUNTEER CORPS
1326746	E	06/30/35	SHANGHAI VOLUNTEER CORPS
1326774	E	06/30/34	SHANGHAI VOLUNTEER CORPS
1326962	E	06/30/34	SHANGHAI VOLUNTEER CORPS
1327031	E	06/30/34	SHANGHAI VOLUNTEER CORPS
1327036	E	06/30/34	SHANGHAI VOLUNTEER CORPS
1327100	E	06/30/34	SHANGHAI VOLUNTEER CORPS
1327103	E	06/30/34	SHANGHAI VOLUNTEER CORPS
1327209	E	06/30/34	SHANGHAI VOLUNTEER CORPS
1327225	E	06/30/34	SHANGHAI VOLUNTEER CORPS
1327755	E	06/30/34	SHANGHAI VOLUNTEER CORPS
1327995	E	06/30/34	SHANGHAI VOLUNTEER CORPS
1328006	E	06/30/35	SHANGHAI VOLUNTEER CORPS
1328090	E	06/30/34	SHANGHAI VOLUNTEER CORPS
1328255	E	06/30/35	SHANGHAI VOLUNTEER CORPS
1328269	E	06/30/34	SHANGHAI VOLUNTEER CORPS
1328275	E	06/30/34	SHANGHAI VOLUNTEER CORPS

SER. NO.	MODEL	MM/DD/YY	USAGE
1328505	E	06/30/34	SHANGHAI VOLUNTEER CORPS
1328537	E	06/30/34	SHANGHAI VOLUNTEER CORPS
1328600	E	06/30/34	SHANGHAI VOLUNTEER CORPS
1328644	E	06/30/34	SHANGHAI VOLUNTEER CORPS
1328649	E	06/30/35	SHANGHAI VOLUNTEER CORPS
1328820	E	06/30/34	SHANGHAI VOLUNTEER CORPS
1328898	E	06/30/35	SHANGHAI VOLUNTEER CORPS
1328942	E	06/30/34	SHANGHAI VOLUNTEER CORPS
1329006	E	06/30/34	SHANGHAI VOLUNTEER CORPS
1329028	E	06/30/34	SHANGHAI VOLUNTEER CORPS
1329109	E	06/30/34	SHANGHAI VOLUNTEER CORPS
1329348	E	06/30/34	SHANGHAI VOLUNTEER CORPS
1329398	E	06/30/35	SHANGHAI VOLUNTEER CORPS
1329553	E	06/30/34	SHANGHAI VOLUNTEER CORPS
1329595	E	06/30/34	SHANGHAI VOLUNTEER CORPS
1329609	E	06/30/35	SHANGHAI VOLUNTEER CORPS
1329728	E	06/30/35	SHANGHAI VOLUNTEER CORPS
1329740	E	06/30/34	SHANGHAI VOLUNTEER CORPS
1329786	E	06/30/34	SHANGHAI VOLUNTEER CORPS
1329844	E	06/30/35	SHANGHAI VOLUNTEER CORPS
1329858	E	06/30/35	SHANGHAI VOLUNTEER CORPS
1329947	E	06/30/34	SHANGHAI VOLUNTEER CORPS
1330149	E	06/30/34	SHANGHAI VOLUNTEER CORPS
1330247	E	06/30/34	SHANGHAI VOLUNTEER CORPS
1330309	E	06/30/34	SHANGHAI VOLUNTEER CORPS
1330598	E	06/30/35	SHANGHAI VOLUNTEER CORPS
1330635	E	06/30/34	SHANGHAI VOLUNTEER CORPS
1330769	E	06/30/35	SHANGHAI VOLUNTEER CORPS
1330773	E	06/30/34	SHANGHAI VOLUNTEER CORPS
1330816	E	06/30/35	SHANGHAI VOLUNTEER CORPS
1330914	E	06/30/34	SHANGHAI VOLUNTEER CORPS
1330987	E	06/30/34	SHANGHAI VOLUNTEER CORPS
1330991	E	06/30/34	SHANGHAI VOLUNTEER CORPS
1331009	E	06/30/34	SHANGHAI VOLUNTEER CORPS
1331066	E	06/30/34	SHANGHAI VOLUNTEER CORPS
1331204	E	06/30/34	SHANGHAI VOLUNTEER CORPS
1331334	E	06/30/34	SHANGHAI VOLUNTEER CORPS
1331377	E	06/30/34	SHANGHAI VOLUNTEER CORPS
1331386	E	06/30/35	SHANGHAI VOLUNTEER CORPS
1331406	E	06/30/34	SHANGHAI VOLUNTEER CORPS
1331429	E	06/30/34	SHANGHAI VOLUNTEER CORPS
1331490	E	06/30/34	SHANGHAI VOLUNTEER CORPS
1331553	E	06/30/34	SHANGHAI VOLUNTEER CORPS
1331567	E	06/30/34	SHANGHAI VOLUNTEER CORPS
1331619	E	06/30/35	SHANGHAI VOLUNTEER CORPS
1331656	E	06/30/34	SHANGHAI VOLUNTEER CORPS
1331687	E	06/30/34	SHANGHAI VOLUNTEER CORPS
1331784	E	06/30/35	SHANGHAI VOLUNTEER CORPS
1331868	E	06/30/35	SHANGHAI VOLUNTEER CORPS
1331872	E	06/30/35	SHANGHAI VOLUNTEER CORPS
1331876	E	06/30/34	SHANGHAI VOLUNTEER CORPS
1331922	E	06/30/35	SHANGHAI VOLUNTEER CORPS
1331971	E	06/30/34	SHANGHAI VOLUNTEER CORPS
1331979	E	06/30/34	SHANGHAI VOLUNTEER CORPS
1332074	E	06/30/34	SHANGHAI VOLUNTEER CORPS
1332096	E	06/30/35	SHANGHAI VOLUNTEER CORPS
1332291	E	06/30/34	SHANGHAI VOLUNTEER CORPS
1332309	E	06/30/34	SHANGHAI VOLUNTEER CORPS
1332376	E	06/30/34	SHANGHAI VOLUNTEER CORPS
1332580	E	06/30/34	SHANGHAI VOLUNTEER CORPS
1332592	E	06/30/35	SHANGHAI VOLUNTEER CORPS
1332707	E	06/30/34	SHANGHAI VOLUNTEER CORPS
1332790	E	06/30/35	SHANGHAI VOLUNTEER CORPS
1332801	E	06/30/34	SHANGHAI VOLUNTEER CORPS
1332842	E	06/30/34	SHANGHAI VOLUNTEER CORPS
1332914	E	06/30/35	SHANGHAI VOLUNTEER CORPS
1333003	E	06/30/35	SHANGHAI VOLUNTEER CORPS
1333120	E	06/30/34	SHANGHAI VOLUNTEER CORPS
1333139	E	06/30/35	SHANGHAI VOLUNTEER CORPS
1333193	E	06/30/35	SHANGHAI VOLUNTEER CORPS
1333248	E	06/30/35	SHANGHAI VOLUNTEER CORPS
1333325	E	06/30/34	SHANGHAI VOLUNTEER CORPS
1333420	E	06/30/35	SHANGHAI VOLUNTEER CORPS
1333518	E	06/30/35	SHANGHAI VOLUNTEER CORPS
1333536	E	06/30/34	SHANGHAI VOLUNTEER CORPS
1333542	E	06/30/34	SHANGHAI VOLUNTEER CORPS
1333564	E	06/30/34	SHANGHAI VOLUNTEER CORPS
1333581	E	06/30/34	SHANGHAI VOLUNTEER CORPS
1333597	E	06/30/34	SHANGHAI VOLUNTEER CORPS
1333668	E	06/30/34	SHANGHAI VOLUNTEER CORPS
1334113	E	06/30/34	SHANGHAI VOLUNTEER CORPS
1334115	E	06/30/34	SHANGHAI VOLUNTEER CORPS
1334369	E	06/30/34	SHANGHAI VOLUNTEER CORPS
1334747	E	06/30/34	SHANGHAI VOLUNTEER CORPS
1334747	E	06/30/35	SHANGHAI VOLUNTEER CORPS
1334873	E	06/30/35	SHANGHAI VOLUNTEER CORPS
1335364	E	06/30/34	SHANGHAI VOLUNTEER CORPS
1335521	E	06/30/35	SHANGHAI VOLUNTEER CORPS
1335545	E	06/30/34	SHANGHAI VOLUNTEER CORPS
1335677	E	06/30/34	SHANGHAI VOLUNTEER CORPS
1335743	E	06/30/35	SHANGHAI VOLUNTEER CORPS
1335843	E	06/30/35	SHANGHAI VOLUNTEER CORPS
1335870	E	06/30/35	SHANGHAI VOLUNTEER CORPS
1336156	E	06/30/34	SHANGHAI VOLUNTEER CORPS
1336234	E	06/30/35	SHANGHAI VOLUNTEER CORPS
1336261	E	06/30/34	SHANGHAI VOLUNTEER CORPS
1336320	E	06/30/34	SHANGHAI VOLUNTEER CORPS
1336396	E	06/30/34	SHANGHAI VOLUNTEER CORPS
1336469	E	06/30/34	SHANGHAI VOLUNTEER CORPS
1336529	E	06/30/34	SHANGHAI VOLUNTEER CORPS
1336552	E	06/30/34	SHANGHAI VOLUNTEER CORPS
1336560	E	06/30/34	SHANGHAI VOLUNTEER CORPS
1336765	E	06/30/34	SHANGHAI VOLUNTEER CORPS
1337130	E	06/30/34	SHANGHAI VOLUNTEER CORPS
1337445	E	06/30/34	SHANGHAI VOLUNTEER CORPS
1337687	E	06/30/35	SHANGHAI VOLUNTEER CORPS
1337714	E	06/30/34	SHANGHAI VOLUNTEER CORPS
1337948	E	06/30/35	SHANGHAI VOLUNTEER CORPS
1338262	E	06/30/34	SHANGHAI VOLUNTEER CORPS
1338329	E	06/30/34	SHANGHAI VOLUNTEER CORPS
1338434	E	06/30/34	SHANGHAI VOLUNTEER CORPS
1338609	E	06/30/35	SHANGHAI VOLUNTEER CORPS
1338985	E	06/30/34	SHANGHAI VOLUNTEER CORPS
1339088	E	06/30/35	SHANGHAI VOLUNTEER CORPS
1340245	E	06/30/34	SHANGHAI VOLUNTEER CORPS
1340275	E	06/30/35	SHANGHAI VOLUNTEER CORPS
1340841	E	06/30/34	SHANGHAI VOLUNTEER CORPS
1340877	E	06/30/34	SHANGHAI VOLUNTEER CORPS
1340914	E	06/30/34	SHANGHAI VOLUNTEER CORPS
1341637	E	06/30/34	SHANGHAI VOLUNTEER CORPS
1341799	E	06/30/34	SHANGHAI VOLUNTEER CORPS
1345116	E	06/30/34	SHANGHAI VOLUNTEER CORPS
1347927	E	06/30/35	SHANGHAI VOLUNTEER CORPS
1348785	E	06/30/34	SHANGHAI VOLUNTEER CORPS
1349583	E	06/30/35	SHANGHAI VOLUNTEER CORPS
1350802	E	06/30/35	SHANGHAI VOLUNTEER CORPS
1354058	E	06/99	CMP INVENTORY
1356934	E	06/99	CMP INVENTORY
1396717		11/10/44	USAT MULCRAS-141
1476996		11/09/44	USAT BOPPLES-147
2115594		11/20/44	USAT GEN MACARTHURS-44

Primary References

Government Documents, Books, Manuals ands other Publications

Numerous War Department, Ordnance Department U.S. Army, Navy and Marine Corps and other government documents, reports and correspondence specifically cited in the endnotes.

Colvin, Frederick H. and Ethan Viall, *United States Rifles and Machine Guns – A Detailed Account of the Methods Used in Manufacturing the Springfield, 1903 Model Service Rifle,* Government Printing Office, Washington, D.C., 1917.

Colvin, Frederick H., *60 Years with Men and Machines: An Autobiography,* McGraw Hill Books, New York City, N.Y., 1947.

Crowell, Benedict, Assistant Secretary of War, Director of Munitions, *America's Munitions,* Government Printing Office, Washington, D.C., 1919.

Crozier, Major General William, *Ordnance and the World War,* Charles Scribner & Son, New York, N.Y., 1920.

Green, Constance McLaughlin, Harry C. Thomson, and Peter C. Roots, United States Army in World War II, The Technical Services, *The Ordnance Department: Planning Munitions for War,* Ordnance Department, 1955.

Thomson, Harry C. and Lida Mayo, *The Ordnance Department: Procurement and Supply,* United States Army in World War II, The Technical Services, Ordnance Department, 1960.

United States Army in the World War – 1917–1919: Organization of the American Expeditionary Forces, Vol. 1, Center of Military History, United States Army, Washington, D.C., 1988.

William Bradford Williams, *Munitions Manufacture in the Philadelphia Ordnance District for the U.S. Army in the World War,* A. Pomerantz & Company, Philadelphia, Penn., 1921.

Army Ordnance Doc. No. 1865, *History of Rifles, Revolvers and Pistols, 1917–1919,* Government Printing Office, Washington, D.C., 1920.

Project Supporting Paper – *Rifle U.S. Caliber .30, M1917, 1917–1945,* Small Arms Division, Industrial Service Office, Chief of Ordnance, August 14, 1945.

Project Supporting Paper Relating to Bayonets, Knives and Scabbards 1917–August 1945, Small Arms Division, Industrial Service, Ordnance Department, Declassified, September 27, 1958.

No. 1917, *Description and Rules for the Management of the United States Rifle, Caliber .30, Model of 1917,* October 8, 1917; Revised January 16, 1918; Revised May 7, 1918, Government Printing Office, Washington, D.C., 1918.

Supplement to No. 1917, *Description and Rules for the Management of the United States Rifle, Caliber .30, Model of 1917,* Government Printing Office, Washington, D.C., 1918.

Handbook of Ordnance Data, November 15, 1918, Ordnance Department, U.S. Army, Government Printing Office, Washington, D.C., 1919.

History of Trench Warfare Material, Army Ordnance 1917–1919, Government Printing Office, Washington, D.C., 1920.

FM 23-6, War Department, Basic Field Manual, U.S. Rifle, Caliber .30, M1917 (Enfield), August 3, 1942.

Small Arms and Small Arms Ammunition, Record of Army Ordnance Research and Development, Book 1, Vol. 2, Office of the Chief of Ordnance, Washington, D.C., January 1946.

TR 1320-B Fire Control and Sighting Equipment, Small Arms and Special Infantry Weapons, April 2, 1927.

Ordnance Base Shop Field Service Data, U.S. Rifle, Cal. .30, M1917, Prepared under the direction of the Chief of Ordnance, Rock Island Arsenal, February 1943.

The Flaming Bayonet – Designing Tactical Surprise, Dr. James W. Williams, Historian, U.S. Army Chemical School, 1987.

Chemical Warfare Monographs, Volume 43, Arthur B. Ray, War Department – Chemical Warfare Service, Research Division, American University Experiment Station, Washington, D.C., May 1919.

Flaming Bayonet, Partial Report No. 3 on Investigation #40, Asire, Lt. H.W., Chemical Warfare Service, 1919.

Secondary References

Books

Brophy, Lt. Col. Williams S. (Ret.), *The Springfield 1903 Rifles*, Stackpole Books, Harrisburg, Penn., 1985.
Brown, Capt. Sevellon, *The Story of Ordnance in the World War*, James William Bryan Press, Washington, D.C., 1920.
Campbell, Clark S., *The '03 Springfield Rifles' Era*, Campbell Books, Richmond, Vir., 2003.
Canfield, Bruce N., *U.S. Infantry Weapons of the First World War*, Andrew Mowbray Publishers, Lincoln, R.I., 2000.
Canfield, Bruce N., *An Illustrated Guide to the '03 Springfield Service Rifle*, Andrew Mowbray, Inc., Lincoln, R.I., 2004,
Canfield, Bruce N., *U.S. Military Bolt Action Rifles*, Andrew Mowbray Publishers, Woonsocket, R.I., 2010.
Canfield, Bruce N., *U.S. Small Arms of World War II*, Andrew Mowbray Publishers., Woonsocket, R.I., 2020.
Crossman, E.C., *The Book of the Springfield, Small Arms*, Technical Publishing Co., Georgetown, S.C., 1951.
Cunningham, Gary M., *U.S. Knife Bayonets & Scabbards*, Scott A. Duff Publications, Export, Penn., 2015.
Daum, Anthony C. and Charles W. Pate, *U.S. Military Arms Inspector Markings*, Andrew Mowbray, Inc., Woonsocket, R.I., 2016.
Ezell, Edward Clinton, *The Great Rifle Controversy*, Stackpole Books, Harrisburg, Penn., 1984.
Ferris, C.S., *United States Rifle Model of 1917*, Scott A. Duff Publications, Export, Penn., 2004.
Freidel, Frank, *Over There – The Story of America's First Great Overseas Crusade*, Temple University Press, Philadelphia, Penn., 1990.
Goodwin, Mark G., U.S. Infantry Weapons in Combat, Scott A. Duff, Publications, Export, Penn., 2005.
Hatcher, Maj. General Julian S., Hatcher's Notebook, The Stackpole Company, Harrisburg, Penn., 1947.
Henshaw, Thomas, The History of Winchester Firearms 1866–1992, Winchester Press, Clinton, N.J., 1993.
Hicks, Major James E., U.S. Military Firearms, Borden Publishing Co., Alhambra, Calif., 1962.
Houze, Herbert G., *Winchester Bolt Action Military & Sporting Rifles 1877 to 1937*, Mowbray Publishing Co., Lincoln, R.I., 1998.
Johnson, George B. and Hans Bert Lockhoven, *International Armament, Volume I*, International Small Arms Publishers, Cologne, Germany, 1965.
Keith, Elmer, *Hell, I was There*, Petersen Publishing Co., Los Angeles, Calif., 1979.
Lawrence, Joseph Douglas, *Fighting Solider – The AEF in 1918*, University of Colorado Press, Boulder, Colo., 1985.
Marcot, Roy, *Remington – The Official Authorized History of Remington Arms Company*, PRIMEDIA, Special Interest Publications, Peoria, Ill., 1998.
Mercaldo, Luke, *Allied Rifle Contracts in American 1914–1918*, Wet Dog Publications, Greensboro, N.C., 2011.
Morrison, Daniel J., *The U.S. Enfield Bayonet*, Breed's Hill Publishing Co., Demarest, N.J., 2017.
Sawyer, Charles Winthrop, *Our Rifles*, The Cornhill Publishing Co., Boston, Mass., 1920.
Senich, Peter R., *The Complete Book of U.S. Sniping*, Paladin Press, Boulder, Colo., 1988.
Skennerton, Ian, *The U.S. Enfield*, published by Ian D. Skennerton, Margate, Australia, 1983.
Stratton, Charles R., *The Pattern 1914 and U.S. Model 1917 Rifles*, North Cape Publications, Inc., Tustin, Calif., 2000.
Tulkoff, Alec S., *Equipping the Corps: 1892–1937*, R. James Bender Publishing, San Jose, Calif., 2010.
Webster, Donald B., *Military Bolt Action Rifles 1841–1918*, Museum Restoration Service, Alexandria Bay, N.Y., 1993.
Whitman, John W., *Bataan, Our Last Ditch*, Hippocrene Books, N.Y., N.Y., 1990.
Williamson, Harold F., *Winchester – The Gun that Won the West*, Combat Forces Press, Washington, D.C., 1952.
Zeglin, Fred, P.O. *Ackley- America's Gunsmith*, Krause Publications, Iola, Wis., 2017.

Articles

U.K. Historic Arms Research Centre, "Pattern 1913 Enfield Rifle,"1997.
Gorelick, Marc, "The U.K. Pattern 1913, Pattern 1914 and the U.S. Model of 1917," Virginia Gun Collectors Association.
Krisko, Jonathan, "Selling the Model 1917 – John T. Thompson and the Rifle Demonstrators," *American Rifleman*, February 2022.
Marcot, Roy and Joe Poyer, "The Story of Eddystone," The Remington Society of America.
Kurt Sellers (Maj. U.S. Army, Ret.) "Eddystone Rifle Plant – History of the Last Remaining Building," www.worldwar1centennial.org

Endnotes

CHAPTER 1

[1] Johnson, George B. and Hans Bert Lockhoven, *International Armament – Volume 1*, International Small Arms Publishers, Cologne, Germany, 1965, p. 231, henceforth referred as "Johnson and Lockhoven."

[2] www.researchpress.co.uk/index.php/firearms/gunmakers/gibbs/gibbs-metford, courtesy of John Spangler.

[3] Skennerton, Ian, *The Lee-Enfield*, Ian D., Skennerton, Margate, Australia, 2008, p. 450.Henceforth referred to as "Skennerton."

[4] *Ibid.*, pp. 439–447.

[5] *Ibid.*, p. 149

[6] *Ibid.*, p. 27

[7] U.K. Historic Small Arms Research Centre, Rifleman.org. a.m. – Pattern 13 Enfield Rifle.

[8] Skennerton, p. 27.

[9] Morrison, Daniel Jay, *The U.S. Enfield Bayonet*, Breed's Hill Publishing Co., Demarest, N.J., 2017, p. 27.

[10] Mercaldo, Luke, *Allied Rifle Contracts in America, 1914–1918*, Wet Dog Publications, Greensboro, N.C., 2011, p. 132. Henceforth referred to as "Mercaldo."

[11] *Ibid.*, p. 134.

[12] Williamson, Harold F., *Winchester – The Gun That Won the West*, Combat Forces Press, Washington, D.C., 1952, p. 226. Henceforth referred to as "Williamson."

[13] File memorandum, Remington Ilion Works, Ilion, New York, August 17, 1917. Subject – World War Production of Remington Ilion Works. By J.E. Brennan, Vice President. Courtesy of Roy Marcot.

[14] Mercaldo, p. 134.

[15] *Op cit.*, File memorandum, Remington Ilion Works.

[16] *Ibid.*

[17] Houze, Herbert G., *Winchester Bolt Action Military & Sporting Rifles, 1877 to 1937*, Andrew Mowbray, Inc., Lincoln, R.I., 1998, p. 117.

[18] Mercaldo, pp. 136–138.

[19] *Ibid.*

[20] Skennerton, p. 63.

[21] *Ibid.*

[22] *Ibid.*, p. 59

[23] John Spangler, email correspondence, Dec. 2022.

[24] Skennerton., pp. 58–61.

[25] Stratton, Charles R., *The Pattern 1914 and U.S. Model 1917 Rifles*, North Cape Publications, Inc., Tustin, Calif., 2000, p. 23. Henceforth referred to as "Stratton."

[26] *Ibid.*, p. 153.

[27] *Ibid.*, p. 140.

[28] Skennerton, p. 111.

[29] John Spangler, email correspondence, Dec. 2022.

[30] Skennerton, p. 125.

[31] Morrison, Daniel Jay, *The U.S. Enfield Bayonet P1913, M1917, M1917 (Vietnam)*, Breed's Hill Publishing Company, Demarest, N.J., 2017, pp. 31, 32. Henceforth referred to as "Morrison."

[32] Mercaldo, pg. 155.

[33] *Ibid.*, p. 143.

[34] Skennerton, p. 62.

[35] Mercaldo, pg. 143.

[36] *Ibid.*

CHAPTER 2

[1] Crowell, Benedict, Assistant Secretary of War and Director of Munitions, *America's Munitions 1917–1918*, Government Printing Office, Washington, D.C., 1919, p. 183. Henceforth referred to as "Crowell."

[2] Colvin, Fred H. and Ethan Viall, *United States Rifles and Machine Guns*, reprinted by Wolfe Publishing Co., Prescott, Arizona, 1984.

[3] Colvin, Fred H., *60 Years with Man and Machines*, McGraw Hill Books, New York, N.Y., 1947, p. 184. Henceforth referred to as "Colvin."

[4] *Ibid.*, pp. 184, 185.

[5] *Ibid.*

[6] *Ibid.*, p. 187

[7] Crowell, p. 179.

[8] *Ibid.*, pp. 180, 181.

[9] Office of Chief of Ordnance memo to the Adjutant General of the Army, April 11, 1917, courtesy of Archival Research Group.

[10] *Ibid.*

[11] Crowell, p. 181.

[12] Krisko, Jonathon, " 'Selling' the Model 1917: John T. Thompson & the Rifle Demonstrators," *American Rifleman*, February 12, 2022.

[13] Houze, Herbert G., *Winchester Bolt Action Military & Sporting Rifles, 1877–1937*, Mowbray Publishing, Lincoln, R.I., 1998, p. 139. Henceforth referred to as "Houze."

[14] "Newton D. Baker and the Genesis of the War Industries Board, 1917–1918," *Journal of American History*, Vol. 52, No. 1, June 1965, Oxford University Press.

[15] *Arms and the Man*, "Two Experts Praise the U.S. Enfield, Part 1," Vol. LXII, No. 18, July 28, 1917, pp. 343, 344.

[16] *Arms and the Man*, "Two Experts Praise the U.S. Enfield, Part 2," Vol. LXII, No. 19, August 4, 1917, pp. 365, 366.

[17] *Ibid.*

[18] Minutes transcribed during an April 4, 1917, meeting attended by officials of Winchester Repeating Arms Co., Remington-UMC Co. and the Remington/Delaware/Eddystone entities.

[19] *New York Times*, December 29, 1916, p. 12:4.

[20] April 1918 report from Remington, WWI Ordnance Notes, U.S. Martial Arms Collector, Jan. 1994, 67-12-67-13, courtesy of Frank Mallory.

[21] Crowell, p. 182.

[22] *Ibid.*, p. 183.

[23] Colvin, p. 188.

[24] *Ibid.*, p. 190.

[25] *Ibid.*, pp. 191, 192.

[26] Crozier, Major General William, *Ordnance and the World War*, Charles Scribner & Son, New York, N.Y., 1920, pp. 60, 61.

[27] Crowell, p. 183.

[28] National Archives research per Frank Mallory.

[29] Williamson, Harold F., *Winchester, The Gun That Won the West*, Combat Forces Press, Washington, D.C., 1952, p. 238.

[30] Crowell, p. 182.

[31] Green, Constance M., Harry C. Thomson, and Peter Roots, *The Ordnance Department: Planning Munitions for War*, United States Army Technical Services, Office of the Chief of Military History, Department of the Army, Washington, D.C., 1955, p. 26.

[32] Colvin, p. 196.

[33] Crowell, p. 186.

[34] Report for week ending May 3, 1919, from Lt. Col. Herbert O'Leary to Maj. C.C. Williams, Chief of Ordnance.

[35] Crowell, p. 179.

[36] Crossman, Edward C., "British Rifles for the American Army," *National Service*, November 1917.

[37] Crowell, p. 181.

CHAPTER 3

[1] War Department, *Basic Field Manual, U.S. Rifle, Caliber .30, M1917 (Enfield)*, Government Printing Office, Washington, D.C., August 3, 1972, pp. 1, 2.

[2] Skennerton, Ian, *The U.S. Enfield*, Ian Skennerton, Margate, Australia, 1983. Henceforth referred to as "Skennerton."

[3] Williams, William Bradford, *Munitions Manufacture in the Philadelphia Ordnance District for the U.S. Army in the World War*, A. Pomerantz & Company, Philadelphia, Penn., 1921, p. 516.

[4] Stratton, Charles R., *The Pattern 1914 and U.S Model 1917 Rifles*, North Cape Publications, Tustin, Calif., 2000, p. 37. Henceforth referred to as "Stratton."

[5] Hatcher, Maj. Gen. Julian S., *Hatcher's Notebook*, The Stackpole Company, Harrisburg, Penn., 1947, p. 16. Henceforth referred to as "Hatcher."

[6] *Description and Rules for the Management of the United States Rifle, Caliber .30,*

Model of 1917, Government Printing Office, Washington, D.C., 1918, p. 22.
[7] *Ibid.*, p. 92
[8] Hatcher, pg. 14.
[9] Crowell, Benedict, America's Munitions 1917-1918, Government Printing Office, Washington, DC, 1919, pg. 185, henceforth referred to as "Crowell."
[10] Hatcher, p. 14.
[11] Stratton, pp. 59, 60.
[12] *Op cit., Description and Rules*, p. 13.
[13] *Ibid.*, p. 14.
[14] Stratton, p. 48.
[15] *Ibid.*, p. 52.
[16] *Ibid.*, p. 70.
[17] *Op cit., Description and Rules*, p. 28.
[18] Stratton, pp. 68, 69.
[19] *Ibid.*, p. 72.
[20] Sellers, Major Kurt, U.S. Army (ret.), *Eddystone Rifle Plant, History of the Last Remaining Building.*
[21] John Spangler, email correspondence with author Jan. 25, 2023.
[22] *Op cit., Description and Rules*, pp. 28, 29.

CHAPTER 4

[1] *History of Rifles, Revolvers and Pistols, 1917–1919*, U.S. Army Ordnance publication No. 1865, U.S. Government Printing Office, Washington, D.C., 1920, pp. 15, 16.
[2] *Ibid.*, p. 17.
[3] Crowell, Benedict, Assistant Secretary of War and Director of Munitions, *America's Munitions 1917–1918*, Government Printing Office, Washington, D.C., 1919, p. 185. Henceforth referred to as "Crowell."
[4] Mallory, Frank, U.S. Martial Arms Collector and Springfield Research Service, No. 79, January 1997, p. 79-2.
[5] September 18, 1918, Executive Committee minutes of the Victor Talking Machine Co., courtesy of John Spangler.
[6] Williamson, Harold F., *Winchester – The Gun That Won the West*, Combat Forces Press, Washington, D.C. ,1952, p. 15, Henceforth referred to as "Williamson."
[7] Mercaldo Luke, *Allied Rifle Contracts in America 1914–1918*, Wet Dog Publications, Greensboro, N.C., 2011, p. 81. Henceforth referred to as "Mercaldo."
[8] Houze, Herbert G., *Winchester Bolt Action Military & Sporting Rifles 1877–1937*, Andrew Mowbray Publishing, Lincoln, R.I., 1998, p. 139. Henceforth referred to as "Houze."
[9] Mercaldo, p. 144.
[10] Winchester Repeating Arms Company, letter from J.E. Otterson to General W.C. Crozier, April 3, 1917, World War I – Letters & Memos Relative to Winchester Producing Enfield Rifles for the U.S. Army, Winchester Arms Collection Archives.
[11] Williamson, p. 237, 238.
[12] Houze, p. 140.
[13] *Ibid.*
[14] Crowell, p. 182.
[15] Williamson, p. 238.
[16] Houze, p. 138.
[17] *History of Rifles, Revolvers and Pistols – 1917–1919*, U.S. Army Ordnance publication No. 1865, U.S. Government Printing Office, Washington, D.C., 1920, pp. 16, 17.
[18] Williamson, p. 243.
[19] *Ibid.*
[20] Ferris, p. 16.
[21] War Department, Office of the Chief of Ordnance memo to the Engineering Bureau, dated April 18, 1918, NARA, courtesy of Archival Research Group.
[22] Memorandum from Col. T.L. Ames, Ordnance Office, Supply Division to Control Bureau, Ordnance Office, dated March 4, 1918, courtesy of Archival Research Group.
[23] Memorandum from Supply Division, Ordnance Office to the Engineering Division, dated June 25, 1918.
[24] Memorandum dated June 27, 1918, from Maj. J.S. Hatcher, Machine Gun-Small Arms Section, to Engineering Division, Ordnance Department, courtesy of Archival Research Group
[25] Memo dated June 29, 1918, from General Headquarters, A.E.F.
[26] Ferris, p. 16, 17.
[27] *Ibid.*, p. 91.
[28] Ferris, p. 91.
[29] *Ibid.*, p. 89.
[30] *Ibid.*, p. 18.
[31] Letter dated June 19, 1918, from Col. John Simpson, Estimates and Requirements Division, to the Requirements Section of his Division, File 474.1/947, courtesy of C.S. Ferris.
[32] Jordan, Col. Harry B., Acting Chief of Ordnance, A.E.F., May 3, 1918, NARA, courtesy of Archival Research Group.
[33] Winchester Repeating Arms Co., Report on Status of Government Contracts, January 23 & 24, 1919, World War I – Government Contracts January 1919–December 1919, Winchester Arms Collection Archives, courtesy of Herbert G. Houze.
[34] Crowell, p. 227.
[35] Houze, p. 142.
[36] Rock Island Auction Company description for Lot #1820, September 11, 2015, auction.
[37] Crowell, p. 186; Skennerton, p. 67.
[38] Winchester factory polishing shop ledger, courtesy of Herbert House.
[39] Houze, p. 141.
[40] Ferris, p. 26.
[41] Skennerton, Ian, *The U.S. Enfield*, Ian D. Skennerton, Margate, Australia, 1983, p. 67. Henceforth referred to as "Skennerton."
[42] Marcot, Roy, *Remington – America's Oldest Gunmaker*, Primedia Special interest Publications, Peoria, Ill., 1998, p. 11. Henceforth referred to as "Marcot"
[43] *Ibid.*, pp 93–98.
[44] *Ibid.*, p. 108.
[45] Mercaldo, p. 33.
[46] *Ibid.*, p. 185.
[47] Crowell, p. 183.
[48] File memorandum, Remington Ilion Works, Ilion, New York, August 17, 1917. Subject – World War Production of Remington Ilion Works. By J.E. Brennan, Vice President. Courtesy of Roy Marcot.
[49] *Ibid.*
[50] *Ibid.*
[51] *Ibid.*
[52] Contract War-Ord 14063 dated July 12, 1917, between United States Government and the Remington Arms Union Metallic Cartridge Company, Inc., for 300,000 U.S. rifles, dated July 12, 1917. Courtesy of Roy Marcot.
[53] Memorandum from General William Crozier, Chief of Ordnance, to I. Betts, Vice President, Remington-UMC, dated November 13, 1917. Courtesy of Roy Marcot.
[54] *Ibid.*
[55] *History of Rifles, Revolvers and Pistols – 1917–1919*, U.S. Army Ordnance publication No. 1865, U.S. Government Printing Office, Washington, D.C., 1920, p. 16.
[56] Ferris, p. 100.
[57] Rifles formerly in the Peter Wainwright collection and the author's collection.
[58] Cowan Auction description for M1917 rifle from the Peter Wainwright collection, with the experimental rear sight.
[59] Ferris, pp. 125, 126
[60] *Ibid.*, p. 127.
[61] *Ibid.*, p. 41.
[62] *Ibid.*, p. 128.
[63] *Ibid.*, pp. 128, 129.
[64] Crowell, p. 186.
[65] Report dated week ending May 3, 1919, from Lt. Col. Herbert O'Leary to Chief of Ordnance, Maj. Gen. C.C. Williams containing daily reports of rifle manufacture after production of the M1917 rifle ceased, Ferris, p. 29.
[66] Marcot, Roy and Joe Poyer, "The Story of Eddystone," *Remington Society of America Journal*, 4th Quarter, 2010. Henceforth referred to as "The Story of Eddystone."
[67] *Ibid.*
[68] *Ibid.*
[69] Williams, William Bradford, *The British Enfield & U.S. Model 1917 Rifles made by the Midvale Steel & Ordnance Co., Munitions Manufacture in the Philadelphia Ordnance District for the U.S. Army in the World War*, A. Pomerantz & Co. Publishers, January 1921.
[70] Ferris, p. 91 and email correspondence with author and researcher Hayes Otoupalik.

[71] Williams
[72] "The Story of Eddystone."
[73] *Ibid.*
[74] Springfield Research Service, U.S Martial Arms Collector, 67-12.
[75] Skennerton, pp. 59, 63.
[76] Ferris, p. 101.
[77] Former Peter Wainwright Collection description.
[78] Rock Island Auction Company catalog description, September 12, 2020, Lot #1424.
[79] Rock Island Auction Company catalog description, September 8, 2019, Lot #3730.
[80] Crowell, p. 186, Skennerton, p. 67.
[81] *Story of Eddystone.*
[82] Ferris, p. 26.
[83] *Story of Eddystone.*

CHAPTER 5

[1] *History of Rifles, Revolvers and Pistols – 1917–1919*, U.S. Army Ordnance publication No. 1865, U.S. Government Printing Office, Washington, D.C., 1920, pp. 14, 15.
[2] Ordnance Department memo from Major J.T. Kenyon to Brig. Gen. Charles B. Wheeler, dated April 9, 1918.
[3] Ordnance memo to "Mr. Marshall" dated February 11, 1918, courtesy of Archival Research Group.
[4] Kreidberg, Marvin A. and Merton G. Henry, *History of Military Mobilization in the United States Army, 1775–1945*, Department of the Army Pamphlet 20-212, Washington, D.C.: Government Printing Office, 1955, p. 221.
[5] Correspondence dated November 24, 1917, from Major General William Crozier to the Adjutant General of the Army.
[6] Ferris, C.S., *United States Rifle Model of 1917*, South Greensburg, Cty., Greensburg, Penn., 2004, pp. 36–40. Henceforth referred to as "Ferris."
[7] Report of Captain James H. Kidder to Col. John T. Thompson, August 31, 1918, Ordnance Office File #ES474.1/1706.
[8] Krisko, Jonathon, "Selling the Model 1917 – John T. Thompson and the Rifle Demonstrators," *American Rifleman*, February 2022.
[9] *Ibid.*
[10] *Description and Rules for the Management of the United States Rifle, Caliber .30, Model of 1917*, Government Printing Office, Washington, D.C., 1918, p. 38.
[11] Memo dated Jan. 27, 1919, from army base hospital No. 45 to the Commanding Officer of the Justice Hospital Group, courtesy of Archival Research Group.
[12] Memo dated February 1, 1919, Col. Walter D. McCaw, Medical Corps, courtesy of Archival Research Group.
[13] March 27, 1919, memo from the Chief of Ordnance, Second Army to the Deputy Chief of Ordnance, G.H.Q. ,A.E.F., NARA, courtesy of Archival Research Group.
[14] Administrative Memorandum No. 71, "February 23, 1919, from Headquarters, Second Division, American Expeditionary Force, NARA, courtesy of Archival Research Group.
[15] *Op. cit.*, Krisko.
[16] *Op cit.*, Ferris, p. 42.
[17] War Department, Training Circular No. 5, Infantry Training, 1918, 19-22.8, Bach, Christian A. and Henry Nobel Hall, *The 4th Division: Its Services and Achievements in The World War.*
[18] *Ibid.*
[19] AEF cable dated February 28, 1918.
[20] March 2, 1918, letter from the adjutant general to the chief of ordnance, file O.O. 474.1/300.
[21] *Op. cit.*, Ferris, p. 56.
[22] *Op. cit.*, Krisko.
[23] Memorandum from American Expeditionary Force, Office of Chief of Ordnance, to Commanding Officers, *Gievres, Is-sur-Tille, Mehun and Jonohery*, dated Sept. 19, 1918, NARA, courtesy of Archival Research Group.
[24] A.E.F Report of a small arms situation, dated April 10, 1918, by Maj. C.C. Sheppard, Ordnance Department, NARA, courtesy of Archival Research Service.
[25] Letter dated June 19, 1918, from Col. John Simpson, Estimates and Requirements Division, to the Requirements Section of his Division, File 474.1/947, NARA, courtesy of Archival Research Group.
[26] Memo dated August 9, 1918, to A.E.F. Headquarters from 1st Lt. G.R. Pennington, Ordnance Officer, 42nd Division, NARA, courtesy of Archival Research Group.
[27] Memorandum dated August 8, 1918, to the Office of Chief of Ordnance, A.E.F. from the Ordnance Officer, 79th Division, NARA, courtesy of Archival Research Group.
[28] Memorandum dated August 10, 1918, from the Chief of Ordnance, A.E.F. to the Chief of Ordnance, Washington D.C., NARA, courtesy of Archival Research Group.
[29] Memorandum dated August 24, 1918, to Chief Ordnance Officer, A.E.F. to the C.O. Advance Unit Army Ordnance Park, NARA, courtesy of Archival Research Group.
[30] October 18, 1918, memorandum from the Office of the Construction and Maintenance Division, Ordnance Department to the Supply Division, NARA, courtesy of Archival Research Group.
[31] October 22, 1918, memorandum from Maj. William C. Young, Commanding Officer, Advance Ordnance Depot, to Chief of Ordnance, A.E.F., NARA, courtesy of Archival Research Group.
[32] AEF Memo dated October 12, 1918, from Camp Ordnance Officer, Camp Hunt, France to the Chief Ordnance Officer, Base Section, NARA, courtesy of Archival Research Group.
[33] Springfield Research Service, M1903-Era Ordnance Notes, p. 54-2.
[34] Springfield Research Service, courtesy of Frank Mallory.
[35] Memo from Chief of Ordnance dated October 5, 1918, to the Chief Ordnance Officer of the A.E.F, NARA, courtesy of Archival Research Group.
[36] Memorandum dated October 25, 1918, the War Department, Office of the Chief of Ordnance, by Col. T.L. Ames to Col. Wells, Director of Storage, NARA, courtesy of Archival Research Group.
[37] *Ibid.*
[38] Ordnance Department memorandum to Col. Little, dated October 26, 1918, courtesy of Archival Research Group.
[39] Tulkoff, Alec, *Equipping the Corps*, R. James Bender Publishing, San Jose, Calif., 2010, p. 147.
[0] January 24, 1918, memorandum from the Director of Gunnery Exercises and Engineering Performances to the Bureau of Ordnance, courtesy of Alec Tulkoff.
[41] Memorandum dated March 7, 1918, from the Office of the Chief of Ordnance, Supply Division, NARA, courtesy of Archival Research Group.
[42] War Department, Office of Chief of Ordnance document titled "Sold to Navy and Marine Corps." Form 18658, June 7, 1919.
[43] War Department, Office of the Chief of Ordnance memo to the Engineering Bureau, dated April 18, 1918, NARA, courtesy of Archival Research Group.
[44] Email correspondence with Andrew Stolinski, Chief Researcher, Archival Research Group, April 3, 2023.
[45] Springfield Research Service, Serial Numbers of U.S. Martial Arms, courtesy of Frank Mallory
[46] *History of Rifles, Revolvers and Pistols – 1917–1919*, U.S. Army Ordnance publication No. 1865, U.S. Government Printing Office, Washington, D.C., 1920, p. 18.
[47] *Arms and the Man*, "1917 National Matches Canceled after U.S. Declares War on Germany," December 1917.
[48] Roberts, J.B. Jr., "Enfield Rifles at the National Matches," *NRA Shooting Sports USA*, March 8, 2018.
[49] *Ibid.*
[50] *Op. cit.*, Roberts.
[51] Headquarters, Fourth Depot Division, France, Division Ordnance Office, dated October 6, 1918. NARA, courtesy of Archival Research Group.
[52] Ordnance Officer, Fourth Depot Division, October 14, 1918, NARA, courtesy of Archival Research Group
[53] Stallings, Laurence, *The Doughboys: The Story of the AEF 1917–1918*, Harper & Row, 1963, New York, N.Y.
[54] Rose, Alexander, *American Rifle*, Delta Trade Publications, Division of Random House, New York, N.Y., p. 288.

CHAPTER 6

[1] *History of Rifles, Revolvers and Pistols, 1917–1919*, U.S. Army Ordnance publication No. 1865, U.S. Government Printing Office, Washington, D.C., 1920, p. 18.
[2] Memorandum dated December 24, 1918, from Col. J.C. Heckman, Supply Division to the Col. H.B. MacFarland, Office of the Chief of Ordnance, courtesy of Archival Research Group.

[3] Memo from Ordnance Office, Administration Division by Col. W.W. Gibson to General Staff, Statistics Branch, June 21, 1919.

[4] Memorandum to the Adjutant General of the Army by Major General Henry Jervey, Assistant Chief of Staff, April 19, 1919.

[5] Keene, Jennifer D., *World War One*, Westport, 2006, pp. 182–187; Sparrow, John C., *History of Personnel Demobilization in the United States Army*, Washington D.C., 1952, pp. 11–19.

[6] Neumann, Brian F. and Shane Makowicki, *Occupation and Demobilization, 1918–1923*, Army Campaigns of World War I, CMH Pub. 77-9, 2019.

[7] AEF memo to Tours, France, Ordnance Depot, NARA, courtesy of Archival Research Group.

[8] November 25, 1918, memorandum from the Chief of Ordnance.

[9] Letter dated June 19, 1918, from Col. John Simpson, Estimates and Requirements Division, to the Requirements Section of his Division, File 474.1/947, NARA, courtesy of Archival Research Group.

[10] Memorandum dated November 11, 1918, for Brig. General J.B. Rice by Major J.A. O'Brien, Ordnance Department, NARA, courtesy of Archival Research Group.

[11] Memorandum dated January 9, 1919, from Col. Earl McFarland, Ordnance Department to Small Arms Supply Section, courtesy of Archival Research Group.

[12] Various Ordnance Reports, courtesy of Archival Research Group.

[13] Report from Camp Kearney, Calif., to Army Ordnance Field Services dated May 15, 1919, courtesy of Archival Research Group.

[14] Springfield Armory Report on the Status of Small Arms and Machine Guns, May 17, 1919, courtesy of Archival Research Group.

[15] Springfield Amory Annual Report, Fiscal Year 1919.

[16] *Ibid.*, Fiscal Year 1920.

[17] *Ibid.*, Fiscal Year 1921.

[18] Springfield Amory Annual Report, Fiscal Year 1942.

[19] *A History of Rock Island and Rock Island Arsenal, Volume II 1898 to 1940*, U.S. Army, courtesy of Kevin Braafladt, Army Sustainment Command Historian, Rock Island Arsenal, p. 342.

[20] *Ibid.*, pp. 344, 345.

[21] *Ibid.*, pp. 346, 347.

[22] Memo from War Department, Ordnance Office, Washington, D.C., to Ordnance Supply Officer, Columbus General Depot, Columbus, Ohio, dated December 19, 1933, courtesy of Frank Mallory.

[23] Project Supporting Paper, Rifle, U.S. Caliber .30 M1917 1917–1945, Small Arms Division, Industrial Service, Office, Chief of Ordnance, August 14, 1945, p. 5; Pate, Charles, *The Almost Model 1917 Sporter Rifle.*

[24] *Ibid.*

[25] Memorandum from the National Board for the Promotion of Rifle Practice to the Assistant Secretary of War, February 16, 1933. Courtesy of Archival Research Service.

[26] Frank Mallory, U.S. Martial Arms Collector 139-21.

[27] Project Supporting Paper, Rifle, U.S. Caliber .30 M1917 1917–1945, Small Arms Division, Industrial Service, Office, Chief of Ordnance, August 14, 1945, pp. 3, 4.

[28] Ordnance Department document, courtesy of Archival Research Service.

[29] Letter dated February 21, 1941, from Col. F.C. Endicott to William E. Van Horn, courtesy of Archival Research Service.

[30] Memo dated March 27, 1924, from Charles R. Baxter, Chairman of the Ordnance Salvage Board to Col. J.P. McAdams.

[31] Telegram dated March 16, 1929, to the Commanding Officer of San Antonio Ordnance Depot and telegram dated March 22, 1929, to Chief of Ordnance from Fort Sam Houston, Ferris, p. 144.

[32] Johnstone, William Crane, *The Shanghai Problem*, Stanford University Press, Redwood City, Calif., 1983.

[33] Shanghai Volunteer Corps Centenary Dinner Program, April 2, 1954.

[34] Springfield Research Service, courtesy of Frank Mallory.

[35] Project Supporting Paper, Rifle, U.S. Caliber .30 M1917 1917–1945, Small Arms Division, Industrial Service, Office, Chief of Ordnance, August 14, 1945, and Supplement I, p. 6.

[36] Rock Island Auctions catalog description, November 30, 2012, with no identified source for the claim.

[37] Shivell, Arthur, "Garand Critique," *American Rifleman*, October 1940.

[38] *Ibid.*, p. 5.

[39] Letter from the Commanding Officer of Springfield Armory to the Chief of Ordnance dated September 22, 1938, regarding M1917 rifle extractors.

[40] A July 1, 1937, memorandum from Lt. Col. Charles A. Walker, Ordnance Department, to Lt. Col. Kenneth Buckhanan, Office of the Assistant Chief of Staff, C-4.

[41] Memo dated October 26, 1938, from Lt. Col. H.B. Sayler, Ordnance Department to Commanding Officer, Rock Island Arsenal.

[42] July 15, 1936, memo from the Office of the Chief of Ordnance to the Adjutant General.

[43] Memo from the Adjutant General to the Office of the Chief of Ordnance, July 15, 1936.

[44] Project Supporting Paper, Rifle, U.S. Caliber .30 M1917 1917–1945, Small Arms Division, Industrial Service, Office, Chief of Ordnance, August 14, 1945, p. 7.

CHAPTER 7

[1] United States National Archives "Milestone Series – The Lend-Lease Act," Reviewed June 22, 2022.

[2] June 11, 1940, agreement between the British Purchasing Commission and the United States Steel Export Co., Ferris, p. 146.

[3] Green, Constance McLaughlin, Harry C. Thomson, and Peter C. Roots, *United States Army in World War II. The Technical Services, The Ordnance Department, Planning Munitions for War,* Office of the Chief of Military History, Department of the Army, Washington, D.C., 1953, p. 74.

[4] Michael Atkins Military Research monograph.

[5] Clarke, D.M., *Arming the British Home Guard, 1940–1944*, Cranfield University, September 19, 2001, p. 64.

[6] Thomson, Harry C. and Lida Mayo, *United States Army in World War II – The Technical Services – The Ordnance Department: Procurement and Supply,* Office of Chief of Military History, Washington, D.C., 1960, p. 156.

[7] United States National Archives "Milestone Series – The Lend-Lease Act," reviewed June 22, 2022.

[8] Skennerton, Ian D., *The U.S. Enfield*, Margate, Ian D. Skennerton, Australia, 1983, p. 148.

[9] Clarke, D.M., Cranfield University, "Arming the British Home Guard, 1940–1944," Security Studies Institute, Department of Management and Security, Cranfield Defence and Security, PhD Thesis, Academic Year 2010–2011, Professor R. Holmes, Supervisor, December 2010.

[10] Vigneras, Marcel, *Rearming the French – Special Studies – United States Army in World War II Series*, Center of Military History, U.S. Army, Washington, D.C., 1989.

[11] Project Supporting Paper, *Rifle, U.S. Caliber .30, M1917, 1917–1945*, p. 7.

[12] *Ibid.*

[13] *Ibid.*

[14] *Ibid.*

[15] 503PRCT – Corregidor.proboards.com

[16] Whitman, John W., *Bataan: Our Last Ditch*, Hippocrene Books N.Y., N.Y., 1990, p. 33.

[17] *Ibid.*

[18] Dunlap, Roy F., *Ordnance Went Up Front*, R&R Books, Livonia, N.Y., 1948, p. 218.

[19] *Ibid.*, p. 301.

[20] "Cabanatuan Camp," Philippine Defenders.lib.wv.us

[21] *Op. cit.*, Skennerton.

[22] Eastman, Lloyd E., *The Nationalist Era in China, 1927–1949*, Cambridge University Press, Cambridge, UK, 1986.

[23] *Op. cit.*, Vigneras.

[24] Lohnstein, Marc, *Royal Netherlands East Indies Army 1936–42*, Osprey Publishing, Oxford, UK, 2018.

[25] History, army.mil.eindies.htm

[26] Edgecombe, David Walter, *Defending the Dominion: Canadian Military Rifles 1855–1955*, 1940.

[27] Cullen, Stephen M., *In Search of the Real Dad's Army*, Pen & Sword Books, Ltd., Barnsley, UK, 2011.

[28] Naval Supply Depot, Report on Rifles, Pistols, Revolvers, Shotguns and Submachine guns on hand at this depot and their condition for week ending February 14, 1942.

[29] Supplement I to "History of U.S. Rifle, Caliber .30, M1917," Small Arms Division, Industrial Service, Ordnance Department, June 25, 1945.

[30] Memorandum dated September 8, 1942, to the Chief of Ordnance from Brig. Gen. Lucius D. Clay, Assistant Chief of Staff for Material.

[31] Project Supporting Paper, *Rifle, U.S. Caliber .30, M1917, 1917–1945*, pp. 8, 9.

[32] Hatcher, Maj. Gen. Julian S., *Hatcher's Notebook*, The Military Service Publishing Company, Harrisburg, Penn., 1947, p. 16.
[33] Project Supporting Paper, *Rifle, U.S. Caliber .30, M1917 – 1917–1945.*
[34] "History of Rock Island Arsenal, 1919 through World War II," Section I, p. 150.
[35] Ferris, C.S., *The United States Rifles Model of 1917*, Scott A. Duff Publications, Export, Penn., 2004, p. 154.
[36] Rock Island Arsenal, "Small Arms Overhauled March 1942–June 1958," courtesy of Kevin Braafladt, Army Sustainment Command Historian, Rock Island Arsenal.
[37] Leinicke, Kris Gayman, Curator of Collections, Rock Island Arsenal Museum, "Keep 'em Shooting," *Man at Arms*, 1997, Vol. 19, No. 3, pp. 38–42.
[38] Springfield Armory Annual Report, FY 1940.
[39] *Ibid.*, FY 1942.
[40] Project Supporting Paper, *Rifle, U.S. Caliber .30, M1917, 1917–1945.*
[41] Zeglin, Fred, *P.O. Ackley – America's Gunsmith*, Krause Publications, Iola, Wis., 2017, pg. 28.
[42] Requisition Form for Small Arms to Naval Supply Depot, Norfolk for stock and issue, dated January 19, 1942.
[43] Naval Supply Depot Report on Rifles, Pistols, Revolvers, Shotguns and Submachine guns on hand at this depot and their condition for week ending February 14, 1942.
[44] Springfield Research Service, courtesy of Frank Mallory.
[45] *Ibid.*
[46] Cunningham, Gary M., *U.S. Knife Bayonets and Scabbards*, Scott A. Duff Publications, Export, Penn., 2015, p. 35.
[47] Goodwin, Mark G., *U.S. Weapons in Combat*, Scott A. Duff Publications, Export, Penn., 2005, pp. 17, 18.
[48] *Ibid.*, p. 58.
[49] *Ibid.*, p. 198.
[50] Pottsman, Chuck, "A Rookie Rifleman in 1943," *GCA Journal*, Vol. 8, No. 3, Summer 1994.
[51] *Op. cit.*, Goodwin, p. 167.
[52] Gruening, Sen. Ernest, Introduction to *Men of the Tundra: Alaska Eskimos at War* by Muktuk Marston, October House, 1969, pp. 2, 3.
[53] Marston, Muktuk, *Men of the Tundra: Eskimos at War*, October House Publishing, 1969.
[54] George, Lt. Col. John B., *Shots Fired in Anger*, a publication of the NRA, Washington, D.C., 1981, p. 262.
[55] Project Supporting Paper, *Rifle, U.S. Caliber .30, M1917, 1917–1945*, pp. 9, 10.
[56] *Ibid.*, pp. 10, 11.
[57] *Ibid.*, p. 11.

CHAPTER 8

[1] Memo dated September 20, 1945, from Breton Schervell, General Commanding, to the Undersecretary of War.
[2] Memo to Chief of Ordnance by Brig. General E.A. Regnier.
[3] Springfield Armory Annual Report for Fiscal Year 1947.
[4] *A History of Rock Island and Rock Island Arsenal, Volume II, 1898 to 1940*, U.S. Army, courtesy of Kevin Braafladt, Army Sustainment Command Historian, Rock Island Arsenal.
[5] Firearms/USA/wwiiafterwwii.wordpress.com/2015/07/24/m1917-rifle-in-21st-century-greenland
[6] *A History of Rock Island and Rock Island Arsenal, Volume II, 1898 to 1940*, U.S. Army, courtesy of Kevin Braafladt, Army Sustainment Command Historian, Rock Island Arsenal.
[7] https://www.forgottenweapons.com/norwegian-k98kf1-repurposed-mauser
[8] https://www.gunboards.com/threads/understanding-a-us-norwegian-danish-model-of-1917.798257
[9] Ferris, C.S., *The United States Rifles Model of 1917*, Scott A. Duff Publications, Export, Penn., 2004, p. 163.
[10] *American Rifleman*, October 1947, p. 41.
[11] Director of Civilian Marksmanship Sales brochure, courtesy of *American Rifleman.*
[12] War Department, Office of the Chief of Ordnance memorandum dated June, 11, 1946, from Brig. Gen. E.L. Ford, courtesy of Archival Research Group.
[13] Hatcher, Maj. General (Ret.) Julian S., H*atcher's Notebook,* The Stackpole Co., Harrisburg, Penn., 1962, p. 247.
[14] *Op. cit.*, Ferris, p. 167.
[15] Memo dated May 28, 1947, from J.D. Frost, Assistant, Rock Island Ordnance Depot to Chief of Ordnance, Washington, D.C., courtesy of Kevin Braafladt, Army Sustainment Command Historian, Rock Island Arsenal.
[16] Memo dated June 24, 1947, from Rock Island Ordnance Depot to the Chief of Ordnance, Washington, D.C., courtesy of Kevin Braafladt, Army Sustainment Command Historian, Rock Island Arsenal.
[17] Springfield Armory National Historic Site Museum, Courtesy of Alex MacKenzie, Curator.
[18] *American Rifleman*, September 1949, p. 43.
[19] U.S. Inflation calculator.
[20] 2022 Civilian Marksmanship Program sales catalog.
[21] *Ibid.*

CHAPTER 9

[1] May 25, 1918, letter from Winchester Repeating Arms Co. to Engineering Bureau, Ordnance Department, File ES 474.4/437.
[2] Campbell, Clark S., *The '03 Springfield Rifles' Era*, Campbell Books, Richmond, Va., pp. 39–41.
[3] *Handbook for Ordnance Data*, U.S. Army Ordnance Department, Government Printing Office, Washington, D.C., 1919, pp. 330, 331.
4 Senich, Peter R., *The Complete Book of U.S. Sniping*, Paladin Press, Boulder, Colo., 1988, p. 64,. Henceforth referred to as "Senich."
[5] Ordnance Supply Division, 3rd Indorsement, dated June 6, 1918 (File 474.4/477, Ferris, p. 53.
[6] Winchester Repeating Arms Co., Report on Status of Government Contracts, January 23 and 24, 1919, World War I – Government Contracts January 1919–December 1919, Winchester Arms Collection Archives, courtesy of Herbert G. Houze.
[7] *Ibid.*
[8] Crossman, E.C., *The Book of the Springfield*, Thomas G. Small-Arms Technical Publishing Co., Georgetown, S.C., 1951, p. 348.
[9] *Ibid.*
[10] Email correspondence with Danny Michael, curator of the Cody Firearms Museum, 2021.
[11] Springfield Armory National Historic Site Catalog Description, courtesy of Alex Mackenzie.
[12] Senich, Peter R., *U.S. Marine Corps Scout-Sniper,* Paladin Press, Boulder, Colo., 1993, p. 7.
[13] Canfield, Bruce N., *U.S. Infantry Weapons of the First World War*, Andrew Mowbray Publishers, Lincoln, R.I., 2000, p. 110.
[14] *Handbook for Ordnance Data*, U.S. Army Ordnance Department, Government Printing Office, Washington, D.C., 1919.
[15] Laemlein, Tom, "Sniping from Below: Periscope Rifles in World War, " *American Rifleman*, April 2018.
[16] *Ibid.*
[17] *Ibid.*
[18] February 9, 1918, memo Major R.C. Scott, Requirements Section, Control Bureau to Engineering Bureau, Small Arms Section, courtesy of Archival Research Service.
[19] Springfield Research Service, courtesy of Frank Mallory.
[20] Canfield, Bruce N., U.S. *Infantry Weapons of the First World War*, Andrew Mowbray Publishers, Lincoln, R.I.
[21] Cablegram dated July 31, 1918, from General Headquarters, A.E.E., signed "Pershing," to Chief of Staff.
[22] Albert, David, "The U.S. M1917 Carbine: Mystery Solved," self-published article, 2014.
[23] Houze, Herbert, *Winchester Bolt-Action Military & Sporting Rifles, 1877–1937*, Andrew Mowbray Publishers, Lincoln, R.I., 1998, p. 158.
[24] *Ibid.*, pp. 158, 159.
[25] Marcot, Roy, *Remington, America's Oldest Gunmaker,* Primedia Special Interest Publications, Peoria, Ill., 1998, p. 189.
[26] Email correspondence with John Spangler, March 2023.
[27] *Op. cit.*, pp. 205-, 206.
[28] Spangler, John, "Guns to Save Lives: An Introduction to Line Throwing Guns," *American Society of Arms Collectors Bulletin 111*, 2015.
[29] Brophy, William S., *The Springfield 1903 Rifles*, Mechanicsburg, Penn., 1985, pp. 150–153. Brophy states that the conversions were done by three firms, W.H. Reisner Mfg. Co., Hagerstown, Md.; Van Karner Chemical Arms Corp., New York (USN); and Price Bros. Inc., Frederick, Md.

[30] U.S. Coast Guard, .30 Caliber Shoulder Line Throwing Equipment, Oct. 10, 1935, Washington, D.C., 1935, pp. 1, 2.
[31] U.S. Coast Guard, Pamphlet No. D30911-0, Line Throwing Guns and Pyrotechnics, Yorktown, Vir., Jan. 1980, p. 21.
[32] Ryan, M.J., "Life Line Projectile," U.S. Patent 2069276, February 2, 1937.
[33] USCG Solicitation Number: HSCG40-12-Q 89 -B830100045 dated Feb 2, 2012; USCG Solicitation Number: HSCG40-15-Q-60200 dated Feb 29, 2015.
[34] John Spangler, email correspondence with author April 25, 2023.

CHAPTER 10

[1] Skennerton, Ian, *The U.S. Enfield*, Ian D. Skennerton, Margate, Australia, 1983, p. 27. Henceforth referred to as "Skennerton."
[2] *Ibid.*, pp. 31, 32.
[3] Mercaldo, p. 155.
[4] Crowell, Benedict, *America's Munitions 1917–1918*, Government Printing Office, Washington, D.C., 1919, p. 227. Henceforth referred to as" Crowell."
[5] Ferris, pp. 173–175.
[6] Cunningham, Gary, *Bayonet Points*, July 2003. Henceforth referred to as "Cunningham."
[7] *Project Supporting Paper Relating to Bayonets, Knives and Scabbards 1917–August 1945*, Small Arms Division, Industrial Service, Ordnance Department.
[8] Hardin, Albert N. Jr., *The American Bayonet 1776–1964*, Albert N. Hardin Jr., Pennsauken, N.J., 1964, pp. 162, 163.
[9] Cunningham.
[10] *Ibid.*
[11] *Ibid.*
[12] Canfield, Bruce N., *Complete Guide to United States Military Combat Shotguns*, Mowbray Publishers, Inc., Woonsocket, R.I., 2007.
[13] Crowell, p. 227.
[14] Cunningham.
[15] Springfield Research Service / U.S. Martial Arms Collector, 67-12.
[16] *Project Supporting Paper Relating to Bayonets, Knives and Scabbards 1917–August 1945*, Small Arms Division, Industrial Service, Ordnance Department.
[17] *Ibid.*
[18] *Ibid.*
[19] *Op. cit.*, Canfield, *Combat Shotguns*, pp. 261–265.
[20] *Ibid.*
[21] Williams, Dr. James W., Historian, U.S. Army Chemical School, The Flaming Bayonet – Designing Tactical Surprise, 1987.
[22] *History of Trench Warfare Material, Army Ordnance 1917–1919*: Government Printing Office, Washington, D.C., 1920, pp. 217, 218.
[23] *Chemical Warfare Monographs, Volume 43*, Ray, Arthur, B., War Department – Chemical Warfare service, Research Division, American University Experiment Station, Washington, D.C., May 1919.
[24] Flaming Bayonet, Partial Report Number 3 on Investigation #40, Asire, Lt. H.W., Chemical Warfare Service, 1919.
[25] Canfield, Bruce N., *U.S. Infantry Weapons of the First World War*, Andrew Mowbray Publishers, Lincoln, R.I., 2000, p. 239.
[26] *Ibid.*, pp. 239, 240.
[27] *History of Trench Warfare Material 1917–1919*, U.S. Army Ordnance Department, Government Printing Office, Washington, D.C., 1920, p. 213.
[28] *Ibid.*, p. 230.
[29] Crowell, Benedict, Asst. Secretary of War, *America's Munitions 1917–1918*, Government Printing Office, Washington, D.C., 1919.
[30] Canfield, Bruce N., *U.S. Infantry Weapons of the First World War*, Andrew Mowbray Publishers, Lincoln, R.I., 2000, p. 240.
[31] Crowell, Benedict, Asst. Secretary of War, America's Munitions 1917–1918, Government Printing Office, Washington, DC, 1919.
[32] *Ibid.*
[33] *History of Trench Warfare Material 1917–1919*, U.S. Army Ordnance Department, Government Printing Office, Washington, D.C., 1920.
[34] *Ibid.*, pp. 211, 212.
[35] *United States Army in the World War, 1917–1919, Organization of the American Expeditionary Forces, Vol. 1*, Center of Military History, United States Army, Washington, D.C., 1988.
[36] Lawrence, Joseph Douglas, *Fighting Solider – The AEF in 1918*, University of Colorado Press, Boulder, Colo., 1985.
[37] Freidel, Frank, *Over There – The Story of America's First Great Overseas Crusade*, Temple University Press, Philadelphia, Penn., 1990.
[38] U.S. Army 89th Infantry Division dispatch, November 1918.
[39] *Project Supporting Paper Relating to Grenade Launchers 1917–1945*, Small Arms Division, Industrial Section, Ordnance Department, January 15, 1947, p. 5.
[40] *Ibid.*
[41] *Ibid.*, p. 10.
[42] *Ibid.*, p. 11.
[43] *Ibid.*, pp. 14, 15.
[44] *Ibid.*, p. 14.
[45] *Ibid.*, p. 13.
[46] *Ibid.*.
[47] Brophy, Lt. Col. (Ret.) William S., *The Springfield 1903 Rifles*, Stackpole Books, Harrisburg, Penn., 1985, p. 435.
[48] *Ibid.*, p. 436.
[49] *Ibid.*
[50] Mallory, Frank, *Springfield Research Service Newsletter*, SRS No. 19.
[51] *Ibid.*
[52] *Ibid.*
[53] Brophy, Lt. Col. (Ret.) William S., The Springfield 1903 Rifles, Stackpole Books, Harrisburg, PA, 1985, p. 436.
[54] Stratton, Charles R., *The Pattern 1914 and U.S., Model 1917 Rifles*, North Cape Publications, Tustin, Calif., 2000, p. 107.
[55] Krisko, Jonathon, "Selling the Model 1917: John T. Thompson & the Rifle Demonstrators," *American Rifleman*, February 2022.
[56] Winchester Repeating Arms Co., Report on Status of Government Contracts, January 23 and 24, 1919, World War I – Government Contracts January 1919–December 1919, Winchester Arms Collection Archives, courtesy of Herbert G. Houze.
[57] Contract War-Ord. 14063 dated July 12, 1917, between United States Government and the Remington Arms Union Metallic Cartridge Company, Inc. for 300,000 U.S. Rifles, dated July 12, 1917. Courtesy of Roy Marcot.
[58] Ferris, C.S., United States Rifle Model of 1917, Scott A. Duff Publications, Export, Penn., 2004, p. 184.
[59] *Handbook of Ordnance Data*, U.S. Army Ordnance Department, Government Printing Office, Washington, D.C., 1919, p. 354.
[60] *Description and Rules of Management of the United States Rifle, Caliber .30, Model of 1917*, Government Printing Office, Washington, D.C., 1918, p. 49.
[61] *Ibid.*
[62] Brophy, Lt. Col. (Ret.) William S., *The Springfield 1903 Rifles*, Stackpole Books, Harrisburg, Penn., 1985, p. 410.
[63] *Ibid.*
[64] *Ibid.*
[65] *Arms and the Man*, May 26, 1917.
[66] *Ibid.*
[67] Brophy, Lt. Col. (Ret.) William S., *The Springfield 1903 Rifle*s, Stackpole Books, Harrisburg, Penn., 1985, p. 412.
[68] John Spangler, email correspondence with author, May 2023.
[69] *Ibid.*, p. 415.
[70] *Handbook of Ordnance Data*, U.S. Army Ordnance Department, Government Printing Office, Washington, D.C., 1919, p. 332.
[71] Brophy, Lt. Col. (Ret.) William S., *The Springfield 1903 Rifles*, Stackpole Books, Harrisburg, Penn., 1985, pp. 415, 416.
[72] Fire Control and Sighting Equipment, Small Arms and Special Infantry Weapons, April 2, 1927.
[73] Brophy, Lt. Col. (Ret.) William S., *The Springfield 1903 Rifles*, Stackpole Books, Harrisburg, Penn., 1985, p. 417.

Index

A

B

C

H

I

J

K

L

M

S